Computer - Integrated Manufacturing Technology and Systems

MANUFACTURING ENGINEERING AND MATERIALS PROCESSING

A Series of Reference Books and Textbooks

SERIES EDITORS

Geoffrey Boothroyd

*Department of Mechanical Engineering
University of Massachusetts
Amherst, Massachusetts*

George E. Dieter

*Dean, College of Engineering
University of Maryland
College Park, Maryland*

1. Computers in Manufacturing, *U. Rembold, M. Seth and J. S. Weinstein*
2. Cold Rolling of Steel, *William L. Roberts*
3. Strengthening of Ceramics: Treatments, Tests and Design Applications, *Henry P. Kirchner*
4. Metal Forming: The Application of Limit Analysis, *Betzalel Avitzur*
5. Improving Productivity by Classification, Coding, and Data Base Standardization: The Key to Maximizing CAD/CAM and Group Technology, *William F. Hyde*
6. Automatic Assembly, *Geoffrey Boothroyd, Corrado Poli, and Laurence E. Murch*
7. Manufacturing Engineering Processes, *Leo Alting*
8. Modern Ceramic Engineering: Properties, Processing, and Use in Design, *David W. Richerson*
9. Interface Technology for Computer-Controlled Manufacturing Processes, *Ulrich Rembold, Karl Armbruster, and Wolfgang Ülzmann*
10. Hot Rolling of Steel, *William L. Roberts*
11. Adhesives in Manufacturing, *edited by Gerald L. Schneberger*
12. Understanding the Manufacturing Process: Key to Successful CAD/CAM Implementation, *Joseph Harrington, Jr.*
13. Industrial Materials Science and Engineering, *edited by Lawrence E. Murr*
14. Lubricants and Lubrication in Metalworking Operations, *Elliot S. Nachtman and Serope Kalpakjian*
15. Manufacturing Engineering: An Introduction to the Basic Functions, *John P. Tanner*
16. Computer-Integrated Manufacturing Technology and Systems, *Ulrich Rembold, Christian Blume, and Ruediger Dillmann*

OTHER VOLUMES IN PREPARATION

Computer - Integrated Manufacturing Technology and Systems

ULRICH REMBOLD
CHRISTIAN BLUME
RUEDIGER DILLMANN

University of Karlsruhe
Karlsruhe, Federal Republic of Germany

INPRO Innovationsgesellschaft für
fortgeschrittene Produktionssysteme
in der Fahrzeugindustrie mbH
- Bibliothek -
Signatur 269

MARCEL DEKKER, INC.   New York and Basel

Library of Congress Cataloging-in-Publication Data

Rembold, Ulrich.
  Computer-integrated manufacturing technology and systems.

  (Manufacturing engineering and materials processing ; 16)
  Includes bibliographies and index.
  1. Computer integrated manufacturing systems.
  I. Blume, Christian. II. Dillmann, R., [date].
III. Title. IV. Series.
TS155.6.R46   1985   670.42'7   85-16052
ISBN 0-8247-7403-5

COPYRIGHT © 1985 by MARCEL DEKKER, INC.   ALL RIGHTS RESERVED

Neither this book nor any part may be reproduced or transmitted in any form or by any means, electronic or mechanical, including photocopying, microfilming, and recording, or by any information storage and retrieval system, without permission in writing from the publisher.

MARCEL DEKKER, INC.
270 Madison Avenue, New York, New York 10016

Current printing (last digit):
10 9 8 7 6 5 4 3 2 1

PRINTED IN THE UNITED STATES OF AMERICA

Preface

Manufacturing creates between 60 and 80% of the real wealth of the major industrialized countries. For this reason most of these countries have numerous industry- and government-supported programs to increase manufacturing productivity. Within the last two decades there have been many important innovations in the development of new design aids, manufacturing processes, engineering materials, and manufacturing systems. The digital computer has contributed to the most significant changes. This so-far almost untapped resource probably has greater potential for increasing manufacturing productivity than any other invention throughout the industrial revolution. With conventional production know-how it becomes increasingly difficult and expensive to improve manufacturing processes but the computer offers possibilities for improving manufacturing technology in many areas. Computers can directly control production and quality control equipment and quickly adapt them to changing customer orders and new products. They make possible the instant evaluation of data to assess the flow of information in the plant and initiate immediate corrective action. Thus they optimize the manufacturing process by their capability to integrate the entire manufacturing system. This decision-making ability will contribute to the conception and design of flexible manufacturing systems (FMSs) that can be reconfigured to changing marketing requirements and new manufacturing processes.

This book discusses the role of the computer in planning and controlling the manufacturing process. Computer-integrated manufacturing (CIM) comprises product design, production planning, production control, production processes, quality control, production equipment, and plant facilities. The creation of the product starts in the design department, where its function, physical design, and manufacturing methods and processes are established. Upon completion of the design activities, 70% of the manufacturing costs have been predetermined. This fact stresses the importance of combining design and manufacturing in a comprehensive CIM system. Thus, an overview of a manufacturing system is first presented to show how its subsystems are interconnected. Then the subsystems are discussed in detail.

Chapter 1 discusses the evolution from fixed to programmable automation. It explains how the computer has gradually been incorporated into manufacturing and the various control functions that it has assumed. Manufacturing requires planning, scheduling, controlling, machining, and material movement operations. Within the context of this book the first three functions are of major interest and the current state of programmable automation in these areas is covered.

Chapter 2 gives an overview of a manufacturing organization including its functions and subfunctions. It explains how different types of computers are interconnected to form a powerful control system to aid management functions at different hierarchical levels. Information processing is supported by the implementation of a comprehensive management information system. It contains all information necessary to conceive, design, and build a product. For this reason, all manufacturing data are stored in this information system and can be retrieved by or distributed to the user.

Typical automation tools needed to build a CIM system are discussed in Chapter 3. A computer-controlled facility can be successfully conceived only if manufacturing hardware is available that was designed for programmable automation. Computer, machine tool, and manufacturing know-how are required to implement a modern production system. The components of such a system are quite numerous and depend on the product to be made and the piece rate. For this reason it is not possible to discuss them all. In this chapter a brief attempt is made to give the reader not familiar with software and hardware used in manufacturing a short overview of typical system components, such as machine tools, material conveyance equipment, computers, and software packages.

Typical computing equipment needed for the control of the manufacturing process is described in Chapter 4. Quite a range of control computers is available. The simplest type is the programmable controller. Micro- and minicomputers are more complex and are employed where the control functions are more demanding (e.g., where a process is optimized). Since individual computers are an integral part of a complex manufacturing system, they are interconnected with the help of efficient data communication systems. The higher control functions of a plant are the domain of large control or business computers.

Computer aids for product design are illustrated in Chapter 5. The reader will obtain an overview of the CAD system needed to design a product and how it is used. In addition to hardware, software aspects are treated. Several different design principles are discussed, and an overview is given of several advanced CAD systems. Some of these already provide the interface between CAD and CAM, whereby in addition to design data, manufacturing information is generated from a description of the product.

Process planning and manufacturing scheduling are the topics of Chapter 6. From a description of the product and a knowledge base containing information about the manufacturing process and the available equipment, it is possible to generate automatically a process plan. This activity interfaces directly with CAD and has access to the CAD data base. Process planning is based on an efficient coding system and uses group technology methods, the most important of which are discussed. The process plan, together with the manufacturing due dates, comprise the basis of scheduling. There must be enough lead time to produce the parts on time and to utilize the plant equipment efficiently. Thus, this chapter shows how operations research methods are used for scheduling of machines and balancing of loads.

Numerical control (NC), direct numerical control (DNC), and computer numerical control (CNC) technologies are the heart of modern manufacturing. They are discussed in Chapter 7. Different control principles and NC languages are explained. It is shown how control data for machining a workpiece can be obtained directly from the CAD data base. In more advanced manufacturing an attempt is made to maximize the utilization of resources; therefore, the necessary adaptive control methods and sensors to optimize the operation of machine tools are additional subjects of this chapter.

Chapter 8 is devoted to assembly by robotic devices. Most assembly operations are currently done by human beings because these functions are very difficult to automate. For assembly to be performed by machine, there are two necessary prerequisites. First, the product developer has to learn how to design a product for assembly. Second, robotic devices have to be conceived which have advanced sensors and some human intelligence. The computer will play a major role in fulfilling the second prerequisite. This chapter describes robotic devices, their computer controls, programming language requirements, and intelligent sensors. Machine vision to determine the identity and angular position of an object is covered in detail. Vision needed to direct assembly operations and to perform a quality check after completion of the assembly task is also discussed.

Layout of a modern manufacturing and a material conveyance system is usually preceded by a computer simulation. With this tool it is possible to investigate various manufacturing configurations, to optimize the machine tool requirements and the material flow system, and to check for bottlenecks. The subject of simulation is covered in Chapter 9. Simulation of discrete manufacturing processes can be done with general-purpose languages such as GPSS or with special languages designed for specific applications (e.g., for the layout of a flexible manufacturing facility). The chapter discusses how to structure the model for the simulation, and how to perform it. It includes examples of applying the simulation to the real world.

Computer-aided quality control is discussed in Chapter 10. Quality control has to be planned within a plant-wide systems approach. The standards for quality are set by the customer. These standards have repercussions on all functions of a manufacturing system. The topics of this chapter are quality control planning, quality control methods and test procedures, the architecture of computerized measuring systems, and hierarchical computer concepts for quality control. Particular attention is given to problems that arise when quality control operations are done on moving assembly lines. Methods to enable a computerized system to learn its own test tolerances are also discussed. The application of coordinate measuring machines is covered in more detail.

Chapter 11 is devoted to the programmable factory. The computer concepts of the previous chapters are incorporated in such a factory to obtain a true FMS. Such a system consists of a group of machine tools and/or production equipment interconnected by an automated material handling system. A computer, or usually a hierarchical computer system, plans, executes, and controls the production process. The components and various configurations of a FMS are described. The chapter concludes with an illustration of several FMSs used by different industries.

Computer-Integrated Manufacturing Technology and Systems is based on a CAD/CAM course for students at the University of Karlsruhe. The course has also been presented at universities in the United States and in other countries in answer to the need of the industrial community for young engineers and computer scientists who have a basic understanding of the complex functions of computers to control manufacturing processes.

The authors would like to thank the publisher for help in creating this book and patience in waiting for its completion. Several people helped with the writing; they deserve our sincere thanks. The numerous drawings and graphs were made by Mrs. Hannelore Neeb and Mr. Bernd Kister. Mrs. Angela Jahn-Held and Miss Bärbel Seufert typed the manuscript and the tables.

Ulrich Rembold
Christian Blume
Ruediger Dillmann

Contents

Preface		iii
1.	Introduction	1
	1.1 General Discussion	1
	1.2 Development History of Computer Use	2
	1.3 Manufacturing Control System	3
	Reference	6
2.	Role of the Computer in Manufacturing	7
	2.1 Introduction	7
	2.2 Management Information Systems	8
	2.3 Manufacturing Data Base	13
	2.4 Hierarchical Computer Control Concept	19
	2.5 Functions of a Manufacturing Organization	29
	2.6 Excursion into a Computer-Aided Manufacturing Plant	45
	References	47
3.	Hardware and Software Components for Computer Automation	48
	3.1 Criteria for Selecting Automation Tools	48
	3.2 Components of a Manufacturing System	55
	3.3 Design and Development	55
	3.4 Manufacturing Planning and Control Modules	59
	3.5 Machining Modules	63
	3.6 Material Flow and Storage Modules	71
	3.7 Assembly Modules	83
	3.8 Quality Control Modules	86
	3.9 System Control Modules	91
	References	95

4. Advanced Computer Architectures Used in Manufacturing — 96

- 4.1 Introduction — 96
- 4.2 Hierarchical Control of Manufacturing — 99
- 4.3 Computing Equipment and its Hierarchical Assignment — 102
- 4.4 The Communication Network — 114
- 4.5 Examples of Hierarchical Computer Systems for Manufacture — 119
- 4.6 Advanced Computer Architectures (Fifth Generation) — 140
- References — 148

5. Computer-Aided Design — 149

- 5.1 Introduction — 149
- 5.2 Hardware — 151
- 5.3 Input Peripherals for Computer Graphic Systems — 159
- 5.4 Special Display Screens — 166
- 5.5 Graphic Output Systems — 171
- 5.6 Software — 173
- 5.7 Computer Internal Representation of Graphic Data — 213
- 5.8 Design Phases — 225
- 5.9 CAD Systems — 235
- 5.10 Discussion of Realized CAD Systems — 245
- References — 284

6. Manufacturing Systems — 287

- 6.1 Introduction — 287
- 6.2 Group Technology — 288
- 6.3 Process Planning — 318
- 6.4 Production Planning and Control — 356
- References — 383

7. Control of Manufacturing Equipment — 386

- 7.1 Introduction — 386
- 7.2 Numerical Control — 388
- 7.3 Control System of the NC Machine Tool — 428
- 7.4 Adaptive Control — 443
- 7.5 Sensors for Computer-Controlled Machine Tools — 457
- 7.6 NC Part Programming — 464
- 7.7 CNC Systems — 485
- 7.8 DNC Systems — 492
- References — 499

Contents ix

8. **Computer-Controlled Parts Handling and Assembly** — 500

   8.1 Introduction — 500
   8.2 Structure of Industrial Robots — 510
   8.3 Kinematics — 512
   8.4 The Effector — 519
   8.5 Sensors — 522
   8.6 Drives to Operate the Robot Joints — 532
   8.7 Control — 533
   8.8 Programming Languages and Programming Systems for Assembly Robots — 548
   8.9 Interfacing of a Vision System with a Robot — 584
   8.10 Optical Workpiece Recognition — 597
   References — 658

9. **Simulation of Manufacturing Processes** — 662

   9.1 Introduction — 662
   9.2 Simulation Methods — 662
   9.3 Iterative Character of the Simulation — 663
   9.4 Advantages and Limitations of Simulation — 665
   9.5 Progression of a Simulation Project — 667
   9.6 Initialization State — 668
   9.7 Validation Problems — 668
   9.8 Concepts of Simulation Systems and Programming Languages — 669
   9.9 Standard Tools of a Simulation System — 669
   9.10 Higher Programming Languages — 673
   9.11 Statement or Block-Oriented Simulation Languages — 674
   9.12 Simulation of a Flexible Manufacturing System — 685

10. **Quality Control** — 689

    10.1 Introduction — 689
    10.2 System Concept for Quality Control — 690
    10.3 Quality Control Planning — 692
    10.4 Function of Quality Control — 693
    10.5 Quality Control Methods and Test Procedures — 695
    10.6 Architecture of Computerized Measuring Systems — 699
    10.7 Equipment Configuration for Quality Control — 699
    10.8 Hierarchical Computer Systems for Quality Control — 717
    10.9 Implementation Problems with Quality Control Computer Systems — 723
    10.10 Computer Language for Test Applications — 734
    References — 740

11.	The Programmable Factory		742
	11.1 Introduction		742
	11.2 General Aspects of Flexible Manufacturing Systems		743
	11.3 Classification of FMSs		748
	11.4 Components of a FMS		753
	11.5 A Factory of the Future		770
	References		776

Index 777

Computer - Integrated Manufacturing Technology and Systems

1
Introduction

1.1 GENERAL DISCUSSION

With the beginning of mass production, a hitherto unknown market potential opened up to industry. Suddenly, a variety of high-quality products were within the reach of many customers. This demand stimulated the design of very complex and inflexible manufacturing systems, which constituted large investments and which were very expensive to modify when the end of the life cycle of a product made a production change necessary.

During the period of fixed automation a large number of manufacturing processes and methods were developed and became a standard for industry. These pioneering tasks were accomplished by persons with a broad manufacturing experience who were persistent and inventive. A formal engineering education was not essential; however, to be a good craftsman was an absolute necessity.

When the market for many mass-produced items was saturated, a change in customer preference in the direction of more individually designed products became noticeable. This made necessary the introduction of flexible manufacturing systems. With these most of the basic production processes could still be used; however, they had to be made adaptable to handle different products and models.

The original emphasis to increase flexibility was directed toward machining and eventually lead in 1949 to the conception of numerical control (NC) technology. The introduction of discrete and integrated digital circuits made possible the development of controls for complex machine tools that could perform different machining operations. With these and the aid of other primitive programming methods, a wide spectrum of workpieces could be produced.

It soon became apparent that programming was a formidable task for workpieces of intricate design, and an improved method to prepare NC programs had to be found. This lead to the development of the automatically programmed tools (APT) concept, which necessitated the use of a computer for program preparation. Thus the basic concept of programmable automation was born.

During the same period the electronic business computer was introduced into the manufacturing organization to support forecasting, accounting, scheduling, allocation, and control functions. In most cases the automation efforts with this computer were very successful after a rather short period. Forecasting, scheduling, and allocation could be greatly enhanced with the aid of mathematical tools developed by the operations research activities at many universities.

1.2 DEVELOPMENT HISTORY OF COMPUTER USE

The process control computer began to appear in the production facilities of several automotive and aircraft manufacturers in about 1964. Originally, it was used for factory data acquisition, quality control, and monitoring of machine functions. The manufacturing engineer quickly realized that the basic data that this computer acquired from the production floor provided the foundation for the management information system. Data could easily be transferred via communication lines to the business computer for further processing. This lead to the conception of distributed computer systems for factory control. With the appearance of mini- and microcomputers, these distributed computers also came within the reach of many smaller companies. Development eventually lead to hierarchical computer control concepts for manufacturing organizations. A good example of such a hierarchical computer control system was introduced with direct numerical control. Here planning, scheduling, and programming are done by a computer at an upper hierarchical level, and the execution of machining operations is directed and supervised by a computer at a lower hierarchical level.

Because of its real-time capabilities, the process control computer was also used for early interactive computer-aided design (CAD) activities. The development of useful computer-aided design methods took quite a long time since both the hardware and software are very complex. However, within recent years this technology is being accepted very rapidly by progressive design departments. Through use of the computer creative designers can improve their productivity at least fourfold over that using conventional design methods. It is expected that computer-aided design will be introduced by many manufacturing organizations.

One of the areas in which the computer has had little effect so far is assembly. However, the computer's impact will also be felt here within the near future. There are presently robots and special assembly machines being developed which will be equipped with the advanced

Introduction 3

sensors and vision systems that are needed for complex assembly tasks.

In the introductory discussion it was shown that there are isolated computer-aided technologies available today with which we are able to increase manufacturing productivity. Within the years to come it will be the task of the manufacturing engineer to integrate these technologies fully into a useful computer-aided design/computer-aided manufacturing (CAD/CAM) concept. Looking at the present state of the art, it is already possible to design a product with the help of a computer and to store the information describing it in a central graphical data bank. With the help of this information, the bill of materials can be generated, which after the drawing, is the most important document needed for the manufacture of a product. These documents can be used to do automatic process planning and to implement a production schedule. Upon release of the design to manufacturing, the material movement will be activated to supply the machine tools with raw material. The program to generate the workpiece will also be sent by computer to the machine tool. Data from the machining operation and material movement will be sent back to the computer system and evaluated to perform automatic process adjustments. The finished workpiece will be measured in a computer-controlled inspection machine, which receives the test program for the product from the central computer. The supply of the raw material for storage and process buffers will also be initiated and supervised by the computer.

The system described above will be very complex and will require the combined skills of the craftsman and the highly trained engineer. The future factory can be visualized as an organization where management is supported by a hierarchy of interconnected computers. It will be the task of the computer to perform repetitive short-range planning and control work, thereby giving management free time to be more creative in long-range planning, product development, and the conception of new processes and more effective manufacturing organizations. In this control concept the computer will never be able to replace human beings, since the processes are too complex to be left to a machine. It must also be realized that the computer is capable of handling routine work economically. However, in general, it is very costly to control process exceptions automatically. In the future factory these exceptions will still be within the human domain. In addition, it is inconceivable that a factory organization that consists of many individual persons can ever be operated efficiently by a programmed computer system.

1.3 MANUFACTURING CONTROL SYSTEM

There are many different methods used to represent a manufacturing system. The most important asset the firm possesses is manufacturing know-how. It requires planning, scheduling, controlling, machining, and material movement operations as well as resources to convert the knowledge into a product. In the context of this book the first three functions are of major interest. The manufacturing machines are of secondary importance. The interaction of the various functions can be represented by a four-plane concept (Fig. 1.1). All manufacturing activities start with planning and lead to a schedule. This is done at different hierarchical levels, where information processed at a preceding level is refined to information needed for the succeeding level, thereby gradually shortening the planning and scheduling horizon. The schedule is the input to control. It furnishes all parameters to initiate the actual manufacturing operation. It is also activated by feedback information when manufacturing problems or bottlenecks are encountered.

Figure 1.2 shows a scheme for a manufacturing control system. It contains planning, scheduling, order release, control, and verification functions for the material flow and manufacturing processes. These are controlled by a complex feedback system that has many interconnected loops. At the present time most controlling functions of this control concept are done by human beings.

In the factory of the future, repetitive control tasks will be assigned to the computer. A model system for such a factory is shown

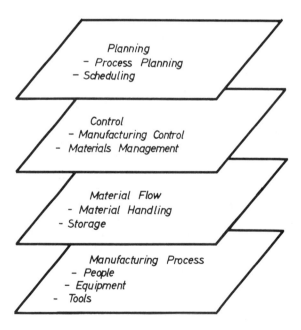

FIG. 1.1 Four-plane concept of manufacturing. (From Ref. 1.)

Introduction

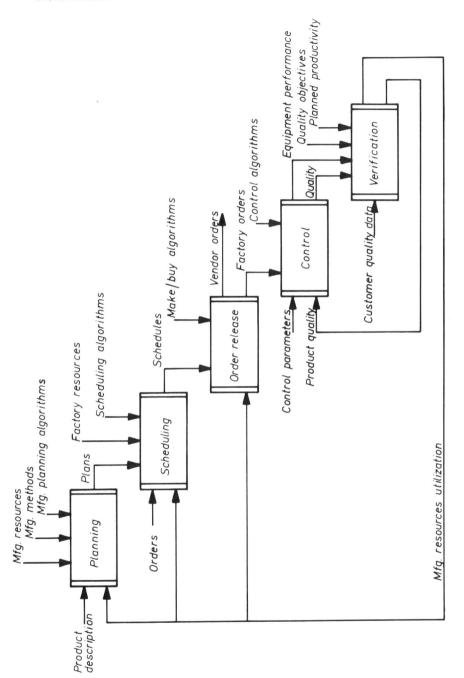

FIG. 1.2 Control loop of a manufacturing system.

in Fig. 1.2. It begins with planning. Planning draws its input from the human knowledge data base of manufacturing resources, manufacturing methods, and planning algorithms. Usually, this knowledge data base is kept stable over a long time frame. Changes are introduced by technological innovations or changes in the resource supply. The product description will change even more dynamically. It needs data from the market, the customer, competing products, and new inventions. The output of planning is input scheduling. There input data, together with knowledge of the factory resources and customer orders, are converted to factory schedules with the help of scheduling algorithms. Order release is activated with the help of this schedule. At this point a make-or-buy decision is made, which depends on cost data, factory load, manufacturing problems, order due dates, and other parameters. The factory orders activate manufacturing control. The manufacturing process is supervised with the help of control algorithms, control parameters, and feedback knowledge of the quality produced. Machine control functions are usually of a repetitive nature and are therefore the domain of automatic control. Control has two outputs: equipment performance and the product quality parameter. These outputs, together with performance standards and customer quality data, are the input for verification. The output of this function is the product quality and manufacturing resource utilization. The former parameter is brought to control and the latter to planning, scheduling, and order release. This closes the control loop.

In this book an attempt is made to present the current state of programmable automation, with emphasis on the interaction of computer-aided design and computer-aided manufacturing. The individual subjects covered in detail are design, process and material planning, material flow control, numerical control, and assembly. In addition, the possible architecture of computer systems for future factories is discussed.

REFERENCE

1. D. L. Judson, Integrated Manufacturing Control Material Management (IMC-MM), Society of Manufacturing Engineers Technical Paper MS79-841, Dearborn, Mich., 1979.

2
Role of the Computer in Manufacturing

2.1 INTRODUCTION

With the invention of the electronic computer it was quickly realized that this tool had an enormous potential for becoming the focal point in future automation endeavors. Conventional automation was based primarily on sophisticated mechanical machinery controlled by cams and levers or electrical switching gear. The equipment in general was conceived to perform fixed manufacturing assignments. The degree of automation that can be achieved with these tools is limited since for more demanding tasks the controls become so complex that they cannot be justified economically. The original automation endeavors were directed primarily toward the improvement of the machining capabilities of the manufacturing equipment. However, the average time a workpiece spends in a machine tool is only about 5% of the total time it spends in a factory (Fig. 2.1). It is apparent that the long in-shop time results in a high inventory of unfinished and finished parts. If this idle time could be reduced, large amounts of capital would be available for more productive tasks. It is also well known that the average machine tool is used only 6% of the time available. The productive time could be improved by automatic loading and fixturing of parts, the use of group technology, and the employment of second and third shifts.

Whereas early automation endeavors were concentrated on the improvement of machining operation, at present attention is focused on the 95% nonproductive moving and waiting time. As a matter of fact, most present research effort in manufacturing is concerned with the possibility of reducing this idle time, thereby increasing machine utilization and productivity. Since the task is very difficult to perform with conventional automation tools, the computer plays an ever-increasing role.

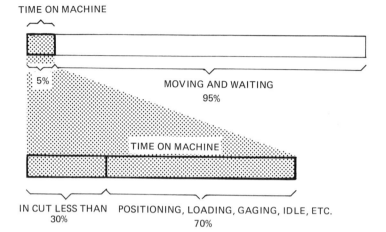

FIG. 2.1 Life of the average workpiece in the average (batch-type production) shop.

2.2 MANAGEMENT INFORMATION SYSTEMS

By "management information system" we mean a computer-based information processing system which supports the decision-making functions of management to control and operate a plant. The control of a manufacturing organization involves many different activities. With ever-increasing national and international competition, the flood of information that must be processed can reach astronomical proportions, even for a small manufacturing organization. Whereas in former times a company may have supplied only one product to a small local market with a limited number of customers, it now has to provide an entire product line to an international market. In addition, a rather short product life cycle will make long-, medium-, and short-range planning necessary. A comprehensive management information system for the firm of the future is shown in Fig. 2.2. One can easily see that if this information system is to be effective, it has to build on a large data base containing a wide spectrum of external information, which eventually will lead to the conception of a product and the market to be serviced. Detailed knowledge of the product and available resources and their properties is necessary for the operation of the plant. Here again, the amount of data to be stored and processed can be quite formidable. Just looking at the parts spectrum, it is common for a medium-sized company to administer over 50,000 parts. In addition, for example, one part may be assembled from five to seven other parts.

The Computer in Manufacturing

However, the number of final models in a product line may be much smaller. To satisfy the customer's desire to own products of different colors or functions, it may be necessary to generate 10 or even 100,000 variations of a single basic product.

From the introductory discussion we have seen that many different types of data have to be collected, prepared, and processed in an ongoing manufacturing organization. These activities involve not only the day-to-day transactions necessary to support accounting, payroll, sales, and so on, but they also provide managerial aids for making decisions for the present and the future. This means that a management information system must comprise data processing and decision-making activities at all managerial levels of an organization. Since most information maintains its value over a longer period, it is also necessary to provide ample storage facilities for data. There are usually different functions performed at different levels of a

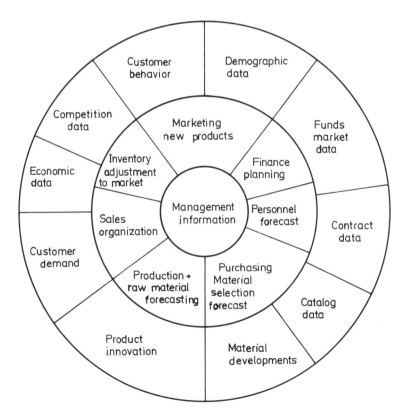

FIG. 2.2 Comprehensive management information system of a firm (according to John Niebold).

manufacturing organization. At the highest level, planning and decision making are of a strategic nature. At an intermediate level there will be tactical and operational planning for the near future. At the lowest level, the daily transactions are processed and actual control of the plant is performed. For this reason a management information system will be hierarchical in structure. The amount of data to be processed by a modern manufacturing company makes the use of a computer mandatory. Thus a management information system will be a comprehensive human-computer system able to aid all operation and decision-making functions necessary to manage an organization. This definition implies that human beings maintain their dominant role and that many subordinated control functions will be assigned to the computer. It is actually the power of the computer that makes the management information system such an efficient tool in conducting the business of an organization.

The operation of a modern plant is based on an efficient management information system supported by a hierarchical computer network. The information supplied by this system has to be accurate and timely. For this reason, management information systems are founded on a computerized data base which may have a centralized and distributed structure. This data base contains all essential aspects of a company's operation together with facts about its environment. The data base activities begin as soon as a customer has placed an order. With the entry of the order, under ideal conditions it would be possible automatically to trigger all manufacturing activities needed to process the order and to follow the order through the factory until it is ready for delivery to the customer (Fig. 2.3). For this, the management information system contains a central data base and the programs to control each manufacturing operation. The activities are coordinated by a supervisor whose operational strategy may be altered dynamically depending on the type or number of products to be manufactured, an order's priority, the availability of resources, and so on.

It is usually not possible to implement such a management information system all at once. The individual activities are entered one by one over a long period. The interconnection of these activities to a functional entity requires careful specification of interfaces, protocols, and data formats. Quite often this is a very difficult task, since the implementation of an entire system may take one or two decades. During this period, computer architectures, programming languages, and data processing procedures may change considerably. There is also the possibility that with advanced technologies, a previously conceived control system will become obsolete. Usually, activities 1 to 4, 11 and 12 are entered first into a management information system (Fig. 2.3). For their programming and execution, general data processing computers are commonly used.

Today a wide variety of packaged programs are available which are based on the assumption that the fundamental activities (e.g., production planning and control) are very similar from one factory to

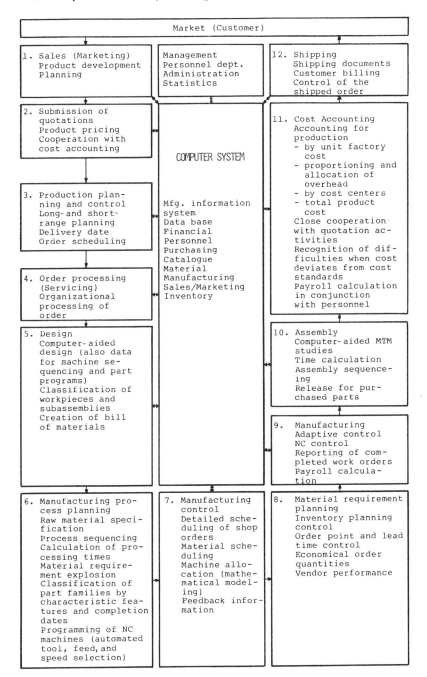

FIG. 2.3 Functions of the computer in a manufacturing organization. (From Ref. 1. Courtesy of VDI-Verlag, Düsseldorf, West Germany.)

the next. The ease of adapting these programs to a plant's function depends on the degree of similarity that exists between the application to be implemented and the underlying concept of the program.

For activities 5 to 7, 9, and 10, it is generally more difficult to conceive a packaged control program. The reasons for this are as follows:

1. Similar activities usually vary considerably among firms and among products.
2. Interaction among different types of computing equipment, such as control computers and business computers, is required.
3. Efficient algorithms to perform functions or subfunctions are not available.
4. Application and system programs are very difficult and have not been developed satisfactorily.
5. Breakthroughs in research are needed for the solution of many problems.

Despite these barriers, many companies are attempting to develop custom-designed solutions for their activities. The most successful are those who manufacture simple products, such as electric motors or standard parts such as ball bearings.

2.2.1 Example of Information Flow in a Firm

The customer initiates information flow with an inquiry or an order (Fig. 2.3). There are, in principle, three types of orders that can enter the system. They are for (1) a new product, (2) a standard item from an existing product line, and (3) a product purchased from another manufacturer. With the help of Fig. 2.3 the information flow to process the order of a new product will be discussed. As an exercise the reader may want to try to construct the information path for the two other alternatives. It is assumed that all the information about the activities of the firm is located in a centralized data base and that the management information system contains all decision algorithms and programs to process the part.

When the sales department obtains an inquiry for a product, a quotation is prepared with the help of product development and manufacturing. The customer then obtains the specification of the product, a price, and a delivery date. If a similar product had been produced previously, more accurate cost accounting data from the centralized data base will support the preparation of the quotation.

When an order is placed, the number of items to be purchased, an order number, and a customer number are issued and recorded in the data base. A credit check is made on the customer. Product planning schedules the order and order processing is initiated.

The specification of the product is then given to the design department. Here the product is conceived and its design and physical properties are selected. The manufacturing documents, in the form of a drawing, bill of materials, and processing specifications, are generated. The product is tested for performance and endurance and released for production. The design department may also classify the components of the product so that they will be manufactured according to group technology considerations. All these activities are supported by the computer. The degree of automation needed to process the order depends on the complexity of the product and the sophistication of the design department.

The next activity is manufacturing process planning. Here the raw materials, machine tools, and machining sequence are selected. In addition, the processing times are calculated and the numerically controlled (NC) programs are generated. This activity may be highly automated.

Purchasing obtains from the manufacturing documents the raw material requirement, the part number, and the number of products to be built. A make-or-buy decision is made and the parts are either ordered from a vendor or released for production.

The vendor delivers the parts and an invoice to the plant. A check is made to verify that the order is complete and that the parts received comply with the specifications. In case of a problem, new parts are ordered. Accounting is notified to pay the vendor.

When all raw material and parts are available, the order is released and manufacturing control gets into action. Machine tools are selected and in the case of delivery data problems, production runs are rescheduled. The parts are manufactured under computer control and delivered to assembly.

Upon completion of manufacturing, cost accounting begins its work. Fixed and variable costs are calculated for each product. Costs are distributed among cost centers and payroll calculations are made. The cost components of the final product are sent to the quotation department to be used for new quotations.

Shipping crates the product and sends it together with the bill of lading and the invoice to the customer. This function also controls the shipping procedure.

2.3 MANUFACTURING DATA BASE

The large number of data to be stored and processed in a manufacturing information system makes it necessary to give careful attention to the design of the data base. As the computer applications grow in a factory, there will be an ever-increasing amount of data entered which may be interrelated through very complex ties. In manufacturing there are basically two groups of data to be processed. The first group comprises:

1. Structural information about the product, contained in the bill of materials
2. Information about process capability and manufacturing operations, which are contained in the process planning file
3. Information describing the manufacturing machines available

All this information is relatively permanent in nature and will be changed only if a product is entered into or deleted from a production plan or if production equipment is changed.

The second group of data is used for order processing. Typical information belonging to the second group is:

Planning and processing an order
Machine selection and process sequencing
Scheduling of shop orders
Material scheduling
Payroll and other calculations

This information is assembled to perform processing of the part and will be deleted upon completion of the operation. There are four different types of data bases that can be used:

A collection of independent data bases
A centralized or solitary data base
An interfaced data base
A distributed data base

Each of these data base types has advantages and disadvantages, and it is often necessary to use a combination of them in a comprehensive manufacturing data base.

When selecting the data base, the following criteria should be considered:

Type of data to be stored
Number of data to be stored
Number of data files
Access speed
User of data
Ease of updating and changing of the file
Flexibility of the file
Redundancy of storage
Access control and security
Maintainability of file

The Computer in Manufacturing

The various types of data bases and their principal advantages and disadvantages are discussed next [2].

2.3.1 Collection of Independent Data Bases

Historically, an independent data base was created for each application [e.g., for engineering calculations, accounting, sales, etc. (Fig. 2.4)]. Each data base was designed for the convenience of the user. Often, little consideration was given to a common data base interface, data formats, expandability, maintenance, and so on. The various files usually had their own method of data organization and their own formats. The following problems are encountered with this approach:

1. The data bases are difficult to combine into a single manufacturing information system. There is little consideration given to evolutionary growth.
2. A change in the data base organization or data format makes it necessary to change all programs for which the file is used.
3. A number of redundant data files have to be implemented.
4. Application program and data base maintenance is expensive.
5. Exchange of data between files is difficult.

FIG. 2.4 Collection of independent data bases.

Despite these difficulties, a dedicated data base has advantages for standalone applications where access to programs and data is limited to a particular user. For example, in computer-aided design the data file that is needed for stress analysis of a component is used only by the designer. However, the bill of materials that is created for the same component is of interest to many users in a factory. Similarly, many control computer installations will have their own data base. In this case, however, it may be necessary to exchange vital operating parameters with other data bases, making it necessary to provide standard data formats, interfaces, and communication protocols.

2.3.2 Centralized Data Base

The idea of a centralized data base was developed when problems were encountered with the concept of individual data bases. An attempt was made to bring into a centralized collection of files all information and data generated, stored, and processed (Fig. 2.5). The high cost of early computer equipment and storage devices helped to lead to this development. The advantages of this concept are that all the information on a manufacturing activity is located in one storage medium and that access is provided to all persons who need to query the system for specific information. Particular problem areas with this type of data base are as follows:

1. The administration of a centralized system is too unwieldy when a large number of data must be handled and processed.
2. The access time to retrieve specific information may be too long.

FIG. 2.5 Centralized data base.

The Computer in Manufacturing

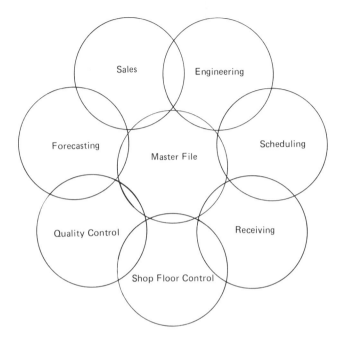

FIG. 2.6 Interfaced data bases.

3. Programming and maintenance of such a complex system becomes very difficult.
4. A large amount of information is of interest to only a limited number of users and does not need to be on such a large file.

A centralized data base is of particular value in forecasting, payroll and financial activities. It is also useful to support quality assurance activities and centralized part programming where a large number of NC machines have to be supplied with programs.

2.3.3 Interfaced Data Base

A data base is said to be "interfaced" (Fig. 2.6) when information for multiple manufacturing functions is entered only once into a system and then distributed automatically to different users. Each user entity has its own individually controlled data base, but its data may be requested and loaded down from another data base. This approach can be used to advantage where different manufacturing functions need a large portion of common data in addition to their specific data. Such information is typically provided by the bill of materials, which is an important document for most manufacturing operations. As the processing of an order proceeds from one operation to another, different specific sets of information are required. With the reduced information in the user files, the timeliness and accuracy of the data processing are greatly enhanced. The specific problems with this type of data are the following:

1. The entire system is very involved and requires expertise to conceive.
2. As numerous data files at various locations are required, they are difficult to control and to maintain.
3. Common data formats, interfaces, and communication procedures must be provided.
4. There must be system programs available which automatically provide each user group with the information they need.

2.3.4 Distributed Data Base

In the distributed data base, common data essential to many users are kept in a master file and specific data are kept in local files. The user group is able to manipulate data in both the master file and in its own file. Figure 2.7 shows a typical application where the information describing a design is kept in a master file. Groups involved in system design, tool design, and stress analysis have access to this master file to retrieve common information, but they also have their own files needed to perform their individual assignments. There are two possibilities for using the master file: It can be copied and loaded down or it can be accessed directly. In the latter case, provisions must be made that the file not be altered without unauthorization. A distributed data base is used with many hierarchical distributed control computers (e.g., in quality control and numerical control systems). In modern computer numerical control (CNC) concepts a large control computer supervises the operation of many direct numerical control (DNC) machines which contain their own computer and data storage. NC programs are generated on the control computer and are kept in its master file.

A part program is loaded down to a computer of the DNC machine as soon as the machine is readied for machining a part. The storage device of this computer may contain several part programs, thereby making the factory operation insensitive to failure of the larger computer. Many of the problems with this interfaced data base are similar to those encountered with the distributed data base. The design of a distributed data base is modular in nature and therefore should be easier to implement and maintain.

The Computer in Manufacturing 19

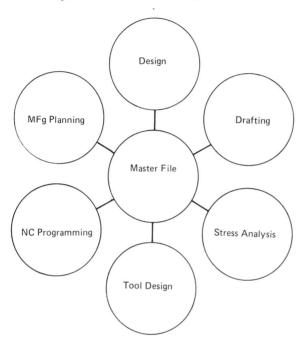

FIG. 2.7 Distributed data base.

2.4 HIERARCHICAL COMPUTER CONTROL CONCEPT

The first large computers to control a manufacturing operation were quite expensive and an attempt was made to utilize fully the available computing power. Thus there were many control tasks of different types assigned to the computer. It was quickly realized that to plan and install such a system was almost an overwhelming task. With increasing installation size, complexity increases and reliability decreases. In case of a computer failure, the entire control system can break down and seriously affect communication within the plant or even cause a total plant shutdown. The software overhead also increases rapidly with increasing size of the control task. In addition to the complex application programs, these computers have to support large operating systems to schedule all processing tasks. Service requests are answered sequentially, and when the computer is utilized to handle thousands of input and output requests, the control task may slow down considerably. In an overloaded system, data may get lost because it is not possible to answer all input requests. Eventually, the system deteriorates. When this occurs the computer has reached its capacity and is no longer expandable. It was also shown that systems which are loaded nearly to capacity are cumbersome to expand since there is a large amount of reprogramming to be done when additional software is to be added. Usually, patched-up software decreases the efficiency of the computer since changes were not originally planned or anticipated and little provision for expansion has been made. With a solitary computer system the entire plant is controlled by a huge master control program, consisting of many subprograms. A change in one subprogram often affects another task, since many programs are interrelated. Also, the mere size of the program and of the system make troubleshooting time consuming and discouraging.

The initial investment is very high in solitary systems, and in general it is difficult to justify an installation of such size. Since computer technology is new, the required capacity of a system is generally underestimated, and it does not take long to reach full capacity when new hardware and software additions are required. Many system designers have learned from this experience and tend to overspecify the hardware requirements.

Despite the many inherent disadvantages of a solitary installation, for some applications, such computers have advantages. For larger computers, there is usually more and better software available than there is for a small system, such as error diagnostic routines, compilers, operating systems, and so on. Larger machines have a definite advantage when complex algorithms used by mathematical models of the control system have to be executed. In particular, for process optimization, which requires many calculations and a sizable memory, the large computer will be more suitable. Also, the operating system in general is more powerful than that of the smaller computer. This includes the capability of allowing multiuser access and to work in both a foreground and a background mode. In the case of process control, large computers tend to have better hardware and priority-interrupt structures.

With the advent of mini- and microcomputers, tools became available which, when interconnected to a network, could make these computers as powerful as a large mainframe computer. Control functions requiring very basic calculations are often much better accomplished by small systems. Many of the tasks in manufacturing involve processing of digital input and output signals. These tasks are perfect for a minicomputer. When computer designers began to analyze complex plant control systems very carefully, it became apparent that many of these systems could be divided into modular subsystems. Mini- and microcomputers could be used to assume responsibility for controlling these subsystems by making one processor responsible for one or more modules of the total control task. It was also found that most control systems could be represented by a hierarchical structure; thus the

The Computer in Manufacturing

concept of hierarchical computer control system was developed (Fig. 2.8).

Hierarchical control systems are similar to the hierarchical structure of various levels of management in a manufacturing organization. Hierarchical systems are usually designed with two, three, or in some cases four levels of control. Figure 2.9 shows the interconnected control loops for part of a manufacturing operation. Typically, the control loop for the highest level—in this case for the sales function—will reside on the corporate computer and the next control loop—for production scheduling—on the plant host level, and so on. The data exchange between the control loops makes it necessary that handshake procedures, data formats, protocols, and interfaces be the same. Synchronization can be done in either the polling or the interrupt mode. The following rules should be observed when selecting computing equipment for and assigning control tasks to different hierarchical levels:

1. Higher levels control a larger part of the manufacturing system.
2. Higher levels require slower reaction times. Their information processing cycles are slow and decisions are made at longer intervals.
3. The decision process at higher levels is more complex and is based on fewer structured data. The formal presentation of a problem becomes increasingly difficult.
4. With increasing hierarchical level, the speed of information flow within an individual control loop decreases.
5. At higher levels real-time requirements decrease.
6. At higher levels the number of tasks to be performed by the computer increase.

The responsibility for the individual levels of control is discussed next.

2.4.1 Control at the Corporate Level

At this level the computer will assist corporate management to operate the business and to perform strategic tasks such as forecasting, marketing, and competitive analysis (Fig. 2.8). The data base from sales, purchasing, finance, and the manufacturing operation will be used to assist executive personnel in strategic planning and in all other areas of the business to be competitive in the marketing environment. Large simulation models and control schemes are very important tools to support work at this level. The central computer system will consist of a large data processing machine and several large peripherals which are typically used for business applications.

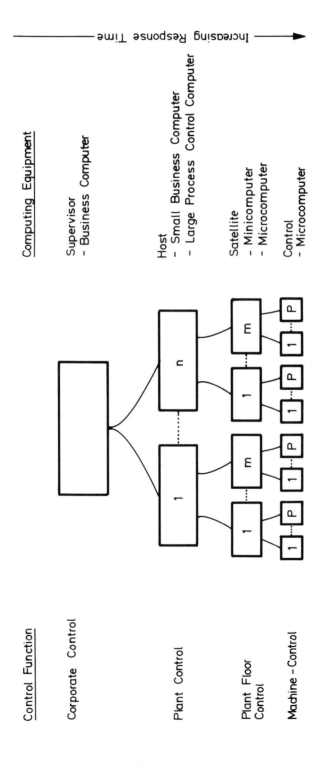

FIG. 2.8 Control functions of a firm and the supporting computing equipment.

The Computer in Manufacturing 23

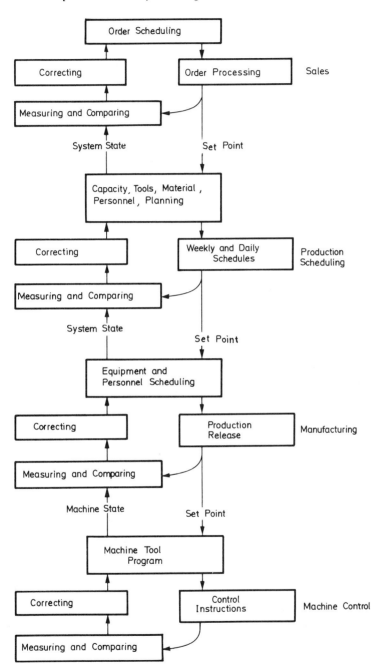

FIG. 2.9 Interconnected control loops in a hierarchical control system.

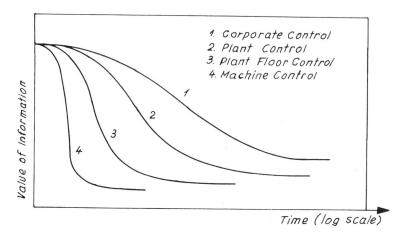

FIG. 2.10 Decline of the value of information over the time for different hierarchical levels.

It is important that these installations have a fast throughput to handle all control tasks in the business environment. This strategic computer operates in a longer time frame. Typically, information is processed and reported on a weekly basis. At this level, information retains its value for a long period (Fig. 2.10). The computer usually does not have to operate in the real-time mode. Reports are issued for the various sections of the business. Information from the manufacturing process is sent from the lower levels to the central computer via minicomputer and host computer. This information enables management to form the data base needed to operate the plant efficiently as part of the comprehensive corporate system.

2.4.2 Plant Control

There are usually several computers at this level, supporting the work of the plant management in operating the manufacturing system. The host computers perform supervisory functions such as scheduling orders for daily production runs, following the orders through the manufacturing facility, and reporting on production status and production problems. This control system provides middle management with the information necessary to supervise the plant efficiently on a daily basis. The information flow from machining centers, material handling equipment, and receiving and inspection stations will be tied together at this level. In addition, important information will be fed back to the various plant sections. This level of control helps the plant to function as an entity. The second level of control will be

The Computer in Manufacturing 25

connected to the corporate computer system; it assists the communication between plant and corporate management. The host computer communicates at the lower level with one or several mini- and microcomputers and coordinates the operations of these units. The hardware configuration of the host computer consists of a smaller business computer or a higher-priced control computer. Typically, the life cycle of the information at this level is 1 day—in many cases only several hours (Fig. 2.10). For this reason the supporting computer has to have real-time capabilities.

2.4.3 Plant Floor Control

Computers at this level operate directly with the process (Fig. 2.8). Their task is to supervise and control production runs and quality activities. Usually, a number of small mini- or microcomputers are located in several sections of a plant in close proximity to the equipment or process to be controlled. Each computer is assigned a limited control task, such as monitoring and supervising several machine tools or a smaller process. Control functions include scanning of process inputs from different sensors, counting, evaluation of critical parameters, some data reduction, and limited control of the process. At this level, the process needs continuous attention and immediate correction when out-of-control conditions are encountered; thus the computer will be communicating continuously with the process control elements and needs real-time capabilities (Fig. 2.10). Communication links have to exist between the machine and plant control level.

2.4.4 Machine Control

Computers at this lowest level are embedded directly in the machine tools and processing equipment. They perform the function of the real-time controller. Control programs and parameters are provided by the computer of the next-higher level. The advantages and disadvantages of large solitary and hierarchical systems are summarized in Table 2.1.

2.4.5 Future Trends in the Development of Computer Systems for Manufacturing Control

Computers for use at both the corporate control and plant control levels are still quite expensive. Also, the multiuser software that is needed to make this equipment a versatile and useful tool may become very difficult to conceive and design, in particular when the manufacturing information system becomes very large. System researchers will try to solve this problem either by standardizing individual manufacturing control functions or by conceiving flexible manufacturing operations with the aid of computer hardware and software.

TABLE 2.1 Centralized Computer System Versus Satellite System

Central computer	Satellite computer
High hardware cost	Low hardware cost
High software cost	Low software cost
Large software overhead	Low software overhead
High in-plant wiring cost	Low wiring cost
Software is complex and difficult to change	Software is modular and easy to write
Low service speed (operates in sequence)	Fast service speed (control task divided among many small but fast computers)
Difficult to expand	Easy to expand
When the computer fails, the entire plant shuts down	When one computer fails, only a section of the plant shuts down
Expensive backup	Inexpensive spare computer can be kept as standby

A large control system can be assembled from modules, where the individual control functions are divided into smaller entities and implemented on small computers interconnected to a distributed computer system (Fig. 2.11). A topic of current research is implementation of the system and user software to operate the individual computers in firmware, thus simplifying the configuration of a distributed computer architecture. For the operation of such systems, distributed data bases are needed. Such data base concepts are still very new and a considerable amount of research has to be performed to make them practical.

2.4.6 Implementation of Hierarchical Computer Systems

The conception and implementation of hierarchical plant control is a long evolutionary process. Usually, each computer installation is directly tailored to a particular manufacturing organization. This even often holds true for factories that produce the same product but are located at different sites. In general, management methods and organizational structure of similar plants differ so much that no common control system can be conceived. In some cases smaller subsystems may have very close similarities and can be used by several organizations.

Computer installations will render satisfactory service only if a plant-wide or even corporate-wide systems approach is taken. All building blocks of an installation must be carefully planned with all departments that will be sources or end users of data. The data flow and interaction between departments must be known to design an efficient control system. In addition, the required degree of automation must be determined. This is a very important step since it determines the cost of an installation. The implementation of the computer system may be started with a simple application to get production personnel familiar with computers, or the entire plant may be automated, with all manufacturing operations supervised by a computer. The latter approach is seldom taken since the risk of startup problems is very high. It is necessary to examine the limitations of mechanizing discrete processes. In continuous processing plants, a very high degree of mechanization is often obtained. Here the entire processing plant can be put under close computer control, starting with incoming raw material and ending with the preparation of the final product for shipment. Because of the nature of discrete processes, total mechanization is not always possible. Some gaps between individual processes will always exist which have to be bridged by manual assembly operations or conventional material handling methods. However, it should be pointed out that it is often possible to streamline discrete manufacturing by viewing the total manufacturing plant as a continuous operation. Optimized process automation can be obtained by linking computer mechanization with systemization. Usually, the first step to automation is to mechanize individual subsystems of a process. After this has been done, the individual components can be tied together at a plant level.

The control system of a new plant will differ from that of an older one. Here advantage can be taken of all the capabilities that computer technology has to offer. A new plant is usually planned with a higher degree of automation, and the outcome of such an endeavor will stand to gain a higher degree of success. Most of the major bad decisions that add to the inefficiency of a plant are made during the early planning stage. Therefore, it is important to spend enough time and effort to develop a good control strategy. It can be quite difficult to design computer control for an older plant, as they have not been built up with the computer in mind. Many facilities have expanded through a natural evolutionary process. The computer then is assigned the additional task of patching up mistakes made in the past. Companies that operate very large manufacturing facilities or a multitude of plants can be faced with considerable inefficiency if computer installations are not planned on a corporate-wide basis. Provision must be made to standardize data formats, interfaces between individual hierarchical levels, and communication protocols.

The communication task can be quite difficult if each installation uses the computer of a different manufacturer. Usually, communication software and computer hardware vary among manufacturers. Software and hardware standardization can provide a considerable advantage. Even during the development of a computer system, standardization can be important. In many cases, software packages and interface hardware have to be developed only once and can be used for different applications. Computer engineers can be transferred easily from one installation to another, and documentation can be simplified. Computer maintenance service may also prove to be much simpler if standardized equipment is used.

There are two principal strategies by which a computer system can be installed: the bottom-up and the top-down methods. The bottom-up approach starts with the computers at the lowest level of the hierarchy. It has the following advantages:

1. Many engineers can start simultaneously with several installations.
2. The elements of the completed computer system can be started up and operated individually.
3. Developments from other installations can easily be adopted and implemented.

The disadvantages of this method are:

1. It may be difficult to add the second control layer since data formats, communication protocols, and interfaces may be different.
2. The interaction of individual computers at the same level may not have been specified in sufficient detail and therefore may cause severe difficulties.
3. Programming the computer installation may become very expensive because computers at the same level may not be compatible.

The top-down approach starts with the computer at the highest hierarchical level. Its advantages are:

1. Data formats, communication protocols, and interfaces for the lower levels are uniformly defined.
2. This approach is a natural way of first defining the overall system and then gradually breaking it down into modules and submodules.
3. Each module at the lower level can be easily interfaced to a module at the higher level.

Its disadvantages are:

1. It is usually difficult to obtain ready-made submodules from other projects which can be adapted to the computer system.
2. The control structure modeled may not provide the requirements specified for the process and thus a necessary change may involve the implementation of an entirely new control strategy.

The Computer in Manufacturing

Generally, the design of an actual system, the two strategies are used simultaneously. There are computers installed at the top and at the bottom and the manufacturing control system slowly grows together from both directions.

2.5 FUNCTIONS OF A MANUFACTURING ORGANIZATION

A computer-aided manufacturing operation can be defined as a production facility in which all activities to manufacture a product are planned, guided, and controlled with the aid of a computer. The computer system may consist of a single or several interconnected computers which contain production plans and control algorithms to operate the plant efficiently. This definition does not exclude the possibility of partial control by human beings, or human intervention if production problems arise. It is not possible for automated production plans and control algorithms to be able to handle all exceptions and random interferences with the process.

The individual functions of a manufacturing organization and their interactions can be explained with the help of Fig. 2.11. The activities shown are typical for a company manufacturing products in small and medium-size production runs. At the top of the figure the plant activities are grouped together in categories: strategic goal setting, technological planning, time scheduling and organizational planning, monitoring and control, and accounting. The time horizon is shown at the bottom of the figure, and the flow of funds, materials, and information is indicated by individual flow lines and arrows.

The terminology used for the individual functions may not conform to the one to which the reader is accustomed. This is due primarily to the fact that there is no standardized terminology for most activities. In addition, this concept of the manufacturing organization may not be representative of all firms. The type of product, the product mix, and the size of the production runs greatly influence the structure of a manufacturing facility and its organization. The individual functions of the firm are described next.

2.5.1 Marketing Research

Marketing research is defined as the systematic gathering, recording, and analyzing of data about problems related to the marketing of goods and services. In modern companies, such research activities to scan the market, predict its future potential, and collect and analyze current developments and the company's marketing performance have become mandatory. This activity produces reports on sales forecasts, new geographic market potentials, market movements, the end of a product's life cycle, and customer preference as to a company's own

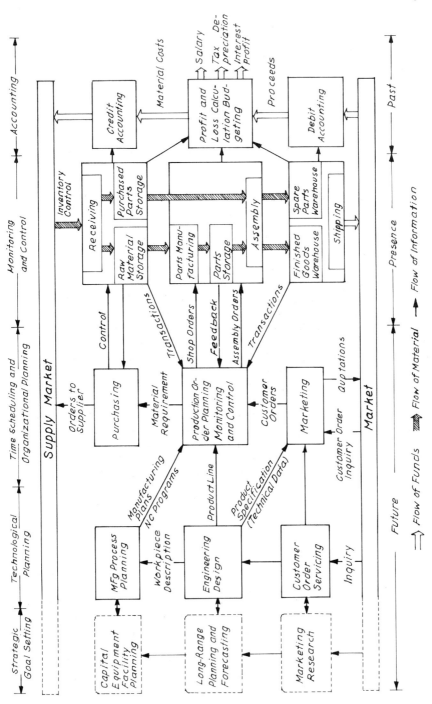

FIG. 2.11 Basic functions of a manufacturing facility for small and medium-size production runs. (From Ref. 3. Courtesy of VDI-Verlag, Düsseldorf, West Germany.)

and competitive products. The information collected is used by long-range planners and forecasters. Marketing research relies heavily on the use of powerful scientific or commercial computers to process and interpret marketing data. The mathematical tools to assist these activities are derived from statistics, Bayesian decision theory, and operations research.

2.5.2 Long-Range Planning and Forecasting

Forecasting involves assessment of a company's past performance and projection into the future. The survival and growth of a company depend on its ability to foresee the long-range performance trend of its products, customer behavior, and market developments. Forecasting must include all segments of a manufacturing organization. A product will be successful only if it is in high demand, of high quality, and is reasonably priced. To obtain these objectives, a forecast must be made for all areas of the business, including the product, the market for the product, the location of the factory, and distribution centers, the manufacturing processes, the labor market, the energy and material, supplies, inventory, and competition.

Typical forecasting activities are shown in Fig. 2.12. They heavily rely on modeling, statistical evaluation, and the use of a commercial computer. The sources of the information going into forecasting usually include marketing research, the market, design, and other activities. The information is needed primarily by product planning, capital equipment and facility planning, design, and marketing. The principal forecasting activities are summarized as follows:

   Forecast of the product line
   Forecast of annual sales budget
   Establishment of profit objectives

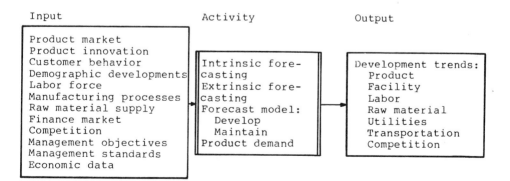

FIG. 2.12 Long-range planning and forecasting activities.

Preparation of corporate and factory budgets
Forecast of product standard costs
Drafting of operating profit or loss plans
Establishment of capital investment programs and appropriation schedules
Assessment of capital and cash requirements
Forecast of the resource and labor supplies

2.5.3 Capital Equipment and Facility Planning

This activity is needed to plan new or improved production facilities. It may start with site selection for a new plant, whereby the market, availability of labor, material supply, utilities, tax structure, and many other factors are considered. Planning can be supported by simulation programs to investigate different site alternatives and the influence of changing resources (Fig. 2.13). In a succeeding step, processing equipment is investigated and the optimal layout of the floor plan is obtained with the help of a simulation program. In addition, a return-on-investment analysis is performed for various plant alternatives. As an output of this activity, plans for a new or improved plant and for the layout of machine tools, processing equipment, and transportation system are obtained.

2.5.4 Customer Order Servicing

The function of customer order servicing is to process an order through a plant quickly and efficiently (Fig. 2.14). Upon receipt of an order,

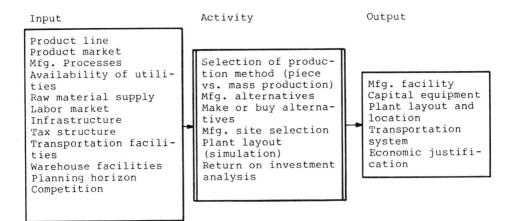

FIG. 2.13 Capital equipment and facility planning activities.

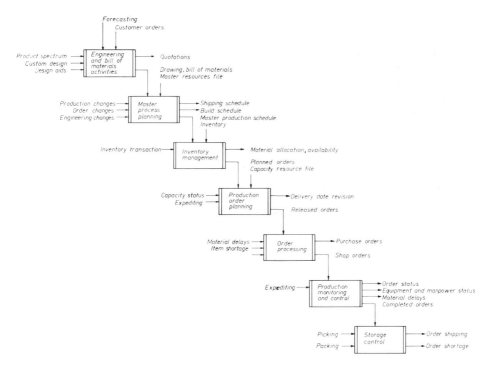

FIG. 2.14 Customer order servicing comprises many manufacturing activities.

the customer and the order are identified and a check for correctness is made. Special order requirements and engineering work to be performed are determined. An inquiry is made to determine if the order can be filled from inventory or if it has to be scheduled for manufacturing. As the output of this activity, the customer obtains a delivery date. A customer's inquiry about the status of an order can be obtained, and the customer may be informed about manufacturing delays, part supply problems, or necessary engineering changes. Customer order servicing will also be activated if the order processing must be accelerated.

2.5.5 Engineering and Design

The product to be marketed will be conceived, designed, and tested by this activity. The design phases are product definition, conception of the operating principle, design, and drafting. In most cases a prototype will be built and then will be engineering and field tested.

The engineering department is also responsible for generating the bill of materials, process plans, and description of the machine tools used.

There are a large number of data to be processed in the engineering department (Fig. 2.15). The sources of the principal data are customer order servicing, marketing, quality control, and short- and long-range planning. The outputs of the engineering department are the fundamental data needed by all manufacturing activities. The most important manufacturing documents are:

1. The engineering drawing, describing the product, its dimensions, tolerance, shape, surface, material, and so on
2. The bill of materials, containing the record on the product structure and the basic components of a part
3. The process sheet, describing the processes to manufacture the product
4. The processing equipment file, containing information as to department, machine tools, cutting tools, and fixtures to be used for manufacturing of the part

In addition to these documents part programs, machine loading plans, line balancing sheets, and quality control instruction can be generated automatically. The engineering department keeps a record of all product changes and product history and is responsible for the design of the product instruction and service manuals.

2.5.6 Manufacturing Process Planning

In this activity the operation plan and the NC programs for a workpiece are prepared (Fig. 2.16). Machining operations, assembly procedures, process sequences, machine tool alternatives, and fixturing methods are kept in the form of decision tables. There is also information on standard parts, part alternatives, and part families. When a part is to be manufactured, a code is entered and the computer performs a part search. The output of the manufacturing activity planning is a comprehensive process sheet on the manufacturing operations, manufacturing sequences, NC programs, and assembly instructions.

2.5.7 Marketing

Marketing actually comprises the entire business organization; it analyzes, organizes, plans, and controls and company's resources, policies, and activities to satisfy the needs and wants of the customers. Because of this broad definition, it is very difficult to give a brief description of this activity. Figure 2.17 shows the principal functions of the sales activities, which are only a part of marketing. The sales

Inputs	Activities	Outputs
Workpiece-related	Functional analysis	Drawings
Identification number	Product definition	Bill of materials
Name	Study of alternatives	Product documents
Where used	Value analysis	Product manuals
Change status	Engineering calculations	System data base
Reference to origin	Simulation	
Position number	Product layout	
Status	Detailed drafting	
Number required	Dimensioning	
	Customer assurance	
	Processing of drawing changes	
Drawing-related	Engineering changes control	
Drawing number	Product history file maintenance	
Designer	Creation of CAD data file	
Date	Part family analysis	
Scale	Part classification	
Microfilm	Product costing and pricing	
Drawing size	Bill of materials processing	
Company	Customer options processing	
Miscellaneous	Engineering testing	
	Field testing	
Geometric data of product		
Shape		
Geometry		
Dimensions		
Tolerances		
Surfaces		

FIG. 2.15 Product design functions.

Inputs	Activities	Outputs
Geometric data of product (continued)		
Standard shapes		
Variants		
Hole patterns		
Tolerance classifications		
Physical properties of product		
Specific weight		
Strength		
Elongation		
Hardness		
Heat transfer		
Fluid dynamics		
Aerodynamics		
Corrosion resistance		
Electrical resistivity		
etc.		
Product requirements		
Function		
Efficiency		
Capacity		
Fuel consumption		
Operating environment		
Speed		
Load		
Service life		
Weight		

Size
Stability
Shape
Noise emission
Cost
etc.

Manufacturing requirements

Machining capabilities
Machining processes
Machining sequences
Tooling
Part commonality
Manufacturing alternatives
QC standard
QC requirements
Fixed and variable costs

Process selection
Machine tool selection
Process planning
Part programming
Machine loading
Assembly line balancing
Process capability study
Tooling analysis
Tool selection
Time standards selection
QC procedures selection
QC limits selection
Product pricing

Master production plan
Manufacturing activity plan
Manufacturing alternatives
Manufacturing simulation results
Machine loading plans
Line-balancing plans
Cost analysis data
Cost and time studies
Part programs
QC limits
QC test descriptions

Material supplier

Vendor file, raw material
Vendor file, components
Vendor performance file

Vendor selection

Material requirements plan
Material purchase orders

FIG. 2.15 (Continued)

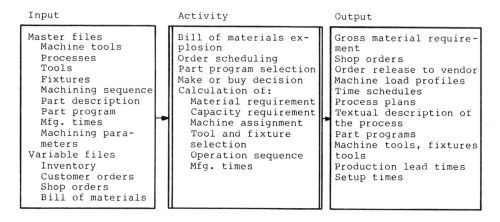

FIG. 2.16 Order planning or process planning activities.

department is the interface between manufacturing and the customer. It sells the products and determines pricing strategies, the model mix, product specifications, quality standards, and so on. The decisions made by this activity usually have a direct effect on all sections of the firm. The sales department is also the feedback channel between marketing, engineering and manufacturing. Competitive behaviors are recorded, customer needs and wants are analyzed, and performance problems with the product are reported.

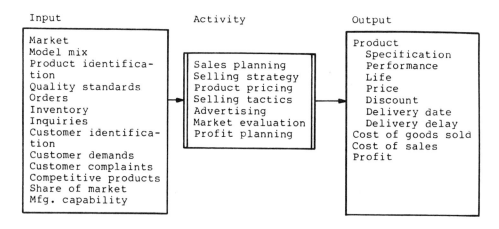

FIG. 2.17 Sales activities.

2.5.8 Production Order Planning and Manufacturing Monitoring and Control

When a workpiece is planned and released for production, it has to compete for the resources of the firm with many other workpieces. If the delivery date of an order is to be met, a release time must be scheduled to allow timely completion of the order. A load profile to detect bottlenecks and waiting queues must be established for each machine. If necessary, alternatives have to be found for either work schedules, processing equipment, manufacturing sequences, overtime schedules, and so on. Order planning can be supported with the help of operations research methods. With these tools it is possible to optimize or suboptimize the utilization of the manufacturing equipment.

Order planning actually starts with the release of the bill of materials (Fig. 2.15 and 2.16). With the help of this document material and tool requirements are determined, after which a make-or-buy decision is made for components. Parts are either released for in-house or supplier manufacturing. There are usually two planning schedules, one short range and one medium range. A machine load profile is established and machine assignments are made to meet the time schedule established.

The primary function of manufacturing monitoring and control is to assure that schedules are met and that activities are coordinated. When the order is released, the NC programs have to be administered and the manufacturing process has to be controlled. There are usually a host of computer activities performing this function. NC programs have to be downloaded and distributed to machine tools, and control data have to be supplied to the shop floor to initiate execution of the order, to direct the material flow, and to administer the distribution of tools and fixtures.

Status information about the process is collected from workstations and machine tools and sent to a host computer for evaluation. Apparent production problems are recorded and steps for corrective action are taken. An effective monitoring and control system should supervise all functions of a plant. Figure 2.18 shows the concept of a work-in-process data collection system that monitors the material flow through a manufacturing facility. Such a system will easily detect production difficulties and bottlenecks during all processing stages. It will be part of a centralized data processing installation. Because of cost considerations, many manufacturers will not install such a comprehensive plant-wide system. The functions of a typical scaled-down plant monitoring and control system are shown in Fig. 2.19. They include attendance reporting, job assignment, expediting and material handling control. In Fig. 2.1 it was shown that the time a workpiece resides in a factory is quite long compared to its actual processing time. A well-conceived factory monitoring and control

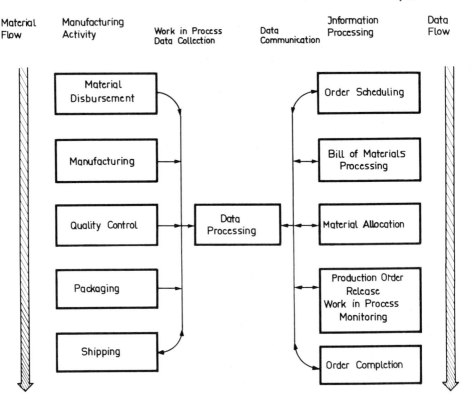

FIG. 2.18 Work-in-process data collection: data and material flow. (From Ref. 5.)

system should be able to reduce this idle time substantially. Thus it should be possible to utilize machine tools, processing equipment, and conveyor lines more productively.

2.5.9 Purchasing and Receiving

Purchasing is responsible for the acquisition of raw material and parts manufactured by outside sources (Fig. 2.20). Criteria for selecting a supplier are the price and quality of a product, delivery conditions, and past performance of the contractor. The factory's material requisition is received by the purchasing department and requests for quotations are sent out to suppliers. The quotations can be initiated automatically or manually. From the quotations submitted, a supplier is selected, who will issue an acknowledgment for the order. Generally, an evaluation file is established for each supplier, ranked on the basis

The Computer in Manufacturing

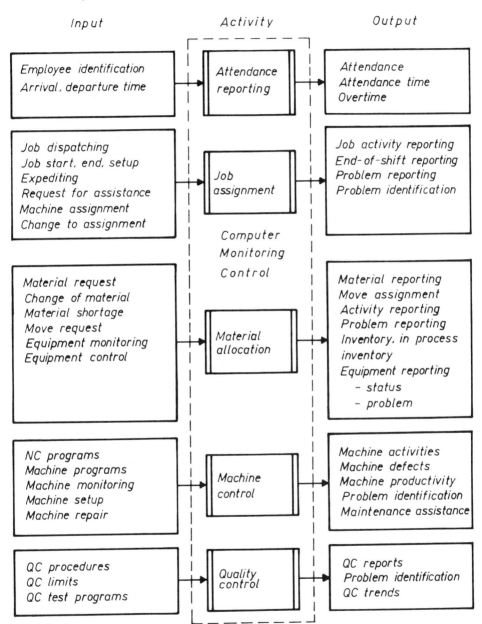

FIG. 2.19 Function of computerized plant floor monitoring and control. (From Ref. 5.)

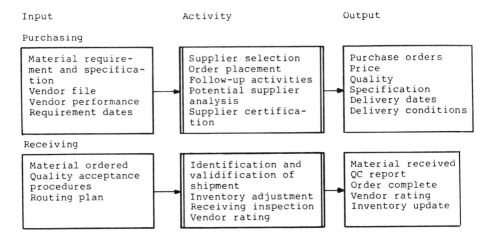

FIG. 2.20 Basic functions of purchasing and receiving.

of price, timely delivery, and the quality of the product manufactured. Often a follow-up of the order is issued to assure that the material will be delivered on time.

Receiving identifies the shipment and checks it against the original order (Fig. 2.20). If discrepancies are discovered, the vendor will be notified to complete the order. The material received will be quality tested and released to the warehouse. The number of parts tested and the thoroughness of testing depend on the past performance of the vendor. The quality of the shipment will be entered in the vendor rating file.

Control computers can be used very successfully for receiving inspection. With this equipment it is often possible to perform 100% testing of all components received. The test program will be initiated by the entry of an identification number. The computer then takes over control of the test, gathers and evaluates the test parameters, and issues a quality report. This equipment may be connected to the plant-wide computer network.

2.5.10 Inventory Management

The material released by receiving inspection is stored in inventory. There are usually two kinds of inventory. Material used immediately will be stored as in-process inventory at special buffers between the processing equipment, as will semifinished parts that are transferred from one machining operation to another. Parts to be used in the future are kept in a central warehouse, from which they are distributed on request.

The Computer in Manufacturing

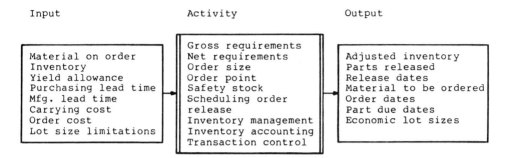

FIG. 2.21 Functions of inventory management.

The computer plays an important part in maintaining an inventory record and in controlling inventory transactions (Fig. 2.21). In larger installations a hierarchy of computers will control the warehousing operation. When an order for retrieval of a part is initiated, a central computer sends the corresponding information to a control computer which is responsible for supervising the operation. It, in turn, issues control signals to a computer or programmable controller to actuate the stacker crane. The part is picked from its bin automatically and brought to the pickup area. From there it will be carried by forklift truck, conveyor system, or other conveyence, again under computer control, to its destination. The central computer updates its inventory file and if necessary issues orders for the purchase of new parts. Since this computer contains an image of the entire warehouse, it is also capable of issuing upon request an on-line inventory report and records on outdated material.

2.5.11 Maintenance

Maintenance plays an ever-increasing role in automated factories. With the introduction of transfer lines and other interconnected processing equipment, equipment breakdown becomes quite a severe problem. Transfer lines often operate at only 60 to 70% of their rated capacity because of the failure of one component. Because of this problem, preventive maintenance is performed on complex machinery. In many cases control computers supervise the functions of important machine components, such as hydraulic cylinders, index tables, relays, pumps, and hydraulic systems. The general role of the computer in maintenance is in scheduling or preventive maintenance, issuing maintenance reports, administration of the maintenance material inventory, purchase-order release of new maintenance material, and real-time supervision of machine components.

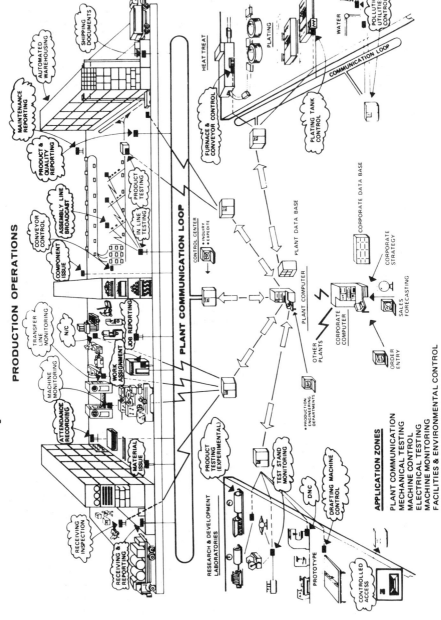

2.5.12 Accounting

Accounting is necessary to establish operating budgets for a firm and to set standards against which are judged the operation of the firm as an entity and those of the firm's individual cost centers. A report is made about what actually did happen as against what was planned. Thus management is provided with a check on its own effectiveness. Sales arising from a given product or from activities in a market segment must be matched with the efforts made in that area. Costs are assigned to departments, divisions, and sections. Cost factors to be considered include labor, material, transportation, utilities, taxes, depreciation, and overhead. A further breakdown is made between fixed and variable costs, although the distinction between these is often very difficult to make. Cost figures for each activity are carefully collected and evaluated. They are represented to management in the form of tabulations and reports, thus permitting a comparison of actual and projected costs.

Accurate cost figures for each product are needed for quotations submitted to customers. In addition, management can obtain information on the profitability of each product and the productivity of manufacturing processes and of each cost center. Previous and present costs are used for cost projections of future sales activities.

2.6 EXCURSION INTO A COMPUTER-AIDED MANUFACTURING PLANT

Figure 2.22 shows the layout of a fictitious plant. In this plant many of the computer applications discussed throughout this chapter are implemented and combined in a comprehensive manufacturing control system. There are computer applications in engineering, plant maintenance, and plant control. The reader will be led through an example involving the movement of a part through the plant, where it is manufactured from raw material, machined, measured, and finally assembled into a product.

The parts or the raw material arrive at the receiving inspection station by truck. The operator identifies the shipment to the computer. From the main computer, information about the number of parts ordered and their quality control procedure and possibly the control program for the automatic test equipment is sent to the receiving inspection department. The number of parts tested are determined with the help of statistical methods which also take into

---

FIG. 2.22 Computer-aided manufacturing. (From Ref. 6.)

consideration the vendor's historical quality performance record. The automatic test equipment measures the parts and sends the quality test together with information on number of parts accepted and rejected to the main computer. Here the vendor performance file is updated and, if necessary, new orders are issued to the vendor to replace defective or missing parts. In the next step the parts are automatically stored in a warehouse, where a storage bin is selected according to such criteria as shortest travel distance, number of transactions for this type of part, and safety provisions. Critical parts may be stored in different isles to avoid part shortages during the possible breakdown of a stacker crane.

When the central computer schedules the part for manufacturing, it releases the part and sends retrieval information to the computer in the warehouse. It, in turn, orders the part to be brought to the pickup area, and a computer-dispatched forklift truck carries the part to the machine tool. The machining instruction and machining sequence for the part have been determined automatically by the central computer and are known to the plant computer. The plant computer sends the NC program for the part to the corresponding machine tool and the part is machined under computer control of the machine tool. Operating data from the machine tool and quality data from the part are returned to the plant computer, evaluated and condensed, and sent back to the central computer. From the operating data of the machine tools and the attendant's identification number, the central computer can calculate the payroll. The part proceeds under computer control from one machine tool to another until it is again automatically stored in a finished parts buffer. The inventory of this buffer is known to the central computer.

When all parts for the product are completed and stored, the central computer schedules the assembly. The plant computer releases all parts needed at the assembly station. They are brought to the individual stations by a conveyor system under computer control. The progress of the assembly is checked automatically at various quality control stations. When the product arrives at these stations, it is identified by a scanner and the computer retrieves its test program and controls the test run. Test data are channeled to the central computer. The finished product is tested again and brought to the finished-goods warehouse. Here it can be grouped together under computer control with other products that are to be shipped to the same destination. In this plant the computer performs scheduling, order release, machine control, machine status reporting, material movement control, warehouse control, attendance reporting, payroll calculation, maintenance scheduling, and other tasks. The conception of possible additional computer applications is left to the reader's imagination.

REFERENCES

1. Electronic Data Processing for Production Planning and Control, VDI Taschenbücher, VDI-Verlag, Düsseldorf, West Germany, 1972.
2. W. L. Howard, Data Base—From Independent to Distributed, Society of Manufacturing Engineers Technical Paper MS79-844, Dearborn, Mich., 1978.
3. Communication Oriented Production Information and Control System, IBM Corporation, White Plains, N.Y., 1973.
4. VDI-Bericht 173, VDI-Verlag, Düsseldorf, West Germany, 1971.
5. W. G. Hoppen, Automated Scheduling of Manufacturing Resources, Society of Manufacturing Engineers Technical Publication EM76-970, 1976.
6. IBM Publication G520-2643-0, IBM Corporation, White Plains, N.Y., 1973.

3

Hardware and Software Components for Computer Automation

3.1 CRITERIA FOR SELECTING AUTOMATION TOOLS

In Chap. 2 an overview was given as to the function of the electronic computer in a manufacturing organization. Computer-controlled automation can be successfully implemented only if production equipment is available which was designed for flexible manufacturing and which can easily be interfaced with computers. In general, it is very difficult to interconnect conventional manufacturing machinery to the computer, since it was not meant for programmable automation. In some cases retrofitting might even be impossible.

Modern control technology does not necessarily antiquate all hitherto known control methods and control equipment. It needs careful skill and experiences to combine old and new control and manufacturing know-how to a well-balanced automation system. Since a successful computer application depends largely on a well-conceived hardware-software approach, it is very difficult to assess the capability and the cost of programmable automation. Thorough software engineering knowledge is a prerequisite when a CAM system is to be implemented. Unfortunately, industry has not yet learned to estimate the cost and time to develop software, in particular for real-time applications. A recent study made on this particular subject has highlighted some of these difficulties [1]. In this research work 165 users of process control computers were queried on their experience with the design and implementation of a control installation. The questions covered selection criteria for a computer system, software responsibility, software and hardware cost, realization time for software and hardware, and so on. Some of the results which are of interest in the context of this book will be discussed. Figure 3.1 shows the realization time for a typical control computer project, and Fig. 3.2 shows the number of person-years required for project completion. It becomes

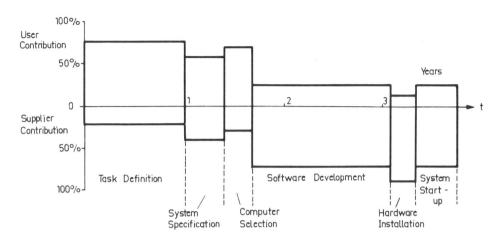

FIG. 3.1 Duration of project phases to install a control process computer system.

evident from this study that the time and effort invested in the project definition and software production are very critical and need careful attention. Most implementers stated that if more time had been used for the project definition phase, considerable reduction of software debugging cost might have resulted. The estimated and actual costs

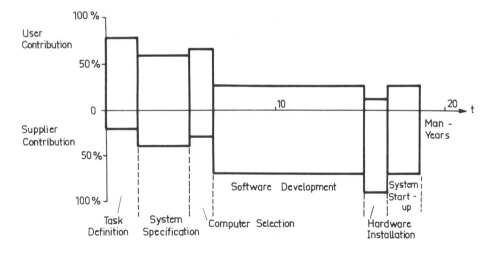

FIG. 3.2 Effort to realize project phases to install a process control computer system.

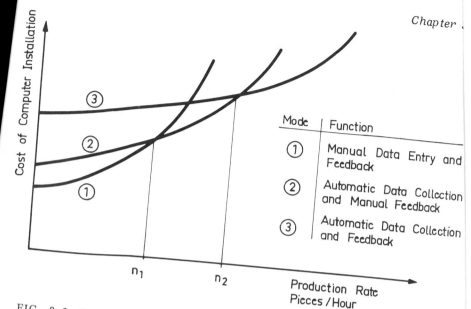

FIG. 3.6 Cost of different control computer installations over production rate.

the process is done by an operator. Because of its simplicity and low cost, this mode of operation will be used for manufacturing processes with small production runs.

Mode 2: Data are automatically collected on-line by sensors under program control and are transferred directly to the computer for evaluation. Feedback control is carried out by the operator. This mode of operation is more expensive and will be useful for medium-size production runs.

Mode 3: Collection and evaluation of data are performed as under mode 2. However, if a correction in the process is necessary, automatic feedback control is done by the computer. This control renders the highest degree of automation; it can be used economically for mass production.

The complexity and cost of the software and hardware increase with increasing mode number. In Fig. 3.6, the cost of computer installations for the three modes of operation is plotted over the production rate. It can be seen that mode 1 is cost-effective only until an output rate of n_1 has been reached. With increasing production mode 2 will render a more economical solution. When the output rate is greater than n_2, a system with computer feedback control will be the best selection for the automation task.

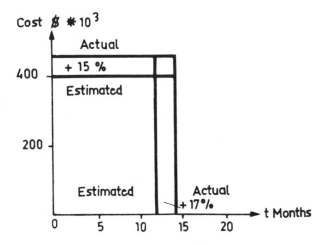

FIG. 3.3 Cost and realization time for control computer hardware.

together with the implementation time for system hardware are shown in Fig. 3.3. The same information for software is shown in Fig. 3.4. In both cases it appeared to be very difficult to estimate the cost and realization times. However, the software imposed most of the problems. For an average software phase it took 53% more completion time than originally estimated and the cost thus increased by 72%. This very typical problem complicated the return-on-investment (ROI) analysis that should have preceded each project. In the study this question was also investigated. It was found that in more than 50% of all installations a proper ROI analysis was not conducted.

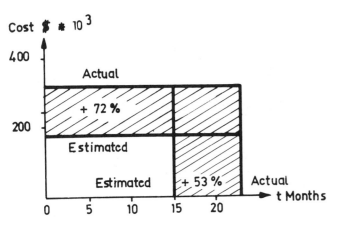

FIG. 3.4 Cost and realization time for software.

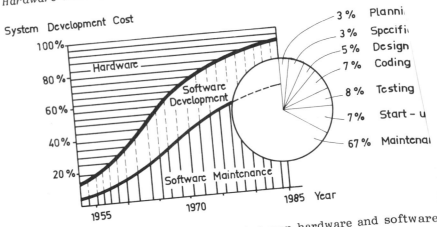

FIG. 3.5 Trend of cost distribution between hardware and software. (From Ref. 2.)

A recent analysis of software cost for automation projects shows that there will be a steadily increasing trend in the future (Fig. 3.5). Thus for the conception of such a project it becomes necessary to find a method to estimate accurately the cost of software and hardware. In addition, techniques have to be found by which the implementer will be able to determine if an automation task can more economically be done by hardware or software control or by a combination of these two technologies.

The hardware and software components needed for automating a manufacturing facility depend on many different aspects, such as:

Size of the plant
Labor market
Cost of labor
Mass, batch, or piece part production
Production rate
Number of models manufactured
Complexity of the product
Available automation technology
Desired degree of automation

When a computer is used to support the automation task, a system architecture may be configured to operate in three different modes (Fig. 3.6):

Mode 1: Data are entered by hand into a data acquisition terminal, evaluated, and processed by the computer. Feedback to correct

Hardware and Software Components

One particular problem should be avoided when computer control is specified. It is very easy to try to overautomate a plant with the help of the computer, in particular when the personnel are not too familiar with conventional automation. In many cases it will be less complex to perform a task with conventional control than with computer control. Overautomation can lead to the situation shown in Fig. 3.7. The cost of computer control typically increases exponentially with increasing degree of automation. The software and control system engineering contribute the most to this problem. Thus it may be possible that the last 5% of an automation task will cost as much as the first 95%.

Figure 3.6 shows that the control technology selected for an installation is a function of the piece rate. Similarly, the piece rate determines the manufacturing technology to be used (Fig. 3.8). NC machines are usually the best selection for small production runs. Flexible automation is more economical for medium-size production runs. For very high production rates, a fixed type of automation will render the best solution. Since the complexity of the part is also a major selection criterion, it is very difficult to determine the transition points between the different types of manufacturing technology. For example, the required piece rate to justify flexible automation is low for very complex parts and high for simple parts.

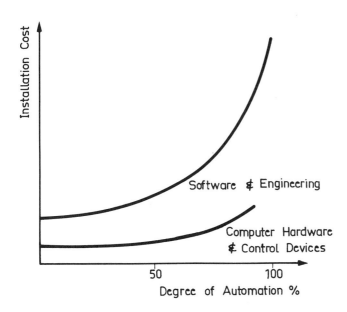

FIG. 3.7 Cost of a control computer installation over degree of automation. (From Ref. 3.)

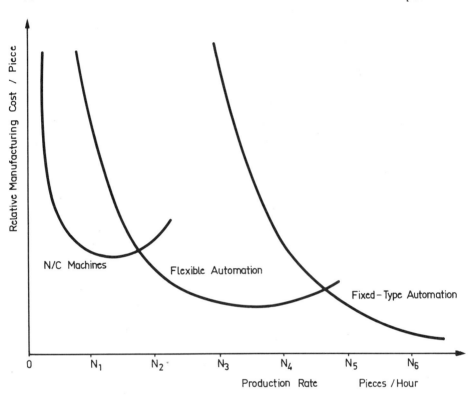

FIG. 3.8 Relative manufacturing cost of workpieces over piece rate for different manufacturing technologies (the value of N_i as an index decreases with increasing complexity of a part).

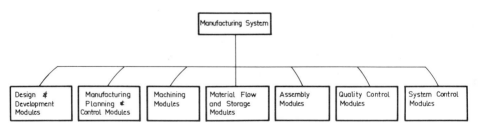

FIG. 3.9 Components of a manufacturing system.

3.2 COMPONENTS OF A MANUFACTURING SYSTEM

The components of a manufacturing system are quite numerous and depend on the product to be manufactured. For this reason it is impossible to itemize them all. However, a brief attempt will be made to give the reader who is not very familiar with software and hardware used for manufacturing a short overview of typical system components. In Fig. 3.9 the automation modules are grouped together under different headings. The following discussion provides an expanded view of each of these groups.

3.3 DESIGN AND DEVELOPMENT

The heart of the automation system is the computer. It will be used for scientific calculations, drafting, and product testing (Fig. 3.10). The designer enters the description of the product with the help of an interactive terminal, a keyboard, a graphic tableau, a light pen, or a digitizer into the central graphic computer (Fig. 3.11). The information is kept in a central graphic data bank and can be accessed for design layout, design changes, and updating. There will also be software modules for calculating physical characteristics of the design, such as strength, deflection, and heat transfer properties. Major calculations such as finite element analysis or product simulation and optimization will be done on the scientific or business computer. Both of these computers can be accessed via a communication line. The bill of materials, NC tapes to produce the workpiece, and, if available, manufacturing plans and schedules will be entered to the central management information file via the business computer. This communication link also serves as an interface between the marketing, forecasting, quality control activities, and engineering. Engineering drawings are produced by plotters or automatic drafting machines. Figure 3.11 shows a possible layout of a CAD system. Several workstations are connected to the central graphic computer, and they can be operated in a multiuser mode.

Engineering will operate its own model shop to manufacture engineering models and prototypes. This function may typically be performed by a NC engine lathe or by milling or boring machines. Figure 3.12 shows an universal two-axis tape-controlled NC engine lathe. The tape describing the rotational part to be machined will be produced automatically by the central graphic computer and read by a tape reader of the lathe controller. With this information the controller actuates the drives of the lead screw and cross lead screw, thereby positioning the lathe carriage. The cutting tool for machining the workpiece contour is fastened via a tool post to the lathe carriage.

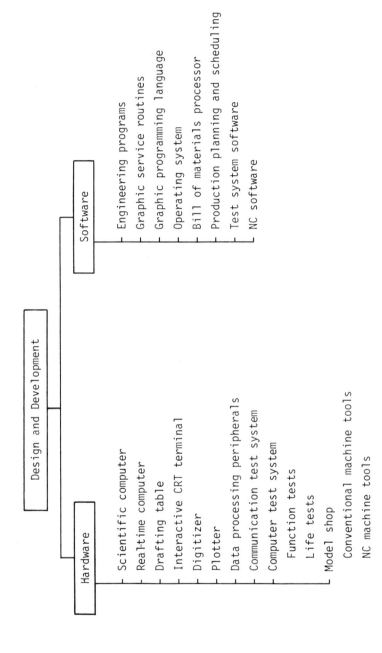

FIG. 3.10 Principal automation modules for design and development.

Hardware and Software Components

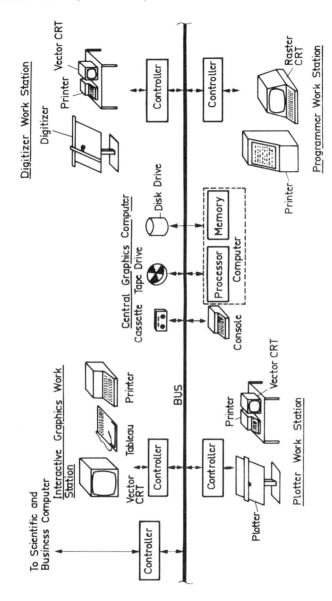

FIG. 3.11 Possible layout of a CAD system. (From Ref. 4.)

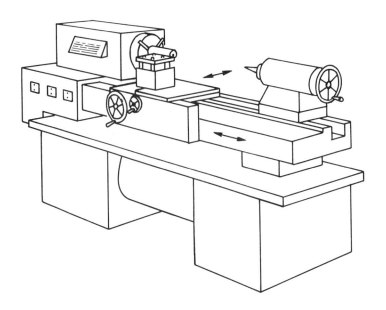

FIG. 3.12 Two axis tape-controlled engine lathe.

FIG. 3.13 Computer-controlled test stand. (Courtesy of Bosch-Siemens Hausgeräte GmbH, Berlin, West Germany.)

Hardware and Software Components

Thus the tool follows the displacement motion of the carriage and produces the desired contour of the workpiece.

The completed prototype of the product will undergo functional and life tests. These operations can be controlled by a mini- or a microcomputer with the aid of a program containing the entire test procedure and test parameters. Figure 3.13 shows a computer-controlled test stand to life test the pump of a washing machine. The computer adds flexibility and accuracy to the test system and thus it substantially improves the productivity of the engineering department.

3.4 MANUFACTURING PLANNING AND CONTROL MODULES

The principal tool for these functions is the central business computer, which strategically controls and operates the management information system. Information from forecasting, marketing, planning, purchasing, accounting, manufacturing, and engineering is entered into this computer by data peripherals which are located in the department where these activities are performed. The central computer processes and stores information that may be required for operation of the plant. In a larger or more modern plant there will be a distributed computer system instead of a central computer (Fig. 2.8). Thus the data processing and storage functions will also be of a distributed nature. Communication with the computer may be done in a batch-type mode with the help of data storage media such as punched cards or in a real-time mode by interactive terminals. The present trend leads toward the interactive mode because it utilizes the computer capabilities most efficiently.

3.4.1 Manufacturing Planning

Manufacturing planning aids are designed to utilize available manufacturing resources efficiently (Fig. 3.14). Planning algorithms may be based on heuristic procedures or on optimization models utilizing operations research tools.

1. *Simulation aids*: There is a large variety of a special-purpose simulation software available, which is of assistance when new facilities are to be planned or ongoing processes are to be investigated or improved. Typical simulation packages are the continuous simulation language DYNAMO to stimulate the behavior of a firm in its market environment or the language GPSS to simulate discrete manufacturing processes (Chap. 9).
2. *Financial modeling and planning*: Such a system would contain an equation-oriented modeling language normally required in planning. The language would include general calculation functions, forecasting and statistical distribution routines, and sensitivity analysis and risk analysis features. A report generator would produce customer reports, plots, and histograms.

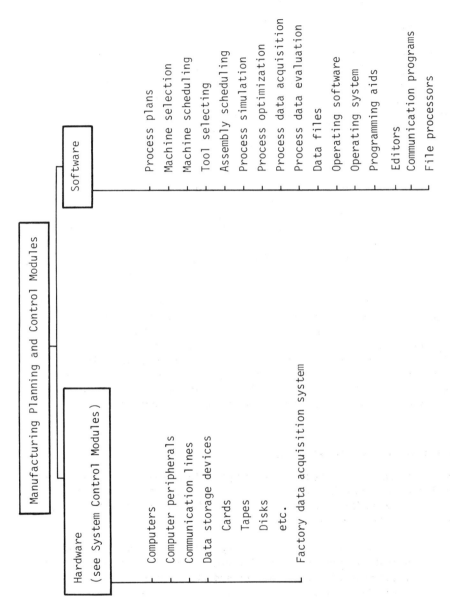

FIG. 3.14 Principal automation modules for manufacturing planning and control.

3. *Financial planning and accounting*: The functions comprise general ledger capabilities, including accounts payable, accounts receivable, payroll, project cost accounting, and billing.
4. *Manufacturing control aids*: Manufacturing control involves numerous functions for which many batch-type and interactive software control packages have been written. Typical functions are:
   a. *Bill of materials processing*: The bill of materials is needed to provide engineering manufacturing and accounting with information on the composition of the product, breakdown of material, labor and overhead cost, manufacturing processes, delivery problems, and so on.
   b. *Inventory management and control system*: Such a module will allow maintenance of proper inventory for manufacturing. It will also assure inventory cost minimization and prevent costly obsolesence of material and parts. Identification of raw material, work in process, and finished goods is essential.
   c. *Material requirements planning*: Material requirements planning is an important manufacturing function. It will assist management to ensure that items are either purchased or manufactured at proper quantities at the correct time.

3.4.2 Manufacturing Control

Manufacturing control collects and evaluates operational data from the ongoing process. This feedback information is needed to perform the following management functions:

   To calculate manufacturing cost and wages
   To check the adherence to schedules
   To detect manufacturing bottlenecks
   To observe the functionality of equipment
   To obtain on-line process inventory

An important computer tool for the collection of operational data is the factory data acquisition system (Fig. 3.15). Intelligent terminals are located at strategic points of a plant and operational data are entered either manually or automatically.

Products or parts are often accompanied by punched cards which are used at a workstation for identification of the object. In this case the operator places the card into a card reader and by means of a keyboard, enters the operational data for the object into the terminal.

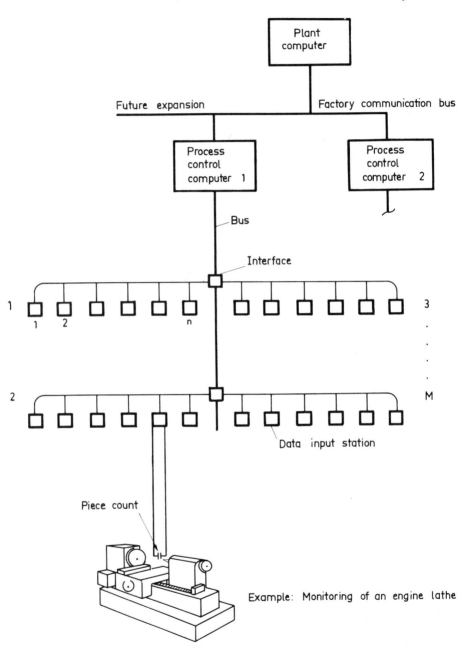

FIG. 3.15 Typical production monitoring system. (From Ref. 5.)

Hardware and Software Components

This process is repeated at each workstation as the product is routed to its completion through the plant. Whenever a problem is encountered at a workstation, immediate corrective action can be taken. In addition, the computer accumulates a complete manufacturing history of the product.

3.5 MACHINING MODULES

In Fig. 3.8 it was shown that the piece rate and complexity of the product significantly determine the manufacturing technology to be used in a plant. There is a full range of different machine tools with different degrees of automation from which a manufacturing system can be built (Fig. 3.16).

The conventional machine tool still has a significant place in low-volume production, in model shop operations, and for custom-designed products. It is also very common to find this tool in highly automated plants to solve special manufacturing problems or to serve as a back-up for automated equipment. The lathe and milling machines are the most versatile of all machine tools. It is possible to produce with these machines most round and flat surfaces at both the inside and outside of the workpiece. The next important tool is the drilling machine, which may be equipped with single or multiple spindles. For sheet metal operations the punch press is an essential tool. It is used for bending, forming, and punching operations and is able to produce high-volume parts to close tolerances.

Significant entries into low-volume production are the NC and DNC machine tools. The machining principles used are similar to those of conventional machine tools. The basic difference is that the machine is controlled with the help of a part program rather then by hand. These machines are classified by the number of axes that are controlled. The basis axis movement can be represented in a cartesian coordinate system. A two-axis engine lathe was shown in Fig. 3.12. Here the movement of the carriage to which the tool post is fastened is controlled in both the x and y directions. A three-axis controlled milling planer typically used to produce aircraft parts is shown in Fig. 3.17. The movement of the machine table is controlled in the x direction and that of the tool post which is mounted to the horizontal cross-rail is controlled in the y and z directions. A four-axis CNC lathe with two eight-station turrets is shown in Fig. 3.18. The movement of each of the eight-station turrets can be controlled in the x and y directions. Different tools are located in the turret heads and are indexed into machining position by a control signal from the parts program. This type of machine tool is very universal and is capable of producing a wide range of precision parts. Figure 3.19 shows a

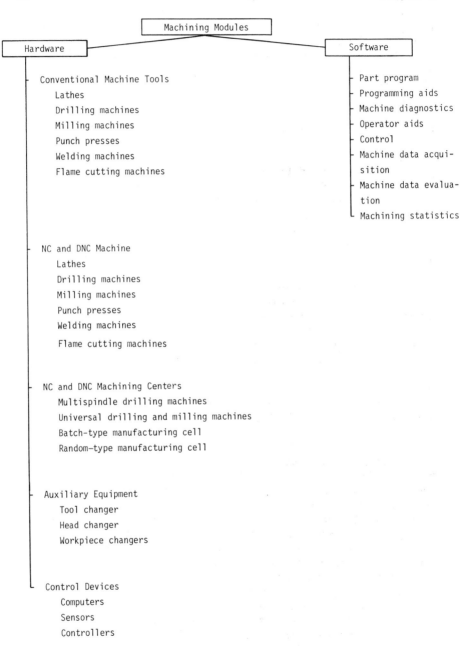

FIG. 3.16 Basic machining modules.

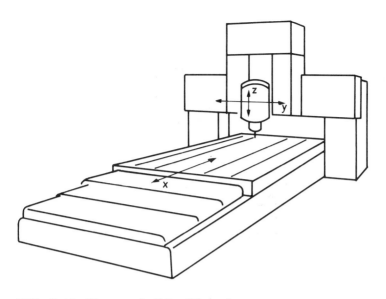

FIG. 3.17 Three-axis NC milling planer.

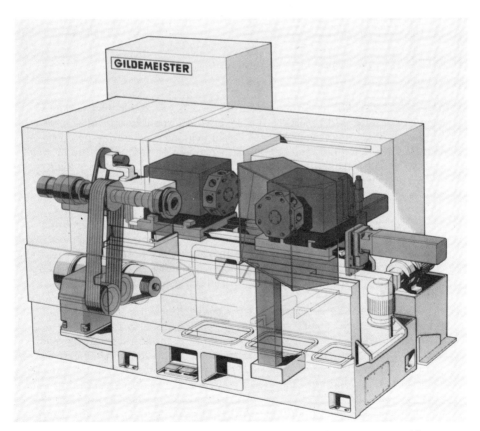

FIG. 3.18 Four-axis CNC lathe with two five-station turrets. (Courtesy of Gildemeister AG, Bielefeld, West Germany.)

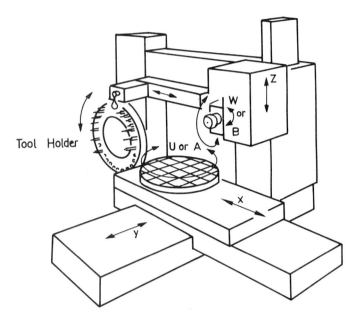

FIG. 3.19 Five-axis controlled machining center.

five-axis controlled machining center. Such a machine tool is mainly used for milling and boring. It may perform specific operations or completely machine a workpiece. Tools are selected under program control from a toolholder. They are returned to the holder upon completion of a work cycle. The designation for the x, y, and z axes are standardized throughout industry. The additional axes do not have such a standard nomenclature. This is indicated in Fig. 3.19.

An increase in productivity can frequently be obtained by the use of multiple-spindle technology. Here various tools are grouped together in one machining head and perform drilling, tapping, boring, reaming, or special operations. To increase the versatility of these types of machine tools (e.g., to process part families), the multiple-spindle heads can be changed by an indexing or other mechanism (Fig. 3.20).

If a well-defined combination of process functions is required, several NC or DNC machines are grouped together to form either a batch-type machining system (flow line) or a random machining cell. The problem of moving the part from one machine to another can be resolved by manipulators (Fig. 3.21), special conveyor equipment, or a combination. Figure 3.22 shows a typical batch-type machining cell. Here the part is mounted on a pallet and moved sequentially from one machining operation to another until it is completed. Here the machine

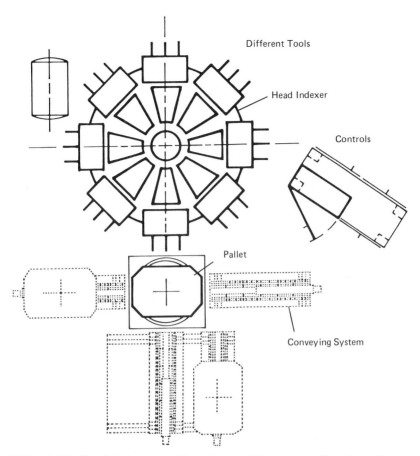

FIG. 3.20 Head indexer. (Courtesy of Kearney & Trecker Corp., Milwaukee, Wisconsin.)

FIG. 3.21 Batch-type machining cell. (Courtesy of Maho, Werzeugmaschinenbau, Pfronten, West Germany.)

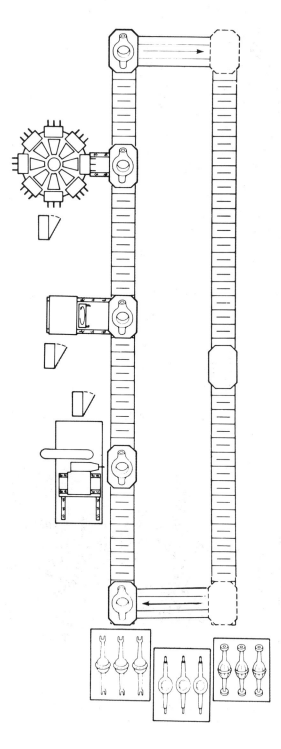

FIG. 3.22 Batch operation: the sequence of machining operations is determined by the nature of the system. (Courtesy of Kearney & Trecker Corp., Milwaukee, Wisconsin.)

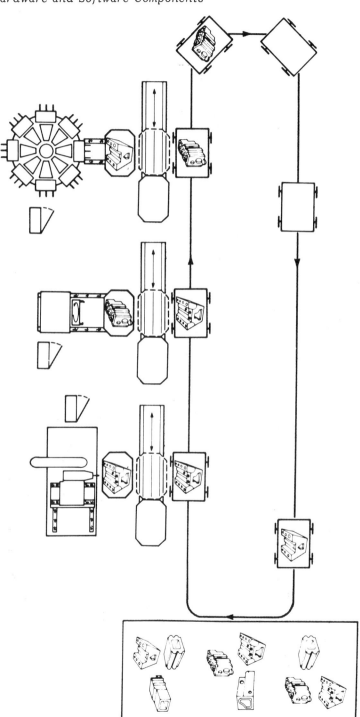

FIG. 3.23 Random operation: the sequence of machining operations is not predetermined and remains a variable in the machining process. (Courtesy of Kearney & Trecker Corp., Milwaukee, Wisconsin.)

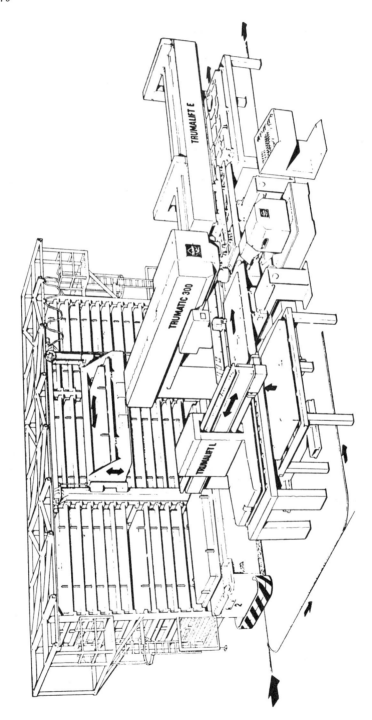

FIG. 3.24 Manufacturing cell for sheet metal fabrication, Trumpf material handling system. (Courtesy of Trumpf GmbH & Co., Ditzingen, West Germany.)

Hardware and Software Components 71

operations is not predetermined and remains a variable in the manufacturing process. With this type of operation an equipment failure will not shut down the system completely. We discuss this subject in more detail in Chap. 11.

The concept of manufacturing cells is used not only for machining, but also leads to very economical solutions when applied to other manufacturing processes. Figure 3.24 shows a manufacturing cell for sheet metal fabrication. In this system the sheet metal stock is brought to an automatic storage system on a guided cart. A hoist automatically unloads the stock and transfers it to a predetermined storage place. From there the sheet metal can be brought under program control to a NC punch press. After completion of the operation, the part is again automatically loaded on a guided cart and shuttled to another workstation.

3.6 MATERIAL FLOW AND STORAGE MODULES

An efficient material flow and storage system is essential for the operation of a manufacturing facility. The application of automatic control will greatly facilitate the movement of discrete parts from receiving to storage, through production, and to finished-goods warehousing. Real-time control of order processing, material handling, production scheduling, and inventory control is done under supervision of the management information system in conjunction with the material flow and storage modules. The equipment and software to be used for these modules is quite varied (Fig. 3.25). Again, the complexity of the material handling system and the degree of automation depend on the type of product to be manufactured, its size, the piece rate, and numerous other factors. In this activity it is also quite common that modules of old and new technology are working closely together.

3.6.1 Part Identification

When a part arrives at a facility it first has to be identified. This may be done by visual or automatic inspection. For visual inspection the name of the part and the number of items delivered are recorded via a computer terminal and sent to the central computer to be stored in a data bank. Automatic identification is possible using a vision system or a code reader. With automated manufacturing it becomes necessary to track a workpiece from the start of the production to its final assembly. The two basic methods used are direct and indirect tracking (Fig. 3.26).

Direct Tracking

With direct tracking the workpiece is identified either by its weight, shape, color, or an attached code carrier. When the workpiece moves

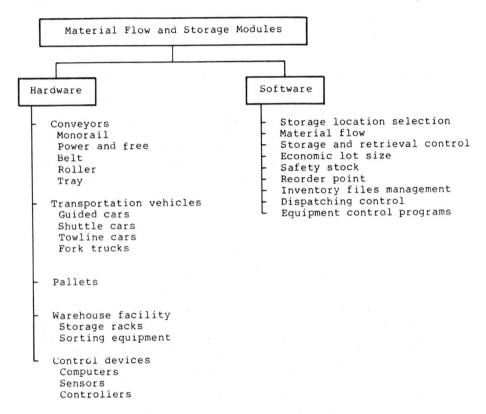

FIG. 3.25 Typical material flow and storage modules.

from one destination to another and a switching station (decision point) detects it, a routing decision is made. In case the station can read the destination address directly from the identifier, we talk about direct coding. If, however, the workpiece is first identified and than a destination address is assigned to it, we talk about indirect coding.

There are numerous identification systems available to recognize a workpiece. The most common principles used are the following:

Light-beam decoding
Bar code decoding
Machine vision
Magnetic decoding
Weighing

The first three principles will be discussed in Sec. 8.10 on optical workpiece recognition. Magnetic code readers are among the most reliable identification systems. With this method the code carrier is attached directly to the part (Fig. 3.27). It is possible to transmit the code free of contact with a write head on the code carrier. When the workpiece passes the read head, the code is identified by the code reader. The principles employed to magnetize metal foils or steel strips are shown in Fig. 3.28. Vertical magnetizing is used predominantly.

A workpiece may also carry a code card (Fig. 3.29). Upon reaching a decision point the code card is taken from the part and inserted into a code reader. This method needs manual handling and is used only where an operator has additional functions to perform, such as

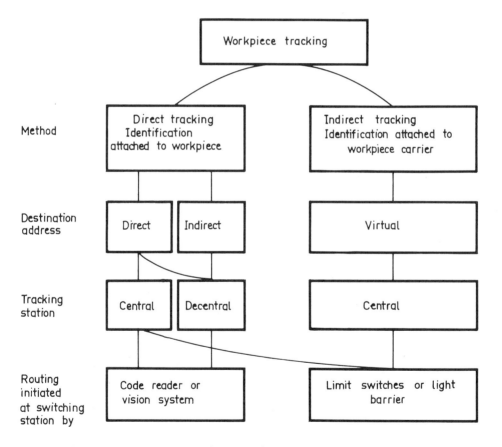

FIG. 3.26 Various methods for tracking workpieces. (From Ref. 6.)

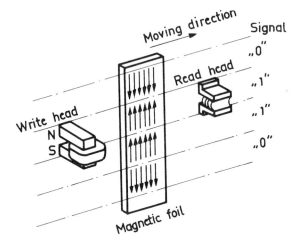

FIG. 3.27 Principle of a magnetic foil code reader. (From Ref. 6.)

assembly or testing. Direct tracking has the disadvantage that it needs a code carrier. In many manufacturing operations, such as heat treating or degreasing, the code carrier may be destroyed, making this method impractical.

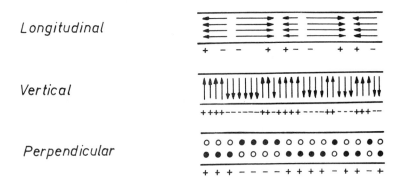

FIG. 3.28 Methods for magnetizing code carriers. (From Ref. 6.)

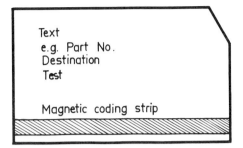

FIG. 3.29 Machine-readable code card. (From Ref. 6.)

Indirect Tracking

With indirect tracking the code is attached to the workpiece carrier, such as the hanger of a chain conveyor or the pallet of a roller conveyor. The principles that are used are the following:

Optical decoding
Light-beam decoding
Magnetic decoding
Ultrasonic decoding
Tracking of the workpiece carrier

The first two methods are discussed in Sec. 8.10 on optical workpiece recognition. For magnetic decoding either permanent magnets or metal flags are set on the workpiece carrier. At the code reader station these devices will be in close proximity to an inductive reader head and the code signal will be transferred. Problems may be encountered when the workpiece carrier oscillates, for example, when a hanger swings back and forth. This may result in several repeated readings for the same workpiece.

Tracking the workpiece carrier. The simplest method uses optical tracking stations along the conveyor system. Whenever a workpiece passes by a tracking station it is recorded. Since the computer knows the number of workpiece carriers between two adjacent tracking stations, it will keep count of the number of carriers passing each station. From the count it can determine when a carrier that has previously left one station has arrived at the next station. In other words, the computer has in its memory a model of the actual conveyor system. This model is constantly updated when a carrier passes a reading station. This method may have a problem with oscillating carriers which interrupt the light beam several times at a station. The computer may lose synchronization.

With another method the conveyor system is divided into contiguous identification sections. There are two tracking guards (e.g., optical sensors) in each section, one at its entrance and one at its exit. Each section can only accommodate one carrier. When the carrier enters a section the entry guard records the move to the computer. The computer now knows the whereabouts of the part. Upon departure from this identification section the computer obtains a signal from the exit guard and the computer records this move immediately. Following the departure notification there must be a signal issued by the entry guard of the next identification section. This method resolves the synchronization problems the method discussed previously may have. However, many more readers are required and the computer is quite busy answering interrupts.

A more complex method is to use an analog system. Here a small analog (physical model) of the actual conveyor system tracks the movement of workpiece carriers. This analog is mechanically connected to the drive of the conveyor system to assure positive synchronization. Whenever a workpiece carrier enters the system it is identified to the analog by means of a code carrier attached to it. From here on the computer tracks this code carrier and is able to follow the workpiece carrier through the plant by tracking the analog.

3.6.2 Material Movement

Common means of transportation are fixed-type conveyors or mobile vehicles. Figure 3.30 shows several different conveyor systems. Belt and roller conveyors are most often used. A part is placed directly on the conveyor and sent to its destination. Often it is necessary to route identical or different parts to different destinations. This makes identification and position tracking necessary. In less complex material flow systems this can be done with the help of light barriers (Fig. 3.30), whereas in complex systems code readers (Fig. 3.27) or vision systems are necessary. To enhance the flexibility of a conveyor system, switch or turn stations are used (Fig. 3.30). In this case positive identification of a part is necessary before the switching operation takes place. In flexible manufacturing systems it is desirable to handle the part as little as possible; therefore, machining or assembly operations are performed with the part located directly on the conveyor system. For this purpose the part must be accurately oriented with regard to the work tool by a positioning fixture (Fig. 3.30). In most cases positioning directly on the conveyor is too inaccurate. For this reason the part is mounted on pallets and is routed on them to the different machining operations. The pallet is positioned automatically before the machining operation starts.

Other types of fixed transportation system are the monorail or power and free overhead conveyors (Figs. 3.31 and 3.32). Parts are placed

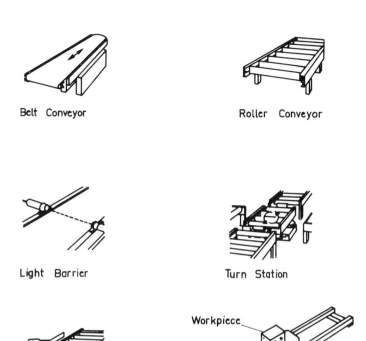

FIG. 3.30 Fixed-type conveyor systems. (Courtesy of Institut für Fördertechnik Universität Karlsruhe, West Germany.)

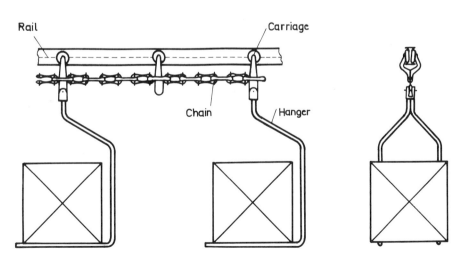

FIG. Chain conveyor overhead system.

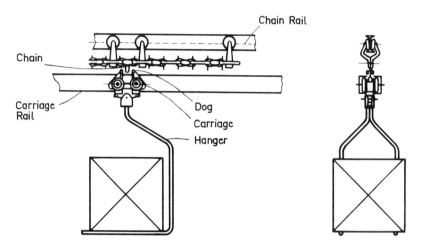

FIG. 3.32 Typical layout of a power and free conveyor system.

on a hanger and routed through a plant along overhead rails by a continuous chain. With the monorail system the hangers are fastened permanently to the chain. This system is inexpensive but inflexible. The power and free conveyor is of advantage where a high degree of flexibility is needed. The hangers are identified by a coding mechanism consisting of small flags or magnets. After identification at a switching station the hangers can be disengaged from its present chain and routed to another location by another chain (Fig. 3.33). The chain-drive principle is also used for ground vehicles (Fig. 3.34). Here the tow car can be engaged to or disengaged from an underground chain. This system does not need the expensive overhead rail; it can also carry heavier loads.

More advanced transportation systems use guided vehicles (Fig. 3.30). These vehicles are self-propelled and are guided along a magnetic induction wire embedded in the floor or along a guide strip painted on the factory floor.

The highest degree of flexibility is obtained with self-propelled ground vehicles which can be dispatched at random to any factory floor location. Here forklift trucks are most commonly used. Their transportation route can either be determined by the driver or by radio communication with the computer.

Hardware and Software Components

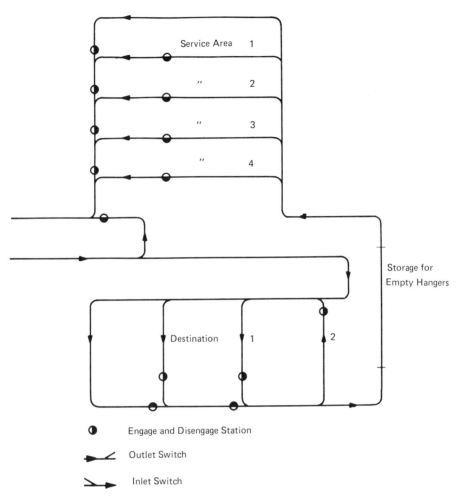

FIG. 3.33 Example of a power and free track layout.

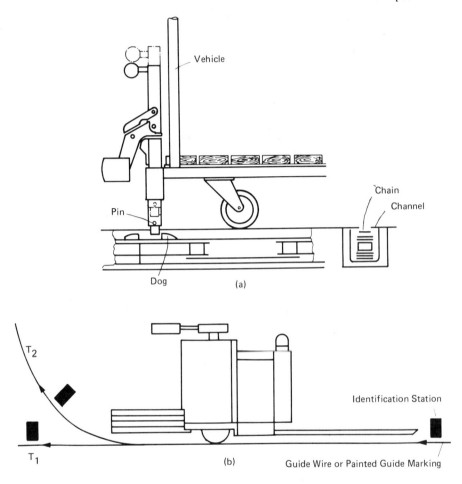

FIG. 3.34 Ground vehicles: (a) chain-driven towline vehicle; (b) guided vehicle.

3.6 Material Storage

A typical warehouse facility consists of a storage rack with a multitude of storage cubicles. Small parts or parts in a low-volume production unit are in general hand-stored and hand-picked. Figure 3.35 shows a storage system in which parts are hand-picked

Hardware and Software Components

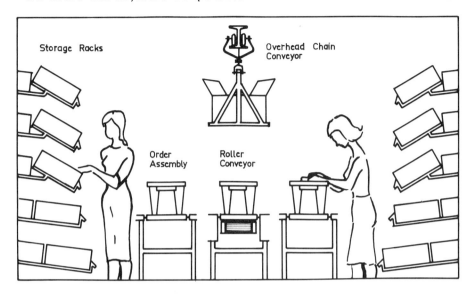

FIG. 3.35 Simple part storage operation. (Courtesy of Krauskopf Verlag, Mainz, West Germany.)

upon issue of a production order and are transported via a roller conveyor to their destination. Empty cartons and material to be discarded are disposed of by an overhead chain conveyor.

A very complex palletizing and warehousing system is shown in Fig. 3.36. Parts or assembled products coming from production are first sorted and then sent to a palletizing station. With the help of an automatic palletizer, individual orders are placed on a pallet. The completed pallets are routed to a high-rise warehouse and distributed to the cubicles. Upon retrieval a stacker crane fetches the parts and delivers them to the conveyor system to be loaded on trucks or boxcars.

Retrieval strategies for parts are determined from production schedules executed by the plant computer. This computer also calculates reorder points for parts, the required safety stock, and economic lot sizes. Material flow, the conveyor system, and storage equipment are controlled with the help of control and microcomputers.

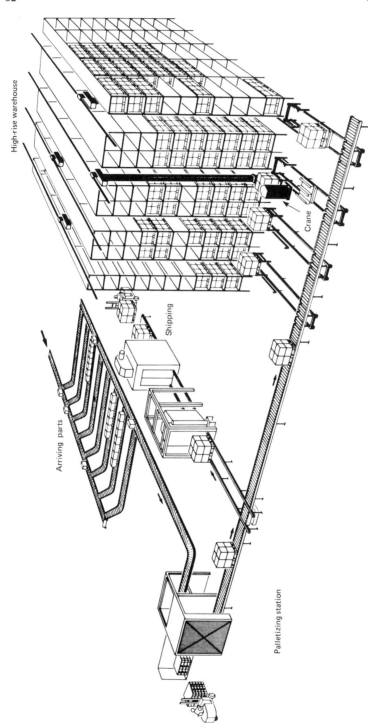

FIG. 3.36 Palletizing a high-rise warehousing system. (Courtesy of Krauskopf Verlag, Mainz, West Germany.)

Hardware and Software Components

3.7 ASSEMBLY MODULES

Automation of assembly lines so far has been one of the most difficult tasks in discrete product manufacturing. The main reason for this is that the majority of assembly operations have been conceived for manual assembly. Human assemblers are equipped with two versatile arms which are capable of performing the most complex assembly operations even in very awkward positions and spaces difficult to access. Our sophisticated sensory system is capable of coordinating easily the work of two arms, to observe the progress of the assembly, and to inspect the completed product. Furthermore, we have an unsurpassed memory capability which allows us to retrieve assembly experiences from other or related products. To date, automatic assembly has been successful only for very simple products. The progress of automation in this area depends mainly on the ability of engineers to design products for machine assembly and on equipment builders to evaluate the human skill and sensory system. Presently, industry uses two basic assembly procedures. In group assembly a product is completely assembled by one or serveral persons, usually in an enclosed assembly area. This type of assembly is very difficult, if not impossible, to automate. In flow line assembly the product is assembled sequentially along an assembly line (Fig. 3.37). Each assembler is responsible for a special assembly operation.

  Automatic assembly systems use the flow line principle. The product may move down the assembly line synchronously or asynchronously. In the synchronous mode there is no provision for in-process buffering. This mode may lead to line stoppages if an assembly operation, frequent product defects, or timely delivery of components to assembly

FIG. 3.37 Manual assembly.

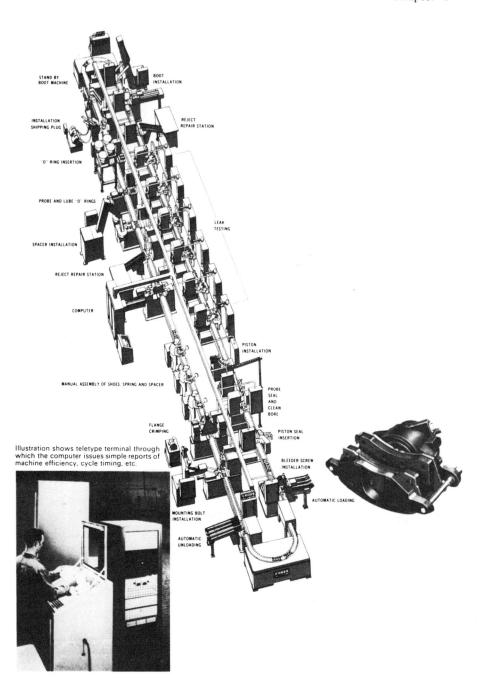

FIG. 3.38 Assembly transfer line for disk brakes. (Courtesy of Cross & Trecker Corp., Bloomfield Hills, Michigan.)

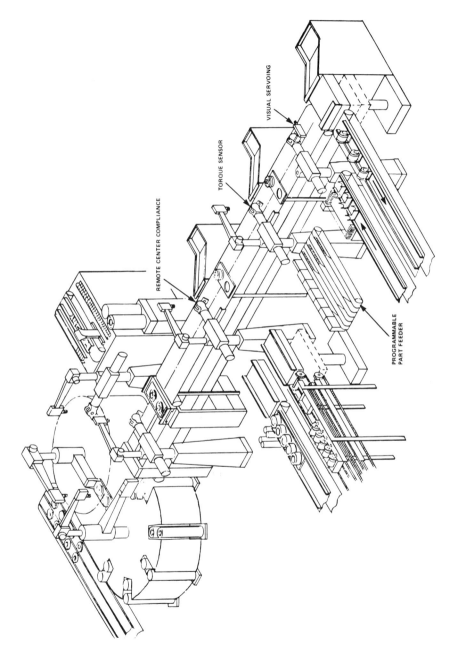

FIG. 3.39 Advanced assembly system for electrical motors using robots and vision. (From *Manufacturing Engineering*, December 1978.)

stations present problems. In this case the asynchronous flow line principle should be used. Here component buffers are provided at strategic locations.

Automatic assembly also requires that the assembly line be paced. This means that the product has to be at a complete standstill and in a fixed position when an assembly operation is performed, whereas manual assembly can be done on moving lines. Figure 3.38 shows an assembly transfer line for automotive disk brake assembly. In this line components are delivered automatically by track or bowel feeders to the various assembly stations. Assembly is done by pick-and-place effectors. There are also automatic test stations to do quality control at different assembly stages. The entire operation is monitored by a computer.

One particular problem imposes feeding of parts. With the present systems, parts have to be preoriented before the pick operation. With modern vision systems this operation may not be necessary any more. They are capable of identifying a part and determining its location and orientation.

Figure 3.39 shows an advanced assembly system for electrical motors in which the computer not only monitors but also controls the operation. Parts are brought by conventional and programmable feeders to the individual workstations and assembled by manipulators as the product proceeds down the assembly system. Recognition of parts and monitoring of the assembly progress are done by different vision modules. There are also remote center compliances devices employed which aid insertion operations of bolts and pegs. Additional inspection is done with torque sensors and visual screening.

The necessity of orienting parts prior to assembly leads to very expensive machine peripherals. For this reason it is of advantage to have sensor devices that allow an assembly robot to pick randomly oriented parts from a bin and to present the parts to the assembly operation in a defined position. Vision systems that are capable of doing this are under development. They also serve for visual inspection of the finished product.

3.8 QUALITY CONTROL MODULES

Computer-guided quality control systems are an integral part of programmable manufacturing operations. There are many activities in a factory that are directed toward the attainment of a high-quality product. Therefore, it is necessary that a systems approach be conducted to integrate all quality control functions. The information flow in a high-quality feedback circuit is shown in Fig. 3.40. It is quite evident from this pictorial that the individual activities affect each other. The entire feedback loop can work properly only when the inputs and

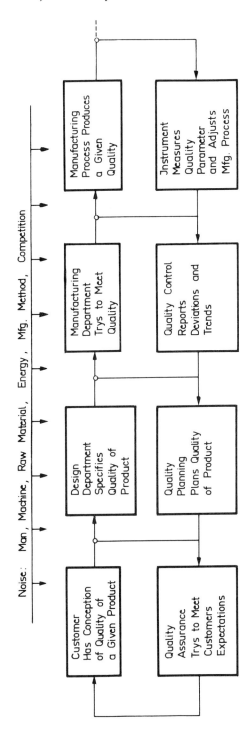

FIG. 3.40 Activities in an integrated quality control system.

outputs of all activities are taken into consideration and properly interfaced. Contrary to older quality control concepts, where effort was concentrated on the possibility of reducing manufacturing problems, the integrated approach includes the customer and the marketing environment. The computer is a valuable tool to use to operate such an elaborate quality control system. Many quality control operations are integrated into manufacturing machines. For example, some modern NC machine tools are capable of measuring a part after it has been produced. We also saw in the preceding section that an acceptance test can be made by a vision system that supervises an assembly operation.

Various quality control (QC) modules are shown in Fig. 3.41. The computer equipment is mentioned only by name. Similar to modern control systems, there are different activities to be performed at different quality control levels (Fig. 3.40). For these functions different components will be used.

At the customer level the commercial data processing system of strategy computer gathers and monitors QC data and looks for QC trends. It compares the QC field data with those obtained from the factory floor. With this information it may be possible to evaluate manufacturing processes and equipment. It is also possible to appraise the quality that was designed into the product. Designers will use the commercial or scientific computer for quality control work. They also need field and factory data. In addition, engineering has to conduct

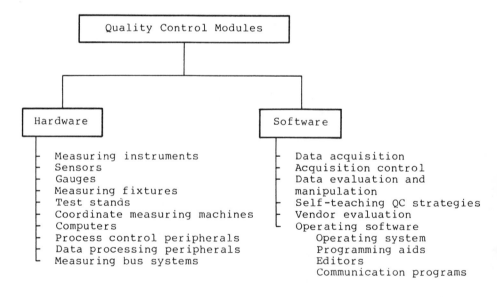

FIG. 3.41 Typical control modules.

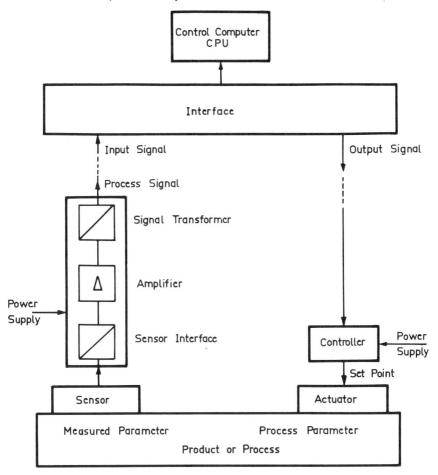

FIG. 3.42 Components of a computerized quality control system.

short- and long-term tests to assure that the product will render an adequate service life (Fig. 3.13). These tests will be supervised by control computers. QC data from the factory floor are gathered with the help of data acquisition systems (Fig. 3.15) and control computers.

The components on an on-line quality system are similar to those of a control system (they will be discussed in detail in Chap. 10). The control computer is a mini- or microcomputer with real-time capabilities. Figure 3.42 shows a schematic drawing of a computerized on-line QC system. Performance or dimensional parameters taken from the product or process are sent to the computer and compared against

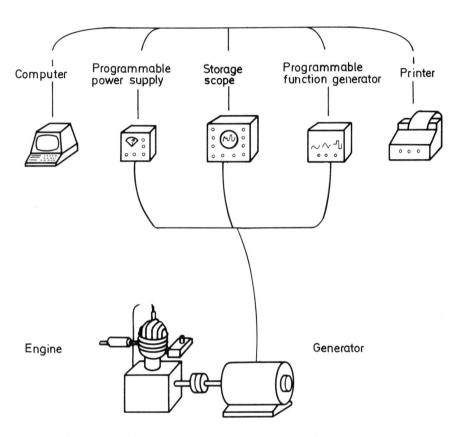

FIG. 3.43 IEEE 488 measuring bus system used for an engine test.

control standards. If necessary, new set-point information is sent back to adjust the process to meet the quality control objectives. In most QC operations the feedback loop is closed by the operator. Trends in the development of low-cost computing equipment have lead to the conception of intelligent and self-learning quality control systems. However, there is still a considerable amount of research to be performed to find a systematic approach to modularize quality control procedures and to tailor from these modules functional systems for individual factory operations.

An important development of computerized measuring equipment is the conception of the instrument bus. Its popular representative is the IEEE 488 bus (Fig. 3.43). Many instrument manufacturers provide

Hardware and Software Components

their equipment with standard interfaces which allow connection of the instrument to the bus by a plug-in device. The entire instrument system can be controlled by a computer acting as a master to the bus.

3.9 SYSTEM CONTROL MODULES

The system control modules contain the entire spectrum of computers, controllers sensors, and communication systems (Fig. 3.44). Only a brief introduction to this subject will be given here, since the application of the computer and its data and process peripherals to control the manufacturing processes is the purpose of the book. More details on the individual building blocks of the control system are given in the chapters devoted to these components. The heart of the control system consists of the data processing, scientific, and process control computers. Table 3.1 shows the principal features of these computers.

The data processing computer is usually a large commercial number-crunching machine. It supports the forecasting, marketing, accounting, financial, and administrative functions. Calculations to be performed are of a rather simple nature. The emphasis is placed on throughput and processing of a large number of input and output data. In older systems the batch-type operational mode predominates; however, with the availability of improved computer architectures and data peripherals, a trend to real-time data processing is imminent. The data base for the management information system is located on large mass storage peripherals such as disk and tape systems. This

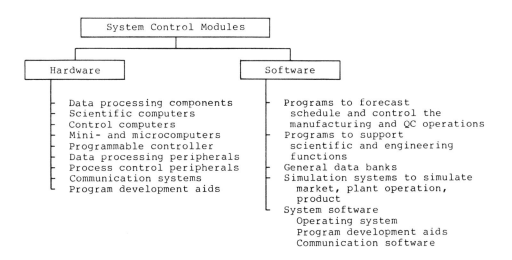

FIG. 3.44 Typical system control modules.

TABLE 3.1 Important Features of Basic Computers

Feature	Computer		
	Commercial	Scientific	Control
Mode of operation	Batch type Scheduled program execution Few priority levels	Batch type Scheduled program execution No priority levels	Input/output of data is possible in real time Many priority levels
Principal functions	Simple calculations Data gathering and updating Large number of input/output data Large data banks	Complicated calculations Large programs Small number of input/output data	Complicated calculations Measurement and control data Large number of input/output control data
Data input/output	Alphanumeric data	Alphanumeric data	Many different types of signals from measurements and for control
Operating system	Cards, tapes, terminals For batch-type operation Few real-time applications	Cards, tapes, terminals For batch-type operation	For real-time operation; often very complex
Peripheral storage	Large disk and tape systems	Large disk and tape systems	Small but fast disk, drum, and cassette systems

computer must also be equipped with communication software and hardware to be able to communicate with system components at the same or a lower level of the hierarchical control system.

The scientific computer is used primarily by engineers and forecasters. This unit must be capable of performing very complex calculations at high speed. This function may, however, be performed by the business computer equipped with a fast processor and a large work memory. The peripherals are used with the scientific computer similar to those of the business computer. Input and output requirements for data are usually small. Typically, a list of program parameters is read in, used in a large program, and output results which may contain only one number are printed.

The control computer will be a mini- or microcomputer with real-time capabilities. A real-time clock and DMA will enhance its operation. Of importance are adequate interface circuits and communication systems for signal and information transfer with the process, operator, and computers at the same or other hierarchical levels. Interfacing is done primarily via standardized interface circuits (Fig. 3.42). Such circuits may even contain their own processor to speed up protocol adaption and preprocessing of data. There will be numerous independent control application within a plant where by a single computer will supervise the operation of one or several machines. Sequential control and logic operations can be done by programmable controller. These elements have the advantage that they are easy to program and to implement. For larger factory control applications distributed control will dominate. This concept greatly simplifies an installation since separate modules are easier to implement, operate, and maintain than are large solitary systems. Figure 3.45 shows a distributed control system for a continuous process. Controllers of modular design are connected to the bus by a plug-in principle. A central processor supervises and coordinates the operation of the different modules. Process optimization can be done by a larger computer via a communication line. The computer system of Fig. 3.45 can in principle also be used for the control of a discrete manufacturing system. Several machine tools and measuring systems could be connected to a central computer. This computer could perform daily scheduling of production runs, distribute part programs to the machine tools, collect operational data, and do failure diagnostics.

Typical application and system software is listed on the right side of Fig. 3.44. The task of numerous software packages will be explained elsewhere in the book. System software needed for the operation of the different types of computers and for plant communication will not be discussed. Inexperienced readers should consult the special literature to familiarize themselves with this subject matter. There will also be no introduction to the operating principles of computers.

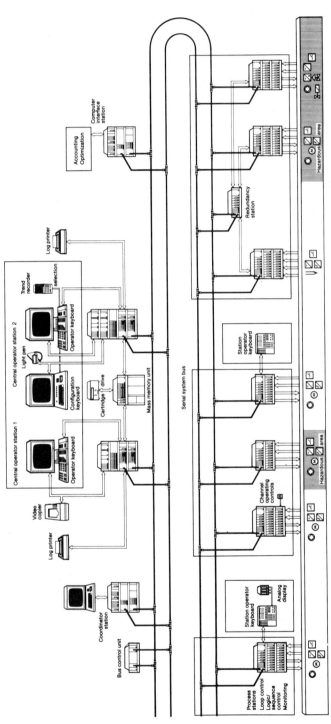

FIG. 3.45 Process control system, Centronic P.

REFERENCES

1. R. Hoffman, A Comparative Study of Implemented Process Control Computer Systems, Dipl.-Arbeit, Institut für Informatik III, University of Karlsruhe, Karlsruhe, West Germany, 1977.
2. B. W. Boehm, Software and Its Impact: A Quality Assessment, *Datamation* Vol. 19, No. 5, May 1973.
3. U. Rembold, *Process and Microcomputer Systems*, R. Oldenbourg Verlag, Munich, West Germany, 1979.
4. H. Flegel, The Use of Modern Aids for Planning and Design, Transmatik Conference '81, Karlsruhe, West Germany, 1981; published in the proceedings of the meeting.
5. S. A. Berry, Techniques in the Application of Computers to Industrial Monitoring, in *CAD/CAM and the Computer Revolution*, Society of Manufacturing Engineers, Dearborn, Mich., 1974, pp. 221–239.
6. F. Thomas, Material Flow Control for Flexible Material Flow Systems, Transmatik Conference '81, Karlsruhe, West Germany, 1981; published in the proceedings of the meeting.

4

Advanced Computer Architectures Used in Manufacturing

4.1 INTRODUCTION

An introductory discussion regarding the necessity of using hierarchical computer architectures in manufacturing processes was given in Sec. 2.4. It was shown that the conventional hierarchical management pyramid of an organization can be supported by a multilevel computer architecture in combination with a highly structured data base concept. The computers of each level aid a corresponding management layer. The computer hierarchy may consist of two or several tiers. Figure 2.8 shows a four-level computer architecture. To simplify the discussions in this section, we will limit ourselves to a three-level computer system (Table 4.1 and Fig. 4.1). Each level by itself may be structured as a modular hierarchy.

The highest level supports the management information activities where management data are processed. Forecasting, marketing, production scheduling, and operational management comprise a subsystem of the highest control level. From this level, horizontal communication to other factories and plants and vertical communication to local production control levels are coordinated. The intermediate control level is for local plant supervision and plant coordination, the weekly production processes, assembly and quality assurance, and the supervision of the lower control levels. Production data and recourse status information are used on this level to aid production decisions. Horizontal communication with adjacent control areas and vertical communication to lower and higher control levels in the hierarchy is performed via high-speed serial data links. The lowest level consists of modules that control technological and geometrical operations. It also contains the machine-specific control algorithms (e.g. direct digital control, DDC algorithms).

Reliability, extendability, and flexibility of the control system can be obtained if there is a clear separation between all levels (micro- and

Advanced Computer Architectures

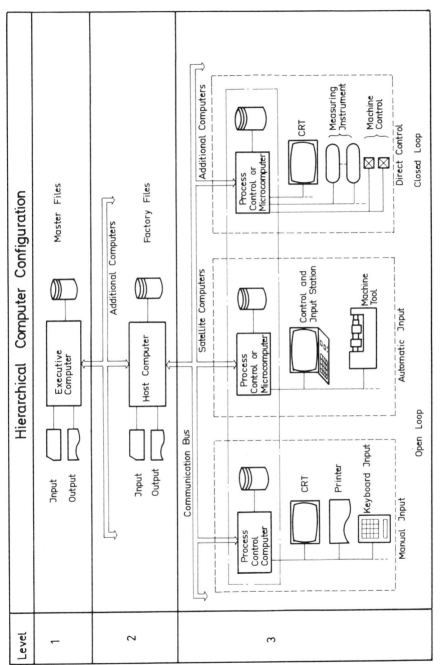

FIG. 4.1 Hierarchical computer configuration for factory control.

TABLE 4.1 Hierarchical Structure of a Three-Tiered Computer System

Computer	Manufacturing unit	Managerial level
Strategy Supervisor (commercial computer)	Corporate office Autonomous production unit Department	President General manager Director
Plant or manufacturing cell supervisor (commercial or control computer)	Factory or manufacturing cell	Plant manager Superintendent
Control (mini- or microcomputer)	Manufacturing system Manufacturing process Machine	Manufacturing engineer Foreman

macrostructure). This modularity will simplify the system implementation. In the near future, numerous development activites will be necessary to conceive standards for communication networks, programming languages, data bases, and very large scale integrated (VLSI) elements. These standards will greatly simplify the installation of computer control hierarchies.

Future supervision systems will be characterized by increasing intelligence located on the different hierarchical control levels. Adaptive controls [adaptive control optimization (ACO), adaptive control constraint (ACC)] for machine tools and self-optimizing and learning controls will be part of the hierarchical control functions.

Advanced Computer Architectures

Information contents	Data processing activity	Task of computer
Knowledge base Market Customer Competition Product line Forecast Performance data Productivity	Central data file Administration Evaluation Updating Concentration Searching Reporting Retrieving	Management Goals Objectives Standards Control Correcting Broadcasting
Factory information Production process Production status Productivity	Operating data Administration Evaluation Updating Concentrating Searching Reporting	Planning Scheduling Controlling Coordinating Optimizing
Basic data Product Personnel Raw material Equipment Energy	Data lists Acquisition Checking Comparing Reporting	Control Open loop Partially open loop Closed loop Input Operating parameters Output Set points Limits

There are different types of computers used at each hierarchical level (Sec. 2.4). The discussions in this chapter assume that the reader is familiar with business and control computer concepts. For the interconnection of a hierarchical computer installation, special interface and communication equipment is necessary. Most of these are explained in dedicated textbooks. Thus only important new developments are discussed here in detail.

4.2 HIERARCHICAL CONTROL OF MANUFACTURING

The basic idea behind the hierarchical control concept is the fact that a complex control task cannot be handled efficiently with a solitary controller (Sec. 2.4). For this reason the control task is subdivided into various control modules according to a hierarchical control concept. It has to serve many different manufacturing functions. Several fundamental hierarchies have to be considered when a computer control system is conceived:

1. *Structural hierarchy of the factory*: A factory consists of multiple subsystems which are located in the same or in different areas (e.g., an assembly system may be structured as component and main assembly).
2. *Organizational hierarchy*: Specific parts of a distributed subsystem comprise an organizational unit (e.g., a fabrication unit of one factory may, from the organizational point of view, belong to another factory).
3. *Management hierarchy*: The management of a factory operates on different control levels.
4. *Information hierarchy*: Decisions are made on the basis of information and data obtained from different control levels.
5. *Structural hierarchy of manufacturing equipment*: The layout of manufacturing equipment is structured as an entity (e.g., it may consist of individual machine tools which are grouped together to machining centers, and they in turn may be combined with a programmable manufacturing system).
6. *Control hierarchy*: The controllers of a factory are interconnected to a hierarchical control installation.

It is the manufacturing engineer's task to form a functional control system from these fundamental hierarchies. The reader can visualize that this is an extremely complex problem. There are no strict rules that can be applied, since even for the same product factories may differ considerably. In addition, some of the requirements for one fundamental hierarchy may contradict those of others.

So far, most of the progress for hierarchical control systems has been made for hardware. Numerous concepts have been published within recent years. Almost every manufacturer of control devices for processes offers its own hierarchical control system. However, the basic concepts of most of these systems are similar. They apply modern control strategies in conjunction with user-friendly control peripherals. Unfortunately, the computers, communication systems, sensors, controllers, and software used are generally very different. For this reason there is hardly any compatibility between the control systems of different manufacturers. Some of the most typical features of the systems are as follows:

1. Strict division of the hierarchical tiers into administration, supervision, and control levels

2. Use of process computers at the upper level and microcomputers at the lower levels
3. Slow communication bus systems at the upper level and fast ones at the lower levels
4. If possible, use of standard components that can be connected by plug-in devices
5. The capability to reconfigure the system on-line with the aid of interactive CRT terminals
6. The substitution of conventional control panels with black-and-white or color picture screens
7. Continued operability of lower levels on failure of a higher level
8. Increasing use of the flying master principle to minimize the potential of system disturbance
9. Simple programmability of the system with the aid of a keyboard, light pen, or graphic terminal
10. Display of schematic pictorials of the process (can be exploded from a higher level to a lower one)
11. User guidance with the aid of error search routines to locate defects
12. Upon disturbance, automatic switchover to a backup
13. Access to any process data of any hierarchical level
14. Selective presentation of process trends for each hierarchical level; free selection of historical data to compare present and past process behavior; simulation of control strategies

Most of these features are included in the majority of advanced control systems. However, often it will be necessary to implement only a few of them, depending on the control task and the justifiable degree of automation.

The integration and implementation of control strategies of each hierarchical level requires unique programming languages, standardized communication data highways, distributed data base systems, and a compatible computer family. The installation has to assure the operation of each manufacturing unit of the plant at its highest possible efficiency. Scheduling and supervision functions have to provide precise synchronization between all manufacturing units. Any malfunction on the operational unit level must be correctable by a supervisory control unit. Disposition and coordination operations have to assign the correct control tasks to each subsystem so that all units work together in the plant as a single entity. The system has to be able to react systematically to emergencies (e.g., to a decrease in quality). For the plant to produce the product at minimum cost by using time, material, labor, and energy efficiently, the overall manufacturing system must be able to carry out its scheduling and control functions. The required reliability, availability, flexibility, and expansibility of the control installation make a modular fault-tolerant hierarchical computer system architecture mandatory. It must be capable of detecting any faults, locating the problems, and correcting them. For this reason, redundant computer components may be necessary.

When computers are integrated into a hierarchical control system, it is advantageous to use the same type of processor from the same family horizontally on each level. This has the advantage that software and hardware can be standardized and maintained more easily. It is also of benefit to use processors of the same family vertically, thereby facilitating software development. If, for example, a stripped-down computer is used at a lower level, the software for it can be developed on a higher level. At a higher tier, the computers are usually equipped with more memory and peripherals, which greatly simplify program development. The lower-level processor may have an instruction repertoire which is merely a subset of that of the higher level. However, this does not matter as long as the programmer observes this restriction during software development.

4.3 COMPUTING EQUIPMENT AND ITS HIERARCHICAL ASSIGNMENT

4.3.1 Supervisory Level (Corporate Level)

An introduction to the highest hierarchical level (corporate, supervisory, strategy) was given in Sec. 2.4.1 (Fig. 2.8). In most organizations the activities of marketing, production planning, order planning, process planning, inventory management, material control, cost accounting, and customer order servicing are supported by this level (Table 4.2).

The strategic management uses a central data base which contains current information about the company, including data on production, machine tools, material handling equipment, assembly facilities, and quality control. Access to all lower data bases is performed via the hierarchy of control computers.

The short- and long-range production schedules are drawn up and their proper execution monitored at this level. In case of production problems, automatic rescheduling is initiated. If scheduling is done properly, the company will efficiently use resources, minimize production cost, and speed up the fabrication of the product. This activity will be supported by mathematical optimization techniques. Optimization runs may be done only once or twice a week because they are usually very time consuming.

Another activity that has to be coordinated and supervised is quality assurance (Chap. 10). The quality assurance loop comprises the customer, engineering, and manufacturing. Quality data are collected at this level and compared to quality standards. If necessary, corrective

TABLE 4.2 Characteristic Data for the Supervisory Computer

Hardware	Software	Task
Processor	Operating software	Data processing
Very large business computer, number cruncher, designed to handle large I/O data streams	Multiuser operating system	Business
Multiprocessor system	Complete set of program development aids	Statistical
Distributed computer system	Diagnostics routines	Competitive strategies
	Communication routines	Central data file management
		Limited engineering calculation
Peripheral	Application programs	Communication with host
Large disk storage devices	Forecasting	Downloading of weekly schedules
Large tape storage devices	Marketing	Collection and evaluation of operating data
Fast printers	Customer servicing	Functional testing of hosts
Interactive terminals	Manufacturing planning	Troubleshooting
	Quality assurance	
Communication hardware for data transmission with:	Maintenance	Communication at same level
	Inventory management	Synchronization with other corporate activities
Same level	Material flow control	Standby operation
Lower level	Purchasing receiving	
	Engineering	Operator communication
	Equipment facility planning	Dialogue operation
	Optimization	Operator guidance and assistance
	Make-or-buy decisions	Input of programs and data
		Input of sales request and manufacturing schedules
		Input of operating parameters

action is initiated. Quality reports are made available to engineering and manufacturing to determine if the design of the product or a fabrication method has to be improved.

This control level has to assure that all factory activities are precisely synchronized, and that any production problem is reported immediately, and that corrective action is taken. There must be alternative production machines and computers available to maintain the operation during an equipment breakdown. The computer will have diagnostic capabilities to check itself as well as the performance of lower hierarchical controllers and that of production machines. If a standby computer is used, it will be updated constantly.

Particular attention must be paid to the user friendliness of the system. Interactive graphic terminals will allow the display of production stream, the location of products, and the inspection of statistical production data and cost data. Fast access to the production data base is a significant parameter in efficient computer use.

A data base containing current information on the manufacturing process, machine tools, the material handling system, and assembly lines makes available the information necessary for possible rescheduling of production. Future strategy computers will be able to solve their own production problems with the aid of decision rules and knowledge of the production process. Such expert systems will have a high-level interactive human-machine interface to support planning and controlling operations. Strategy computers are usually big commercial number-crunching machines with a standby system to ensure data integrity. Self-checking and diagnostic routines assure equipment reliability. Only higher-level languages are used for application programming.

4.3.2 Operative Level (Plant Control)

The most important task of this control level is the execution of actual production schedules for the operation of a manufacturing entity (Table 4.3). It has to interact with all manufacturing cells and with material flow control. The production schedule generated by the management strategy computer on the higher control level has to be adapted to the manufacturing status of the local plant. Efficient allocation of machine resources and the capability of a dynamically responding to production problems on lower levels permit cost optimization.

Frequently, the control and synchronization tasks will be of such a magnitude that individual manufacturing cells with different machine tools, conveyors, robots, and other material handling equipment will have their own operative computer. This unit will also collect and process production data and solve functional problems at lower hierarchical levels. The distribution of control data and the optimization of all unit tasks are performed on this level. Any emergencies from a lower level

TABLE 4.3 Characteristic Data for the Host (Plant) Computer

Hardware	Software	Task
Processor	Operating software	Communication with strategic computer
Small business computer with limited real-time capabilities	For host	
Large minicomputer	Real-time operating system foreground/background	Data processing
Super microcomputer	System configuration program	Technical
Peripherals	Compiler	Business
	Assembler	Statistical
Large-capacity disk storage	Development software	Control strategies
Tape storage	System library	Optimization
Printer	Communication routine	Data administration
Interactive terminal	Diagnostics routine	Communication with satellite
Process I/O's	For satellite	Downloading of daily schedules
Communication hardware for data transmission with:	Cross-compiler	Downloading of software
	Cross-link editor	Data acquisition and concentration
	Real-time simulator	Downloading of operating parameters
Upper level	Application program	Functional testing of satellite
Same level	Plant monitoring control	Troubleshooting
Lower level	Data acquisition and reduction	Communication at same level
	Control algorithms	Synchronization of manufacturing operation
	Optimization programs	Standby operation
	Diagnostics routines for satellite and manufacturing equipment	Operator communication
		Dialogue operation
		Operator guidance

TABLE 4.3 (Continued)

Hardware	Software	Task
		Operator communication (Continued)
		Operator assistance
		Input of programs and data
		Program changes
		Program development
		Input of operating parameters
		Initialization of computing and manufacturing equipment

control or from this level will initiate correction actions to maintain the production schedule. Quality data acquisition and quality assurance operations have to be performed on-line. An additional assignment will be to diagnose and update standby systems.

Operational control necessitates a real-time control system that must be operable under real-time multitasking mode and batch-type operation. The computer also has to communicate with the supervisory and plant levels. To assure economic software development, the implementation languages to be used may be real-time FORTRAN or BASIC, and in the near future will be ADA. Problem-oriented languages greatly simplify programming and can easily be learned by the manufacturing engineer under user guidance.

The computer at this level will have a large working memory and a high-capacity disk memory. It will also be employed for program development for the computers at the lower level. This can be done directly if the satellite is instruction compatible; otherwise, cross-software has to be used. The completed program is downloaded on the satellite. The required communication functions with the lower level are as follows:

1. Turning the satellite on and off
2. Erasing the memory and register contents of the satellite
3. Retrieving operating software and application programs on its own disk and downloading these to the satellite
4. Initialization of the satellite
5. Initialization of the time and date
6. Periodically checking that the satellite is operable
7. Collecting operating data from the satellite
8. Locating problems
9. Disconnecting the satellite in case of an irreparable defect
10. Connecting standby computer equipment
11. In case of system expansion, interfacing new satellites without the need to reconfigure its own operating system
12. Providing program development capacity in background operation without interfering with the control program that is currently running

These functions may be done in a manual or an automatic mode. It is necessary that the operator can interact directly with the satellite if this is required during an emergency situation.

The computer at the operative level will act as the master of the communication system. In other words, the computer is the supreme commander and has the authority to set priorities and to activate and deactivate satellite functions. Since in most cases the satellites will be stripped-down computers, they will use the data peripherals, such as the disk memory unit of the host. This means that the running programs on the satellite must be independent of the disk to operate production equipment. Otherwise, during a possible breakdown of the host, the control system may become inoperable.

A standby for the host may be necessary if the production process is very critical. There are several mechanisms used to assure continuous operation if the master fails:

1. The standby will be activated only when the failure is detected.
2. The master operates in parallel to the standby.
3. A conventional standby control system is activated.
4. The satellites can carry on the operation under reduced control functions.
5. The flying master principle is used, whereby upon failure of the master, another host computer takes on the function of this master.
6. Backup is provided only for the system components that have the highest failure rate. This typically holds true for the disk memory.

This control level is the domain of the minicomputer or of a business computer with real-time capabilities. Since manufacturing problems will become increasingly complex in the future, the "supermini" will predominate. It is usually designed with a data length, a bus, and registers to handle 32-bit words, which allow the user to address a large memory space, to index into long arrays, and to specify records in large files. In general, superminis have a capacity and speed advantage over microcomputers. Typical performance data of such mainframes are shown in Table 4.4. The large address space makes the mainframe very useful to direct the operation of distributed computer systems. In addition, it is possible to control manufacturing systems with the help of large process models.

The future process computer will have several dedicated processors (e.g., for I/O, to perform mathematical calculations, and to operate on data banks). For many applications it is necessary to improve the reliability of a computer network. For this purpose computers with redundant processor configurations will be available. Today, only a few computers offer this as a standard feature.

A further innovation will be the associative data bank for minicomputers, where the key to locating data is not an identifier but its contents. This data bank will be integral part of the future supermini. The operating system of this powerful mini will have capabilities similar to those found in today's large computers. Some of the particular features will be multiuser and multiprogramming capabilities and a virtual memory. UNIX-like operating systems specifically tailored toward real-time application will be the predominant system software.

TABLE 4.4 Typical Performance Data of Superminis

Number of instructions per second	500,000 to 850,000
Single-precision whetstones (FORTRAN)	600 to 1200
Main memory (megabytes)	1 to 128
Mass storage (megabytes)	Up to 4700
Maximum workstations	1 to 128
Typical system software	Time sharing Transaction processing Demand paging Data base management Word processing Electronic mail
Networking	SNA, X.25, and in most cases vendor-designed network
16-bit compatibility	Not provided in all cases
Languages	APL BASIC COBOL FORTRAN PL/1 PASCAL ADA

Particular attention will also be paid to the man-machine interface. Graphic interactive terminals will allow the display of macroscopic and microscopic pictorials, such as flow diagrams of any manufacturing process. This will aid the control of the process and the production problem. There will be graphical operator training and guidance.

4.3.3 Control Level (Plant Floor)

On the control level of the manufacturing process, a wide variety of equipment, such as machine tools, material handling systems, and assembly lines, are directly supervised by dedicated micro- or minicomputers. The microcomputer plays a predominant role. It is embedded in (1) programmable controllers (PCs), (2) numerical controllers (NCs), and (3) direct numerical control (DNC). In addition, a variety of special-purpose controllers are used.

The functions at this level may be in open-, semiopen-, or closed-loop control (Table 4.5). The control algorithms are implemented in software or firmware, and in general cannot be changed by the user. It is possible to employ one type of controller for different equipment just by providing different programs. Usually, there is an interface available for upward communication. This makes possible data acquisition of process data and supervision of the function of the controller via a higher-level computer. In more sophisticated installations, as in flexible manufacturing systems, it is also necessary to change the program when a product change is made.

The computer in general supports only process and no data peripherals. Communication with the operator can be done via the host computer or via a mobile terminal or a simple dedicated keyboard. Process data are accumulated and stored over a predetermined time period or until the working memory is full and then sent upward to the disk memory of the host. The following capabilities have to be provided for a comfortable host-satellite configuration:

1. The satellite must contain a completely independent program for process control and should be operable if the host fails.
2. In case of a failure, it should be able to make the latest process status available to the operator.
3. The operator should have provisions to interact directly with the satellite when the host has problems.
4. The satellite should be able to self-diagnose itself and the equipment it controls. It may have to get the diagnostics program from the host.
5. Sequential or random access to the disk memory of the host should be possible in a transparent mode.
6. The satellite must be able to initiate the transfer of process data to the host if its memory is full.
7. The satellite must be programmed so that it can open, maintain, and close a file on the disk memory of the host. The file must be protected from unallowable access, and it must be possible to inquire about the status of the contents of the file.
8. It must be possible to initiate loading of the operating system and the application programs from the satellite.

At this level the computer will contain an 8- or 16-bit processor and for special purposes, a 32-bit processor.

There has been a continuous trend toward relocating many of the conventional tasks of the processor into the interface components. Numerous input and output modules are available which are capable of operating parallel to the main processor. Typical representatives are digital input/output channels and control units for floppy disks.

TABLE 4.5 Characteristic Data for the Control Computer

Hardware	Software	Tasks
Small CPU; if possible, upward-compatible instruction set	Operating software	Computational assignments
Relatively small working memory	Simple operating system	Determination of operating limits
	Error diagnostics routines	Statistical evaluations
	Communication software	Simple control instructions
		Trend analysis
		Plausibility tests
Process peripherals	Application programs	Communication with:
Transducers	Control algorithms	Host computer
Controllers	Data acquisition and reduction routines	Process
Multiplexer		Operator
A/D and D/A converters		
Digital static and dynamic I/O's		Simple program changes
Counter		Control of process
Clock		
Pulse generator		Inquiry
Signal display		Recognition of interrupt
Communication hardware		Testing of functions of process peripherals
Data peripheral (if required)		Output of control parameters
Keyboard/printer		Emergency shutdown of equipment
Disk memory		
		Operator access
		Input of control limits
		Initialization of equipment
		Initialization of diagnostics routines
		Turn on and shutdown of entire or partial process

Many manufacturers have developed one-chip processors that users can configure to their specific applications. These processors are instruction and bus compatible with the main processor and can easily be integrated into a multiprocessor system.

For the operation of automation systems, the speech analyzer and speech synthesizer are increasingly gaining importance as peripherals for the control computer. With these devices the machine operator can give verbal instructions to the machine; for example, the operator can command a lathe to cut a workpiece to dimensions specified by the designer. There are numerous speech analyzers already available. However, to date their vocabulary is very limited and the hit rate is unacceptable for most applications. Another problem is the necessity to have a specific vocabulary for every instructor. In some cases, even the tired voice of the same instructor may lead to recognition problems.

The speech synthesizer has been developed to a high degree of sophistication. When installed in control devices, it can inform the operator about the task a machine tool is presently performing, or about difficulties with the control system, and if needed, it can give repair instructions. Natural speech processing systems are proposed for the future.

Implementation of complex control algorithms (DDC algorithms) can be realized with the aid of 8- and 16-bit dedicated microprocessors. Provided with fast memories and arithmetic processors, efficient I/O interfaces, and multichannel timers, they are well suited for the solution of many control tasks. They have a high accuracy, are very flexible and economical, and render good solutions when the central processing unit (CPU) and other control circuits are integrated on one chip.

Control programs may be generated on a higher level and can be written in high-level languages such as ADA, PASCAL, FORTRAN, PEARL, APT, or BASIC or their dialects. Compilation, code generation, and program tests are done on the higher level, which transmits executable programs to each manufacturing unit. To increase the efficiency of a program translated from a compiler language, it may have to be optimized when it is used on a small microcomputer. Where a fast execution speed is required, the program will be written in assembly code or a block-type language. To permit shop floor programming and servicing, each unit can be equipped with a RS-232 interface as part of the programmer console. Efficient low-cost color displays with simple graphic capabilities will greatly aid manual process guidance.

4.3.4 Some Future Trends for Peripherals

The development of process peripherals will be greatly influenced by microelectronics. The trend leads to autonomous devices which are capable of processing independently signals in close proximity of the manufacturing equipment. In many cases the microprocessor will be incorporated into the sensor.

The cost to improve the resolution, speed, and noise immunity of present converters would be exponential. For this reason, suppliers will direct more attention to custom-designed circuits which can readily be manufactured with available design tools and manufacturing systems.

The analog devices will be conceived from predesigned functional blocks. This will be made possible by efficient use of CAD/CAM technology. The analog cell concept takes advantage of the possibility of storing design data for converters, operational amplifiers, multiplexers, and memory circuits in the CAD computer. The digitized circuit design data can be configured by the engineer to different combinations and can be verified by a simulation program to check adherence to the customer's specification. It will also be possible to check the economics of the design and to output yield data obtained from stored information about process geometry, number of masks, diffusion, and the size of the wafer and chip.

CMOS (complementary metal-oxide semiconductor) technology, previously used primarily for digital circuits, will have a great impact on analog devices. This will be made possible through the use of the dual polysilicon gate process in conjunction with dual-layer metallization. This manufacturing method allows high yield production of operational amplifiers, voltage references, latches, and data converters.

4.3.5 Software and System Development Aids

With the majority of real-time computer installations, the software is the most expensive component. Despite the availability of higher programming languages, on-line control systems are at present implemented principally with the aid of assembly languages. In Germany, PEARL is becoming a popular language for control and microcomputer systems. In the near future, ADA will be emerging as a stiff competitor. ADA is a highly standardized language. For this reason, it is registered as a trade name. Translators for ADA must undergo stringent validation. Thus it can be expected that ADA will be highly machine dependent, which is necessary for good portability. With this feature software houses will be able to offer a diversity of application programs for manufacturing. Thus the customer will have the opportunity to purchase reasonably priced software.

Another trend that can be observed is the implementation of software functions in firmware or even in hardware. This will be true in particular for the kernel of operating systems and for dedicated data acquisition and control functions. It will also be possible to configure firmware with the aid of an operating system.

There are presently hundreds of operating systems for microcomputers available. They range from stand-alone to real-time and multiuser systems. The stand-alone systems are mainly derivatives of the Digital Research CPM and Microsoft MSDOS developments. With increasing use of microcomputers for real-time applications, the new operating systems will become similar to Bell Laboratories' UNIX system. With the help of this operating system, many future microcomputer installations will become more powerful than, for example, the VAX 780 system.

There will be new software systems available to aid the entire software development cycle. This trend will be stimulated by research results obtained from computer science and artificial intelligence. There are numerous ongoing research activities trying to conceive software development tools for many different applications. Some of them follow the concept of automated software generation; others are only trying to support specific phases of the software development cycle. The first interactive workstations that will allow computer-aided software engineering are being offered on the market. They usually provide graphic capabilities to construct hierarchical system diagrams. The EPOS system, developed by the University of Stuttgart in Germany, renders development aids for the design of real-time software [1]. The SARS system of the University of Karlsruhe in Germany provides tools for the specification phase of software and hardware for automation systems [2].

4.4 THE COMMUNICATION NETWORK

The efficiency and flexibility of the hierarchical manufacturing control system depends on the availability of a unique communication network which connects the distributed processing units. Efforts have been made to standardize process control communication systems. Different protocols and a great number of communication bus systems have been developed and are in use by industry [3]. Typical bus configurations are shown in Figs. 3.43 and 3.45. Most of them employ a protocol hierarchy with different communication levels, including:

   The interprocess communication protocol
   The bus access protocol
   The transmission protocol
   The logical link level protocol
   The physical link level protocol

These protocols are implemented partially in hardware and partially in software. Early communication networks are implemented with a star configuration (Fig. 4.2). Today's networks are configured as ring, star, or open bus systems. Communication is either bit parallel or bit serial. Several buses may be connected to hierarchies with interface

Advanced Computer Architectures

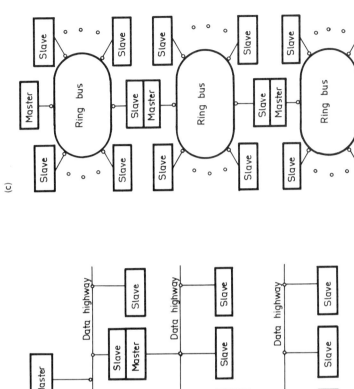

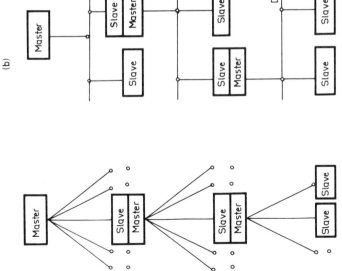

FIG. 4.2 Different structures of plant communication networks: (a) star structure; (b) open bus structure; (c) ring structure.

and control elements. The nodes in the network can be active or passive or both (sender, receiver of data). Most of the actual plant communication networks operate in a single master mode. This means that only one node has control of the network at one time. In some implementations this master function can be shifted from one node to another according to the flying master principle. Ring configurations with bit-serial transmission lines are becoming popular. They use coaxial cables or fiber optics with interface elements to support logical and physical link levels. Transmission rates of 5 and more megabytes per second are possible. Most microcomputer manufacturers supply their systems with communication chips that control access to the bus. They also generate protocols for different link levels, encode and decode signals, and perform master or slave functions.

The objects of standardization for an interface are the hardware connectors, their number of pins, the pin placements, and so on. Also, the hierarchy of protocols has to be defined, including the code, message and data formats, status words, and the method of error detection and correction. For further discussion, three plant communication networks are selected, which are all in an advanced state of standardization [1]: (1) the IEEE 488 bus, (2) the PDV bus, and (3) the Ethernet. They all have unique features and were designed for different applications. The IEEE 488 bus was developed for communication over short distances (maximum 20 m) at a high data rate. The communication is performed byte serially via a bidirectional data bus. The PDV bus is a ring bus for long distances (1 to 3 km) with standardized interfaces and bit-serial data transfer. The Ethernet, one of the most popular systems, is based on a coaxial cable for bit-serial data transfer over long distances and has standardized interfaces.* The efficiency of a bus system depends on the following parameters:

Data rate, transmission distance (maximum)
Time delay to respond to interrupts and data requests
Additional hardware and software needed for expansion
Reliability, fault tolerance, and availability
Unique logic structure, standard plug-in principle
Possible geographic distribution of communication process
Cost of the system components

4.4.1 IEEE 488 Bus

The IEEE 488 bus is a bit parallel/byte serial data transmission system with a transmission rate of 1 megabyte/sec (Fig. 3.43). The area of

---

*The General Motors Corp. is presently specifying, within the MAP program, the communication protocol for a factory bus, which may become a standard.

application for this bus is local acquisition of process and quality control data. Most instrument manufacturers support their instruments and signal processors with standard IEEE 488 interfaces (25 pins, eight data lines, five control lines, three lines for acknowledge signals, eight twisted-pair lines). The nodes may have master functions and can be listeners or talkers or both. The system master controls data transfer between talkers and listeners. For data communication, ASCII code is used. A unique programming language is not defined for the IEEE 488 bus. Software is usually written in BASIC, FORTRAN, PEARL, or PASCAL. The transmission speed is 200 to 400 kilobytes/sec.

If communication over a distance of 20 m is necessary, a second bus system has to be used. This second bus system connects multiple distributed clusters of IEEE 488 bus-controlled measuring instruments. It transfers the data of the lower level to the next control level (Fig. 10.25). A typical bus system for local distributed processes is the PDV bus.

4.4.2 PDV Bus

The bit-serial PDV bus operates with a transmission rate of 100 to 500 kilobytes/sec over distances of up to 3 km. The bus is capable of hosting 252 stations and is qualified to interconnect hierarchical computer systems. The system has been developed for the industrial environment and was sponsored by the Ministry of Research and Technology (BMFT) of West Germany. Typical applications are the interconnection of control computers of a hierarchical control system for manufacturing equipment. Real-time and high safety requirements were an important criterion for the design of this bus. Presently, the bus is offered for industrial use by several manufacturers in fast and slow versions (Fig. 3.45).

The fast version is used for direct process communication with process peripherals, and the slower version interconnects computers at a higher hierarchical level. Process and data peripherals can be connected with the bus by the plug-in principle via an interface control unit. The controller is available as a PDV bus chip (Valvo MEE 3000). The bus is independent of the communication technology. It can accommodate an electrical or a fiber-optics communication cable. There are interrupt capabilities for 24 privileged participants. Service requests for other interrupts are processed by periodic polling.

4.4.3 Ethernet Bus

The Ethernet is presently gaining popularity with local area networks. It is supported by Intel, Digital Equipment, and Xerox. There will be integrated circuits or transceivers available to interface computers and peripherals with the bus. The communication media is a 50-Ω 10-mm coaxial cable. It allows a transmission rate of 10 megabits/sec. The

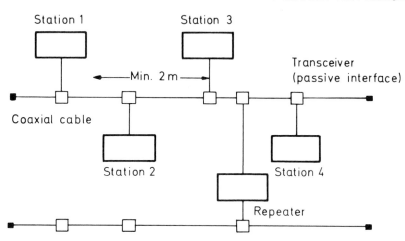

FIG. 4.3 Principle of the Ethernet bus.

maximum length is about 2.5 km and there is a possibility of interconnecting 1024 participants. Figure 4.3 shows the principle of the bus system. A physical layer was standardized. This includes cable diameter and impedance, pin arrangement and plug dimension, the coding method, the voltage level, and the noise margin. The next layer is the data link layer. Here the frames, the protocol, address coding and decoding, and error detection are subject to the standard. All following layers are to be specified by the user. With this it is possible to interconnect existing network architectures and communication systems with the Ethernet. With the aid of a transceiver, the user can connect different Ethernet modules. The bus does not employ the master principle. Each participant can independently request access to the bus.

The Ethernet was intended primarily for use in automation, distributed data processing, terminal access, and other situations requiring an economical connection to a local communication medium. Originally, the Ethernet was not designed for real-time computer systems. However, this application was not specifically excluded. The lack of an interrupt possibility limits the bus to the control of slow or noncritical processes.

Advanced Computer Architectures 119

4.5 EXAMPLES OF HIERARCHICAL COMPUTER SYSTEMS FOR MANUFACTURE

Different manufacturing applications require the availability of different control systems. For this reason, a hierarchical computer architecture has to be tailored toward its application. Typical factors that determine the architecture of a distributed computer installation are as follows:

Degree of automation
Size of an installation
Physical layout of the plant
Number of control levels
Organizational structure of management
Task of the control system
Data volume and data rate
Availability and Safety considerations
Real-time requirements

Three different computer architectures will be discussed in this section, one for the control of NC machines, one for factory data acquisition, and one for plant data processing and communication. Several other hierarchical controls are presented elsewhere in the book. To obtain an overall knowledge in this area, it is suggested that the following sections be studied:

Section 2.4	The Hierarchical Computer Control Concept
Section 3.8	Quality Control Modules
Section 5.9.1	CAD Hardware
Section 7.7	CNC Systems
Section 7.8	DNC Systems
Section 8.7	Hierarchical Control for Robots
Section 10.8	Hierarchical Computer Systems for Quality Control
Section 11.4.3	Control of FMS

4.5.1 Modular Multiprocessor Control System for Machine Tools (MPST)

System Architecture

The MPST hierarchical control system was developed by the University of Stuttgart [4]. It was funded by the Ministry of Research and Technology of West Germany. Originally, the system was conceived as a universal computer architecture to control machine tools, measuring systems, and robots. However, it appears to be best suited for machine tools. A schematic layout of the concept is shown in Fig. 4.4. A word

serial bus connects several processors to operate in parallel as a
fast control system. Each processor has a solitary function which
may consist of one or more control tasks. A typical assignment of
functions for a NC machine tool is shown in Table 4.6. In this case
there are five different processors provided: one for central data
processing, one for input/output control, one for NC program scheduling, one for geometric data processing, and one for processing technological data. The central processor or master controls the access to
the bus and oversees the data communication between the participants.
Each active participant can request access to the bus by sending a
request signal to the master. When access is granted, communication
with another participant can be done in the DMA mode. An active
participant may take over the function of the master to coordinate
the administration of the bus and the central control tasks. There
are also passive participants such as memory, I/O module, and position control circuits. They are assigned to active participants.

The system is designed to accommodate 16-bit microprocessors which
may be of different design. A standardized interface including pin
assignment and bus procedures assures compatibility with the bus.
Each processor contains a 12K EPROM (erasable programmable read-only
memory) and a 4K RAM (random-access memory). 2K of the RAM

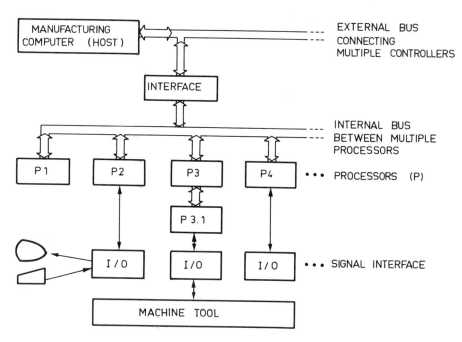

FIG. 4.4 Modular multiprocessor control system.

TABLE 4.6 Assignment of System Functions to Different Processors of a NC Control

1. Central processing unit

    System initialization
    Central interrupt processor and bus scheduler
    Switching of operating mode
    System error processing
    Initialization and execution of diagnostics functions

2. Input and output of control and service data

    Driver programs for peripherals
    Operator work shop programming (manual)
    Execution of diagnostics programs
    Monitor and editor functions for program support and control of other bus participants

3. Scheduling, processing, and storing of NC programs

    NC program correction and optimization
    Work cycles and subtasks
    Distribution of control data to interpolators and PCs

4. Geometric information processing for multiple axes

    Interpolation
    Geometric corrections (e.g., correction for pitch errors of a lead screw)
    Adaptive control (e.g., on-line adaption of cutting parameters)
    Information storage

5. Processing of technological information for I/O (function control)

    Logic operations with input signals
    Decoding
    Storage
    Interlocking
    Text processing
    Simple positioning tasks
    Housekeeping of memory storage for tool and workpiece information

memory serves as a data transfer buffer which can be addressed via the bus by any participant in the read or write mode.

The bus consists of 16 parallel data lines, 16 address lines, five control lines, and two acknowledge lines (Fig. 4.5). In addition, there are three interrupt lines. Interrupts are processed by the daisy chain principle with the help of a common interrupt line, a interrupt freeze line, and an acknowledge line (Fig. 4.5).

The flexibility of the control system is enhanced by information interfaces. For a desired control function, the interconnection of resources within the system is coordinated via these interfaces. Figure 4.6 shows the general characteristics of this concept with the help of an undirected graph. The edges represent data paths and the nodes functional NC blocks. Typical operating phases that can be obtained with this concept are shown in Fig. 4.7. The interconnection of the individual functional blocks is done via data transfer buffers in the virtual address space of the memory.

System Software

The program to operate the machine tool is of hierarchical structure (Fig. 4.8). A central scheduler coordinates the cooperation of the different functional blocks. The number and type of functional

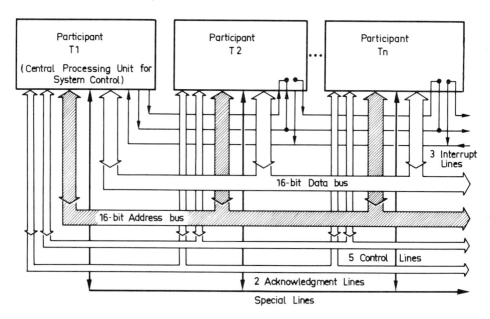

FIG. 4.5 Bus structure of the microprocessor control system.

Advanced Computer Architectures 123

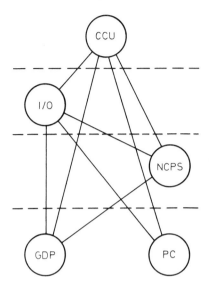

FIG. 4.6 CNC information flow structure: CCU, central control unit; I/O, operator and control data I/O; NCPS, NC program scheduling and processing; GDP, geometric information processing; PC, programmable controller.

blocks depend on the application. Each block contains its own scheduler, which supervises the execution of the individual software modules. Typical modules are for position control, slope generation, fine and course interpolation, and so on. A functional block may contain one or several software modules.

The operating system consists of a system-specific hardware scheduler and an application-specific supervisor (Fig. 4.9). Its task is system initialization, switching of operating modules, interrupt handling, bus scheduling and control, and error processing and exception handling.

The initialization is done wherever the system is turned on. It resets all participants in the ready state. First a timer is turned on and a status check is made. After 50 msec a test is done to assure that all participants are functional. Each participant enters its relative starting address for parameter transfer into the transfer buffer. After initialization, each functional block sends an interrupt signal together with its address and logical name to the master. From this information the master calculates the absolute addresses for the transfer regions in the buffer. A ready signal is issued when all functional units are able to exchange data and the complete address field has

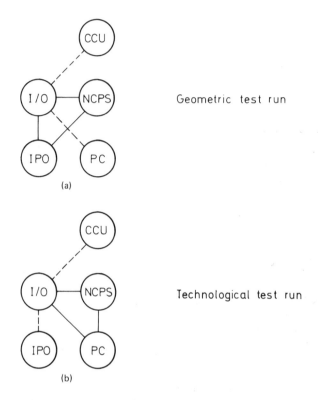

FIG. 4.7 Information flow during different operation phases: (a) geometric test run; (b) technological test run; (c) automatic or program operation, test run of NC programs; (d) data storage; (e) setup operation, stepwise, and conventional NC; IPO, interpolator.

been set up in the transfer buffer. Now each participant knows the address of the transfer fields of the other participants. In case of a missing entry, an error message is issued.

A change of an operating mode is initiated by an interrupt activated through the operator terminal. The participants will assume a new mode of operation and activate the new desired system function. The master supervises the changeover and activates or deactivates all modules for the new assignment.

The recourse management coordinates, supervises, and synchronizes the participants. It monitors the bus when an interrupt occurs, reads the status word, decodes it, and initiates work routines. The

Advanced Computer Architectures

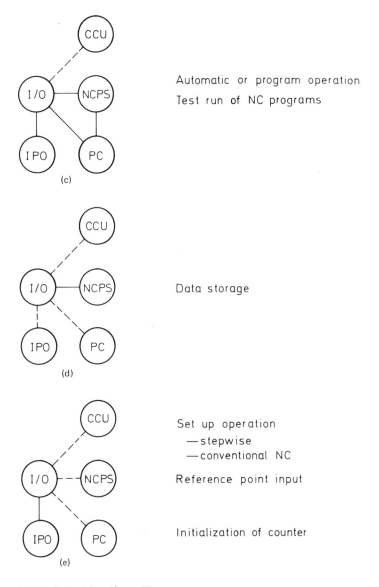

FIG. 4.7 (Continued)

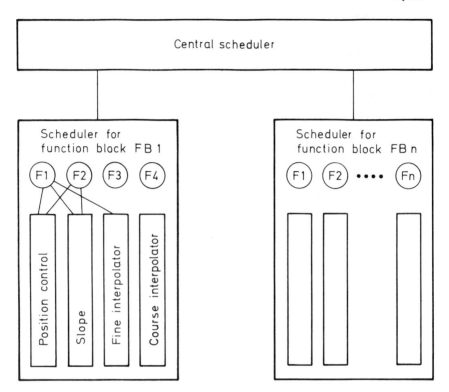

FIG. 4.8 Hierarchical software structure of the MPST system. Example for block FB 1: F1, go to point; F2, stop feed.

participants may issue interrupt requests for read and write, operating mode change, ready for operation, and system errors.

Error processing and exception handling use error-reporting mechanisms and error-handling strategies. It is possible to diagnose errors reported by participants and to oversee timing constraints. Errors may be printed out on a terminal or may lead to a defined off-state of a participant.

Figure 4.10 shows an example of a MPST system configured for a grinder. It has two active and three passive participants.

4.5.2 Example of a Hierarchical Data Acquisition System

System Architecture

The basic function of a data acquisition system is to collect operational data and to process these data. The system acquires data about

Advanced Computer Architectures

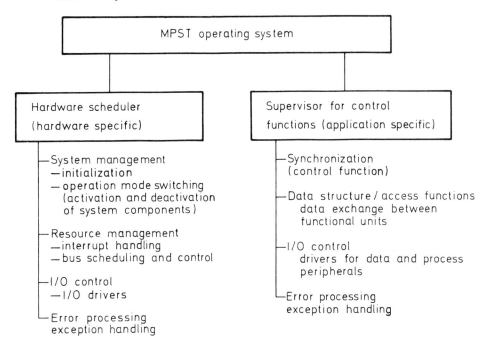

FIG. 4.9 Operating system of the MPST computer system.

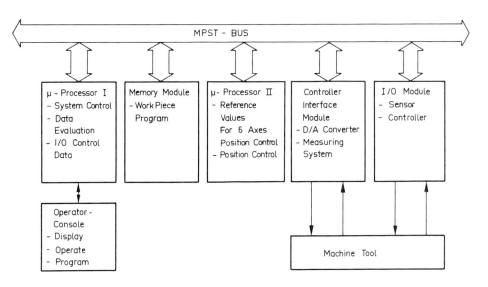

FIG. 4.10 Multiprocessor control system for a grinder.

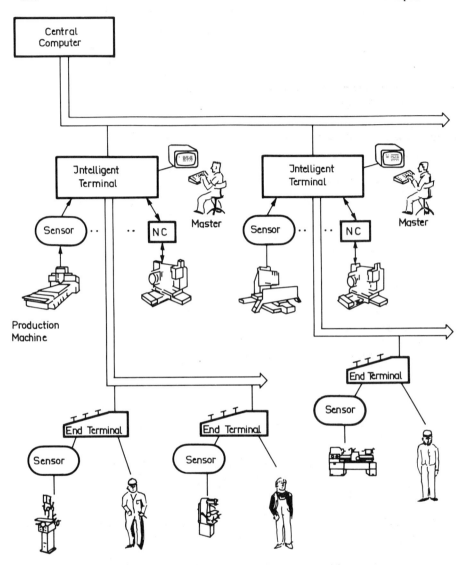

FIG. 4.11 Different types of factory data acquisition functions. (From Ref. 5.)

fabrication, storage, assembly, and quality control (Fig. 4.11). It also communicates with product engineering, production scheduling, and control. The different types of data acquisition systems are (Fig. 4.12):

Basic data acquisition system
Manufacturing information system
Factory control system

Most installations are conceived as manufacturing information systems. Typically, the architecture of a data acquisition system forms a hierarchy of interconnected terminals with different intelligence (Figs. 3.15 and 4.11). In general, advanced systems use three tiers.

Level 1. At the lowest level operational factory data are acquired by end terminals (Fig. 4.13). Their task is to collect data from keyboards, punched cards, badge readers, document readers, magnetic code carriers, and input devices for digital signals. The keyboard has three types of keys: alphanumeric, enabling, and function keys. The output of data from this terminal is visual information via a screen

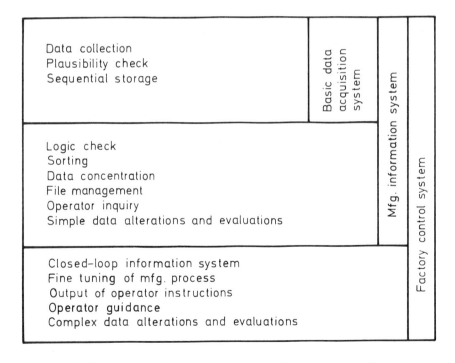

FIG. 4.12 Scheme of a factory data acquisition system. (From Ref. 5.)

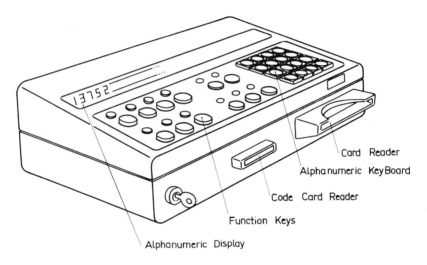

FIG. 4.13 Data acquisition terminal. (From Ref. 5.)

(display) or via indicator lights or digital signals. If requested, the output can also be obtained on a printer. Simple peripherals are used as end terminals. Some end terminals can be installed as stand-alone acquisition devices. However, their usefulness is questionable since they only collect data and do not use the capability of the computer to perform calculations.

Level 2. Usually, several end terminals are connected to the intelligent terminal of the next level of the hierarchy. It will obtain programs for supervising the operations of the end terminals. It will handle and preprocess the raw data from the lower tier and store them in a local memory or in a remote file at the central computer. In contrast to the end terminals, intelligent terminals can communicate directly with each other (cross communication). The peripherals connected to these terminals usually have complex functions such as dialogue or printing capabilities. The intelligent terminal at this level may also be a full-grown computer. In this case it may serve as a management information system operating in the open-loop mode (Fig. 4.12).

Level 3. Finally, all intelligent terminals are connected to the administrative computer or the master. Its principal task is to aid management to control the plant. Central factory data processing and storage is done at this level. Also, the activities of all party lines are monitored. This master function can be transferred to a predetermined intelligent terminal if the master station is not available. Connection to a large planning computer is also possible. Smaller systems can operate

Advanced Computer Architectures 131

without intelligent terminals; in this case the end terminals are connected directly to the central computer (Fig. 4.14). A data acquisition installation controlled by a master may be configured as a factory control system (Fig. 4.11). With this architecture the manufacturing process is supervised in a closed-loop control mode.

The data communication link topology is of nonswitch multipoint design. The intelligent terminal obtains messages from the end terminals by polling, and in turn, the central computer polls the intelligent terminals. Two types of buses may connect the terminals:

1. A fast bus that interconnects the intelligent terminals with the center; it operates at about 200 kilobaud.
2. A slow bus that connects the end terminals with the intelligent terminals; it operates at about 10 kilobaud.

Both may be of the PDV bus design. The activity on the party line is controlled by the PDV data communication procedure.

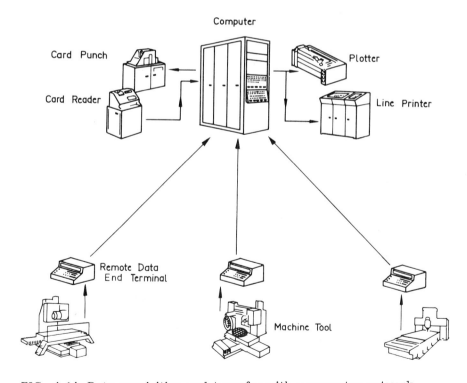

FIG. 4.14 Data acquisition and transfer with a computer network.

Transaction Language

The dialogue with the end terminal and that with the intelligent terminal will differ considerably. Operator communication at the end terminal will be done via a display. It allows the intelligent terminal to formulate tasks such as instructions to the operator at the end terminal or requests for shift reports. The dialogue of the operator with the end terminal is characterized by a rather rigid procedure issued by the intelligent terminal. This dialogue is referred to as operator guidance. A problem-oriented language for manufacturing control including commands for terminal tests and for operator guidance is used to perform the dialogue between the master computer and the operator. The language can handle digital in-process control and human-computer communications. It is based on a transaction concept [6]. There are preplanned instruction sequences for a machine or an operator. They are of logical design. To reduce the error rate, the commands of this language are of simple structure. For example, a display may show the following sentence: "Turn on lathe 3, clamp part, report any difficulties."

A distinction is made between two elementary types of transactions: They are of internal and external nature. The internal transactions are used for the control of the transaction interpreter, which is located in the end terminal. Examples of this kind of transactions command from the intelligent terminal to reset the end terminal or to revocate a transaction. The external transactions are further subdivided into two classes: mask and I/O transactions.

The mask transaction based on a menu technique is used for the human-machine dialogue via a display terminal. The image shown on the screen is divided into output and input fields. The output fields are used to guide the operator at the end terminal and cannot be changed. The input fields can be filled randomly. Mask transactions are used for the generation of forms.

The I/O transactions activate the additional peripherals connected to an end terminal. This can be done in strict sequential or in parallel mode (default). The parallel I/O transactions include concurrent operations by several devices, initiated by one transaction or the use of one device activated by several transactions. Typically, 12 I/O commands can be identified. The most important ones (eight commands) are listed below.

1. *Input*: This command initiates the data collection from input devices. The devices are specified by their full names.
2. *Move*: This command allows to transfer data from input devices to output devices.
3. *Check*: This command offers the possibility to verify the input data. It is always combined with an input command. If the

check result is true, the input data are transferred to the private buffer of the input device. If the result is not true, this part of a transaction is marked as erroneous and the transaction is continued by a case statement.
4. *Output*: This command is similar to the input command. Data received from the intelligent terminal are placed on output.
5. *Collect*: All input data of a transaction are transferred to the intelligent terminal, initiating the transaction.
6. *Buffer*: This command is used to collect all input data of a transaction which did not pass a check and to send them to an intelligent terminal.
7. *Exclusive*: This command allows a transaction to request exclusive use of a peripheral device (input, output) and releases it after use.
8. *Sequence*: This command constrains the different operations of a single transaction to a strict sequential execution.

Data Handling Schedule and Data Structures

A data acquisition system, for example, one used for quality control (QC), collects information from different local subsystems (stations) and must contain a distributed data base that can be accessed by different users, such as manufacturing or assembly personnel. Many basic functions are very similar for both systems. These are:

1. Dialogue-oriented data manipulations (e.g., form generation)
2. List-controlled data acquisition (the process of taking information from an order list)
3. Order processing
4. Graphical protocols

The solution of the problems can be done in a similar fashion in both applications. Three levels of software control hierarchy can be implemented on the intelligent terminals (operating systems). The highest level is the command tier. It receives and decodes orders coming from a dialogue device or from the control computer and initiates appropriate actions. The next-lower tier is defined by the user program. It guides all terminals according to a unique schedule. Here the semantics of the acquired operational data are ignored. The data evaluation program represents the lowest tier in this control hierarchy. It is activated if a transaction is prepared and it is specific to an application. The handling of operational data coming from the factory floor is performed by a four-level approach (Fig. 4.15).

The rate of information arriving from the end terminals is very high. For this reason the data cannot be instantly sorted according to their origination and information contents, so the data are first stored temporarily in a buffer. In the following sorting process (compression,

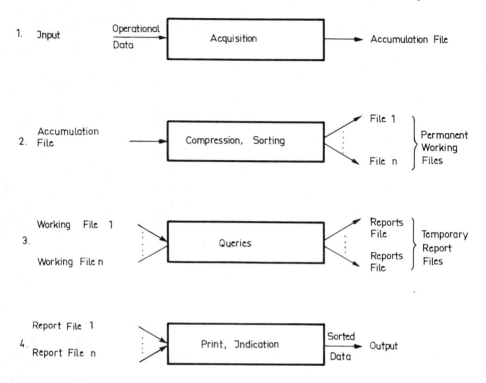

FIG. 4.15 Data handling schedule.

distribution) the input data are transmitted as a sorted package to a permanent working file. Operator queries transfer these data into temporary report files. Finally, a printer, a screen, or a similar output device is used to display the sorted data.

Every transaction is divided into a header and a body (Fig. 4.16). The data structure that is constructed on an end terminal as a response to the request of an intelligent terminal has the same form as that of a transaction. The response header characterizes the type of the response and identifies the transaction. The response body is made up from the usable information of the answer and has four parts, which describe the data, and the data themselves. A permanent data structure is located in the intelligent terminal (permanent working files) to aid storage of the responses of the supervised end terminals. It is also subdivided into a header and a body. The header carries the identification and the control information that defines a unique assignment of the response data. The body saves the response data, which are evaluated by a special program.

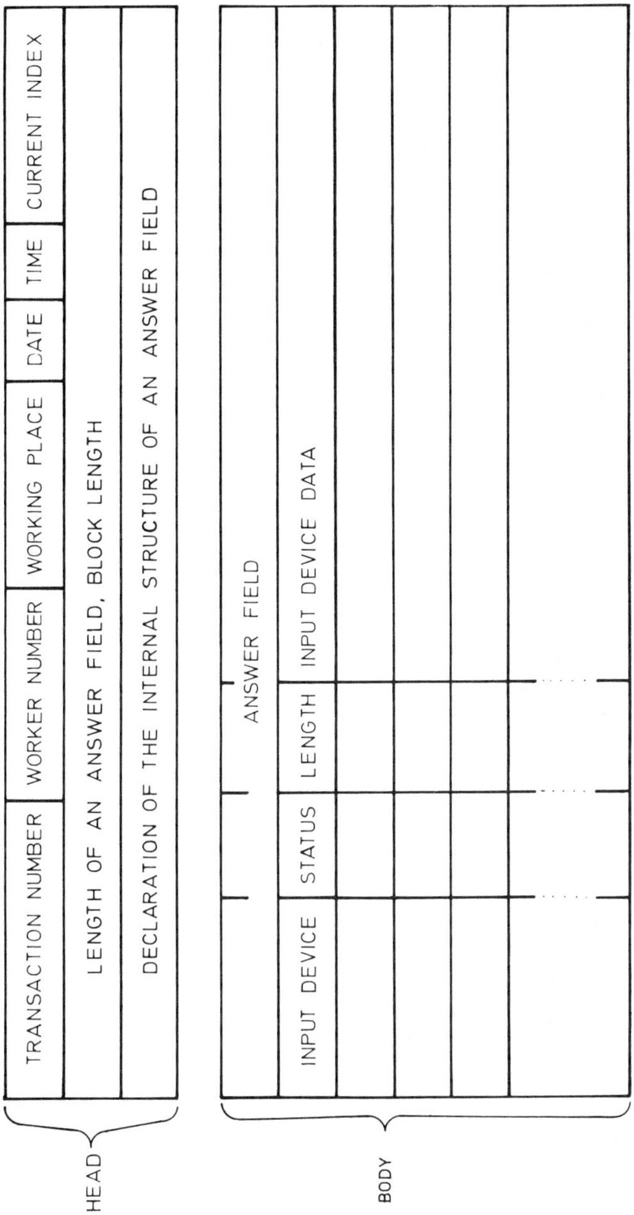

FIG. 4.16 Structure of a permanent working file.

Operating Systems

A data acquisition system typically is a distributed system. For this reason, two operating systems for the two types of terminals have to be implemented. The basic requirements for the structure of these two operating systems are the same. To be able to handle asynchronous events, both systems must have real-time capabilities and they must support multitasking. These two requirements can be obtained by dividing the operating systems into a kernel and a shell. The same kernel is used for the end terminal and the intelligent terminal. The shell for the end terminal accomplishes the following basic tasks: scheduling of the peripherals, communication, buffer management for messages, transaction interpretation, and clock management for wake-up requests. The shell for the intelligent terminal has in addition to these features dialogue device management, data volume management, and a virtual file system. The control hierarchy mentioned before can be implemented on the outer shell (e.g., user guidance).

4.5.3 Local Area Network

General System Considerations

In this section a local area network (LAN) for a medium-size manufacturing firm will be described. A LAN is defined as a bit-serial communication system that interconnects independent computers or peripherals. Usually, it is limited to the company's premises and uses a broadband communication line. Important parameters are:

1. The length of the line is in general below 2 km; however, it may extend to 10 km.
2. The data rate is from 10 kilobits/sec to 10 megabits/sec.
3. The response time lies below 10 mec.
4. Several hundred participants may be connected having identical access mechanisms.
5. All participants have decentralized access rights; there usually is no master controller.
6. Each type of network uses its own data format.
7. There is no specific technology defined for the communication line.

The computer network of the plant is divided into several sections. There is one LAN each for administration, manufacturing planning and control, engineering, and marketing (Fig. 4.17). All functions of the plant have access to the central computer and to the different sections. The communication line between the computers is a broadband coaxial cable. It is configured as an Ethernet (Sec. 4.4). The central computer

Advanced Computer Architectures

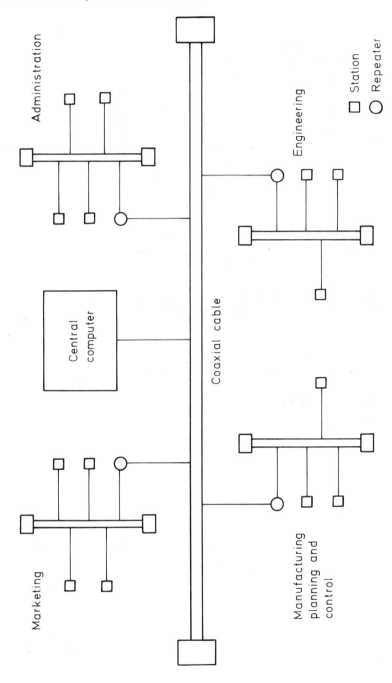

FIG. 4.17 Plant communication system using four local area networks.

is used to execute large programs at high speed. It also contains a central management information system whith uses an interfaced database (Sec. 2.3.3). The computers at each section have their own data base and can access that of the central computer. If desired, it is possible for a station (participant) to configure a data path to any other station. The stations may consist of computers or intelligent terminals. The basic addressable device, however, is a computer. Terminals are usually connected to a terminal controller which provides access to the network. The controller of a station is designed as a transmitter and receiver; it represents a set of hardware circuits and algorithms which manage the access to the bus. A station may place an information package on the Ethernet when it is not occupied. An addressed participant will accept a package if it is correct and will discard it if it is in error. It may happen that a package will not be recognized by a receiver. For this reason there should be a supervisory function on the bus which will recognize such problems. This is of great importance in manufacturing; otherwise, it may happen that the communication to, and control of, a part of the system may get lost.

After this general introduction, the function of manufacturing planning and control will now be explained in detail.

User Functions

Since the discussed system is installed in a medium-size plant, the computer network will serve many different functions (Fig. 4.18):

Production scheduling
Process planning
Material control
Facility monitoring
Machine control
Quality control
Word processing
Electronic mail

For production scheduling, process planning, and material control, several small business computers (personal computers) are installed as intelligent terminals. They can be used in the stand-alone mode or they may access the resources of the central computer. For example, the production engineer can load a scheduling program and part of the master scheduling file into a computer and devise a weekly production schedule. If the engineer desires to draft a long-term schedule, which usually requires a considerable amount of calculation and memory, he or she can do this from a terminal in the time-sharing mode on the central computer. The engineer may need production

Advanced Computer Architectures

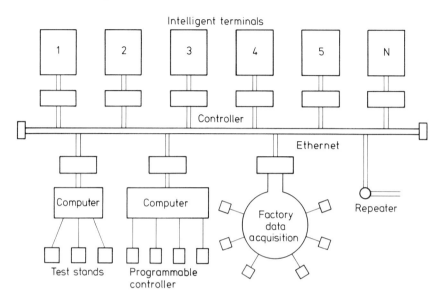

FIG. 4.18 Local area network for manufacturing planning and control.

data and can obtain these by setting up a communication path to the factory data acquisition system. Communication with other terminals is also possible. In addition, the Ethernet may be used to distribute electronic mail.

To perform secretarial service, intelligent terminals may serve as word processors and/or may perform additional functions, such as the following:

Typing of letters, forms, etc.
Mail handling
Scheduling of meetings
Planning of trips
Keeping of calender
Calculating
Searching through files

Very typical, for example, is the arrangement of joint meeting schedules. The secretary may query the computer to find a common date for the various participants of the meeting. If a date is found, the computer will set up the meeting and enter it into the different calendars.

The company uses for quality control a process computer that supervises three test stands. This computer obtains the test program and quality limits from the central system. It supervises the test, acquires performance data, and stores these in memory. In case of a quality problem, the operator is notified immediately. At predetermined intervals, the central computer polls the test system and transfers the test data into its memory.

The material flow system is also supervised by a control computer. The functions of the conveyors and the part feeders are directed by PCs which are connected to the process computer. Here again, the programs needed to control the product to be manufactured are loaded down to the material flow system and operate the installation. The PCs gather performance data from the material flow equipment and send these to the central computer via their supervisor. It is also the task of the process computer to synchronize all units of the material flow equipment with each other and with production machines. The factory data acquisition device constantly monitors all machine tools and accumulates performance and machine status data. This information is also brought periodically to the central computer.

The local area network is connected to the central bus via a repeater station. Through this station all participants of the entire plant computer installation can be addressed. In addition, all data that are transferred to and from other systems pass through the repeater.

4.6 ADVANCED COMPUTER ARCHITECTURES (FIFTH GENERATION)

Advanced architectures of computer systems of the future (fifth generation) are presently being developed in several countries [7,8]. These are expected to be realized in the 1990s and will process knowledge-based information. The following technologies and academic disciplines will influence their development:

VLSI technology
Decentralized parallel computing
Very high level programming languages
Knowledge-based expert systems

The combined software and hardware solutions of the fifth computer generation will provide many benefits to the user, such as (1) intelligent interfaces, (2) knowledge-based management, (3) problem solving, and (4) inference capabilities. Communication with this new computer generation will be possible via natural graphical, or textual languages. They allow the use of innovative programming methods to describe implicitly the manufacturing problem to be solved. This is in contrast to traditional programming methods, in which each step is explicitly entered into the computer. Implicit programming needs a knowledge-based expert system, a problem solver, and an inference module. They assist the user to solve a specialized problem efficiently. In the case of CAD/CAM systems, the fifth computer generation will support planning and control of activities at all hierarchical levels to operate a completely automated factory. The components of a fifth-generation computer are outlined in Fig. 4.19 and discussed in the following sections.

4.6.1 Components of Fifth-Generation Computers

The fifth-generation computers consist of an information processing network which can be linked together by a unique programming language (Fig. 4.19). Such a powerful architecture permits solving problems by inference, with the assistance of a knowledge-based management and intelligent input and output components. Each node of the network contains software and hardware modules which perform special assignments. The hardware modules are configured from VLSI processors of high speed and functionality. They can be configured for a specific application to solve any desired problem: for example, to plan and control automatically the fabrication of a product from information entered by sales. A basic module is the intelligent interface which supports communication between the user and the computer by speech, graphics, or another natural language. The components of this I/O module are built from special-purpose VLSI circuits designed for signal and speech processing. They allow communication with the computer in a form that is more natural to human beings.

The problem-solving and inference operations are equivalent to those functions performed by the CPU of a traditional computer. It will be possible to perform 10^7 to 10^9 logical inferences per second (LIPS; 1 LIPS is equivalent to nearly 1000 instructions on a conventional computer).

The knowledge-based management unit consists of a main memory, a virtual memory, and a file system. It has a storage capacity of 100 to 1000 gigabytes and is capable to access within seconds the knowledge needed for inference.

When a manufacturing problem is presented to the fifth-generation computer, the inference function gets active. It accesses from the knowledge base prior experience stored from earlier production experiments and deduces from these, by logical reasoning, a suitable manufacturing plan. It will be possible to contain the entire data base for a manufacturing operation in the knowledge base. All activities such as design, manufacturing planning, and manufacturing control use the same data base and can thus be linked together very easily.

4.6.2 Applications of Fifth-Generation Computers

The communication with the knowledge-base module that contains the manufacturing information is done by (1) natural languages, (2) speech,

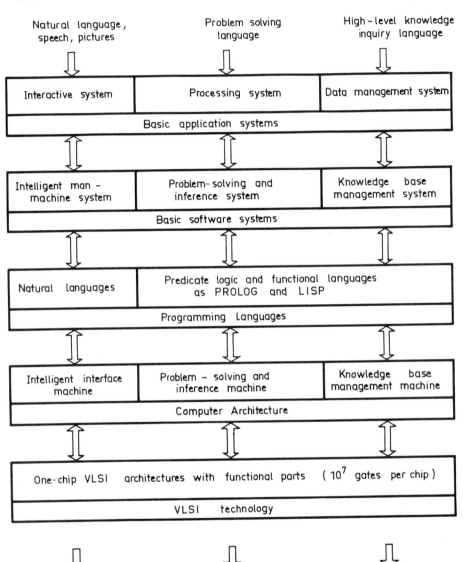

FIG. 4.19 Components of a fifth-generation computer system.

(3) pictures, or (4) images. The natural language processing system is shown in Fig. 4.20. It is subdivided into different modules which contain knowledge of the language (syntax, semantics, pragmatism), the problem-solving domain, and the interactive dialogue interface. A basic function of the semantical and pragmatical analysis is to extract structures from the problem description and to build the syntax tree [9]. This internal tree allows an analysis of the context of the input. From it the problem description and its structure are extracted. If the problem description is not complete, a dialogue will be initiated to request more input information (detailed questions) until the problem

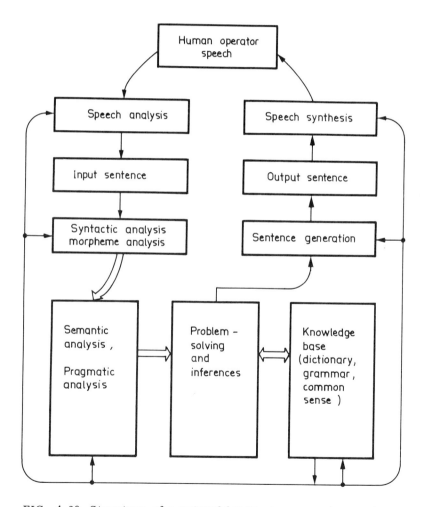

FIG. 4.20 Structure of a natural language processing system.

is totally defined. To perform this task, knowledge about the problem domain (e.g., the manufacturing task), its effective use, and the ability to learn new attributes and methods has to be available. The answer to the problem definition can then be generated by referencing the application problem solver and the knowledge base. The result is transformed from its internal structure into a speech signal using a sentence generator. If pictures or images are used as input, an analogous procedure is performed to generate the answer. Examples of the applications of fifth-generation computers in the area of computer-integrated manufacturing (CIM) systems are knowledge-based expert systems to aid scheduling and production planning as well as to perform local and global plant control. Software engineering, design, and material processing will also be aided by expert systems. Knowledge of the system should include control rules for every level of a hierarchical manufacturing system (operational control, coordination level, disposition, scheduling, etc.).

4.6.3 Basic Software System and Programming Languages

The basic software system supports the three components of the expert system (Fig. 4.19):

    The intelligent interface system
    The problem-solving and inference system
    The knowledge-based management system

General functions of the system software are basic information processing, interactive processing, and management. The intelligent interface software system has to operate on data derived from natural language, speech, picture, and image processors [9]. The problem solver includes a semantic analyzer, a meta inference system, a dictionary, a grammar, an image data base, and a common sense data base. For generating the answer of a problem definition, the question-answering system (intelligent answer generator) uses a picture and sentence generator. To solve the problem definition, reference must be made to the application knowledge data base system. It contains a basic program file, a computer architecture, and a VLSI design file. The system software includes support systems for handling, editing, debugging, acquisition, and consistency checking of knowledge as well as a data base compiler. Besides the application knowledge data base, a system knowledge data base exists, which contains specifications of the system itself (processor operating system and file specifications which facilitate use of the system).

    The programming languages for fifth-generation computers will be of three types [8]:

1. Natural language, speech, and graphical language (pictures), which interact with the intelligent interface
2. The high-level inquiry language, which interacts with the knowledge-based management system
3. The kernel (logic programming) language, which interacts with the problem-solving and inference system

The kernel language is a problem-solving language that is based on predicate logic (PROLOG). A program written in a predicate logic language is a collection of logic statements. Its execution is equal to a controlled logic deduction of the logic statements defined by the program. The inquiry language for the knowledge base and the knowledge representation language can be implemented by using the kernel language. This language will be the machine language of the fifth-generation computer. Other classes of languages, such as the conventional languages, process control languages, functional languages, and object-oriented languages, will also be supported by this computer generation.

4.6.4 Computer Architecture of Fifth-Generation Computer System

The fifth-generation computer architecture consists of three basic machines:

The intelligent interface machine
The problem-solving and inference machine
The knowledge-based machine

Intelligent interface machines are dedicated special-purpose processors that perform the operation on speech, pictures, and signals. The problem-solving and inference machine has the role of a CPU in the fifth-generation computer. It is a logic programming machine with:

Abstract data types
Data flow processing
An innovative von Neumann architecture

Sequential and parallel inference machines are under development using the data flow principle [10]. The knowledge-based machine operates with a relational data base and relational algebra [11]. It holds a large number of knowledge data which are well structured so that each knowledge item can be referenced effectively. When the knowledge-based machine receives a call from the inference machine, it searches and re-

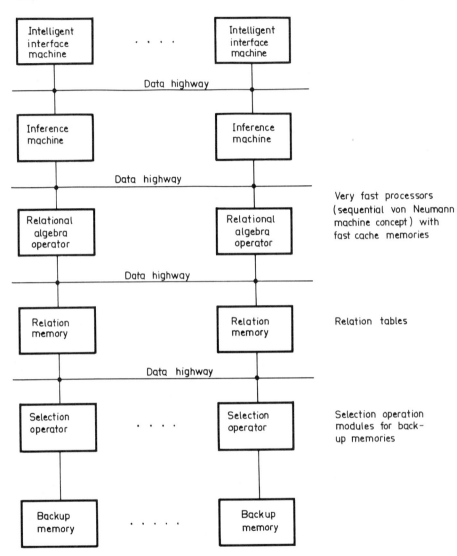

FIG. 4.21 Processor configuration of a knowledge-based machine.

Advanced Computer Architectures

trieves knowledge items and hands them to the inference machine. When the inference sends knowledge data to the knowledge-based machine, the latter compiles and integrates them into the knowledge base. The hardware structure of such a machine is shown in Fig. 4.21. New advanced architectures such as the following will enable the fifth-generation computer to process information at high speed:

- A logic programming machine
- A functional machine
- A relational algebra machine
- An abstract-data-type support machine
- A data flow machine
- An innovated von Neumann machine

With the VLSI design it will be possible to develop the following:

- A high-speed numerical computation machine
- A high-level human-machine communication system
- An efficient network architecture
- A distributed function system
- A data base machine

This will allow fifth-generation computers to operate with a processing rate of 10^7 to 10^9 logical inferences per second.

4.6.5 Conclusion

The architecture of the fifth-generation computer that we have discussed will be available in the early 1990s. Such computers will work as expert systems in CIM for manufacturing planning and control. Its capabilities can be summarized as follows:

1. To perform intelligent human-machine communication, capable of understanding speech, images, and natural language
2. To understand problem definition and requirement specifications
3. To synthesize the information processing procedures
4. To generate response to a problem definition using procedure referencing knowledge

The knowledge bases contain knowledge about the problem areas, the programming languages, and the various machine components of the entire computer.

REFERENCES

1. R. J. Lauber, Development Support Systems, *IEEE Computer*, May 1982, pp. 36–49.
2. W. Epple, M. Hagemann, M. Klump, and U. Rembold, The Use of Graphical Aids for Requirements Specification of Process Control Systems, *Proceedings of COMPSAC '83*, Chicago, 1983.
3. U. Rembold, K. Armbruster, and W. Ülzmann, *Interface Technology for Computer-Controlled Manufacturing Processes*, Marcel Dekker, New York, 1983.
4. U. Spieth, The Design and Operation of a Modern Multiprocessor Control System for Manufacturing, PDV-Bericht 165, Prozeßlenkung mit µ-Rechnern, Kernforschungszentrum Karlsruhe, West Germany, 1978.
5. P. Levi, The System Architecture of Data Acquisition Networks, *Euromicro Journal*, Vol. 5, No. 6, 1979, pp. 350–357.
6. P. Levi, Transaction Language Characteristics and User/Computer Interface in Manufacturing Systems, *Microprocessors and Microsystems*, Vol. 7, No. 1, 1983, pp. 3–17.
7. H. Aiso, Fifth Generation Computer Architecture, *Proceedings of the International Conference on 5th Generation Computer Systems*, Tokyo, Japan, Oct. 1981, pp. 2–29.
8. P. C. Treleaven and J. G. Lima, Japan's Fifth-Generation Computer Systems, *Computer*, Aug. 1982, pp. 79–88.
9. H. Tanaka, S. Chiba, M. Kidore, H. Tamura, and T. Kodera, Intelligent Man-Machine Interface, *Proceedings of the International Conference on 5th Generation Computer Systems*, Tokyo, Japan, Oct. 1981, pp. 3-3-1 to 3-3-11.
10. S. Uchida, et al., New Architectures for Inference Mechanisms, *Proceedings of the International Conference on 5th Generation Computer Systems*, Tokyo, Japan, Oct. 1981, pp. 4-1-1 to 4-2-10.
11. M. Amamiya, et al., New Architectures for Knowledge Based Mechanisms, *Proceedings of the International Conference on 5th Generation Computer Systems*, Tokyo, Japan, Oct. 1981, pp. 4-2-1 to 4-2-10.

5
Computer-Aided Design

5.1 INTRODUCTION

The great potential of the computer for graphical data processing was recognized during the early phase of computer technology. Thus the first applications for drafting and design started to emerge around 1965. During this period an attempt was made to solve very complex tasks with extremely expensive hardware and sparse software, which in most cases led to discouraging results. It took many more years until economic solutions could be demonstrated and trust in the computer was regained.

The research facilities and industry of the United States did the most important pioneering work in computer graphics. They are still the leader in this technology, ahead of Europe and Japan. For example, the company Computer Vision has the leading position in the number of CAD systems installed. They have a worldwide market share of 32%. This outstanding U.S. market position is due to the fact that computer-aided design was introduced very early in space technology and automobile design. These industries are important innovators and the nucleus of many new technologies. They also have the advantage that they can afford the development of exotic technologies supported by a large sales volume. Today, these companies have an edge in CAD over the rest of the industry.

The distribution of design activities in mechanical engineering is shown in Fig. 5.1. Drafting and processing of the bill of materials are most easily automated with the aid of the computer, whereas product engineering has many creative activities that can only be done by human beings.

Computer graphics may be defined as follows: Graphical data processing comprises all techniques that process and generate data in form of lines and figures with the aid of the computer. Thus the input and

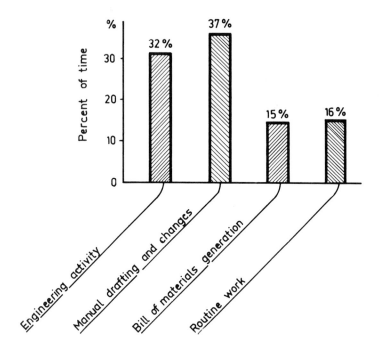

FIG. 5.1 Distribution of design activities. (From Ref. 1.)

interpretation of textual or pictorial data are done with techniques of character and pattern recognition.

Depending on the type of communication with the computer, two graphic methods are distinguished:

1. *Passive graphic data processing*: The generation of graphic output is done here without interaction with the computer. The user enters textual instructions into the computer and obtains a completed solution as output. This mode of operation is called batch processing.
2. *Interactive graphic data processing*: The user solves a problem interactively in close dialogue with the computer. The user can enter instructions and wait for an immediate response from the system. Thus the user is able, in real time, to draw a picture on the screen and do additions and deletions.

In this chapter we first describe the computer hardware needed for computer graphics work. This is followed by a discussion of the software and a description of computer-aided design methods and systems.

5.2 HARDWARE

5.2.1 Display Tube

The display tube is the most important component through which the user communicates with the computer. In most cases it consists of a cathode ray tube (CRT) with a large picture screen (Fig. 5.2). It is similar to the CRT used for television, oscilloscopes, and radar screens. For this reason, many principles have been adopted from these applications and are employed in computer graphics.

The function of the tube will be described briefly. The cathode is heated by current and emits electrons. An electron beam is formed in the electron gun, where it is modulated and focused. From here it travels through the accelerator electrode and another focusing device. The columnated electron beam, which now consists of a small dot, is sent to the picture screen, on which it impacts on a phosphorus coating. At this layer the electrons give up some of their energy and produce light. The x and y deflection of the beam, which is needed to draw characters and figures on the screen, is obtained with the help of a deflection coil. The screen has a certain amount of persistence. It sustains the image for 1 msec when white or green phosphorus is used and up to 1 sec with orange phosphorus. For interactive work using a light pen, the shorter times are preferred. To provide a flicker-free picture the display is constantly refreshed at a rate of 30 to 100 refreshes per second. This rate is determined by the persistence of the phosphorus of the CRT.

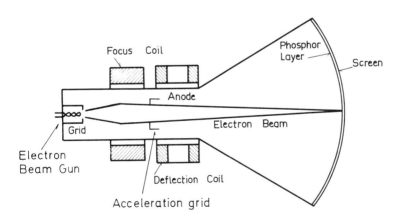

FIG. 5.2 Cross-section of a picture tube.

5.2.2 Deflection System

The deflection electrodes and coils of the CRT are controlled by x and y deflection registers (Fig. 5.3). Two deflection principles are used: the raster method and the vector method.

Raster Screen Principle

This method employs the same horizontal raster principle as that operating a television tube (contrary to this, the radar technology deflects the electron beam vertically). Figure 5.4 shows the principle. The electron beam traverses the screen from left to right in a zigzag pattern, line by line. At the end of each line the beam is turned dark and deflected to the beginning of the next line, where it is turned bright again to draw the next line. Normally, 1024 lines are used.

Advantages include:

1. The entire screen can be used as a graphic display without giving the illusion of picture flicker.
2. The picture can be modified at the refresh rate.
3. Control circuits are rather simple and well developed.
4. The method is well suited for producing shaded picture.

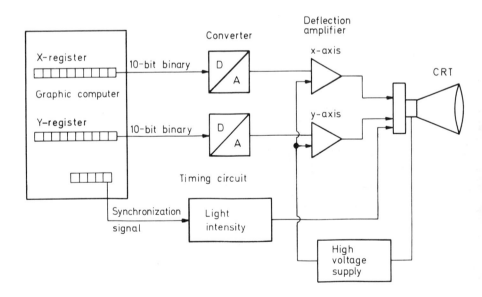

FIG. 5.3 Simplified graphic system.

Computer-Aided Design

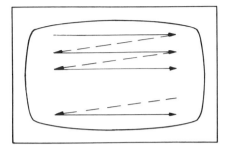

Character B

FIG. 5.4 Principle of the raster display method.

Disadvantages are:

1. The resolution is rather low.
2. It is difficult to manipulate pictures in the raster formate.
3. All lines that traverse the screen at angles greater than 0° and smaller than 90° have a staircase pattern.

Character generator. A character is usually represented by a 5 × 7 matrix of light dots (Fig. 5.4). It is necessary to produce such a matrix for each character (the ISO standard specifies 7 bits to produce characters). To generate a sequence of characters line by line according to this principle would be too cumbersome. For this reason a character generator is used. This device directs the electron beam under hardware control and draws the dot pattern for the character to be displayed (Fig. 5.5). When a character or another sign is to be presented on the screen, it is advantageous to store this information with little or no redundancy in a refresh memory and to instruct the sign generator periodically to draw the character matrix on the screen.

With the dot presentation, for every character 5 × 7 = 35 bits of information would have to be stored. However, with the 7-bit control information (ISO standard), it is necessary to provide only 7 bits to the character generator to display a character. Thus the reduction of redundance for the refresh memory would be 35/7 = 5.

The heart of the sign generator is a ROM (often a PROM or EPROM). It must contain for each character a map of the locations of all dots in the rows and columns which are needed for display. In case of a 128-character set, the ROM has to store 128 × 7 × 5 = 4480 bits of information.

When in operation the character generator is supplied by the refresh memory with 7 input bits, b_1 to b_7, representing the character code (Fig. 5.5). The display lines are controlled by signals placed on inputs, AY1 to AY3 of the line selection logic. The three inputs can select seven character lines and one spacing line. At the outputs

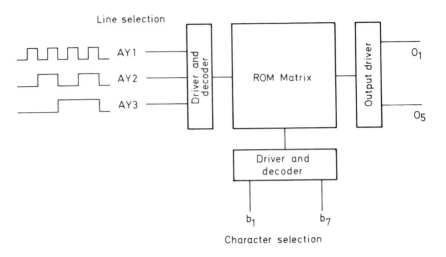

FIG. 5.5 Character generator.

O_1 to O_5, the O/I information for the display of one character line is represented as an input to the deflection coil.

For computer graphics a standard character set is often not sufficient. There may be special symbols to be displayed, such as the symbol for a valve or an electronic gate. These symbols are application oriented and are usually too large to be presented by a 5 × 7 matrix. It is possible to group several generators together or to design a unique symbol generator. A specially designed instruction allows addressing of this generator.

Presentation of a character string. We explain this principle with the help of Figs. 5.3, 5.5, and 5.6. The information to be drawn on the screen is stored in the refresh memory. This storage device can be part of the work memory or a separate entity. With the help of the deflection counters for the x and y directions, the code word for the first character, X1, of the character string is placed on inputs b_1 to b_7 of the character generator. At the same time lines AY1 to AY3 address the first of the seven character lines. This produces at its output the first line of the character. Now, the code word of the second character, X2, is placed on the input of the character generator producing the first line of this character. This process is continued until all first lines of the character string are displayed. Now, the control logic for the line selection switches over to the second character line, thereby repeating the above cycle to produce this line. After the seventh character line has been drawn a spacing line is inserted on the screen by turning the electron beam dark. Now the

Computer-Aided Design

second line of a character string is drawn. This process repeats itself until all the information has been written on the screen.

With the raster screen method it is also possible to produce the vectors needed for computer graphics. In this case it is only necessary to enter into the system the coordinates of the beginning and end points of the lines. The necessary dots between these points are automatically generated. Because of the rather low resolution, the raster screen lines at an angle greater than 0° and smaller than 90° appear as a staircase pattern. Since each point on the screen is presented by one bit in memory, this method needs much memory space. Despite this shortcoming the raster screen is commonly used to present

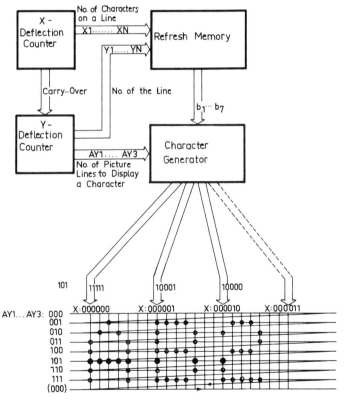

FIG. 5.6 Display of characters by the TV raster method. (Courtesy of PDV Funding Agency, Kernforschungszentrum Karlsruhe GmbH, West Germany.)

statistical pictorials, flow diagrams of processes, and so on, where the low resolution is acceptable.

The Vector Screen

With this method the electron beam can be randomly positioned at any place of the screen. The user has only to specify the coordinates of the beginning and end points of a vector. A new vector may also originate at the end point of another vector; thus only coordinates of one additional point have to be entered. In case part of a figure has to be erased, this can be done with dark vectors or hidden lines. It is also possible simply to indicate coordinate increments Δx and Δy. The resultant end point of this vector is calculated from the coordinates of the end point of the previous vector and the displacement Δx and Δy. This method uses the least amount of memory.

Advantages of the vector screen method are:

1. Only those lines of the screen that are of interest will emit light.
2. This screen has a very high resolution and is well suited to produce detailed drawings of high accuracy.
3. It can be used interactively with a light pen.

Disadvantages are:

1. Positioning of a vector may be time consuming.
2. If the information contents on the screen are very dense, the picture starts to flicker; thus the write speed of the electron beam is limited.

Vector generator. Drawing a vector is done with a vector generator (Fig. 5.7). The majority of vector generators use two integrator circuits, one for the horizontal and the other for the vertical axis. For this purpose, the vectors are classified into different lengths. To produce a vector of a stated length, a capacitance is connected to the integrator. The time constant $\tau = R \times C$ determines the length of the vector. To place a vector on the screen, the control signals x and y are sent to digital-to-analog converters. There is one converter for the x direction and one for the y direction. The output of these converters are brought to the integrators that control the deflection of the electron beam. Thus the length and the direction of the vector are set (Fig. 5.8).

The main problem when drawing a vector is to maintain a constant speed at which the electron beam sweeps the screen. This speed may

Computer-Aided Design   157

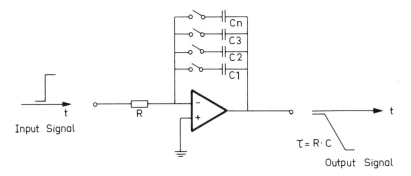

FIG. 5.7 Vector generator using the integrator principle.

change with the position and length of the vector and the amount of information on the screen. Thus the brightness of the picture may change or it may even flicker.

5.2.3 Refresh Memory

With the information presented so far, an improved graphic display system can be conceived (compare Figs. 5.3 and 5.9). One of the main components that was added is the refresh memory. It contains the display code. This is the program that draws the graphic and alphanumeric information on the screen. The display system loops constantly through this program at the refresh cycle speed and maintains the picture on the screen. With this separate memory the system computer is relieved from the refresh routine work.

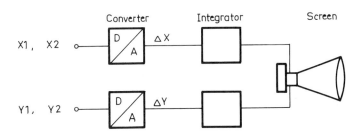

FIG. 5.8 Display of a vector with the help of a vector generator.

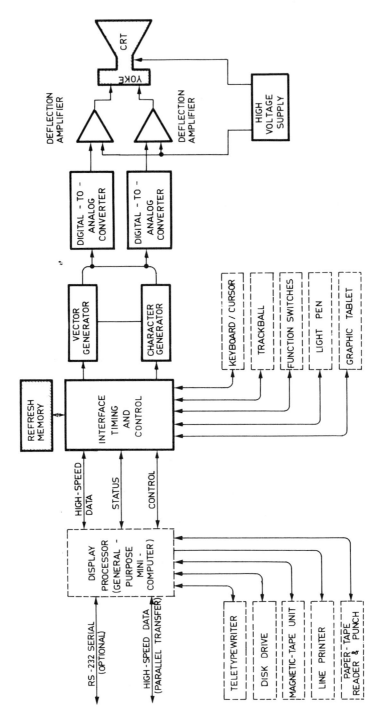

FIG. 5.9 Computer graphics system with random access CRT terminal. (Courtesy of McGraw-Hill, Inc., New York.)

Computer-Aided Design

5.3 INPUT PERIPHERALS FOR COMPUTER GRAPHIC SYSTEMS

Several input/output devices are depicted in Fig. 5.9. The input/output method can be divided into different classes: (1) picture-independent input peripherals, (2) input from tracings, and (3) input from the display tube.

5.3.1 Picture-Independent Data Peripherals

The picture-independent input can be done via a keyboard. This also permits the use of complex instructions. In addition, functional keyboards are available where the keys are assigned to specific instructions. It is possible to make this keyboard flexible by implementing several sets of functions for different applications. The computer only has to know which set of functions is presently used. Most alphanumeric display terminals have these input capabilities.

5.3.2 Picture-Independent Graphic Peripherals

Positioning Wheel

With the positioning wheel two hairlines are moved on the display screen by actuating two potentiometers. There is one wheel for the x axis and one for the y axis. The intersection of both hairlines points to the coordinates to be manipulated.

Joystick and Trackball

The joystick and the trackball both serve the same function. They actuate two potentiometers placed at a right angle to each other. One potentiometer is assigned to the x axis and the other to the y axis. The resistance change of the potentiometer causes a cursor to move along the picture screen. With the help of this cursor points can be added or deleted. Also, entire picture elements can be manipulated. The trackball is a very accurate device. Figure 5.10 shows the use of a joystick.

The Digitizer

Graphic information such as machine drawings, maps, diagrams, and patterns, which require very high resolution, can be entered with the help of a digitizer (Fig. 5.11). Frequently, automatic drafting machines are designed such that they can also be used as digitizers (5.12). The drawing is placed on the board and with the help of a position sensor, the coordinates are entered into the computer. This is done by placing the crossbar cursor over the relevant coordinate

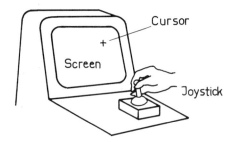

FIG. 5.10 Input by joystick.

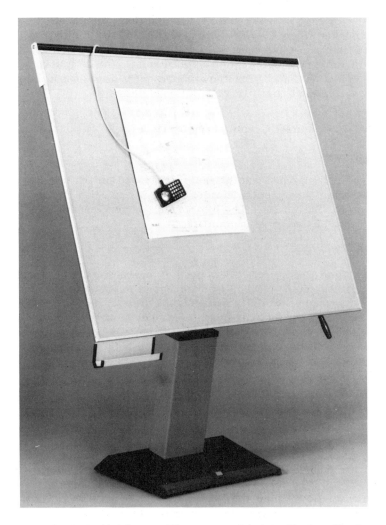

FIG. 5.11 A digitizer. (Courtesy of Aristo-Werke, Hamburg, West Germany.)

Computer-Aided Design

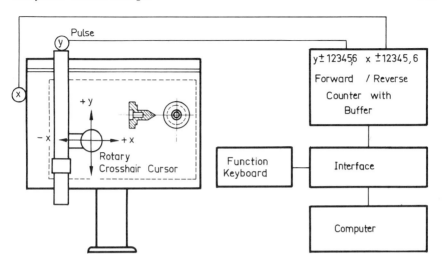

FIG. 5.12 Schematic of a drawing machine and digitizer. (Courtesy of Aristo-Werke, Hamburg, West Germany.)

points and by advising the computer to record its values. The device shown in Fig. 5.12 uses impulse counters to determine the x and y positions. The counters are activated whenever the cursor is moved. There are many other positioning principles used. For example, the digitizer of Fig. 5.11 has embedded in its board a series of wires for the x and y directions. Pulses are sent through these wires at fixed intervals. The cross-hair sensor is moved to its work position. When the pulses from the x wire and the y wire coincide, the cursor picks up this incident as a strong signal and leads it back to a detection logic. This circuit calculates the travel time of the pulses from the instant when they were emitted to the instant when they were detected. Thus the location (the x and y positions) of the cursor is determined. The resolution of a typical 1.5 m × 1 m digitizer is 0.0025 mm.

Graphic Tablet

A graphic tablet is used to enter small sketches, drawings, signs, or instructions into the computer with the help of a stylus (Fig. 5.13). The x and y positions of the stylus are tracked approximately 200 to 500 times per second. This is done with ultrasound, light, a magnetic pickup, a potentiometer, or metallic contact of cross-wires embedded in the tablet. In principle, the tablet is a digitizer with low resolution. In case it is used to enter characters, instructions, or special picture elements, an overlay is placed on it. Users may design their own overlay (Fig. 5.14). With this overlay, macros for

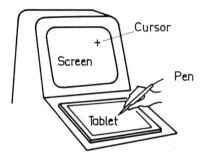

FIG. 5.13 Input by tablet.

raster points, machine screws, filler circles, and other can be called up to display such pictorials on the screen. It is also possible to give instructions to the computer, such as delete, top view, or side view. This information is entered into the computer when the overlay is designed. In this way the function of each field is defined and made known to the computer. When the overlay is used, it is first identified and the macros are called up with the help of the stylus.

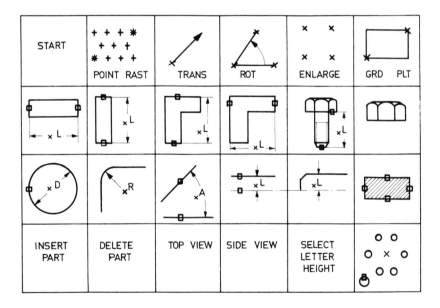

FIG. 5.14 Menu instructions.

5.3.3 Input by Graphic Terminal

In the following section methods of entering graphic data into the computer via a graphic terminal are discussed. This is done with the menu technique using a light pen or a touch-sensitive terminal.

Light Pen

The light pen input technique uses the principle of sequential picture design in a cyclical refresh mode. This means that:

1. The light spot generated by the electron beam traverses the entire area of the screen.
2. Its location is known from the contents of the x and y deflection registers (Fig. 5.3).
3. The contents of only one memory cell of the refresh memory is in the picture buffer at a given time. The current address is known.
4. The picture code is traversed sequentially, including jumps.

Figure 5.15 shows the light pen, which contains a photoelectric cell and a preamplifier. There are also light pens which use fiber optics to conduct the light to an external photoelectric cell and amplifier (Fig. 5.16). With the light pen the user identifies the location of points or picture elements on the screen, and in this way informs the computer that he or she wants to manipulate them. The picture elements belong to a menu, one containing design elements for a pictorial. Thus the user can build up the pictorial from graphic primitives with the aid of the menu. The user can add, delete, move, and interconnect primitives.

The operating principle of the screen is as follows (Fig. 5.17). When the pen is directed at a picture element [i.e., in our case at

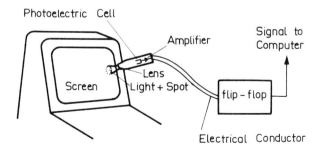

FIG. 5.15 Light pen with integrated photoelectric cell.

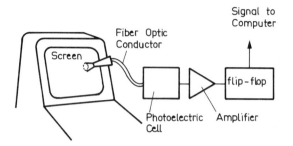

FIG. 5.16 Light pen with remote photoelectric cell.

point (X_5, Y_5)], the photoelectric cell in the pen detects the light of the electron beam at the moment when it traverses this point. The computer knows that at the same time the instruction for the display of this point is being executed. Thus detection of the light pen signal and relating it to the contents of the program counter allow the user to locate the point. By selectively pointing at picture elements, changes may be made to the contents of the pictorial. In this way, program sections in the refresh memory are altered.

Input via a Design Menu

Use of the design menu for input is an extension of the form discussed above. In this case also, the light pen is the input peripheral.

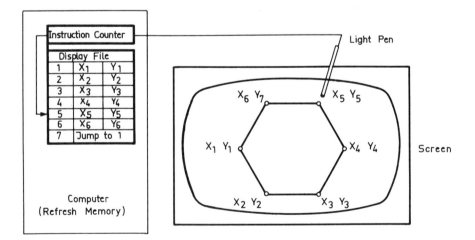

FIG. 5.17 Coincidence of the interrupt signal from the light pen with the instruction counter.

Computer-Aided Design 165

The menu is a list of instructions drawn on the screen: for example, all elements which are needed to draw a circuit. It may have textual and pictorial information. The menu is divided into different fields, each of which can be identified by the light pen. The computer acts on the identification of an instruction. Figure 5.18 shows how a drawing can be assembled from a menu. The almost completed drawing is shown in the center. On the right and left sides of the screen there are textual instructions. In the lower part of the picture the design elements used for the drawing are shown. The elements are activated by pointing at them with the light pen. Thereafter they can be moved into position just by moving the light pen along the screen surface to the desired location. Thus the drawing can be constructed piece by piece.

Touch-Sensitive Terminals

This terminal is a specially constructed display screen. In principle, it functions as a tablet. Touch sensitivity is obtained by three methods.

FIG. 5.18 Interactive graphic work station. (Courtesy of Aristo-Werke, Hamburg, West Germany.)

1. *Infrared*: At opposite sides of the screen an infrared sender and a receiver are located. When an object breaks the light, a control logic determines the location of the interrupt and that of the associated field. This method has high resolution; terminals have been manufactured with 1920 fields.
2. *Capacitance*: A glass plate is coated with a very thin, almost invisible metallic surface. When this plate is touched, there will be a capacitive difference between the area touched and its surrounding. A control circuit identifies this spot and communicates its position to the computer. Contrary to the aforementioned system, this method can be retrofitted to an existing display. Its disadvantage is low resolution and its sensitivity to environmental conditions.
3. *Mechanical*: Two circuits, one having vertical and the other horizontal conductors, are bonded to the screen surface. Upon touch they conduct electrically at this location, providing an analog voltage. This is digitized and the location of the interrupted area is determined. This method also has a high resolution, but is very complex to integrate into a computer system.

The first two methods are very reliable and the corresponding peripherals have a long service life.

In the next section special types of display systems are discussed, as well as some peripherals used to output graphic information.

5.4 SPECIAL DISPLAY SCREENS

5.4.1 Storage Tube

The storage tube allows the user to store pictures in the tube itself on a phosphorus dielectric coating. The refresh memory is replaced by a pair of registers (Fig. 5.19). A high negative potential is placed on the storage grid. The electron beam that hits the storage medium displaces from it secondary electrons and causes the phosphorus to light up. Thus it generates light and dark zones. The image is maintained with the help of electrons emitted from a flood gun which has a potential of -150 V. Usually, the picture can be sustained for about an hour. Thereafter it has to be refreshed. Erasing is done with the help of secondary electrons, which is a relatively slow process.

Advantages include:

1. Random deflection and almost unlimited storage capacity (no dynamic limitations, no picture refreshing, no flicker, no jitter of the light spot)
2. High resolution
3. A relatively high writing rate

Computer-Aided Design

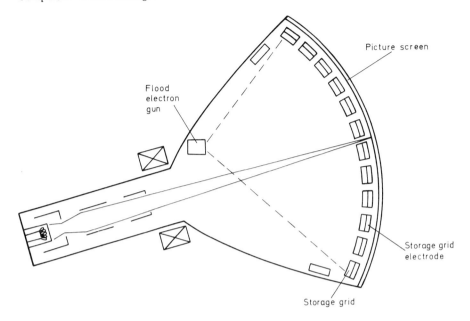

FIG. 5.19 Storage tube. (From Ref. 2. Courtesy of McGraw-Hill, Inc., New York.)

4. Favorable cost, since no refreshing memory and only a simple deflection system are necessary
5. Availability of gray-scale features in newer devices (third dimension)

Disadvantages are:

1. The tube itself ages fast and is expensive.
2. Brightness and contrast are limited.
3. Work with the light pen is almost impossible.
4. No selective erasing of the picture is possible; to modify the picture, it must be completely erased and redrawn.

Despite its limited application, the storage tube is a very popular device for the display of text and simpler graphics. Figure 5.20 shows a possible configuration of a graphic system using a storage tube.

5.4.2 Plasma Screen

The heart of the plasma screen consists of two transparent glass sheets. On each sheet a grid of electrodes is deposited or etched (Fig. 5.21).

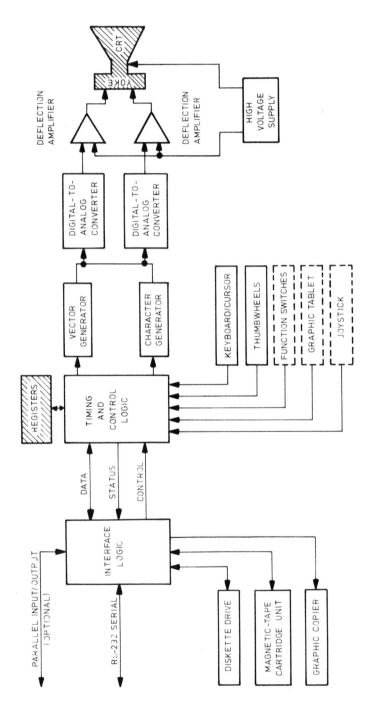

FIG. 5.20 Graphic system using a storage tube. (From Ref. 2. Courtesy of McGraw-Hill, Inc., New York.)

Computer-Aided Design

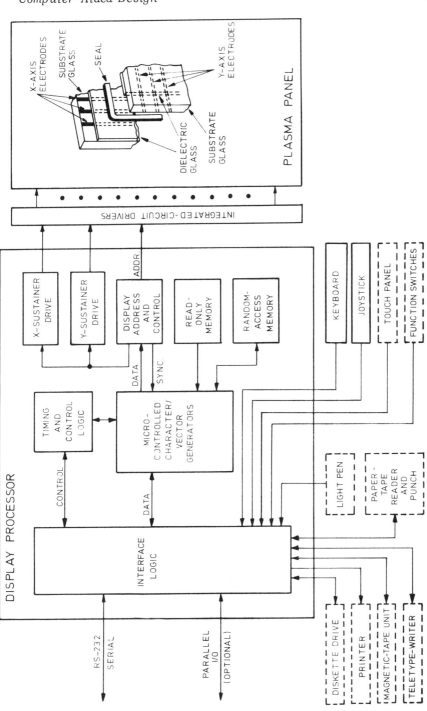

FIG. 5.21 Graphic system using a plasma panel. (From Ref. 2. Courtesy of McGraw-Hill, Inc., New York.)

TABLE 5.1 Capabilities of Various Display Technologies

Capability	Raster	Vector	Storage	Plasma
		Type		
Resolution	2048 × 2048	4096 × 4096	4096 × 4096	512 × 512
Gray-scale display	Yes	Limited	No	No
Color	Yes	No	No	No
Contrast	100 ft-lamberts		8 ft-lamberts	
Price	Inexpensive	High	Fair	
Refresh	30-100 times/sec	30-100 times/sec	Approx. once/hr	
Flicker	Possible	Possible	Not possible	Not possible
Dimensional accuracy	Poor	Fair	Fair	Good
Interactive operation	Yes	Yes	No	Yes
Refresh memory	Yes	Yes	Not needed	Not needed
Memory size	Large	Medium	None	None
Smearing effect	Possible	Possible		
Graphics				
Number of vectors		Limited	Unlimited	
Line straightness	Poor	Excellent	Excellent	Poor
Shading	Good	No	No	No
Rereading of display	Yes	No	No	Yes
Areas	Yes	No	No	Yes

The grid is oriented vertically on one sheet and horizontally on the other sheet, thus forming a matrix. A dielectric layer of magnesium oxide covers the electrodes. The glass sheets are kept parallel to each other by a precision seal, which is also used to contain a mixture of neon and a small amount of xenon (argon) in the glass envelope under a pressure of 0.5 to 1 atm.

Presently, the available resolution is 512 × 512 cells, for a total of 262,144 addressable locations. If a cell is to be lit up, 100 V is placed on the two intersecting electrodes of this cell, starting a gas discharge and causing a glow. A continuous ac voltage (lower than the firing voltage) sustains the image. The glow can be erased by applying an out-of-phase ac voltage to the cell grids. This momentarily reduces the sustaining voltage at this cell and extinguishes the glow.

Advantages of using the plasma screen are:

1. The display is very flat (starts at 7 cm).
2. The picture has high dimensional accuracy.
3. Selective reading and writing are possible.
4. Transparent projection or drawing can be placed directly on the screen and superimposed over the display graphic.

Disadvantages are:

1. Hitherto unconventional technology
2. Rather low resolution
3. Additional control hardware necessary
4. Presently only light and dark presentation of picture; no gray-scale display possible

A summary of the four different display technologies is given in Table 5.1.

5.5 GRAPHIC OUTPUT SYSTEMS

5.5.1 Automatic Drafting Machine

A typical drafting machine is shown in Fig. 5.22. Its operation principle can be depicted from Fig. 5.12. The drawing pen is moved about the drafting table with the help of two motor drives, one for the x axis and the other for the y axis. Automated drafting machines are used as output devices to produce drawings and graphics at high speed and accuracy. The contents of the drawing is created with the help of the CAD system, stored in computer memory, and retreived upon demand. The output of the computer is in the form of control signals issued to the drafting machine controller. Input to the drafting machine can be done by two methods:

FIG. 5.22 Flat-bed plotter. (Courtesy of Aristo-Werke, Hamburg, West Germany.)

1. *On-line*: The CAD system is connected directly to the drafting machine via a data channel.
2. *Off-line*: The output of the computer is placed on a tape or on a disk and manually transferred to an input peripheral of the drafting machine.

Typically, a drafting machine contains several microprocessors to prepare data for output and to control its two axes. For on-line connection to the computer, standard interfaces are usually provided.

The control information from the computer arriving at the drafting machine controller is interpreted. Numerous short vectors are calculated to direct the drawing pen along the paper to produce the tracing. A typical resolution is ±0.005 mm. The size of the drafting table can be up to 1.8 m × 6.8 m. Obtainable drawing speeds in the direction of an axis are 60 m/min and along a diagonal, 40 m/min. This speed can be varied under computer control and optimized to include acceleration and deceleration of the pen. Modern drafting machines have diagnostics capabilities to locate the cause of a malfunction.

Dimensions on the drawing can be specified in terms of absolute or relative coordinates. For absolute coordinates the origin of the coordinate system may be located inside or outside the drafting table. For relative coordinates the new point to be drawn refers to one produced previously.

Characters and special signs are generated by hardware generators. In addition, it is possible to do crosshatching and to vary the thickness of the drawing line. Contortion and mirror operations are possible. The following tools can be used: drawing pen, graving needle, light drawing pen, and dotting needle. There are special drafting machines available to produce masks for integrated circuits with the help of a light drawing pen. Their accuracy is ±0.0025 mm. They allow high acceleration, which is necessary when many short lines are to be drawn.

5.6 SOFTWARE

An important criterion for any data processing system is its user friendliness. A graphic system must be designed such that a user can quickly learn how to operate it. The user should be able to solve a problem in dialogue mode with the computer. It is also necessary to produce computer-independent software so that it can be used for several machines. This will minimize software effort when different computer models are used or when the computer is exchanged for a newer system. In addition, it should be possible for the user to call on programs at different levels without thorough knowledge of the entire system. The following programs are of importance for a CAD system: (1) the operating system; (2) service programs, graphic and nongraphic; and (3) application programs.

5.6.1 Operating System

The operating system is a set of programs that control the operation of the computer and the execution of the application program. The following three tasks have to be performed by this software:

Human-machine communication
Scheduling and coordination of the system function
Control and protocolling of the system operation

To perform these assignments, the operating system contains the following components:

Computer scheduling
  Resource allocation
  Interrupt handling

Task scheduling
Time watching
Synchronization of internal task communication
Memory management
Internal memory
External memory
I/O management
Management of peripheral devices
Management of communication units
Control of data transmission

The power of these components determines the versatility of the CAD system and its available functions. The operating system is normally bought from the CAD system manufacturer.

5.6.2 Service Programs

The following nongraphic service programs are contained in well-conceived computer systems:

Assembler and binder
Translator
Interpreter
Editor

With a CAD system the graphic service programs are very important. They are used to solve the following functions:

1. Implementation, administration, and alteration of data files
2. Search functions
3. Generation and manipulation of graphics
4. Service to the user input and interrupt processing
5. Providing I/O capabilities for the different peripherals

To be efficient, these programs are usually written in the assembly code of the computer mainframe. They are called on by programs written in a higher programming language.

The users do not have to know the exact structure or the code of the operating system and the service programs. They only have to know how to use them. The service programs are usually written by the system manufacturer or a firm that specializes in this type of software.

5.6.3 Application Program

It is the user's responsibility to conceive and write the application program. The program development phases are as follows:

Computer-Aided Design

Problem analysis
Program design
Program implementation
Functional testing of the program
Implementation of the program on the CAD system

Most application programs for CAD are written in FORTRAN. Some machine-dependent routines will be implemented in assembly code.

Since CAD programs usually are very complex, good software engineering practices should be employed; thus the modularity is of extreme importance. The entity of all modules comprises the application program. This point will be discussed with the help of an example (Fig. 5.23).

A CAD program to design a shaft is to be conceived. The input and output parameters are given by the application engineer. The program needs a data file that contains standard data and project specific data about the part. The main program consists of standard modules, such as programs for technical calculations and input/output. These modules call on submodules (i.e., for calculating the load, stresses, strains, and material selection). In this example it is assumed that the shaft can be designed by the variant method. In other words, the designer only has to supply major shaft parameters. A submodule to generate shaft variants will automatically generate the display code for the shaft to be designed.

To interface the modules with submodules it is often necessary to implement a pre- and a postprocessor. This is required when universal software was written in an intermediate language and when it has to be adapted to the user system. In some cases where many similar design concepts exist, it is possible to purchase application software from a systems house. However, most applications are very user specific; thus it is virtually impossible to design general software. The manufacturer of the CAD system usually only furnishes software that contains all the primitives of the basic drafting procedures, such as routines to draw points, lines, circles, parabolas, and so on (Fig. 5.24).

Improved software may contain spline and curve function and routines for three-dimensional graphics. With the basic software the user is able to design application-oriented routines such as dimensionless basic macros to produce drawings of machine elements, parts, or variants. At the next higher level the designer may want to construct from these basic macros other higher-order macros to design subassemblies, such as electromotors or gear cases. These higher-order macros may be used to compile supermacros to design entire machines, such as lathes or cranes.

It is difficult to determine how far the automation effort will go. A manufacturer with a variety of product families may stay at a rather

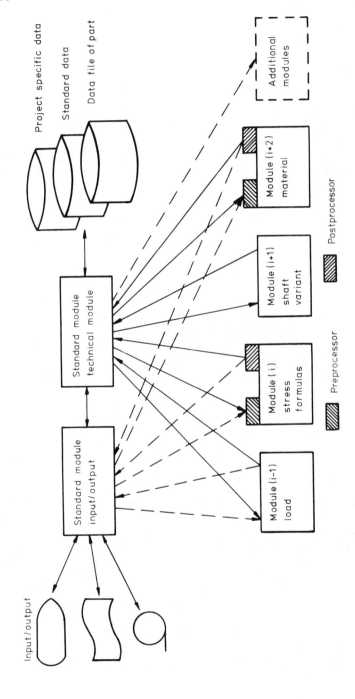

FIG. 5.23 Modular construction of software. (Courtesy of PDV Funding Agency, Kernforschungszentrum Karlsruhe GmbH, West Germany.)

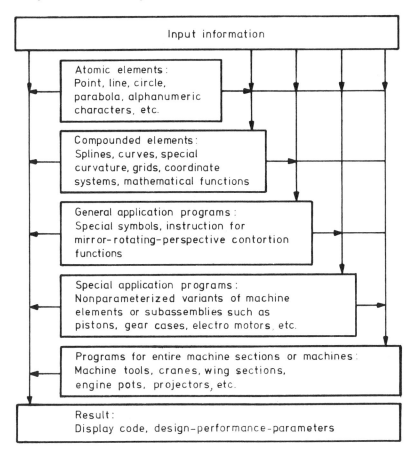

FIG. 5.24 Different stages of CAD software packages. (From Ref. 3. Courtesy of VDI-Verlag, Düsseldorf, West Germany.)

low level, whereas one with only one line of standard products will be highly automated. It is very important that the interface of the individual levels are well described in a user handbook so that the user will be able to easily extend the application of the system software for a variety of applications.

5.6.4 Information Transfer Between Memory and Display Code

The user controls the data communication in the graphics system with the help of the input data and the application program (Fig. 5.25). The output routines search the data base for the display information

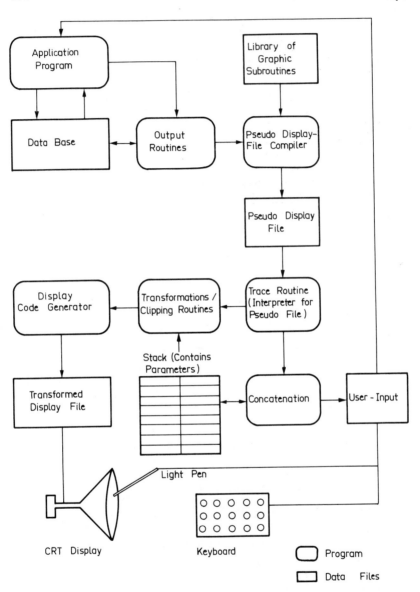

FIG. 5.25 Pseudo-display-file system. (From Ref. 4.)

which describes the graphics to be drawn on the pictures. They take data from the application program and from the data base and transfer these to the pseudo-display-file compiler. Typically, the following information is contained in the data base:

1. Data that describe the graphics; these may include macros of nonparameterized variants.
2. User-defined graphic functions.
3. Calls to graphic subroutines, which have the task to perform transformations with the picture. The subroutines are called by names and are supplied with the necessary parameters.

The graphic subroutines are stored in a program library on an external memory, in general a disk. They can be called upon at any time. The pseudo-display-file compiler links the subroutine calls to library routines and stores the highly structured data in the pseudo display file. The contents of this file can be thought of as an intermediate code which is generated by the translation of a higher programming language. The actual display code, however, contains the information in a form that can be used directly for display. The pseudo display file contains calls to subroutines with the help of which references to data and transformation parameters are transferred. The trace routine traverses the display file, interprets the calls, and gives the necessary instructions to the transformation, clipping, and concatenation routines. A transformation can be a picture translation, a scaling, or a rotation. Clipping is necessary when only parts of a picture are to be displayed on the screen. A more detailed description of these functions will be given later.

Before the transformation routine can be fully activated, the concatenation routine has to perform the following tasks:

1. Call the graphic subroutines and prepare them for execution
2. Perform the concatenation of several individual transformations

To do this task, the routine needs a stack by which parameters are transferred to the graphic subroutines. It also serves as a temporary buffer for the serially connected transformations. In this way each subroutine accesses its own data, changes them if necessary, and again stores them in this buffer. The data or parameter to be executed next are on the top of the stack. The transformation routines retreive and store the information. The use of the stack has the advantage that nested subroutines can easily be handled.

After the trace routine has processed all the pictures, the information is ready for display. It is now sent to the display code generator. This program produces the display code for the particular hardware in use. It has a similar task as a translator (compiler). From here, the transformed display file is stored in the picture file (refresh memory). It is now ready to drive the display screen.

5.6.5 Graphic Transformation/Clipping

If we consider a point (x,y) and it is transformed to the position (x', y'), the different types of graphical operations can be expressed as follows.

1. *Translation*: The point is translated by adding a positive or negative displacement Δx or Δy, respectively, to each of its coordinates.

   $x' = x + \Delta x$

   $y' = y + \Delta y$

2. *Scaling*: Stretching or compressing of a figure relative to the origin of the coordinate system is obtained by

   $x' = x * S_x$

   $y' = y * S_y$

   If the scaling factor is smaller than 1, the figure is compressed; if it is greater than 1, it is stretched.

3. *Rotation*: Counterclockwise rotation by an angle α about the x axis can be described by the following equations:

   $x' = x * \cos \alpha + y * \sin \alpha$

   $y' = y * \cos \alpha - x * \sin \alpha$

4. *Clipping*: Clipping is done by separating a specific section from a picture with the help of a rectangular window.

5.6.6 Display File

This file contains the code that is stored in the refresh memory; it operates the display screen. To obtain a flicker-free picture the information is refreshed by traversing the code 30 to 50 times per second. The structure of this file is explained with the help of Fig. 5.26, in which a circuit is depicted. Figure 5.27 contains the corresponding display file. It is composed of the following constituents: (1) name list, (2) correlation list, (3) a list of the display code, and (4) a table for user data.

Entry into this file is done through a descriptor identifying the graphic object (Fig. 5.27). The correlation table is the interface

Computer-Aided Design

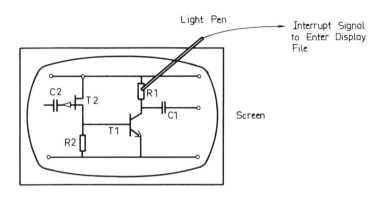

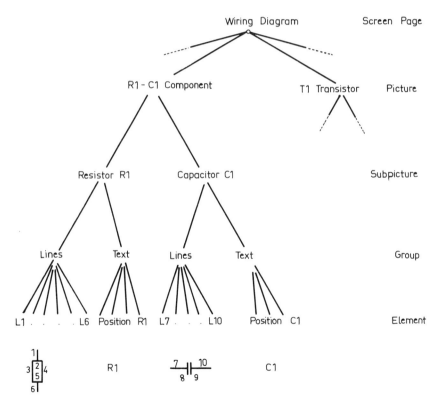

FIG. 5.26 Representation of a drawing by a tree structure. (From Ref. 5. Courtesy of PDV Funding Agency, Kernforschungszentrum Karlsruhe GmbH, West Germany.)

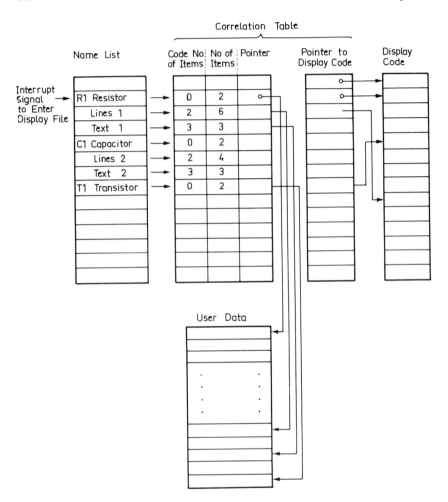

FIG. 5.27 Organization of the display file. (From Ref. 5. Courtesy of PDV Funding Agency, Kernforschungszentrum Karlsruhe GmbH, West Germany.)

Computer-Aided Design

between the user, the data of the application program, and the output of the desired picture. The information contained in a page displayed on the screen is structured as follows:

Picture
Subpicture
Group
Primitive

The primitives can be of the following types:

Point
Line
Character
Symbol generated by hardware

For each entry into the name list the correlation table contains several types of information.

1. Descriptor or type of data (subpicture, point, line, etc.)
2. Number of elements used (number of groups of subpictures, number of elements in a group)
3. Pointer to the parameters entered by the user

If the location of a primitive in the display file is known, the user can access it with the help of the light pen and thus is able to manipulate the picture on the display screen. Since there are pointers addressing the user file, it is also possible to alter user data. The construction of a display file depends on the following factors:

1. The capability of adding or deleting primitives
2. Aids to perform scaling, positioning, duplication, and the change of the intensity of picture elements
3. The frequency of manipulating the picture primitives
4. The necessity of performing sequential changes in the tree-structured picture file

Depending on the method of presenting information, the display code file can be of one of two types: (1) a picture-oriented display file, or (2) an object-oriented display file. The first structure is user friendly and the second is computer friendly.

5.6.7 Working with the Displayed Picture

Interactive work requires that the user is capable to perform changes with the picture on-line. This makes it necessary that the program which contains the code can be dynamically altered. Thus it frequently has to be retranslated in order to accommodate a change. To speed up reconfiguration the program is divided into several segments (Fig. 5.28). With this it becomes possible to recompile only that part of a program to which a change has been made. However, problems may be encountered if this change is done during the execution of the picture code. Thus the newly generated code can no longer be interpreted. This is referred to as: "The display screen is running wild." For example, by such a change meaningless instructions or unintended jumps may be generated, because the old picture had not been completely deleted nor the new one fully reconstructed. There are two methods to avoid this problem.

1. Separation of the segments to be changed during reconfiguration of the new picture. If there are many changes, this may lead to picture flicker (Fig. 5.29).
2. A much better method is the insertion of a new element after it had been recompiled. However, this presumes a nonsequential structure of the data (Fig. 5.30).

In the second case each segment has a pointer to the next segment. The new segment is translated in free memory space. Thereafter, the pointer of the preceding element is directed toward the new element and the old element is moved into the free space. This segmentation method can be used only for object-related data files.

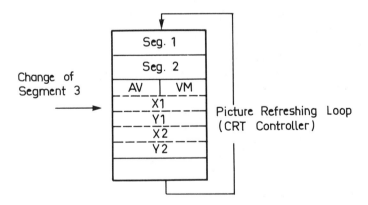

FIG. 5.28 Problem-oriented segmentation of a graphic program. (From Ref. 6.)

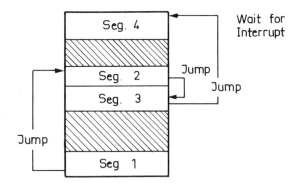

FIG. 5.29 Nonsequential segmentation of a graphic program. (From Ref. 6.)

5.6.8 Data Structures for Graphics

First some of the terms used for data structures are explained.

1. *Data element*: a bit string representing any elementary object which can be stored, retrieved, and processed as a single entity
2. *Block*: a collection of several data elements occupying a consecutive number of memory cells
3. *Component*: a set of consecutive data elements of a data block
4. *Pointer*: a data word containing addressing information
5. *Data file*: a consecutive number of data blocks

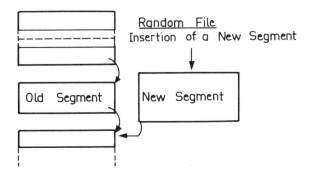

FIG. 5.30 Addition of a new element in a random file. (From Ref. 6.)

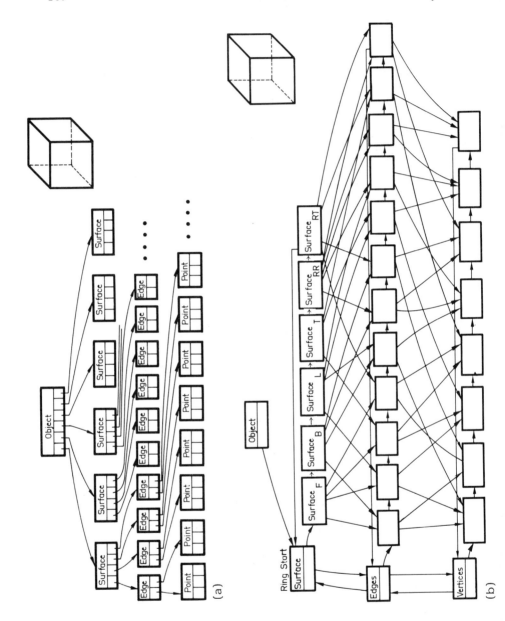

Computer-Aided Design

A data structure must have the following capabilities:

1. It must allow easy and fast access.
2. Changes, additions, and deletions must be easy to perform.

There are three basic methods in use for organizing data files:

1. *Sequential organization*: Blocks are stored in sequential order. Sorting is done with the help of a key.
2. *Direct access*: There is a fixed relationship between a key of a block and its direct memory address: for example, the translation of the key with the help of a hashing table.
3. *Linked list*: With the help of a pointer, linear allocation is possible. Pointers separate the logical and physical organization.

Normally, tree or ring structured lists are used to store graphic data (Fig. 5.31). This figure shows how a cube is represented by both methods. The hierachical structure of both lists is identical. It consists of three levels. At the highest level the six surfaces of the cube are described. The middle level contains the 16 edges, and the lowest level contains the eight vertices. There are pointers directed from the elements of each higher level to the elements of lower level.

By viewing the tree structure it can be seen that the vertices belong exclusively to edges, which means that for each edge there are exactly two vertices. The same holds true for the relationship between edges and surfaces. There is exactly one pointer directed at each element (the root of the tree is excluded). There are 6 data objects of the type "surface," 24 objects of the type "edge," and 48 of the type "vertices." Altogether, 79 data objects are represented. If the ring structure is viewed accordingly, there are only 30 data objects.

The simple and hierarchical structure of the tree and the fact that only one pointer is directed at each element makes additions, deletions, and changes in this list easy. Since identical elements occur several times in this list, it may impose identification problems during a search. The additional memory space this method requires is a disadvantage.

For graphical data processing, the ring structure is most often used. This can be viewed as a linked linear list whose last pointer is directed toward the first element, or the list head. Each ring has exactly one list head and a number of succeeding elements. The list head may belong to several rings. It can represent simultaneously a ring head or another element of the list.

FIG. 5.31 Different graphical presentations of a cube. (a) Tree-structured list; (b) ring-structured list. (Courtesy of the University of Rochester, Rochester, New York.)

An advantage of the ring structure is that the search can start at any element of the list, resulting in a fast search since it does not have to start at the head. Since the ring elements have a greater number of pointers than the elements of the tree, the search for a location of an element is much simpler. There may be complications, however, because several routes can be taken for the location of an element. Each element is unequivocal, but the route to reach for it is not. Additions, deletions, and changes are much more complex. With a clear, well-structured, and modular program, this disadvantage is compensated for by the small memory requirement.

The two-dimensional data structure must be converted into the one-dimensional physical data structure of the computer memory. This means that a structural model has to be derived for each list of the different lists for each figure to be presented on a display screen:

1. *Point list*: Points have names, which are usually sequential numbers. They are the only objects that have absolute coordinates. Points belonging to one figure have forward pointers.
2. *Line list*: Lines have names. They are referred to by point coordinates. The ring list for lines leads to the point list via the figure list.
3. *Figure list*: Figures are denoted by names. Points and lines belonging to a figure can be reached via separate rings.

Rules to manipulate a figure are as follows:

1. *Deletion of a line*: End points of a line may be deleted only if they are not end points belonging to another line.
2. *Deletion of a point*: All lines of a figure in which the point represents an end point are deleted.
3. *Deletion of a figure*: The figure is searched for in the figure list. Its entry in the figure list and all lines and points in the corresponding lists are deleted.

Figures 5.32 and 5.33 show the computer-internal representation of a rectangle and triangle and the various lists needed to describe the two figures. The example of Fig. 5.32 shows the deletion of a line and the example of Fig. 5.33 the deletion of a point. The empty space list is used to keep a record of available memory space. It is left as an exercise to the reader to trace the deletion processes.

5.6.9 Programming Languages for Graphic Data Processing

A graphic language should have the following capabilities:

1. It must support interactive generation, description, and manipulation of a picture.

Computer-Aided Design

2. It must have dynamic flexibility.
3. It must allow graphic analysis of pictures, subpictures, and picture elements.

Since interactive computer graphics and image analysis are two separate fields, the foregoing postulates are difficult to combine into one general-purpose language. The majority of programs for geometric data processing are based on FORTRAN. Table 5.2 shows possible functions that should be included in a language to describe technical drawings. They permit drawing of line segments, circles, and any desired curve. The functions are implemented with the help of subroutine calls. Table 5.3 shows a small selection of typical representatives of such calls. The subroutine calls are divided into three classes:

1. *Administrative calls*: These calls identify the different graphic peripherals and data files and administer the data files. Furthermore, pictures or picture groups of partial pictures can be defined. There are 20 such subroutine calls.
2. *Graphic calls*: With the help of these calls, basic graphic elements and text can be written on the display screen. Coordinate translations and window operations can be performed. Figure 5.34 shows the use of a window operation. There are 50 different graphic calls implemented.
3. *Dialogue calls*: These calls are needed to manipulate groups of pictures, subpictures, and elements on the screen. The different interrupts generated by the light pen are handled with dialogue calls. Figure 5.35 shows the use of a mirror call and Fig. 5.36 that of a rotation call, respectively.

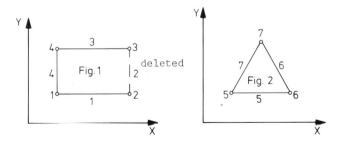

FIG. 5.32 Structure of a ring file for two basic geometric figures. (From Ref. 7. Courtesy of Oldenburg Verlag, Munich, West Germany.)

190   Chapter 5

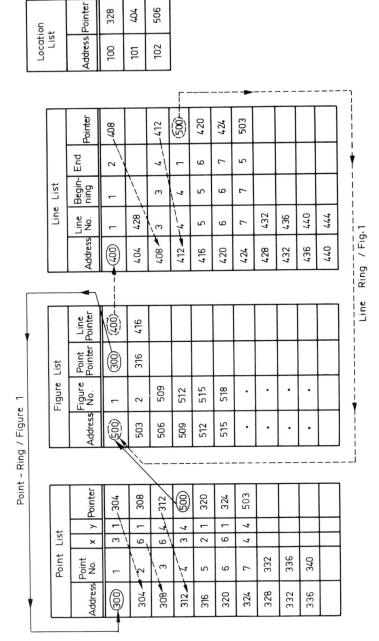

FIG. 5.33 Deletion of point 1 of Fig. 5.32. (From Ref. 7.)

Computer-Aided Design

Empty Location List

Address	Pointer
1	300
2	400
3	506

Line List

Address	Line No.	Start	End	Pointer
400	412			
404	2	2	3	408
408	3	3	4	500
412	428			
428	432			

Figure List

Address	Figure No.	Point Pointer	Line Pointer
500	1	304	404

Point List

Address	Point No.	x	y	Pointer
300	328			
304	2	6	1	308
328	322			

Deletion of point 1

FIG. 5.33 (Continued)

TABLE 5.2 Typical Functions Needed for Computer Graphics

Drawing of a point
    Cartesian coordinate
    Polar coordinate
    Vector
    Position on screen or plotter
    Entered through keyboard
    Center of circle
    Intersection of line with circle
    End point of curve
    Tangent of line to circle

Drawing of a line
    Two points entered through keyboard
    Two points positioned on screen or plotter
    Through a point parallel to a line
    Tangent to a circle from a point
    Tangent to two circles
    Through intersection points of two circles
    Parallel to a line and tangent to a circle

Drawing of a circle
    A point and a radius (diameter) entered through keyboard
    A point and a radius (diameter) positioned on screen or plotter
    Center point and tangent
    Center point and point on circumference

Drawing of a circle (continued)
    Three points on the circumference

Special curves
    Fillet between two intersecting lines
    Ellipse
    Parabola
    Hyperbola
    Spline

Three-dimensional presentation
    Cuboid
    Cone
    Cylinder
    Hemisphere
    Translated polygon
    Rotated polygon
    Spatial curves

Special functions
    Erase Object
    Move object
    Rotate object about point (axis)
    Dark hidden lines
    Delete hidden lines
    Merge pictures
    Group
    Duplicate
    Mirror
    Pattern
    Surface intersection
    Geometric projections
    Window

Source: Ref. 8.

TABLE 5.3 Different Subroutine Calls of the Siemens Graphics System

Administrative calls
CALL BEGIN (IFELD, IANZ)	Open system
CALL INIT (ISEN, IVERS, ITR)	Initialize a peripheral
CALL BEPIC (ISEN, X, Y, ID)	Beginning of a picture
CALL BEGRP (ISEN, X, Y, ID)	Beginning of a group
CALL EGRP (ISEN, ID)	End of a group
CALL BMENU (ISEN, X, Y, ID)	Beginning of a menu field
CALL EMENU (ISEN, ID)	End of a menu field

Graphic calls
CALL LINE (X, Y, IABS)	Draw line
CALL DLINE (X, Y, IABS)	Draw dark line
CALL LINE3D (X, Y, Z, IABS)	Draw three-dimensional line
CALL CIRC (X, Y, RAD, IABS)	Draw circle
CALL LINID (X1, X1, X2, Y2, ID)	Draw line with ID number
CALL TEXT (X, Y, : N)	Write text
: N FORMA ('REXT')	Text field
CALL MODUS (ISEN, INTEN, LTYP, LP, IBLINK'IDIMA)	Set status for graphic calls
CALL LNTYP (ISEN, LTYP)	Set type of display

Dialogue calls
CALL DELETE (ISEN, ID, IZU)	Delete an object
CALL DARK (ISEN, ID, IZU, INTEN)	Darken picture
CALL MOTYP (ISEN, ID, IZU, LTYP)	Change of displayed object
CALL MOVE (ISEN, X, Y, IABS, ID, IZU)	Move an object
CALL ROT (ISEN, ALPHA, BETA, GAMMA, IABS, MODUS, ID, IZU, ITYP)	Rotate three-dimensional objects
CALL ENINT (ISEN, INTM)	Set interrupt mask
CALL SPECI (ISEN'IWAIT, ISPEC, ID, NRGR, IFELD)	Process a given interrupt
CALL ASPECI (ISEN, ISPEC, IFELD, IANZ)	Process several given interrupts

194   Chapter 5

TABLE 5.3 (Continued)

Parameters:
ISEN	Peripheral No.
IFELD	Working memory for task
IANZ	No. indicator needed with different instructions
IVERS	Model No. of peripheral
ITR	Cursor - use or not use
X	User coordinate
Y	User coordinate
ID	Identification number
IABS	Coordinate - relative/absolute
RAD	Radius
:n	Format instruction
INTEN	Brightness
Ltyp	Representation of a line
LP	Light pen interrupt - permitted/not permitted
IBLINK	Blink/not blink
IDIM	Dimension statement
IZU	Sub-group picture
ALPHA BETA GAMMA	Angle of rotation
MODUS	Operation modus for rotation program
INTM	Interrupt mask (light pen, function key)
IWAIT	Interrupt - wait for/do not wait for
ISPEC	Interrupt mask for a specific interrupt
NRGR	Group No. for interrupt

Source: Refs. 9 and 10.

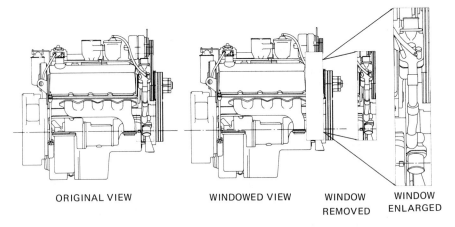

ORIGINAL VIEW   WINDOWED VIEW   WINDOW   WINDOW
                                    REMOVED   ENLARGED

FIG. 5.34 Effect of a window operation. (From Ref. 11. Courtesy of McDonnell Douglas Corp., Huntington Beach, California.)

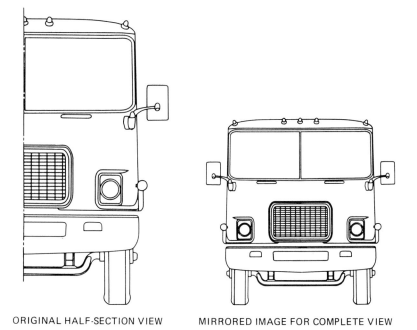

FIG. 5.35 Effect of a mirror operation. (From Ref. 11. Courtesy of McDonnell Douglas Corp., Huntington Beach, California.)

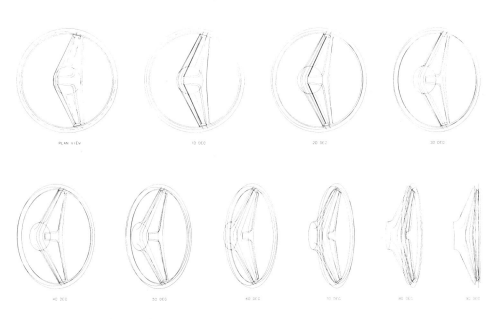

FIG. 5.36 Effect of a rotate operation. (From Ref. 11. Courtesy of McDonnell Douglas Corp., Huntington Beach, California.)

5.6.10 Generation of Drawings

When drawings are produced with the aid of a graphics system, the basic concepts of computer-aided design must be known. The user must be capable of generating, via textual programming, the control signal needed to actuate the drafting peripherals.

The steps in producing a figure on a drawing will be discussed briefly. A subroutine call for FORTRAN to produce a line is

$$\text{CALL LINE } (x_1, y_1, x_2, y_2, p)$$

This call causes the stylus of the drafting table to draw a line between the points $P_1 = (x_1, y_1)$ and $P_2 = (x_2, y_2)$. The additional parameter p defines the type of line (e.g., the width of a line or a centerline). With this call it is possible to generate any drawing.

Now we take a look at the type of information that is contained in a drawing. In principle, the information contents can be divided into two groups:

1. Technological contents
   a. Pictorial presentation
   b. Dimensions
   c. Machining and finishing information
2. Organizational contents
   a. Workpiece-related information
   b. Drawing-related information

A computer-aided design system must be capable of graphically describing the workpiece, the dimensions, and tolerances, and entering text into the drawing (Fig. 5.37). A drawing contains a great amount of information and thus is very cumbersome and expensive to produce. Therefore, a graphics system can be employed economically only if the programs that produce the drawing or its components are used repeatedly.

For economic reasons it is necessary to search for drafting and design methods that are computer oriented. The user can employ a graphics system to advantage only by utilizing its flexibility. The computer is of particular help if the drawing contains similar forms or when changes are often made.

A good example is the layout of seats in an aircraft body (Fig. 5.38). The designer starts with the selection of basic elements, such as seats and rows of seats. The complete drawing is assembled from these elements. This is of particular advantage for airplanes since the seating arrangement differs from one customer's specification to another.

Computer-Aided Design

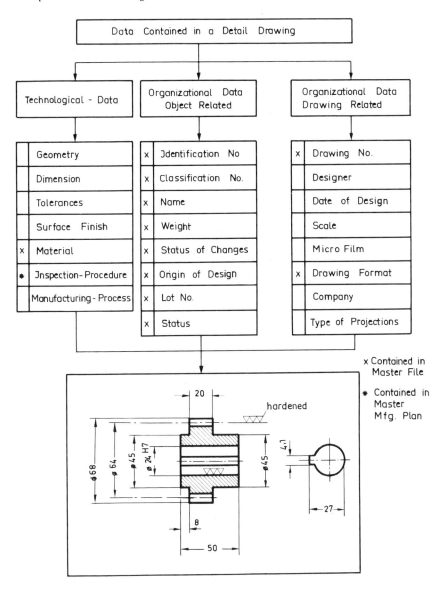

FIG. 5.37 Information contents of a drawing. (From Ref. 12. Courtesy of VDI-Verlag, Düsseldorf, West Germany.)

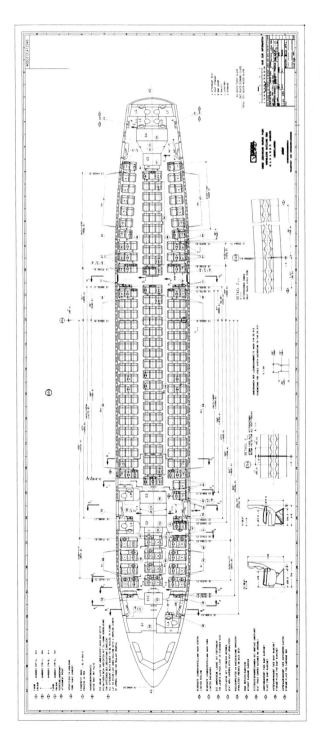

FIG. 5.38 Layout of seats for an airbus. (Courtesy of Messerschmitt-Boelkow-Blohm GMBH, Hamburg, West Germany.)

Computer-Aided Design

There are basically two different methods used to produce a drawing.

1. Generation of drawings from basic geometric primitives or form elements
2. Use of a special catalog containing macros of machine elements, which can be accessed by macro calls

The advantage of the second method is that frequently used parts need to be generated only once and entered into a library. This includes all information used to perform dimensioning, such as letters, dimensioning lines, and arrows.

5.6.11 Generative Principles

Four generative principles are deduced from the described above (Fig. 5.39).

Generation of Drawings from Geometric Primitives

The basic elements are points, lines, dimensioning arrows, circles, and other two-dimensional figures. In addition, macros must be available to draw these elements on the screen. Generally, these basic macros are supplied by the vendor. It is very cumbersome to produce drawings with them. This will be demonstrated with the help of an example. However, these elements are the basic building blocks used to write higher-order macros. In the following example a program to generate the drawing of a gear will be written with the help of a hypothetical language containing only four language constructs.

1. PLACE(x,y) places the drawing pen on the point (x,y).
2. LINE(x,y,n) draws from the present position a straight line connecting the new position (x,y); n = 1 produces a solid line, n = 0 produces a dashed line.
3. POINT(x,y) produces a point at the location (x,y).
4. DECLARE defines a subprogram.

The nonparameterized gear is shown in Fig. 5.40. The dimensions are to be interpreted as follows:

D1	Diameter of the hub
D2	Diameter of the bore
L1	Length of a tooth
L2	Length of hub section 1
L3	Length of hub section 2
D3	Outer radius of the gear
TH	Tooth height

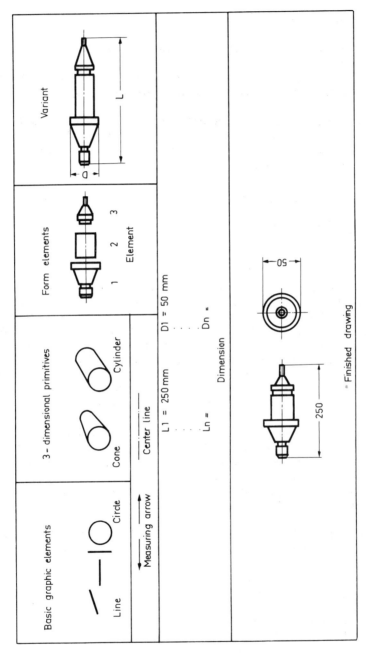

FIG. 5.39 Different graphic generative methods. (From Ref. 13.)

Computer-Aided Design

The coordinate system will be placed on the intersection of the centerline with the left border line of the hub. To draw the gear, several basic dimensions have to be calculated by the computer (Fig. 5.40). Since we are concerned only with two-dimensional projections it is sufficient to consider only the xy plane. Thus we obtain the dimensions given in Table 5.4 and Fig. 5.40.

With the help of previously introduced instructions, the following subroutines can be written:

DECLARE FUNCTION UGEAR (D1, D3, L1, L2, L3, TH)

UGEAR PLACE(0,0) + LINE(0, D1/2, 1) + LINE(L2, D1/2, 1) +
LINE(L2, R, 1) + LINE(L2 + L1, R, 1) +
LINE(L2 + L1, D1/2, 1) + LINE(L2 + L1 + L3, D1/2, 1) +
LINE(L2 + L1 + L3, 0, 1)

DECLARE FUNCTION MARK(D2, D3, L1, L2, L3, TH)

MARK = PLACE(0, D2/2) + LINE(L2 + L1 + L3, D2/2, 1) +
PLACE(L2, R-TH) + LINE(L2 + L1, R-TH, 1) +
PLACE(L2, R-TH/2) + LINE(L2 + L1, R-TH/2, 0)

The complete drawing is generated as follows:

DECLARE FUNCTION (D1, D2, D3, L1, L2, L3, TH)

GEAR = UGEAR(D1, D3, L1, L2, L3, TH) +
MARK(D2, D3, L1, L2, L3, TH) +
UGEAR(-D1, -D3, L1, L2, L3, -TH) +
MARK(-D2, -D3, L1, L2, L3, -TH) +
PLACE(0,0) + LINE(L2, L1, L3, 0, 0);

The last line is the instruction to draw the centerline of the gear. When the seven values for the gear to be drawn are inserted into the program, a call to the subroutine GEAR will actuate the computer peripheral to draw the gear.

A drawing also requires dimensioning instructions and additional text to be inserted into the pictorial. This can be done with the following instructions:

DARROW(x1, y2, x2, y2, 1) will draw a dimensioning arrow between the points (x1,y1), (x2,y2) and also inserts the dimension between the two arrows. The value 1 can indicate a variable or an actual dimension.

For more realistic dimensioning of the gear, additional information, such as pitch, pitch circle, and root circle, has to be generated by the computer. This has been omitted to simplify the example. The

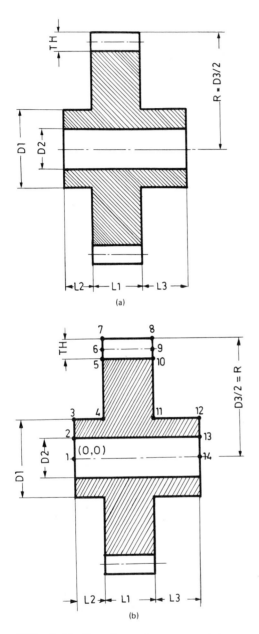

FIG. 5.40 Drawing of a gear: (a) gear to be drawn; (b) marking of construction points. (From Ref. 14. Courtesy of VDI-Verlag, Düsseldorf, West Germany.)

Computer-Aided Design

TABLE 5.4 Calculating the Basic Dimensions of a Gear

Point i	x_i	y_i
1	0	0
2	0	D2
3	0	D1
4	L2	D1
5	L2	R-TH
6	L2	R-TH/2
7	L2	R
8	L1 + L2	R
9	L1 + L2	R-TH/2
10	L1 + L2	R-TH
11	L1 + L2	D1/2
12	L1 + L2 + L3	D1/2
13	L1 + L2 + L3	D2/2
14	L1 + L2 + L3	0

Source: Ref. 14.

computer could also select the gear material and perform the stress analysis calculations.

Generation of a Drawing from Three-Dimensional Primitives

This method uses as basic elements three-dimensional primitives such as cuboids, cones, cylinders, and hemispheres. Figure 5.41 shows an example where a simple workpiece is described by the Computer Oriented Part Coding (COMPAC) method. It is a programming package built on spherical elements. The program will be discussed briefly.

The first instruction assigns the name SHAFT to the workpiece. The second instruction specifies the surface finish. The third instruction defines the material. The fourth instruction specifies the dimension of element K1 in relation to the coordinates x, y, and z; they are 25, 35, and 23 mm. In the next instruction the coordinate system is translated in the z-axis direction and then rotated about the y-axis by an angle of -15°. Upper wedge of the K1 is omitted and a new name K2 is

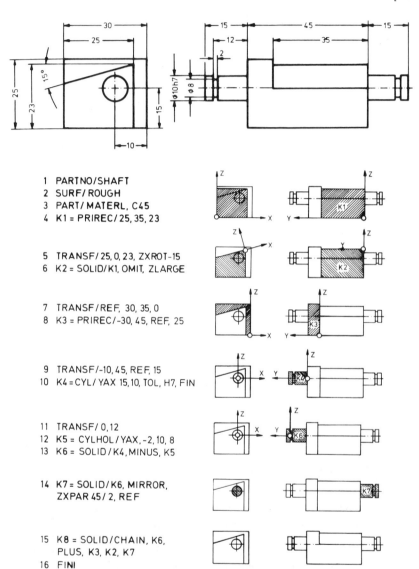

FIG. 5.41 Programming of a workpiece. (From Ref. 15. Courtesy of Rudolf Hauffe Verlag, Freiburg, West Germany.)

Computer-Aided Design

assigned to K1. The additional elements K3, K4, K5, and K6 are specified in a similar manner with the help of several coordinate translations. Element K7 is generated with the mirror instruction 14 about the xz plane and by moving the coordinate system back along the y axis by 45 mm. The instruction 15 assembles the complete workpiece from elements K2, K3, K4, K5, K6, and K7.

The drawing of Fig. 5.40 contains many dimensions. It is desirable to dimension a workpiece properly and to avoid over- or underdimensioning. This dimensioning information can be expressed in simple graph-theoretical terms (Fig. 5.42). Here an xy projection of a simple workpiece is shown, together with a dimensioning tree. The nodes of the graph are the lines L and the branches are the distances between the lines. For a properly dimensioned part there is a unique path between the nodes of the tree. A missing dimension is indicated by an interrupted path, whereas for an overdimensioned part the graph is not a tree. The latter case is denoted by the dashed line F. This dimension would be unnecessary.

Figure 5.43 shows a very complex workpiece generated from three-dimensional facets with the help of the IBM CAD system GDP/GRIN. In principle, this system works similarly to the COMPAC system; it will be described later in more detail.

Generation of Drawings from Simple Form Elements

A high degree of flexibility can be reached by using form elements. This method, however, relies on the principle that workpieces can be constructed from similar form elements, which implies

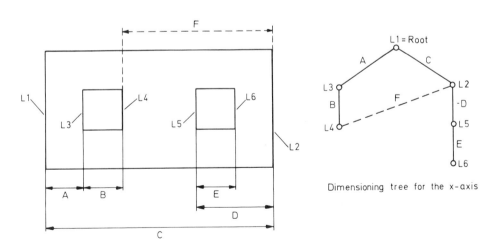

FIG. 5.42 Verification of dimensions with a dimensioning tree. (From Ref. 16. Courtesy of the University of Rochester, Rochester, New York.)

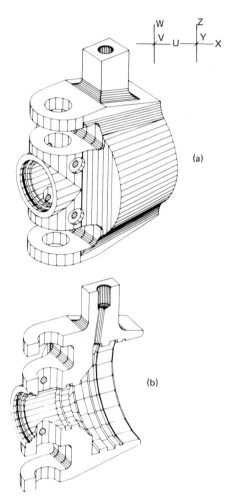

FIG. 5.43 Drawing of a complex part by the facet method. (From Ref. 17. Courtesy of International Business Machine Corporation, White Plains, New York.)

that the workpieces have a certain degree of similarity. The form elements deduced from the workpiece spectrum will be placed in a form element library (Fig. 5.44). The description of the workpiece is done with the aid of an input language or a menu or a tablet or a display screen. The number of form elements used depend on the frequency of their occurrence on the drawing. For economic reasons only macros for those form elements that are used most are entered into the library. Figure 5.45 shows the result of an investigation to determine the

Computer-Aided Design

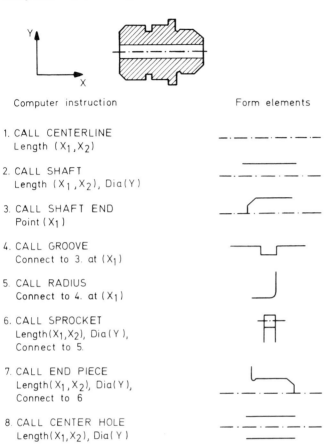

FIG. 5.44 Generation of a drawing from form elements. (Courtesy of VDI-Verlag, Düsseldorf, West Germany.)

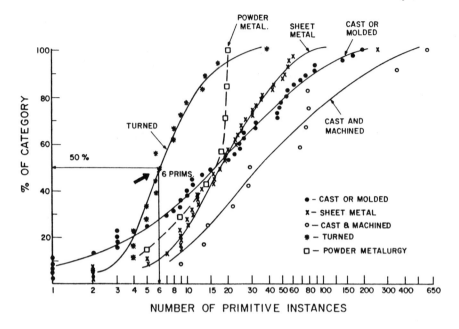
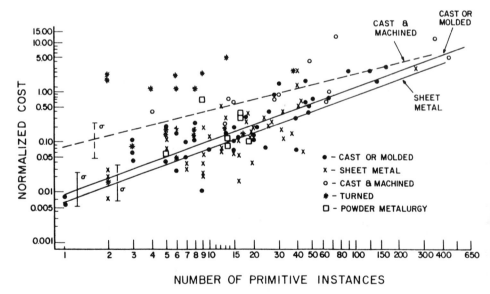

FIG. 5.45 Design primitives for the parts of a copying machine. (From Ref. 18. Courtesy of the University of Rochester, Rochester, New York.)

Computer-Aided Design

number of primitives needed to design a copying machine. A distinction was made between powder metal, sheet metal, cast or molded, and cast or machined parts. In this study it was found that 50% of all parts of the copying machine could be described with six primitives. In the lower part of the figure an attempt was made to relate cost to geometric complexity. These figures show that with a high number of primitives, the form element method becomes very expensive. For this reason it may be more economical to design the components that have seldom used primitives with the help of geometric primitives.

Generation of Drawings with the Help of the Variant Method

In case the workpieces are very similar in design, the variant method is the most efficient alternative (Fig. 5.46). With this method a nonparameterized master variant will be designed and its description will be stored in the computer as a macro. When the user calls this macro, he or she parameterizes it with the desired dimensions. The computer calculates the shape of the gear and produces the drawing. The variant method can be simplified with the help of the two following design methods. However, both have the disadvantage that the drawing may not show the exact shape of the workpiece.

Fill-in-the blanks drawing. With this method the computer contains a ready-made drawing, including the dimensioning arrows. Some of the dimensions that are fixed by the manufacturing process may also be included (Fig. 5.47).

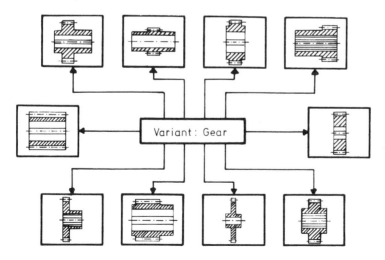

FIG. 5.46 Variant design method applied to gears. (From Ref. 19.)

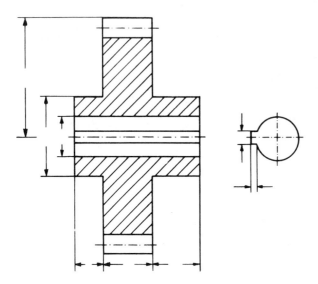

FIG. 5.47 Fill-in-the-blanks drawing.

Table printout method. Here again a nonparameterized drawing is contained in the computer. Upon entry of a part number the computer prints out the drawing and a list of parameter values (Fig. 5.48). The user has to select the dimensions from the parameter value list.

5.6.12 Geometric Modeling of Solids

When a workpiece is modeled with the help of the computer, it is necessary that the algorithm representing the solid be unequivocal. To assure this, a class of subjects in a three-dimensional euclidean space (E^3) is postulated to model solids. The class has the following properties [20]:

1. The configuration of the solid must stay invariant with regard to its location and orientation.
2. The solid must have an interior and must not have isolated parts.
3. The solid must be finite and occupy only a finite space.
4. The application of a transformation, rotation, or other operation that adds or removes parts must produce another solid.
5. The model of the solid in E^3 may contain an infinite number of points. However, it must have a finite number of surfaces which can be described to the computer.

Computer-Aided Design   211

6. The boundary of the solid must uniquely identify which part of the solid is exterior and which interior.

In mathematical terms this implies that suitable models for solids are subsets of E^3 that have the properties: bounded, closed, regular, and semianalytic.

These are several methods to represent solids (Fig. 5.49). They are:

1. *The wire frame*: This model is ambiguous. For example, according to the rules above, the boundary does not identify which part is exterior and which is interior.
2. *The primitive instancing presentation*: With this method objects can be defined as primitives that belong to a family of

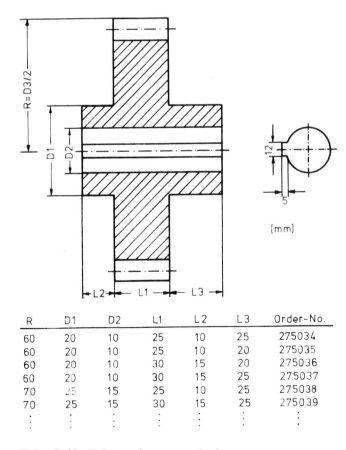

R	D1	D2	L1	L2	L3	Order-No.
60	20	10	25	10	25	275034
60	20	10	25	10	20	275035
60	20	10	30	15	20	275036
60	20	10	30	15	25	275037
70	25	15	25	10	25	275038
70	25	15	30	15	25	275039
⋮	⋮	⋮	⋮	⋮	⋮	⋮

FIG. 5.48 Table pointout method.

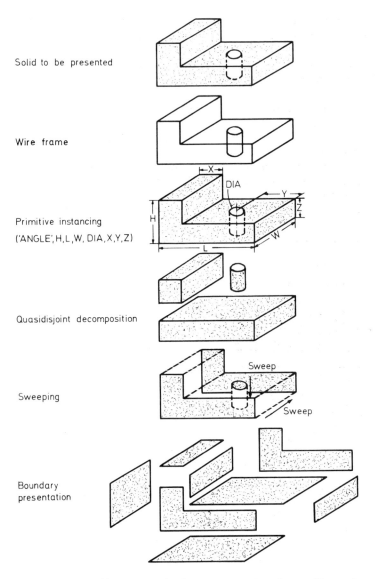

FIG. 5.49 Different methods to present a three-dimensional object.

objects having a finite number of parameters. In our case the angle was defined as a primitive; it is represented by the 8-tuple ('ANGLE', H, L, W, Dia, x, y, z). This method is particularly useful when the variant method is applied for part presentation.
3. *Quasi-disjoint decomposition*: With this method the solid is segmented into smaller solids which have no holes or disjoint interior. The smaller solids may be cubes, sheets, columns, or be of different shape. Particular problems may be encountered when curved objects have to be modeled by this method.
4. *Sweeping*: With this method a plane moving through space produces the model of the object.
5. *Boundary presentation*: In this case a solid is modeled by nonoverlapping faces of its boundaries.

The last three modeling techniques are the most important. Since they all have advantages and disadvantages, several may have to be used to facilitate modeling of three-dimensional workpieces. However, it is necessary that all workpieces can be presented by the system and that the model is unambiguous.

5.7 COMPUTER INTERNAL REPRESENTATION OF GRAPHIC DATA

5.7.1 Phase Objects

The main task of a CAD system is to translate the product description to production information. This is done with the help of algorithms and data. The CAD process is a sequence of translations that starts with the designer's conception of the product, leading to the workpiece production drawing and completing the cycle with instructions to the machine tool to produce the part. Phase objects are the models of the different stages to translate the design idea into the manufacturing drawing. For this purpose, first information has to be gathered on the product and the processes needed to produce it. Then calculations have to be performed to determine physical characteristics. Thereafter an image of the product is drawn on paper and the design is evaluated. The last phase is usually a design change to include improvements or alternatives.

A phase object can be defined more closely as an 8-tuple consisting of identification, classification, spatial definition, technological means, environment, and the state [21]. Thus the spatial definition consists of the shape and the dimensions of the product. The shape may be further divided into geometry and topology. Figure 5.50 shows the different stages of phase objects. For each stage the following applies:

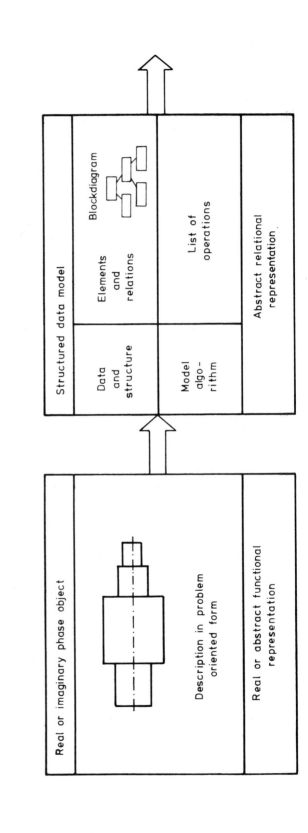

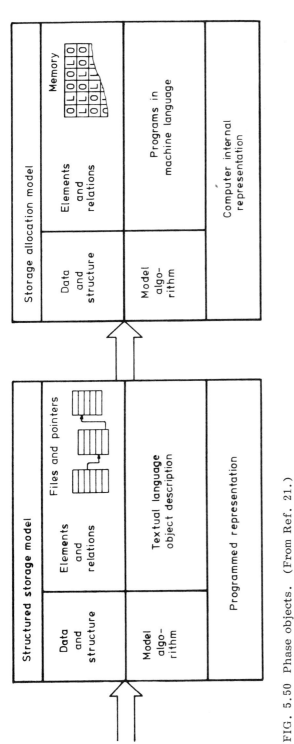

FIG. 5.50 Phase objects. (From Ref. 21.)

1. The semantics must be maintained when one stage is translated into another.
2. The syntax will change with each consecutive phase.

The design of the product goes through the function, principle, shape, and detail phases (Fig. 5.50). Each of these will be discussed later in more detail. There will be phase objects for each phase. They are represented in the computer by complex data structures and are described by combinational logic, graphs, verbally, or by special textual methods. The best method is the use of graphs. In particular, geometric interdependencies can easily be related.

First, the workpiece has to be made known to the computer. In the ideal situation the computer would contain a sufficiently high degree of artificial intelligence to design the part on its own. It would obtain a formal description of the parts environment and its functions. Information typically needed are function, principle, the overall shape, the load, the torque to be transmitted, life expectancy, corrosion resistance, and so on (Fig. 5.50). In practice, however, it is not possible to provide creative design capabilities to the computer. This is the domain of the designer and should be left to the designer. In particular, the "function" and "principle" phases require high creativity. The "shape" and "detail" phases are more routine and can be automated more easily.

The phase objects contain identifying, classifying, geometric, and problem-oriented elements (Fig. 5.51). In the data model there are hierarchical and problem-oriented relations. Hierarchical relations can be described by means of a tree, showing, for example, the relations between a solid and its surfaces, edges, and vertices (Fig. 5.52). To denote the hierarchical order of the elements, they are divided into different classes. Problem-oriented relations are defined by the semantics. For example, they relate that two surfaces have the same finish or that they are interconnected by a dimensioning arrow (Fig. 5.52). This describes a problem-oriented relationship within a class of geometry-dependent elements.

5.7.2 Description of a Machined Cube

To continue the preceeding discussion, we take a cube which represents a simple workpiece and show how it can be described completely to the computer. This example was taken from the dissertation of J. P. Lacoste [22]. Figure 5.53 shows a simplified generalized graph to describe a workpiece. The workpiece is represented at the highest level as a solid, together with its description and physical parameters. At the next level the subcomponents, if present, are listed. Here also the shape and technological properties are shown. One level lower the surface boundaries are depicted, together with information on dimensions,

Computer-Aided Design

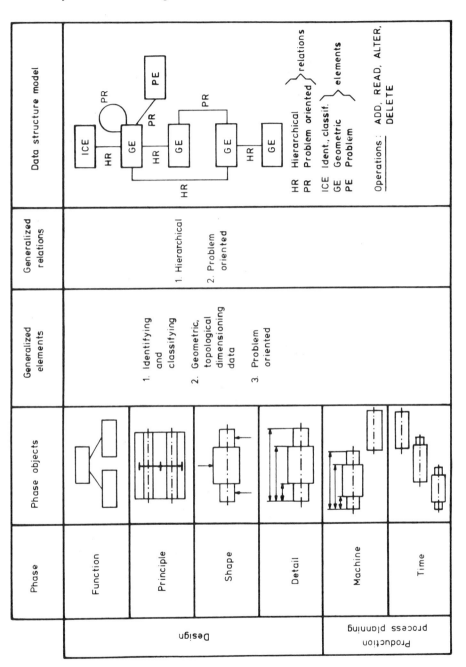

FIG. 5.51 Generalization of a data structure model. (From Ref. 21.)

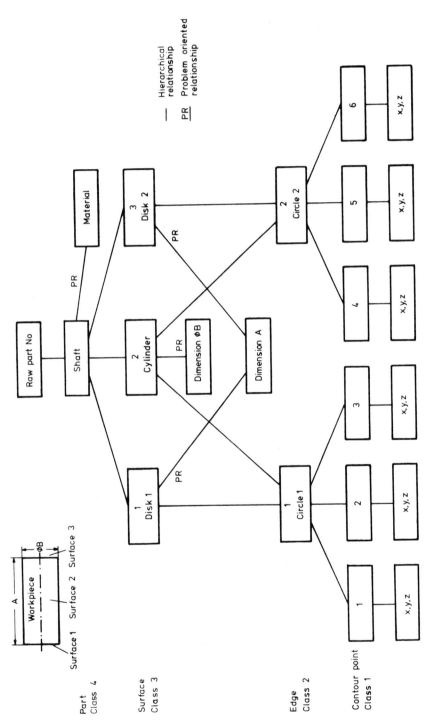

FIG. 5.52 Possible representation of a phase object with the help of a graph. (From Ref. 21.)

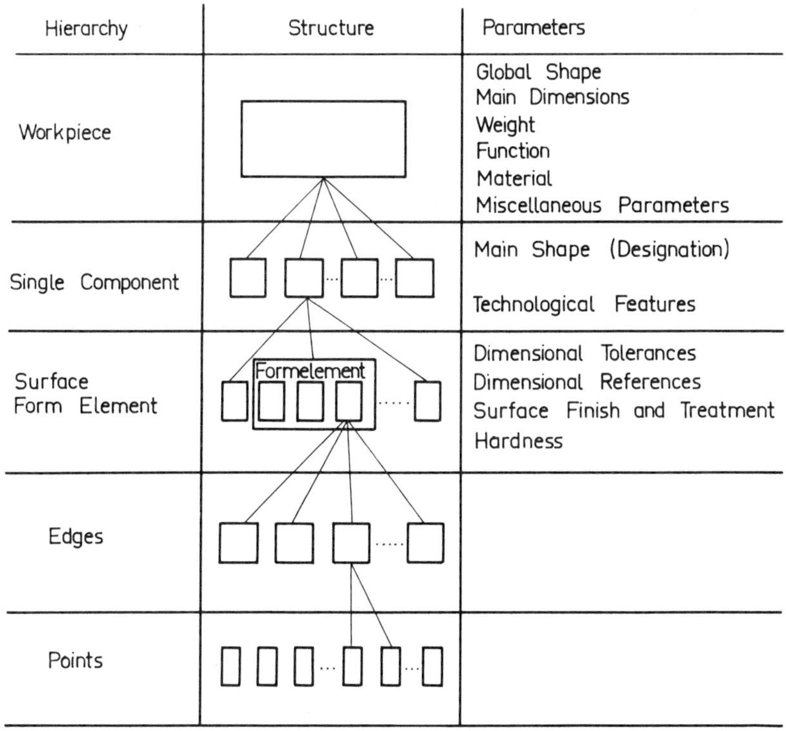

Computer-Aided Design

Hierarchy	Structure	Parameters
Workpiece		Global Shape Main Dimensions Weight Function Material Miscellaneous Parameters
Single Component		Main Shape (Designation) Technological Features
Surface Form Element	Formelement	Dimensional Tolerances Dimensional References Surface Finish and Treatment Hardness
Edges		
Points		

FIG. 5.53 Fundamental structure of a workpiece. (From Ref. 22.)

tolerances, surface treatment, and hardness. The graph branches farther, down to edges, and from there terminates with points. The criteria to assign these relations are, for example, that an edge belongs to a surface. This type of network is called a directed graph. We are able to specify the following relationship graphs: (1) solid/surface graph, (2) surface/edge graph, and (3) edge/point graph. The center part of Fig. 5.53 describes these graphs. The problem-oriented relationships are shown on the right side.

In addition to showing the hierarchical relationships in the graph of Fig. 5.53, there must be element graphs that show the cross-relationships of the individual elements at each hierarchical level (Fig. 5.54). These graphs are (1) solid graph, (2) surface graph, (3) edge graph, and (4) point graph. These are all undirected graphs. They express coincidence relationships, which may also be termed neighborhood functions. This means, in the case of the surface graph (Fig. 5.54), for example, that each surface section is neighboring with four others and not with the fifth.

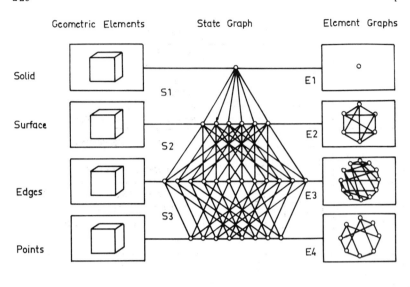

FIG. 5.54 Graph presentation of a cube. (From Ref. 22.)

Mapping the Graph Structure in the Computer

The graph structure has to be described with the help of three components:

1. The symmetric element or coincidence matrix, in which the description of the undirected element graph is stored
2. The asymmetric relationship matrix, which stores the description of the directed element graphs
3. The element table, which contains the description of the properties of the individual elements

These three components will be discussed next in more detail.

Element matrix. The geometric relations between two identical elements are represented by the element graphs. The coincidence relationships are stored in the element matrix (Fig. 5.55). The row and

Computer-Aided Design

column indices of this matrix are the designators of the geometric elements. The surfaces of the cube are given the names a, b, . . ., f. To determine a neighborhood relationship, the matrix has to be entered either from its top or its side and traversed in a straight line. Whenever a 1 is found, a coincidence relationship exists. If there is a 0, no neighborhood exists. For example, if the row c is depicted, the neighbors are a, b, e, and f.

Element table. Since the element matrix contains only coincidence information, an additional element table with element-specific information is constructed. The different files of the workpiece model are of similar design. The description of the element is stored in a data block. Its information contents depends on the characteristics of the element. The contents of the file can be accessed through the row index via a pointer field which is directed toward the address of a data block (Fig. 5.56). The pointer permits random storing of the data block, which facilitates file housekeeping.

A data block consists of a head and a data field (Fig. 5.57). The head allows identification of a block when it is being read. The information is stored sequentially and is always preceded by an identifier. The identifier specifies the type of information and the format of the succeeding parameter (Fig. 5.58 and 5.59). This method assures

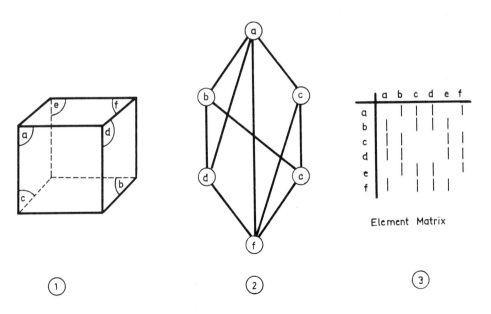

FIG. 5.55 Connection of similar geometric elements. (From Ref. 22.)

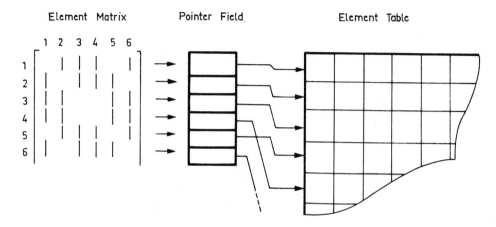

FIG. 5.56 Connection of element matrix and element table. (From Ref. 22.)

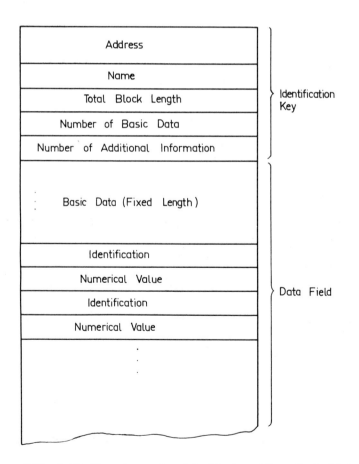

FIG. 5.57 Structure of a data block. (From Ref. 22.)

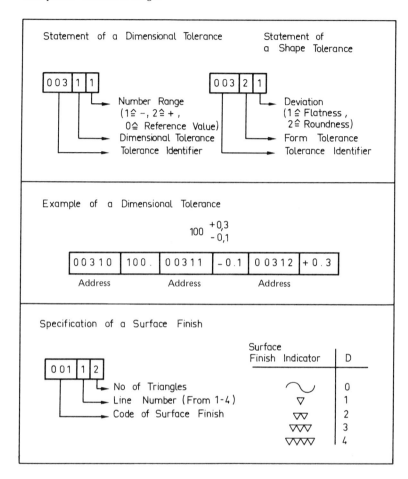

FIG. 5.58 Specification of technological data. (From Ref. 22.)

efficient utilization of computer memory. Additional files would be needed for the coordinate system, component, workpiece identification, edge and point description, and so on.

Relationship matrix. The relationship of elements of different hierarchical levels is described by the individual relationship matrices. Each of these graphs contains information on the relationship between two neighboring hierarchical levels.

Figure 5.60 shows the relationship between the surfaces and edges of the cube. Again a 1 means that there is a relationship and a 0 that it does not exist. For example, surface a is related to

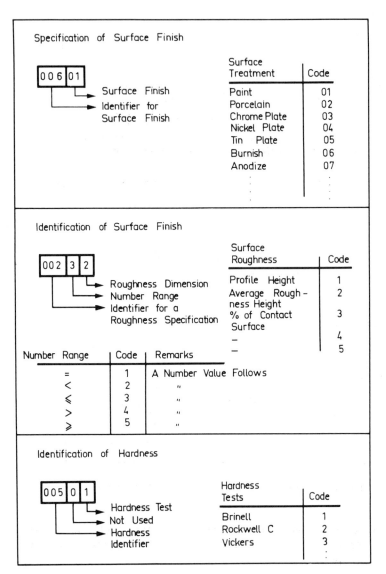

FIG. 5.59 Representation of technological data. (From Ref. 22.)

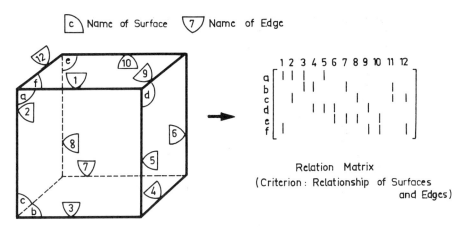

FIG. 5.60 Connection of dissimilar geometric elements. (From Ref. 22.)

lines 1, 2, 3, and 5. Additional relationship matrices would have to be designed for describing dimensioning and tolerance information.

5.8 DESIGN PHASES

The design process comprises all activities from the creative conception of the product to its final documentation in the drawing. Design starts with the concept of a product, which may be the idea of the designer or a result of a marketing research activity. The physical properties of the product are thoroughly evaluated and the product is designed to be functionable and manufacturable. Aesthetic considerations will also be included to please the customer's taste. The image of the final product is drawn on paper and serves as a manufacturing document.

It is extremely important to investigate the different design phases thoroughly to find out where the computer offers the most benefits (Fig. 5.61). In principle, the design process can be divided into four different phases [23]: (1) functional analysis, (2) product definition, (3) layout, and (4) detail drafting. Not every product needs to go through all phases. It depends primarily on whether the product is new or represents a redesign. The tasks of the individual phases are shown in Fig. 5.61. It shows which design methods they use and what results they deliver. The various phases will now be discussed in more detail.

Design phases

	Functional analysis	Product definition	Layout	Detail drafting
Stages:	1. Problem task 2. Sum of all solutions 3. Error critique 4. Evaluation	1. Problem task 2. Sum of all solutions 3. Error critique 4. Evaluation	1. Problem task 2. Sum of all solutions 3. Error critique 4. Evaluation	1. Problem task 2. Creation of solutions
Goal:	Functional concept	Working concept	Principle layout	Manufacturing information
Concerned with:	Problem character	Problem character (Task character)	Task character (Problem character)	Task character
Type of activity:	Mentally - intuitive	Mentally - intuitive (Manually - schematic)	Manually - schematic (Mentally - intuitive)	Manually - schematic
Methods and aids:	Methodology Morphology Intuition	Design systematics Value analysis Morphology Simulation techniques	Layout systematics Calculation methods Value analysis	Manual and machine techniques

FIG. 5.61 Design phases. (From Ref. 23.)

Computer-Aided Design

Functional analysis. The main tools used in this phase are methodology, morphology, and intuition. With their help a functional analysis of the product is made. This usually includes investigation of different alternatives (e.g., there may be an electrical, hydraulic, or a mechanical answer to the solution of the problem). A thorough error critique and evaluation of the concepts is needed to find the optimal functional design. The work requires high intellectual abilities and a thorough knowledge of product design and manufacturing. The result of this phase should be a concept of the product or product line. The material to be used, and information about the manufacturing processes and the tools used, should be known. This phase is in the human domain; a computer cannot be a substitute for a human.

Product definition. In this phase the results of the functional analysis are further refined and several alternative working principles are investigated. From this an optimum is selected (e.g., the result may be an efficient hydraulic pump). This work still requires mental activities; however, there may already be some manual labor involved. A clear picture of the product is created and it may be supported by sketches and functional drawings. The display screen may serve as a vehicle to investigate the functional concept with the help of simulation programs to animate the function of components or that of the entire product. The working tools are design systematics, value analysis, and simulation techniques. The investigation of alternative solutions has an important influence on the functional quality of the product. For this reason there should be good simulation techniques available on a computer graphic system.

Layout. In the machine layout phase, a practical functional solution with, in many cases, aesthetic appeal is created. The activity requires manual skills and can be heavily supported by the computer. Depending on the complexity of the product, the product family, and the production rate, the design aids may differ considerably (Fig. 5.24). The product may be designed from graphic primitives, form elements, or variants (Fig. 5.39). The output of this phase is the layout of the entire product. All individual parts have now been conceived, and their shape, overall dimensions and physical properties are known. In addition, the subassemblies and the main assembly are defined.

Detailed drafting. In this phase the detailed workshop drawings and the bill of materials are produced. This activity is of a predominantly manual nature and should be performed by the computer.

5.8.1 Types of Design

The emphasis on the individual design stages in each phase depends on the product itself. For a new product it is necessary to go through all design phases. Here automation endeavors may be difficult to implement. If a product had been designed before or is a variant of one for which design knowledge is already available, a very high degree of automation can be obtained. Thus we have to distinguish among three different types of design (Fig. 5.62): (1) a new design, (2) a design change, and (3) a variant design. This makes it necessary before design starts to investigate the method to be used very thoroughly. An attempt should be made to use as much previous design knowledge as possible. It is also advantageous to take a look at the entire design spectrum to assure commonality of parts and subassemblies. The more commonality there is, the more easier it is to conceive an efficient design system (Fig. 5.45).

5.8.2 Detail Drafting

In this section we discuss the forth design phase, detail drafting, more closely. It takes the main design effort (Fig. 5.1) and is a candidate for high automation. Depending on the product line and the expectation of the design department, many computer-aided design concepts and aids have demonstrated their usefulness to industry. The integration of CAD with CAM will not be possible or economically feasible in all cases. The individual design systems for the three basic production methods: (1) low-volume production, (2) medium-volume production, and (3) high-volume production, will differ from each other considerably.

In low-volume production the designer will work primarily with graphic primitives (Figs. 5.39 and 5.45), and if the designer has a similar product family, also with form elements or variants. Here the CAD system may be interfaced directly to NC equipment to control machining.

In medium-volume production the form element or the variant method will predominate (Fig. 5.39). Here the designer may also be concerned with part families and part commonality. In many cases direct connection of the CAD output to NC equipment will be possible.

With high-volume production the designer will use mainly graphic primitives and form elements, since there are usually not many design options. Interconnection of the CAD with the CAM system may not be possible, since production equipment uses few programmable control systems.

We may conclude from this discussion that a CAD system will show the greatest benefits for medium-volume production. It will be necessary here to survey the entire part spectrum when the CAD system is installed and to divide this spectrum into part families. For a family with few members, the form element method will probably apply (Figs. 5.39 and 5.45). For each of these form elements a macro will be written and placed in a form element library. With the help of these form

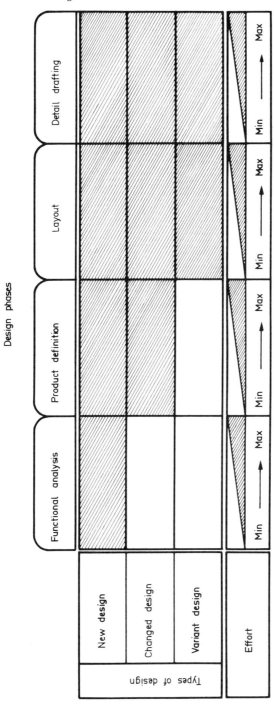

FIG. 5.62 Design phases for different types of designs. (From Ref. 23.)

elements the designer is able to specify complex parts just by adding the macros together. Only different combinations of macros may be allowed, to avoid impractical or nonpermissible designs.

If the part spectrum contains many similar parts, the variant method will be the predominant choice. Here it may only be necessary to call a variant macro and to supply it with parameters. During the initial implementation of the CAM system, this method may be very expensive.

A decision about which of these design methods renders the most economical solution can be determined only after a careful investigation of the entire part spectrum and the processes needed to generate the parts. If there are many parts which are very close in design, redesign of the parts to standard items or variants should be considered. In this way the efficiency of the CAD system can be greatly improved.

The system for automatic detailing should contain all those parts that have the highest probability of occurrence. With the help of this part spectrum, experience can gradually be built up. The capability of automatically generating NC information should be provided. Now let us take a look at the workpiece analysis to select a proper part spectrum.

5.8.3 Workpiece Analysis

The workpiece analysis is done for four phases [24]:

1. Selection of the workpiece spectrum
2. Assignment of parts to product groups
3. Functional analysis of the parts
4. Determination of the frequency distribution of the parts (functions)

The selected workpiece spectrum of the first phase may contain rotational, cubical, or sheet metal parts (Fig. 5.45). The groups of the second phase may be formed for motors, gear boxes, brakes, and so on. In the third phase the selected workpieces will be analyzed for their functions. This will determine for what percentage of workpieces functional variant macros can be conceived. They will be entered in a variant catalog. After this investigation there will be many workpieces left over for which variant macros are uneconomical. These parts are investigated in the fourth phase and common form elements will be selected. Macros will be written for those and entered into the form element macro. There will be some workpieces left over. Those have to be drawn with the help of geometric primitives. This procedure will assure an economic solution.

Input list 1 — Main element

Element	Sketch	Computer input
Cylinder	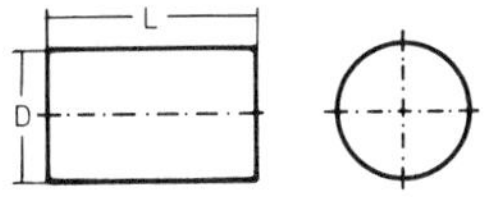 (L, D)	XXX_CYL/L,D /TL*= /TD*= /B1**=...
Cone	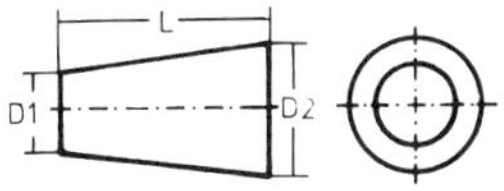 (L, D1, D2)	XXX_CON/L,D1,D2 /TL*= /B1**=...

* Tolerance ** Location of finish indicator

FIG. 5.63 Catalog for main elements. (From Ref. 24.)

Input list 2 — First-order element

Element	Sketch	Input
Key	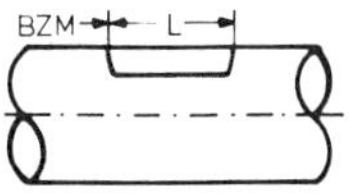 (BZM, L)	XXX_KEY /L,BZM /TL*= /TBZM*=
Teeth	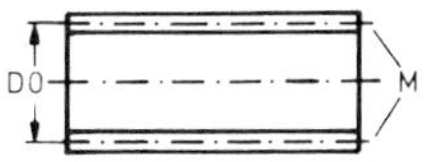 (D0, M)	XXX_TETH /D0 M

* Tolerance

FIG. 5.64 Catalog for first-order elements. (From Ref. 24.)

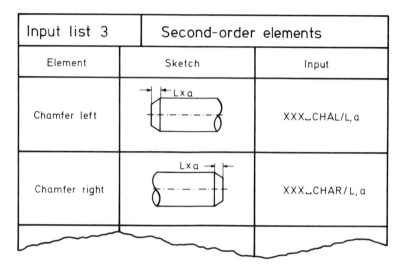

Input list 3	Second-order elements	
Element	Sketch	Input
Chamfer left	⊢L×a	XXX⌴CHAL/L,a
Chamfer right	L×a	XXX⌴CHAR/L,a

FIG. 5.65 Catalog for second-order elements. (From Ref. 24.)

5.8.4 Example of a Workpiece Description

In this part we describe the creation of macros of form elements for rotary parts. In the first step, groups of fundamental form elements are devised. A fundamental form represents a section of a workpiece. It has a uniform surface over its length without interruptions. Its cross section may change according to a continuous function. Typical form elements are cylinders, cones, and cubes (Fig. 5.63). By adding

		Main element						
		CYL	CON	HEX	SQAR	BAR	POL	KEY
First-order element	THRED	■						
	CHAL	■	■	■	■	■		■
	CHAR	■	■	■	■	■		■
	RADAL	■	■					
	RADAR	■	■					

Combinations permitted

FIG. 5.66 Combinational possibilities of main and first-order elements. (From Ref. 24.)

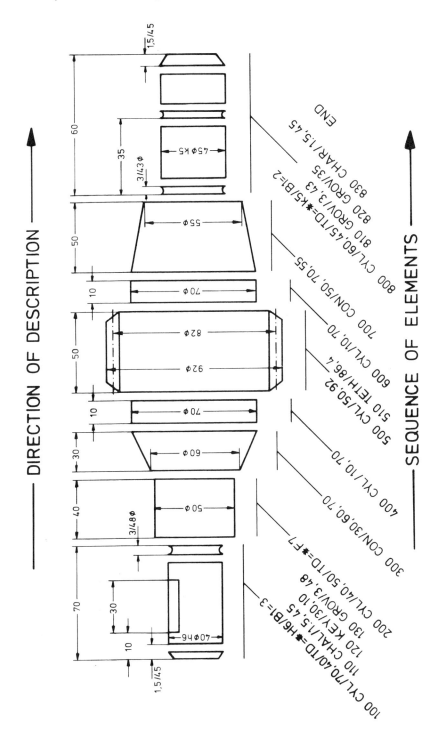

FIG. 5.67 Description of a gear. (From Ref. 24.)

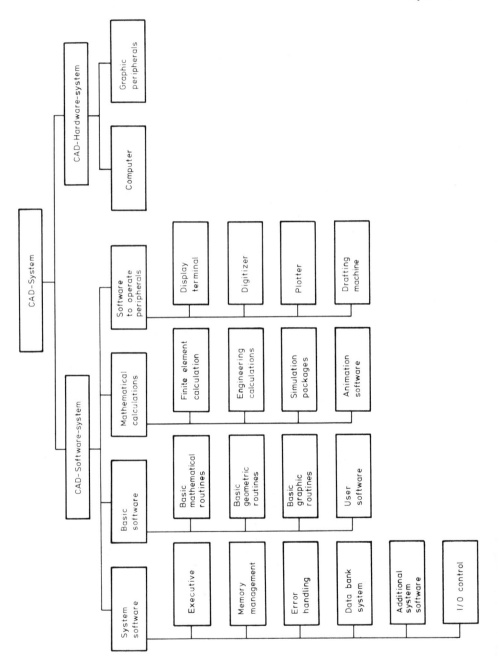

Computer-Aided Design

or deleting a first-order element, the form element may be changed (Fig. 5.64). A first-order element does not change the function of the main element; however, it may change its form over the entire length. These elements are normally included in the layout phase. They may already contain dimensions.

Second-order elements do not change the function of a main element. However, they are usually added during the detail phase (Fig. 5.65). An analysis of workpieces has shown that they are always located at either end of the main element.

Figures 5.63 through 5.65 contain a syntax for describing the form elements to the computer. It is also necessary to check for the permissible combination of principal and other form elements (Fig. 5.66). For example, a cylinder can be combined with any of the first-order elements. In this graph the following abbreviations are used:

THRED	Outer thread
CHAL/CHAR	Chamfer left/right
RADAL/RADAR	Outer radius left/right
CYL	Cylinder
HEX/SQAR	Hexagonal/square
POL	Polygon
CON	Cone
BAR	Barrel
KEY	Key

Figure 5.67 shows how a gear can be programmed with the help of form element macros. This picture is self-explanatory and does not need further discussion. Note also that since the direction of the workpiece description is from left to right, it is not necessary to define new coordinate systems for each element.

5.9 CAD SYSTEMS

In this chapter we give a short and general description of available CAD systems and their application. At the present time the CAD technology is advancing at a very rapid pace. For this reason a detailed discussion of systems currently available would be obsolete in a short period of time.

Figure 5.68 shows a summary of basic components of a CAD system. In principle, the installation is divided into hardware and software components. The majority of these components were covered adequately in previous sections of this chapter. Most of the components shown should be supplied by the vendor of the system. The user is concerned mainly

---

FIG. 5.68 Components of a CAD system. (From Ref. 25. Courtesy of Mark & Technik Verlag, Munich, West Germany.)

with application software, engineering calculations, simulation, and animation. Most of this software will be developed by the user; however, there may also be quite a few standard programs available for the user's applications. Typical user applications are as follows:

Electrical engineering
    Design of transformers
    Design of electromotors
    Layout of wiring harnesses
    Layout of wiring diagrams
    Design of integrated circuits
    Layout of circuit boards
    Investigation of dynamic behavior of circuits

Physics
    Dynamic investigation of physical phenomena
    Mapping of optical and thermodynamic patterns

Mechanical engineering
    Calculation of mechanical, thermodynamic, and fluid-dynamic properties
    Design aid for components and products
    Drafting of machine drawings
    Design of part families
    Styling of surfaces
    Automatic design of major engineering components, such as aircraft wings, fuselages, ship hulls, and gear boxes
    Piping and structural component layout
    Animation of machine functions
    Simulation of mechanical designs
    Plotting of engineering test results

With the acceptance of the CAD system by the designer, it will be used for almost any design function. Results from present systems show that the output of an efficient designer who uses a CAD system increases four- to fivefold over that with conventional design methods.

One of the major drawback of the CAD system is the high capital investment needed to purchase a configuration that contains adequate design functions. It is also very difficult to cost-justify it because there are so many intangible benefits which are not apparent to the average designer. Many conventional design procedures and methods have to be altered when a CAD system is introduced.

The versatility of a CAD system depends greatly on its ability to handle two- or three-dimensional graphics. Figure 5.69 shows typical applications for CAD systems in automobile development. They can be summarized as follows:

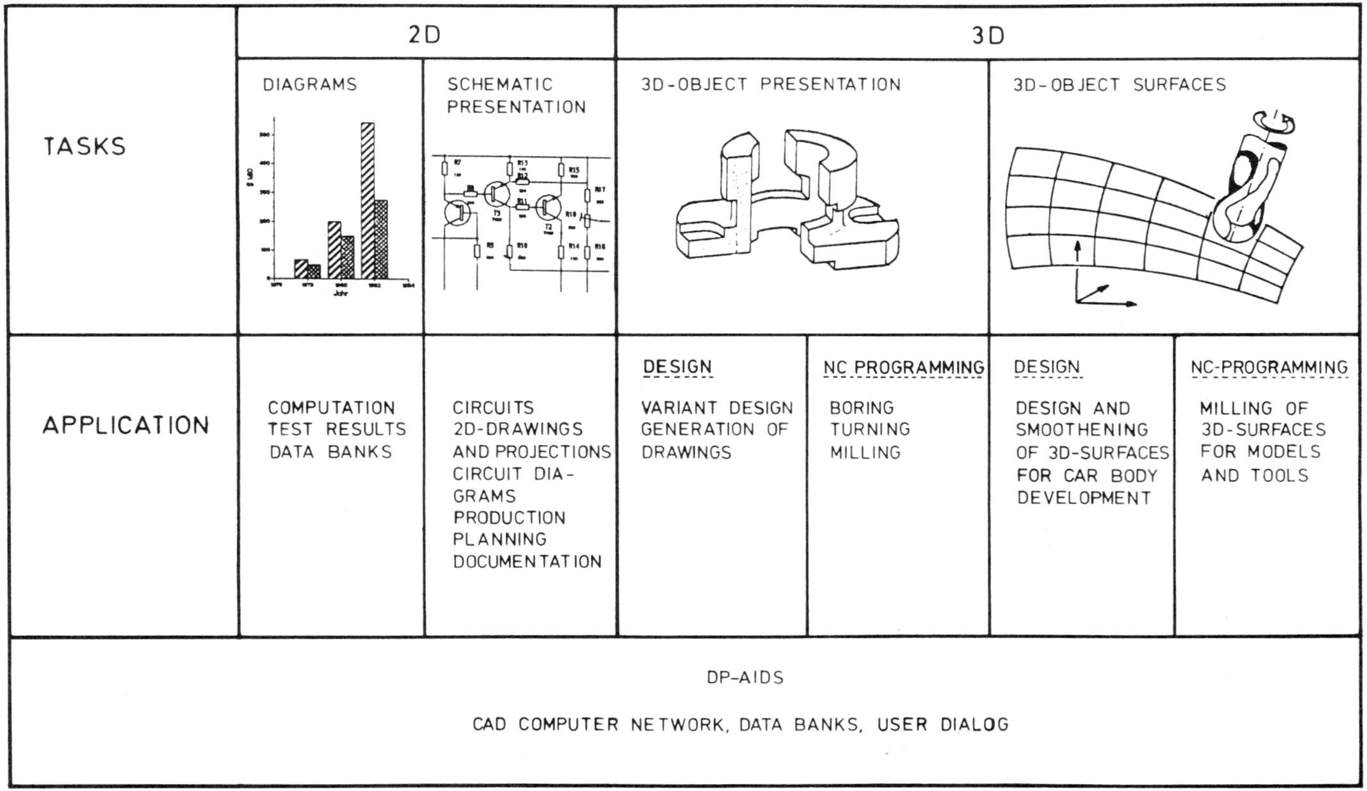

TASKS	2D		3D	
	DIAGRAMS	SCHEMATIC PRESENTATION	3D-OBJECT PRESENTATION	3D-OBJECT SURFACES
APPLICATION	COMPUTATION TEST RESULTS DATA BANKS	CIRCUITS 2D-DRAWINGS AND PROJECTIONS CIRCUIT DIAGRAMS PRODUCTION PLANNING DOCUMENTATION	DESIGN VARIANT DESIGN GENERATION OF DRAWINGS	DESIGN DESIGN AND SMOOTHENING OF 3D-SURFACES FOR CAR BODY DEVELOPMENT
			NC PROGRAMMING BORING TURNING MILLING	NC-PROGRAMMING MILLING OF 3D-SURFACES FOR MODELS AND TOOLS
	DP-AIDS CAD COMPUTER NETWORK, DATA BANKS, USER DIALOG			

FIG. 5.69 Typical applications of two- and three-dimensional CAD systems for automobile development. (From Ref. 26. Courtesy of Bayrische Motoren Werke, BMW, Munich, West Germany.)

Two-dimensional system
> Plotting of diagrams from computational results, simulations, and engineering tests
> Schematic presentation of ladder diagrams for circuits; design of integrated circuits
> Layout of manufacturing facilities
> Two-dimensional generation of engineering drawings
> Graphic balancing of machine loads

Three-dimensional system
> Interactive design of machine elements which are manufactured predominantly by chip-removal processes
> Pictorial simulation of machine element functions
> Simulation of chip-removal processes
> Design of workpiece variants
> Simulation of assembly systems (robots) and assembly alternatives
> Design and smoothening of three-dimensional surfaces for car bodies

For example, for the design of a car body, the following CAD modeling software is needed:

1. *Line model*: This system projects a visual picture of the surface skin of a car body (Fig. 5.70).
2. *Surface model*: This system allows analytical and graphical description of surfaces.
3. *Volume model*: A solid-oriented system makes possible the development of models from solid primitives. An object will be designed from these and altered until the final design fulfills the requirements.

Typical capabilities of two- and three-dimensional systems are shown in Fig. 5.71. The trend in industry is to enhance the capabilities of three-dimensional systems and to obtain flexible and universal software packages. Three-dimensional CAD systems are more complex than two-dimensional systems. For this reason, they are also very expensive. The complexity of a three-dimensional system versus a two-dimensional system will be demonstrated with an example. We are trying to design a square with the help of the two-dimensional system and a cube with the help of the three-dimensional system [26].

Two-dimensional system
> x coordinate of the reference point
> y coordinate of the reference point
> Length of the square
> Height of the square
> Angle about which the square is tilted against the x axis

Computer-Aided Design

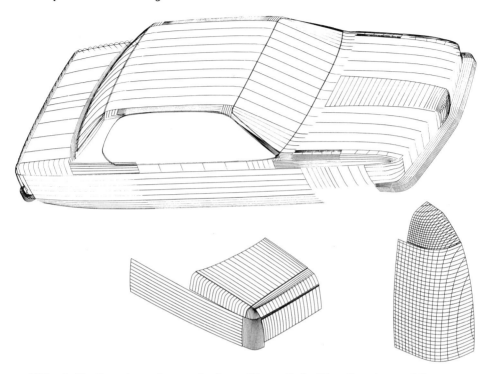

FIG. 5.70 Drawing of a car body. (From Ref. 26. Courtesy of Bayrische Motoren Werke, BMW, Munich, West Germany.)

In this case, five pieces of information are needed:

> Three-dimensional system
> x coordinate of the reference point
> y coordinate of the reference point
> z coordinate of the reference point
> Length of the cube
> Height of the cube
> Width of the cube
> Angle of the cube in regard to the x axis
> Angle of the cube in regard to the y axis
> Angle of the cube in regard to the z axis

In this case there are nine pieces of information needed to draw the object.

The three-dimensional design method requires considerably greater effort than the two-dimensional design. For this reason more qualified

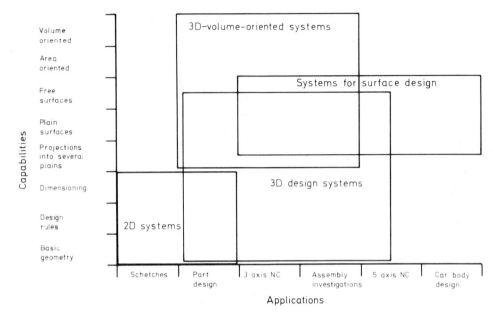

FIG. 5.71 Capabilities of two- and three-dimensional graphic systems. (From Ref. 26. Courtesy of Bayrische Motoren Werke, BMW, Munich West Germany.)

personnel are needed to operate such a system. We end this section with the presentation of a very complex three-dimensional drawing of an airbus (Fig. 5.72).

5.9.1 CAD Hardware

As mentioned earlier, expensive computer equipment usually is needed to do meaningful CAD work. Figure 5.73 shows a typical CAD installation that can be operated in a multiuser mode. In other words, several designers can share the computer and several peripherals.

A workplace may consist of an alphanumeric terminal, a graphic display screen, a tablet, and a hard copy device. At this level there may be personal computers provided to perform simple engineering calculation and to produce simple NC tapes.

At the next level there is a 16- or 32-bit host computer shared by several designers. The peripherals are large disk drives and a magnetic tape system. All the major graphic software packages will operate on the host. The system may also be used to perform engineering calculations or simple simulations. In addition, NC programming and the creation of the bill of materials can be done at this level.

Computer-Aided Design

At the highest level there is the central computer, which may be a business or scientific computer. Only a large engineering department can afford to have its own central computer. In most cases this computer will be a business computer shared with other departments. If such a computer is used by unrelated functions, engineering should be sure that they will get a high enough priority to perform its functions adequately and in a timely fashion. This computer will host the central engineering file, with interfaces to manufacturing and other functions of the business. Typical tasks done at this level are large engineering calculations and simulations. Complex NC programming may also be performed. Process planning, part family work, and bill of materials processing are usually very computer intensive and may have to be done at this level.

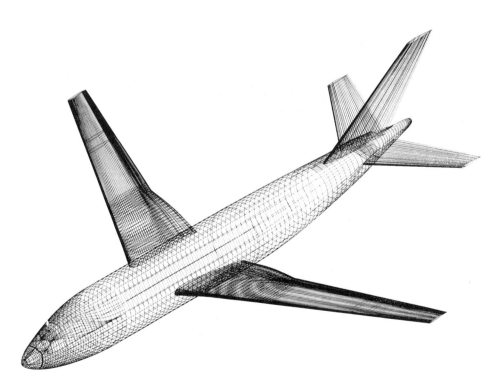

FIG. 5.72 Three-dimensional presentation of an airbus. (From Ref. 27. Courtesy of Messerschmitt-Boelkow-Blohm GMBH, Hamburg, West Germany.)

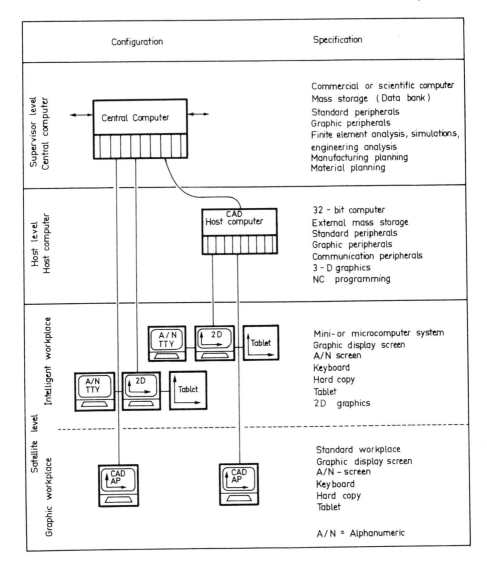

FIG. 5.73 Typical layout of a computer-aided design system. (From Ref. 26. Courtesy of Bayrische Motoren Werke, BMW, Munich, West Germany.)

5.9.2 Standardization of CAD Software

Presently available CAD systems were developed by many independent industries and laboratories. There were no standards used for software, hardware, and computer or user interfaces. For all practical purposes, presently it is not possible to operate software from one type of computer on another type. This problem has been recognized by many users and an effort is being made to standardize at least all interfaces to those functions of a CAD system which are needed for general operations. In the United States the ACM/SIGGRAPH Graphic Standard Planning Committee has made the first standardization attempt with the CORE system. In Germany, a DIN committee developed the Graphic Kernel System (GKS). This kernel system is independent of the programming language, the computer, and the application program. Thus it is only necessary to describe its function and structure. The system is presently being adopted in the United States.

The purpose of the KGS system is to supply to the user a standardized method for graphic programming. It contains the following:

Methods for the abstract description of peripherals
Definitions for basic graphic elements
Definitions for graphic operations
Methods for abstraction of graphic applications
A definition of a user model

Figure 5.74 shows the layer model of this approach. The GKS software system is located between the operating system and the user program. The design of the interface between the operating system and the GKS kernel is the task of the system supplier.

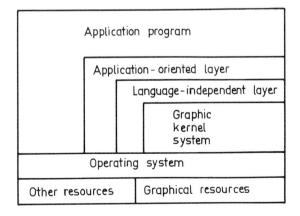

FIG. 5.74 The layer model of GKS. (From Ref. 28.)

The syntactical definitions needed to use the GKS system are left up to the user. All the user has to know is how to interface with the language-independent layer and how to initiate processing of a picture. When the programmer communicates with the computer and wants to use its graphic capabilities, he or she opens (deletes) a GKS process with an OPEN GKS (CLOSE GKS) instruction (Fig. 5.75). This initiates the creation of a workstation process, OPEN WORKSTATION, or changes its operating state or deletes it. A process is an object of limited life; it performs the interaction between the program and the computer. It can be identified and has a certain state and uses computer resources or communicates with other processes. A process may create other processes or delete them or share parts with other processes. Thus a process will be configured for a certain period of time and will use the computer resources to perform its assignment, such as the output of a picture from the display screen on a plotter.

For the manipulation of a partial picture on the screen, the user needs certain basic elements which can be combined under a name. Such partial pictures are called segments. They can be opened, used, and stored. It is also possible to save information in the GKS metafile

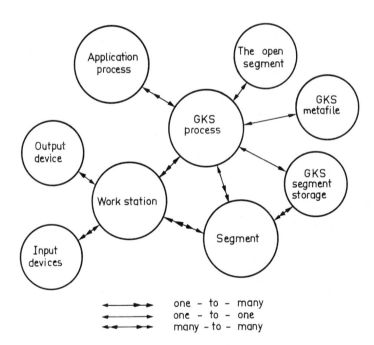

FIG. 5.75 Relationships between the processes of a GKS application. (From Ref. 28.)

Computer-Aided Design

for long-time storage. A variety of GKS functions are available to the user. They may consist of a simple instruction to a plotter or may be very complex, as when a picture is manipulated interactively. Typical functions include the following:

1. *Output primitives*: These allow graphic output of drawings or images on a graphic display or plotter.
2. *Input primitives*: These allow input of graphic information to a graphic display.
3. *Picture manipulation and transformation*: For design or drafting operations it is necessary to perform manipulations, transformations, and editing with pictures.
4. *Picture structure*: When pictures are created on a screen, they may have to be superimposed or changed or gradually built up. For this reason it is necessary to identify subpictures. It is the task of GKS to provide the tools for identifying such partial pictures.
5. *Peripheral allocation*: A graphic system usually has numerous peripherals. These must be allocated to the individual user or process. The GKS system has to schedule the peripherals. The operation of the peripherals is done by dedicated drivers.
6. *Storage*: The storage provision allows users to save graphic information temporarily over a long time frame.

5.10 DISCUSSION OF REALIZED CAD SYSTEMS

5.10.1 COMPAC System

The Computer Oriented Part Coding (COMPAC) system was developed in 1972 at the University of Berlin. This system can be considered one of the pioneering works in the use of computers for mechanical design. The system allows the description of raw parts and finished workpieces. The parts may consist of the following geometric elements:

Plain surfaces
Conical surfaces
Cylindrical surfaces
Spherical surfaces
Torus surfaces

It is possible to add other elements if desired.

Form Elements

In the COMPAC system it is assumed that any workpiece can be described with the help of three-dimensional primitives by addition or subtraction operations. The following primitives are included in the program library:

Cuboid
Sphere
Cylinder
Cone
Pyramid
Hexagonal prism
Triagonal prism
Rectangular prism
Torus

The instruction to describe such an element has the following syntax: symbol = name of the form element/principal axis, dimensions. The form element cuboid (PRIREC), for example, is uniquely specified by the length of an edge (Fig. 5.76). Figure 5.77 (upper part) shows the same cuboid; however, in this case a coordinate transformation has been made. The principal axes are designated as shown in the figure. If one axis has to be changed, it can be done in the following manner:

CENTER   Determines the origin of the coordinate system
XAX   x axis is the principal axis
YAX   y axis is the principal axis   Modifier for an axis
ZAX   z axis is the principal axis

The parameters k_x, k_y, and k_z in Fig. 7.77 represent the lengths of the corresponding edges. They are entered as absolute values. The coordinate assigned to the main axis has to be marked with a plus or minus sign, depending on the orientation of the workpiece. The

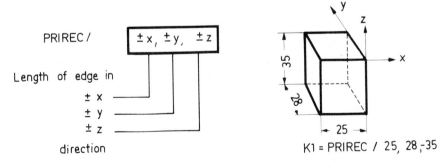

FIG. 5.76 Syntax for a cuboid. (From Ref. 29. Courtesy of Carl Hanser Verlag, Munich, West Germany.)

Computer-Aided Design

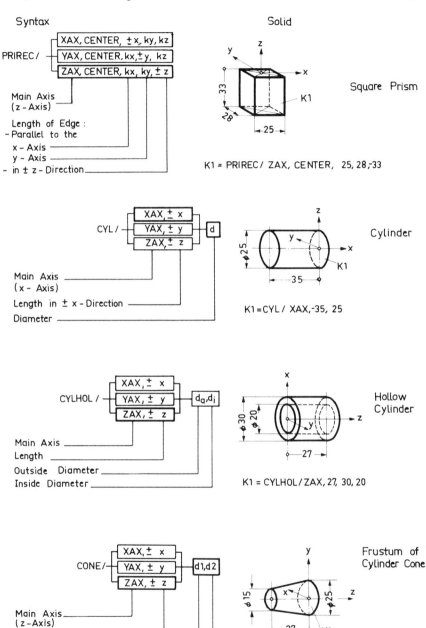

FIG. 5.77 Syntax for describing different solids. (From Ref. 29. Courtesy of Carl Hanser Verlag, Munich, West Germany.)

parameter field of the instruction shows the programming alternatives that are permitted.

In the case of a cylinder, it is of advantage to locate its centerline in the direction of one of the axes of the coordinate system. Furthermore, one end of the cylinder should be placed on the origin. Thus the length of the cylinder can be stated in terms of ± a dimension along the principal coordinate axis. The syntax of some more solids is described in Fig. 5.77.

Technical standard primitives are usually those which occur often in workpieces. However, it may not be possible to describe all of these with the help of simple rules. Typical elements are (1) key, (2) tooth profile, (3) thread, (4) hollow cylinder, and (5) dead-end hole.

Modification of Form Elements

When form elements are described, it is often necessary to change their shape. In Fig. 5.41 it was shown that a wedge had to be taken off form element K1 to generate form element K2. This was possible by moving the coordinate system into a new reference position and by tilting it at an angle of about -15°. The instruction for separating the wedge from K1 was:

$$\text{OMIT}, \begin{Bmatrix} X & \text{LARGE} \\ Y & \\ Z & \text{SMALL} \end{Bmatrix}, e$$

Here LARGE means, for example, in the positive x direction and SMALL means in the negative x direction. The letter e defines a surface specified previously. Another modification is needed for dimensional tolerances. It is placed after the dimension and indicated by TOL.

K1 = PRIREC/30, −40, TOL, .3, .1

Sequentially shown tolerances are interpreted as upper and lower limits. If coded tolerances are used, they are indicated as follows:

K2 = CYL/XAX, −10, 20, TOL, H7

Surface finishes can be described in a similar manner.

K1 = PRIREC/10, −40, ROUGH

K2 = CYL/XAX, −10, 20, FIN

A coordinate transformation is also possible, similar to the APT language. One method is the implicit transformation depicted in Figs. 5.78 and

Computer-Aided Design

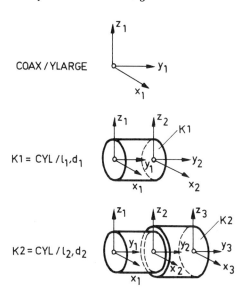

FIG. 5.78 Principle of an implicit coordinate transformation. (From Ref. 29. Courtesy of Carl Hanser Verlag, Munich, West Germany.)

5.67. It is of advantage when a workpiece is assembled from different form elements.

Various instructions to perform an explicit transformation are shown below.

Label, TRANSF/WX, WY, WZ, XYROT, walpha

Label, TRANSF/WX, WY, WZ, YZROT, walpha

Label, TRANSF/WX, WY, WZ, ZXROT, walpha

In this instruction WX, WY, and WZ are translation parameters and walpha is the angle of rotation. XY, YZ, and ZX indicate the plane about which rotation takes place.

The label is used similarly to a symbol. It identifies a defined coordinate system, thus always making it possible in a part program to refer back to a coordinate system used previously. The use of this instruction was shown in Fig. 5.41.

In case the dimensions of a new form element to be added to the workpiece do not refer to the origin of the coordinate system defined previously, the modifier REF may be used:

Transformation constant, REF, Label, . . .

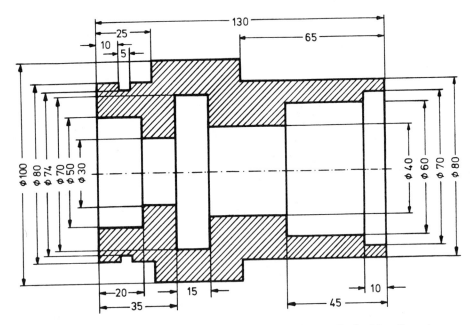

FIG. 5.79 Rotational part to be described. (From Ref. 30. Courtesy of Carl Hanser Verlag, Munich, West Germany.)

The instruction to combine several primitives to a part is:

$$\text{SOLID/Form element 1} \begin{Bmatrix} \text{PLUS} \\ \\ \text{MINUS} \end{Bmatrix} \text{Form element 2}$$

Using the COMPAC System

There are two examples that show the use of the COMPAC language. One is shown in Fig. 5.41. In this example there is also a mirror instruction used, which was not discussed. The second shows a program for a rotational part (Fig. 5.79). Readers should be able to follow this program on their own.

5.10.2 Interactive Graphics for Modelling Solids (GRIN)

This system was developed by IBM and allows three-dimensional modeling of solids. It has a data structure and programs to produce objects interactively from a set of primitives.

Viewing Three-Dimensional Models on a Three-Dimensional Display

There are two methods available to produce three-dimensional objects. One uses the primitives cuboid, cylinder, cone, and hemisphere, and the other allows the generation of solids from translated and rotated polygons (Figs. 5.80 and 5.81).

Table 5.5 contains an example of different commands available to the user. They may be grouped together as follows:

1. Input commands for primitives
2. Commands to produce points and line segments
3. Editing commands to work with objects on the graphic display screen
4. Display command to work with different views and presentations of the object
5. Miscellaneous commands

A method for displaying objects in different modes is shown in Fig. 5.82. The two cylinders are merged together with a merge command. The wire frame model (a) shows the cylinders merged together as solids. In view (b) the hidden lines are removed. This view could also represent two pipes merged together. In view (c) the facet lines are removed, and in view (d) the hidden intersection line between the two cylinders is shown.

Figure 5.83 shows how an object can be observed from different viewpoints. The operator is able to place himself into different viewing positions by specifying to the computer an elevation and azimuth angle. The view can be projected on a plane which goes through the viewer's position perpendicular to the viewing angle. The viewing commands are listed under display commands in Table 5.5. They are easy to use. There is also the possibility of scaling the three-dimensional image. Orthographic projection is used because it gives a clear and uncluttered picture of the object.

Input of Volume Primitives

Complex solids are generated by placing many solid primitives together. The primitives are positioned on the screen by simple commands (Table 5.5). For example, to draw a hemisphere, only two entry points are needed: its center and the distance of the radius. The computer then draws the entire primitive. Similarly, a cuboid can be specified by the direction of a principal axis and three distances.

For a swept volume, first a two-dimensional polygon is entered (Fig. 5.81a). The arc is defined by two end points and a radius. To produce a volume, the two-dimensional figure is either moved along a vector or swept along the arc of a circle (Fig. 5.81b and c).

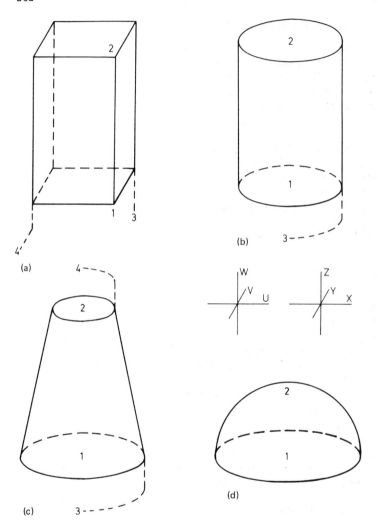

FIG. 5.80 Volume primitives and their entry points: (a) cuboid; (b) cylinder; (c) cone; (d) hemisphere. (From Ref. 31. Courtesy of International Business Machine Corporation, White Plains, New York).

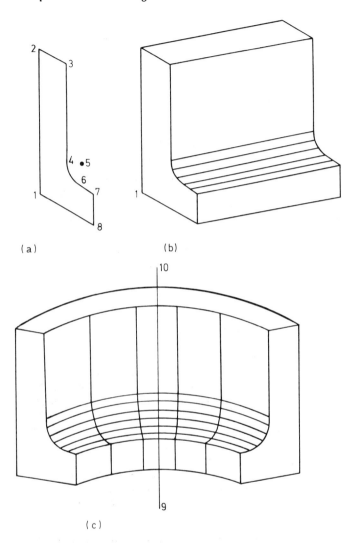

FIG. 5.81 Translated and rotated polygons and their entry points: (a) polygon; (b) polygon translated; (c) polygon rotated. (From Ref. 31. Courtesy of International Business Machine Corporation, White Plains, New York.)

TABLE 5.5 Some GRIN Commands

Input commands	
CB	Enter a cuboid
CO	Enter a cone
CY	Enter a cylinder
HE	Enter a hemisphere
PT	Enter a translated polygon
PR	Enter a rotated polygon
Points and line segments (used within other commands)	
AB x y z	Enter absolute coordinates of a point
CU	Display cursor to select an existing point
a<n>	Move n units in the a direction from current point, where a is an axis (a = X, Y, Z, U, V, or W)
EN	Accept coordinates of current point
RJ	Reject last entered point or arc
ARC	Enter a circular arc in a polygon
Editing commands	
EO	Erase object
MO	Move object
ROX	Rotate object about an axis
ROP	Rotate object about a point
ROS	Rotate and scale object about a point
PO m	Reset polarity mode to m = H for hole, S for solid. For m = R, polarity of selected object is reversed, without changing mode.
F<n>	Set the number of facets for new primitives. If defaulted, permits refaceting of a selected primitive.
RC a<->n	Alter rotated coordinated system. Rotates n degrees (counter)clockwise about the a axis (X, Y, Z, U, V, or W)
Display commands	
RL n, RR n, RU n, RD n	Rotate eyepoint left, right, up, or down n degrees
RI	Rotate to standard isometric angle
RX <->a	Rotate to axis a (X, Y, Z, U, V, or W)
RA n	Rotate to absolute azimuth of n degrees
RE n	Rotate to absolute elevation of n degrees
T	Translate so indicated point is centered on screen
TC	Automatically center picture and scale to fit screen
TL n, TR n, TU n, TD n	Translate eyepoint left, right, up, or down n units

TABLE 5.5 (Continued)

Display commands (Continued)	
DS<n>	Scale smaller or larger by n. (Default is 2.)
DL<n>	
DW	Set wire frame display mode
DH <m>	Set hidden line display mode. For m = d, shows hidden lines dashed; for m = f, shows facet lines of curved surfaces; for m = df, shows both.
DR	Redraw display
DN <m>	Display numeric values of coordinates of a point, or a variety of other data.
Miscellaneous	
CC	Cancel current command
IM	Merge new primitives
TX	Enter text on picture
MATHC	Enter desk calculator mode
FETCH	Fetch a named model from storage
FILE n	File current model under special name

Source: Ref. 31.

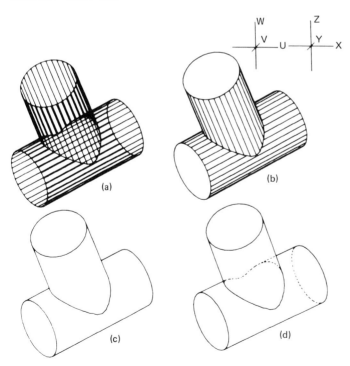

FIG. 5.82 Drawing of two intersecting tubular workpieces. (From Ref. 31. Courtesy of International Business Machine Corporation, White Plains, New York.)

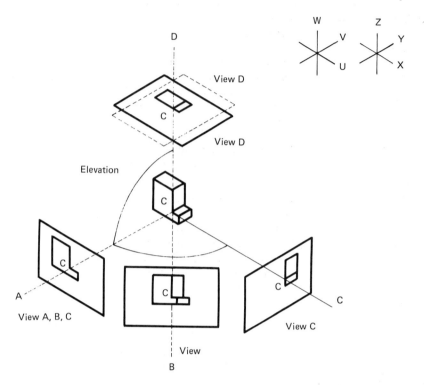

FIG. 5.83 Different views of a workpiece. (From Ref. 31. Courtesy of International Business Machine Corporation, White Plains, New York.)

The number of facets for a rotated polygon can be defined. Facets can also be deleted if desired.

There are two possibilities for using a coordinate system. Normally, the engineer will do design work with the help of a world coordinate system. Often, however, the engineer may want the assistance of a rotational coordinate system, which is defined as the rotation of the object about an axis of an existing coordinate system.

Generation of a Workpiece

As an example, the workpiece shown in Fig. 5.84 will be drawn. To follow the gradual buildup of the workpiece, the reader may first want to study the commands in Table 5.5. An object can be defined as a solid or a negative volume. The latter is needed to remove solid material from the object.

We take advantage of the fact that the part is symmetrical about two planes except for the appendate on the top. For this reason the design starts with a one-quarter section of the part. A four-sided

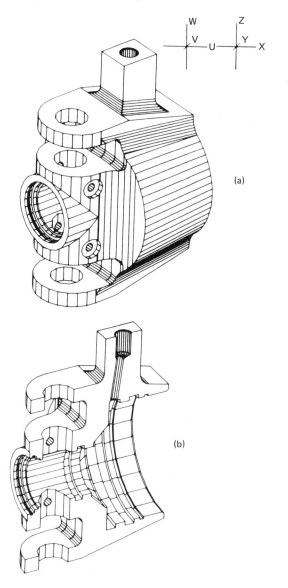

FIG. 5.84 Drawing of a complex workpiece by the facet method. (From Ref. 31. Courtesy of International Business Machine Corporation, White Plains, New York.)

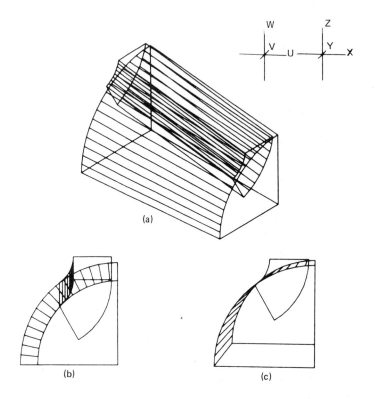

FIG. 5.85 Phase one. (From Ref. 31. Courtesy of International Business Machine Corporation, White Plains, New York.)

polygon is drawn and rotated about a 90° angle to form a truncated cone (Fig. 5.85). A fillet is produced which tangentially blends into the planar surface parallel to the z axis. The fillet is produced by a four-sided polygon attached to the inside of the truncated cone. It is rotated and translated up into the vertical plane perpendicular to the z axis (Fig. 5.85b and c).

Figure 5.86 show the top view of the completed section of Fig. 5.85. Two negative translated polygons are used to trim off the ends. On the left side of the part a fillet is generated. Figure 5.87 shows the addition of a quarter-cylinder with the help of a polygon and a sloped section on the top. The next step is to remove material from the undercut with the help of a negative rotated polygon (Fig. 5.88). The main body is now formed by two mirror operations. In addition, the protruding stud in the center of the workpiece face is generated (Fig. 5.89). The auxiliary holes and the main cavity are generated in a similar fashion (Fig. 5.90). This completes the part.

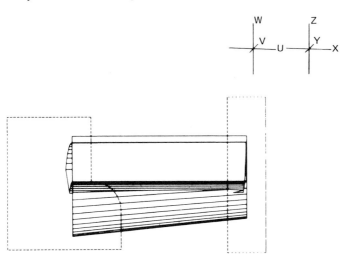

FIG. 5.86 Phase two. (From Ref. 31. Courtesy of International Business Machine Corporation, White Plains, New York.)

Design Feasability Study

To increase the versatility of the system, the programmer may include FORTRAN assign statements in the graphic program. The programmer can use symbolic names and assign a value to these. This feature permits parameterization, branching, and iteration.

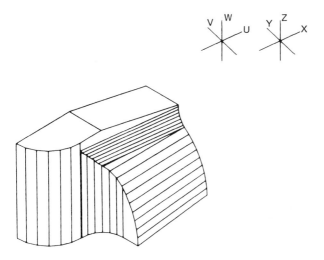

FIG. 5.87 Phase three. (From Ref. 31. Courtesy of International Business Machine Corporation, White Plains, New York.)

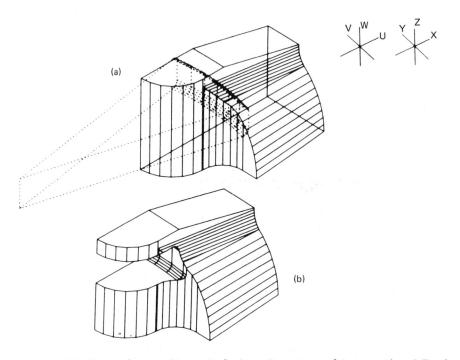

FIG. 5.88 Phase four. (From Ref. 31. Courtesy of International Business Machine Corporation, White Plains, New York.)

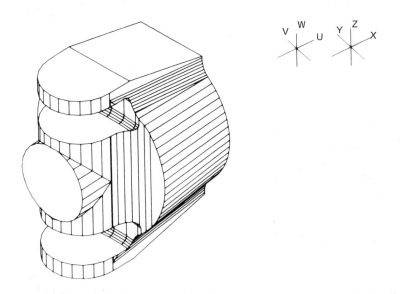

FIG. 5.89 Phase five. (From Ref. 31. Courtesy of International Business Machine Corporation, White Plains, New York.)

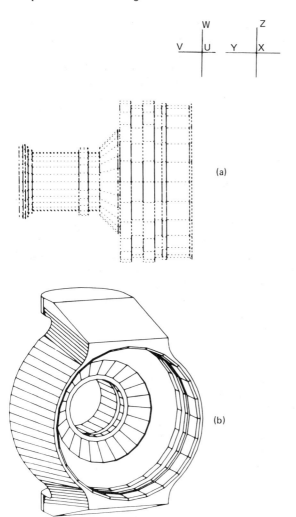

FIG. 5.90 Phase six. (From Ref. 31. Courtesy of International Business Machine Corporation, White Plains.)

New graphic functions such as the intersection of two lines can be interactively entered without the need for a recompilation. This facility will be discussed with the aid of an example.

Figure 5.91 shows studies to investigate the design of a scraper blade of a bulldozer. The user can simulate the lift, tilt, and sway motions of the blade and check the function of the linkage and possible interferences with the drive mechanism and other components of the machine.

5.10.3 Part Assembly and Description Language (PADL)

Early work on the PADL system dates back to about 1972. The system was conceived for the design of mechanical workpieces. A modeling scheme was to be developed to describe the geometry of parts and assemblies. The question of tolerances and surface finishes was also to be included. The designer was to be provided with algorithms and software to describe mechanical designs and to produce production drawings. A future goal was the production of manufacturing plans and command data for machine tools. The PADL system was not conceived for the solution of specific design problems. It simply renders general design aids to automate the design function.

PADL 1

This version of PADL allows the description of objects with the aid of primitive solids [32]. The primitives are the block and the cylinder. The buildup of complex workpieces is possible with the help of combinational operations of the set theory (Fig. 5.92). In this figure several different methods are shown to describe a simple L-shaped workpiece. The operations used are .INT. (intersection), .UN. (union), and .DIF. (difference). The drawing for this workpiece is shown in Fig. 5.93 and the program to describe it to the computer is shown below.

PADL Definition	Explanation
εBASE = B(4.0, H1, 2.0)	Assign the name BASE to a solid block of size 4.0 in the x direction, H1 in y, and 2.0 in z. The left-back-bottom corner of the block is at the origin of the coordinate system.
εLIP = B(1.0, 2.0, 2.0)	Use another block (which happens to overlap BASE) to make the left portion of the part.

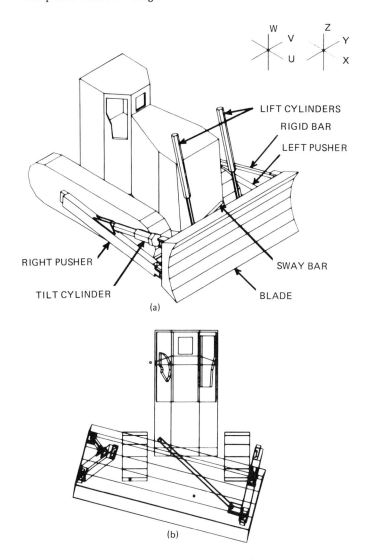

FIG. 5.91 Design of a scraper plate for bulldozer. (From Ref. 31. Courtesy of International Business Machine Corporation, White Plains, New York.)

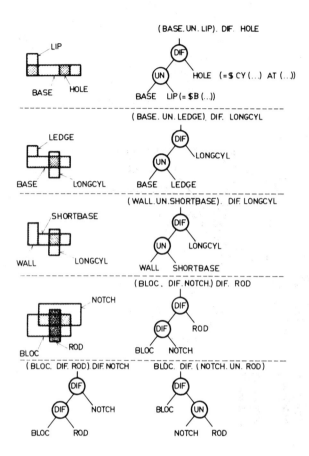

FIG. 5.92 Alternative constructions for a part. (From Ref. 33. Courtesy of Rochester University, Rochester, New York.)

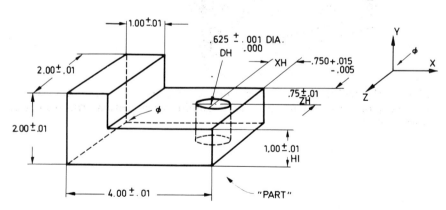

FIG. 5.93 Example of a PADL 1 program for a workpiece. (From Ref. 34. Courtesy of Rochester University, Rochester, New York.)

Computer-Aided Design

PADL Definition	Explanation
εHOLE = CY(DH, H1) AT (4.0-XH, 0.0, ZH)	HOLE is a vertical solid cylinder of diameter DH and length H1, whose bottom face is on the Hzy plane.
εPART = (εBASE.UN.εLIP) .DIF.εHOLE	Combine the solids into a single entity named PART.
DH = .625 : PM (.001, .000)	Assign a value and tolerance to the diameter of the hole.
XH = .750 : PM(.015, .005)	Locate and tolerance the hole in the x direction.
ZH = .75	Locate the hole in the z direction.
H1 = 1.00	Specify the height of BASE and HOLE.
DEFTOL( ) = PM(.01)	Specify the default tolerance.

The system allows the investigation of spatial interference when tool motions are planned, thus avoiding possible collision. The material removal operation can be modeled with the help of the .DIF. operator. A comprehensive program for a more complex part is shown in Fig. 5.94.

PADL 2

The core of the PADL 2 system is a constructive solid geometry (CSG) module together with a boundary representation (B-Reps.) module (Fig. 5.95). The CSG module permits the design of workpieces with the help of primitive solids. With the B-Reps. module, solids are linked together from faces and edges. The boundary presentation may be obtained from the CSG module with the help of a boundary evaluation (BEVAL) system. The PADL 2 system pays particular attention to dimensioning and tolerancing. The primitive solids that can be presented by CSG are the following:

Rectangular blocks
Right corner wedge
Circular cylinder
Right corner circular cone
Sphere
Torus (available to limited applications)

There are 10 classes of nodes available for the user to select from to represent the solids (Table 5.6).

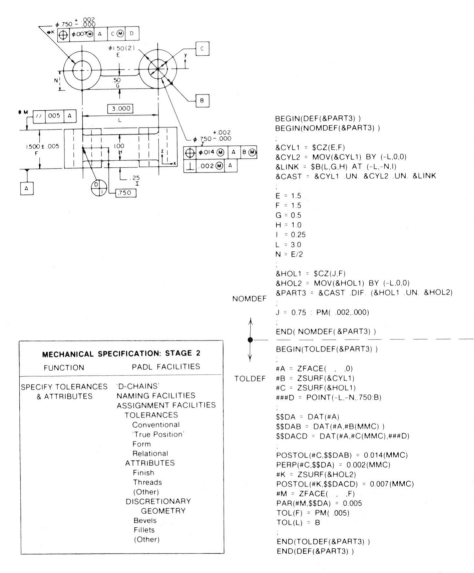

FIG. 5.94 Example of a PADL 1 program for a workpiece. (From Ref. 34. Courtesy of Rochester University, Rochester, New York.)

Computer-Aided Design

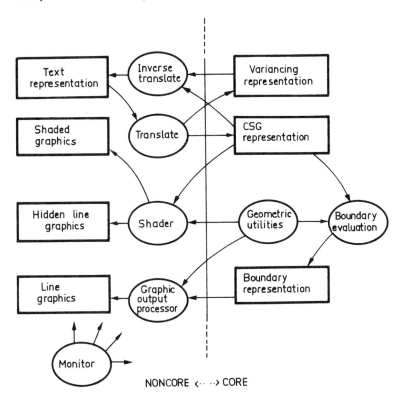

FIG. 5.95 PADL-2 core and noncore representations (□) and processes (○). (From Ref. 35. Courtesy of Rochester University, Rochester, New York.)

A workpiece can be presented by a CSG graph and subgraphs containing the nodes of Table 5.6. The basic unit is the generic solid which is parameterized with regard to its configuration and location. The relations expressed by dimensions and tolerances are described by variancing presentation (Fig. 5.95). Geometric relations are described as relations between real numbers using static expressions. The configuration and the location of solids can be described by real-number-valued expressions. Coordinate systems needed for relational positioning are related to rigid transformations as translations and rotations.

Figure 5.96a shows the design of a hexagonal solid with the help of PADL 2. The program may be conceived as follows:

TABLE 5.6 CSG Node Classes[a]

Semantics	Type	Remarks
Solid	Invocation	The instantiation of a generic or primitive solid with parameters of location and configuration.
	MovedSolid	The application of a rigid motion to a solid.
	SetOp	The combination of two argument solids by assembly or one of three (regularized) set operations: union, intersection, and difference.
Coordinate system	CoordSys	A transformed copy of LAB, the "primitive" reference coordinate system.
Real number	Variable	A string representing a real. If it has no user name, it acts like a constant; otherwise, it acts like a variable.
	RealOp	The application of a monadic or diadic real operation.

Computer-Aided Design

Name value pair list	ParmList	For configuration and location parameters, names of CSG objects in the invoked generic are paired with values of the same type. Any number of configuration pairs is allowed. A single coordinate system-valued pair gives location.
	MoveBy	Six names designate translation and rotation relative to formal X, Y, Z directions. Values are REAL NUMBERs. When the formal X, Y, Z are bound to a coordinate system, the MoveBy specifies an absolute motion in LAB. Any number of pairs is allowed; resulting motions are composed in order.
	Properties	Attachable to any node is a property list which may contain attributes such as color, a text note, or a material code.
Generic header	Generic header	Identifies the name of a parameterizable object and points to its root node.

[a]Nodes are linked in a graph data structure representing solid, coordinate system, and real-number expressions.
(*Source*: Ref. 35. Courtesy of Rochester University, Rochester, New York.)

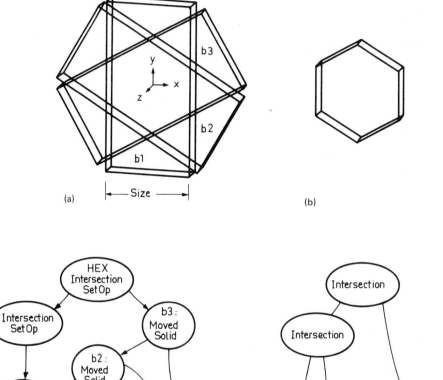

FIG. 5.96 Design of a hexagonal workpiece with the PADL-2 system. (From Ref. 35. Courtesy of Rochester University, Rochester, New York).

Computer-Aided Design

```
          GENERIC HEX(AHEX);
(1)   B0 = BLO(x = (1), y = (2), z = (.25));
(2)   M1 = (MOVX = (-.5, MOVY = (-1));
(3)   B1 = MOVE B0 BY (M1);
(4)   M2 = (ROTZ = ((2 * 3.1416)/3));
(5)   B2 = MOVE B1 BY (M2);
(6)   B3 = MOVE B2 by (M2);
(7)   AHEX = ((B1 INT B2) INT B3);
```

The first command defines a block B with the parameters 1, 2, .25. One corner of the block is designated as the origin by default. Command (2) defines the origin to be at location -.5, -1. Command (3) defines a new block B1 and moves the center of the block to the location specified previously. Command (4) specifies the rotational increment needed for the block to form the hexagonal shape. With command (5) a new block B2 is formed and rotated by 60° about the x axis. Similarly, command (6) defines block B3. Command (7) forms the final object by combining (B1 INT B2) INT B3 (Fig. 5.96b).

The corresponding CSG graph for this solid is shown in Fig. 5.96c. This graph is an unambiguous representation of the solid but does not allow the geometric computations for the display of the part. To be able to draw the hexagonal solid, the user has to issue the command DRAW AHEX. This generates the CSG tree needed to envoke the draw instruction (Fig. 5.96d).

5.10.4 Technical Information Processing System (TIPS)

With this Japanese approach an attempt is made to design an integrated CAD/CAM system. The development can be divided into three parts:

The design of a workpiece description system
The design of complex machine description system
The integration of CAD with CAM

Workpiece Description System

The system was designed for the description of complex workpieces (Fig. 5.97). At the top of this figure the generalized instruction format is shown. The different descriptive methods are:

1. *Canonical pattern*: These are basic geometric elements: the point, the line, the circle, the rectangle, the thread, and so on. There is also the possibility of describing a miller contour.
2. *Contour pattern*: These are more complex contours and three-dimensional primitives. They include revolve and rotate provisions, the plane, complex curves, and solids. There is also the possibility of including specially designed contour patterns.

Symbol / □, Pattern,   □ / □, Constant (Coordinate), Domain, Connection

Free   Mode   Direction   x1, y1, z1 x_1, x_u Label
        P: Positive   I: Inside   x2, y2, z2 y_1, y_u for the
        Q: Negative   O: Outside   x3, y3, z3 z_1, z_u connection
        D: Dummy   R: Right   of product
        T: Transe   L: Left   set
                             U: Upper
                             D: Down
                             E: Equal

                          Routine name
                          for conversion

(1) Canonical
    pattern   LINE,   / □, ($x_1, y_1, z_1, x_2, y_2, z_2$)
    (fixed)   CIRCLE,   / □, r (x_1, y_1, z_1)
                 RECTANGL,/ □, α, l_1, l_2 (x1, y1, z1)
                 ROUND,   / □, r ($x_1, y_1, z_1, x_2, y_2, z_2, x_3, y_3, z_3$)
                 CONTORnn

                 POINT   (x_1, y_1, z_1)
                 HOLE   r (x_1, y_1, z_1)
                 THREAD   r, p, (x_1, y_1, z_1)

                 MILLER   Symbol ($x_1, y_1, z_1, x_2, y_2, z_2$)

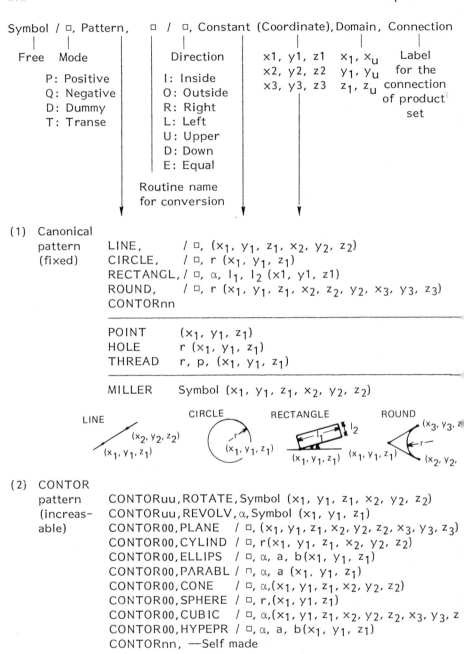

(2) CONTOR
    pattern   CONTORuu, ROTATE, Symbol ($x_1, y_1, z_1, x_2, y_2, z_2$)
    (increas-   CONTORuu, REVOLV, α, Symbol (x_1, y_1, z_1)
    able)   CONTOR00, PLANE   / □, ($x_1, y_1, z_1, x_2, y_2, z_2, x_3, y_3, z_3$)
                 CONTOR00, CYLIND / □, r($x_1, y_1, z_1, x_2, y_2, z_2$)
                 CONTOR00, ELLIPS / □, α, a, b(x_1, y_1, z_1)
                 CONTOR00, PARABL / □, α, a (x_1, y_1, z_1)
                 CONTOR00, CONE   / □, α, ($x_1, y_1, z_1, x_2, y_2, z_2$)
                 CONTOR00, SPHERE / □, r, (x_1, y_1, z_1)
                 CONTOR00, CUBIC   / □, α, ($x_1, y_1, z_1, x_2, y_2, z_2, x_3, y_3, z$)
                 CONTOR00, HYPEPR / □, α, a, b(x_1, y_1, z_1)
                 CONTORnn, —Self made

FIG. 5.97 Description format of the TIPS system. (Courtesy of Society of Manufacturing Engineering, Dearborn, Michigan.)

Computer-Aided Design 273

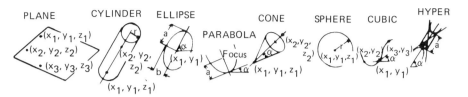

(3) Convertible
pattern POINT, APT, INTOF, L1, L2
(increas- POINT, APT, C1, ATANGL, α
able) POINT, APT, P1, RIGHT, TANTO, C1
 LINE, APT/□, P1, RIGHT, TANTO, C1
 LINE, APT/□, RIGHT, TANTO, C1, RIGHT, TANTO, C2
 LINE, APT/□, P1, ATANGL, α
 CIRCLE, APT/□, CENTER, P1, TANTO, L1
 CIRCLE, APT/□, CENTER, P1, P2
 CIRCLE, LOFTIG/□, P1, P2, P3
 ROUND, CV/□, r, L1/□, L2/□

FIG. 5.97 (Continued)

3. *Convertible pattern*: These are patterns from the APT language. They may facilitate the connection of the design function with manufacturing. An instruction set for points, lines, and circles is given. This may be extended if the designer needs additional functions.
4. *FORTRAN subroutines*: Complex parts often cannot be designed with the generalized instruction set supplied. For this purpose it may be necessary to include a provision for describing contours with the help of mathematical formulas. In the TIPS system the designer has the option to add FORTRAN subroutines to the part description program.

Figure 5.98 shows how a turbine blade is designed with the TIPS language. In the upper part of the program the base of the blade is described from canonical patterns, and in the lower part the blade contour is described with the help of a FORTRAN program.

Complex Machine Description System

For the design of complex parts a processor bank is provided in which FORTRAN subroutines are located. These subroutines, called Processor (P), describe, for example, individual components of the object or they may even generate process planning and manufacturing information. Figure 5.99 shows a processor. The interconnection of

274   Chapter 5

```
*10 TURBINE BUCKET(0.,0.,0.)0.,64.,0.,174.,0.,84.;
R1/P,RECTANGL,/l,0.,10.,70.,(27.,2.);
R2/P,RECTANGL,/l,0.,50.,5.,(7.,67.);
T1/P,LINE,/D,(37.,37.,62.,22.)32.,22.;
T2/P,LINE,/D,(32.,17.,57.,2.)32.,2.;
L1/D,LINE,/E,(37.,0.,37.,10.);
L2/D,LINE,/E,(0.,67.,10.,67.);
L3/D,LINE,/E,(0.,22.,10.,22.);
C1/P,ROUND,CV/I,10.,L1/R,L2/D;
C2/P,ROUND,CV/I,8.,L1/R,T1/U;
C3/P,ROUND,CV/I,4.,L3/D,L1/R;
C4/P,ROUND,CV/I,5.,T2/U,L1/R;
C5/Q,CIRCLE,/0,5.,(47.,27.)47.,80.,22.,32.;
C6/Q,CIRCLE,/0,4.,(45.,6.)45.,80.,0.,12.;
M1/P,MIRROR,,T1,T2,(32.,0.,32.,10.);
M2/P,MIRROR,,C1,C2,C3,C4,C5,C6,(32.,0.,,32.,10.);
R1/P,RECTANGL,/l,0,50.,50.(7.,2.);
B1/P,CONTOR10(),,72.,/72.,:C0;
B2/P,CONTOR20(),,72.,/72.,:C0;
*     EOF
      SUBROUTINE CONTOR(IA,T1LIST,N,M,J,F,XX,YY,ZZ)
      X=XX
      Y=174.0-YY
      Z=ZZ
      IF(Y-76.) 10,10,20
10    IF(Y-72)11,15,15
11    F=-1
      GO TO 100
15    W=4.-SQRT(16.-(Y-72.)**2)
      X0=52+W
      Z0=6-W
      ASINA=1
      ACOSA=0
```

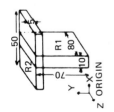 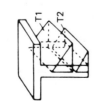 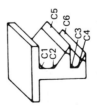

Computer-Aided Design

```
      E=72+2*W
      P=0.25
      B=30/E
      T=(24+2*W)/E
      GO TO 30
   20 X0=303./4.-5./16.*Y
      Z0=5./4.+1./16.*Y
      X1=529./12.+5./48.*Y
      Z1=331./4.-1./16.*Y
      E=SORT((X1-x0)**2+(Z1-Z0))**2)
      ASINA=(Z1-Z0)/E
      ACOSA=(X1-X0)/E
      P=0.25
      B=(335./8.-5./32.*Y)/E
      T=(115./4.-1./16.*Y)/E
   30 D=(X-X0)*ACOSA+(Z-Z0)*ASINA
      Z=-(X-X0)*ASINA+(Z-Z0)*ACOSA
      X=D
      IF(X/E.LT.0) X=ABS(X)+E
      IF(X/E-P)40,40,50
   40 CC=B/P**2*(2*P*X/E-(X/E)**2)
      GO TO 60
   50 CC=B/(1-P)**2*((1-2*P)+2*P*X/E-(X/E)**2)
   60 TT=T/0.2*(0.2969*SORT(X/E)-0.126*X/E-0.3516*(X/E)**2
      /+0.2843*(X/E)**3-0.1015*(X/E)**4)
      IF(IA-10)80,70,80
   70 F=-Z/E+TT+CC
      GO TO 100
   80 F=Z/E+TT-CC
  100 CONTINUE
      RETURN
      END
```

FIG. 5.98 TIPS-1 shape description example for an object including free surface. (From Ref. 37. Courtesy of Society of Manufacturing Engineers, Dearborn, Michigan.)

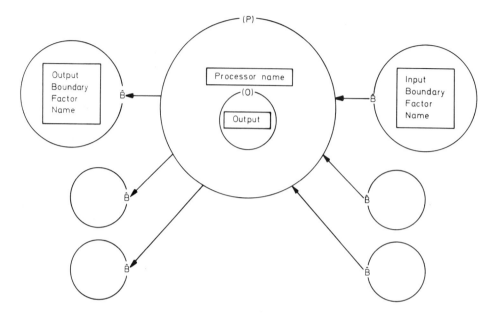

FIG. 5.99 The program concept of the TIPS system. (From Ref. 37. Courtesy of Society of Manufacturing Engineers, Dearborn, Michigan.)

different processors is done with input boundary factors $\check{B}$ or output boundary factors $\hat{B}$. There are also provisions for optional output O. The designer using the processor bank selects the modules needed to solve a problem and interconnects these with the boundary factors (Fig. 5.100). Different symbolic names may be assigned to the same processor; thus it may be used in different incarnations. The presentation of the workpiece described by this method is compiled by an executable program.

Figure 5.101 shows a picture of a punch press. The actual use of the TIPS system will be demonstrated with the help of this example. The example is simplified by reducing the press to eight parts: (1) a die, (2) a piston rod, (3) a crank, (4) a clutch, (5) a brake, (6) a flywheel, (7) a belt drive, and (8) a motor.

The interconnection of these parts with the help of boundary factors $\check{B}$ and $\hat{B}$ is shown in Table 5.7. A graphical presentation of the processors and their boundary factors is depicted in Fig. 5.102. To reconstruct the logic of this example, we first view the die and then work our way up through the pictorial. The inputs to the die are, for example, a 500-kg force, an energy of 50 hp to form the part,

Computer-Aided Design

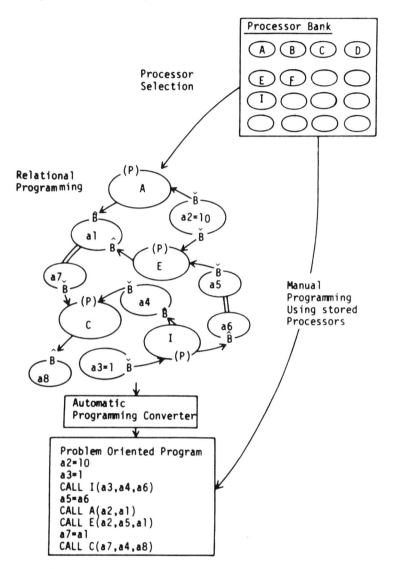

FIG. 5.100 Problem-oriented programming by selecting processors from a processor bank. (From Ref. 37. Courtesy of Society of Manufacturing Engineers, Dearborn, Michigan.)

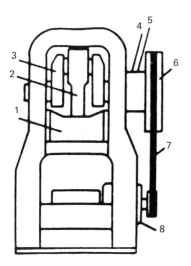

FIG. 5.101 TIPS-2 design description example. (From Ref. 37. Courtesy of Society of Manufacturing Engineers, Dearborn, Michigan.)

TABLE 5.7 Notation for Boundary Factors

	∨B	∧B		∨B	∧B
Press force	P *	*	Energy	E *	
Energy	E *	*	Rotating speed	N *	*
Repetition speed	N *	*	Rotating torque	T *	*
Stroke	S *	*	Brake torque	T *	
Axial force (Max.)	W *	*	Rotating speed	N *	
Stroke	S *	*	Output speed	No *	
Tangent force (Max.)	F *		Input speed	Ni *	
Rotating torque	T	*	Output torque	To *	
Rotating speed	N *	*	Input torque	Ti *	*
Arm length-2	A *		Rotating torque	T *	*
Transmitted torque	T *	*	Rotating speed	N *	*
Rotating speed	N *	*			

Source: Ref. 37.

Computer-Aided Design

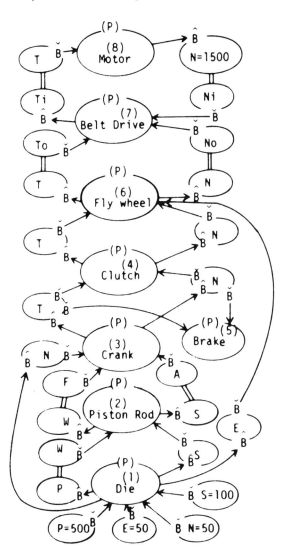

FIG. 5.102 Processors for example of Fig. 5.101. (From Ref. 37. Courtesy of Society of Manufacturing Engineers, Dearborn, Michigan.)

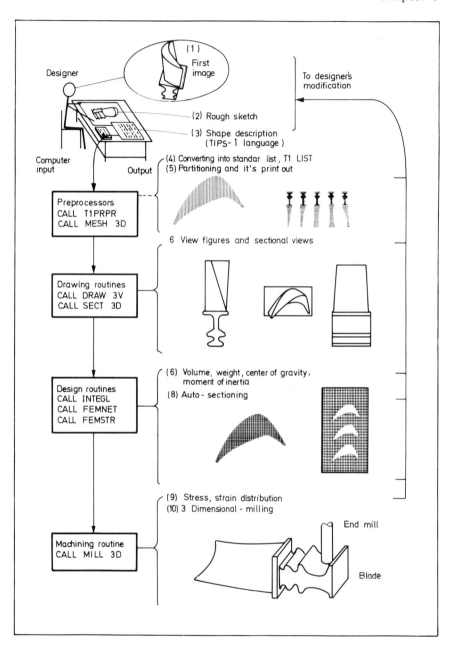

FIG. 5.103 A CAD/CAM example for TIPS. (From Ref. 37. Courtesy of Society of Manufacturing Engineering, Dearborn, Michigan.)

a cycle speed of 50 cycles/min, and a stroke of 100 cm. The die transmits the force to the piston rod, which must have a stroke of 100 cm. The cycle speed is produced by the crank, whereas the energy for forming the part must be supplied by the flywheel. From here on readers may continue on their own. The textual program is shown below.

```
DIE=PRESSDIE.0010
PROD=PISTNROD.0015
CRNK=CRANKSHT.0005
CLUH=CLUTCH-D.0020
BRAK=BRAKE-D.0030
FLYW=FLYWHEEL.0100
BELT=BELTDRIV.0010
MOTR=MOTOR-I.0050
DIE(P=500.,E=50.,N=50.,s=100.)
MOTR(N=1500)
MOTR/BELT<T:TI>
BELT/MOTR<NI:N>
```

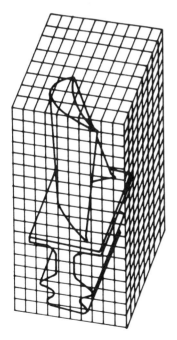

FIG. 5.104 Partitioning. (From Ref. 37. Courtesy of Society of Manufacturing Engineers, Dearborn, Michigan.)

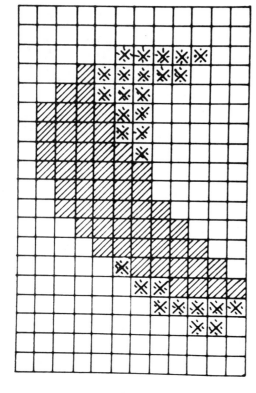

FIG. 5.105 Mosaic processing.

```
BELT/FLYW<TO:T,NO:N>
FLYW/CLUH<T:T,N:N>
FLYW/DIE<E:E>
CLUH/CRNK<T:T,N:N>
BRAK/CRNK<T:T,N:N>
CRNK/PROD<F:W,A:S>
PROD/DIE<W:P,S:S>
CRNK/DIE<N:N>
END
```

The first part of the program contains the definition of the eight processors describing the eight machine parts. In the next two instructions the boundary factors for the die and for the motor are defined. The following instructions show how the different machine parts are tied together.

Integration of CAD/CAM

The goal of this research work is the interconnection of CAD and CAM activities. This can be conceived as a small-scale computer-integrated manufacturing activity. From the conception of the design, all functions to describe the part, to calculate its physical parameters, to draw it, and to produce its NC program are done completely automatically. For this it is necessary to build a program library that contains information on the different activities in form of processors. How such an integrated CAD/CAM system may work is shown with the help of Fig. 5.103.

In this picture the engineer designs a turbine blade by first putting a sketch of the part on paper and then entering this into the computer with the help of the TIPS language (Fig. 5.97). This input is standardized and partitioned by preprocessors. The partitioning is shown in Fig. 5-104. Mosaic processing may follow this step. It renders rough cross sections of the profile of the turbine blade (Fig. 5.105). A

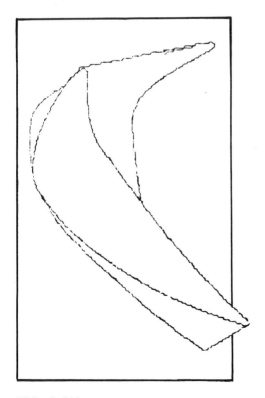

FIG. 5.106

trial-and error procedure smooths the mosaic to produce continuous surface lines (Fig. 5.106). The volume, weight, center of gravity, moment of inertia, and stress and strain distribution across the blade are calculated in following steps.

The work is concluded with the generation of NC tapes for rough and finish milling. The information for producing the surface contour is directly taken from the design file.

REFERENCES

1. H. Opitz, Trends in Manufacturing Technology, *VDI-Zeitschrift*, No. 7, May 1972.
2. L. Turner, Which Data Terminal Display: Plasma panel or CRT?, *Electronics*, Feb. 17, 1977.
3. Data Processing in Design, Methods and Aids: Automatic Generation of Drawings, VDI Guidelines 2211, Düsseldorf, West Germany, 1973.
4. J. Shopiro, Survey of Computer Graphics, Technical Report 1, Production Automation Project Report TR-1-A, University of Rochester, Rochester, N.Y., 1974.
5. M. Burmeister, et al., Introduction to Computer Graphics, Bericht KfK-PDV 92, Gesellschaft für Kernforschung, Karlsruhe, West Germany, 1976.
6. G. Krüger, Lecture Notes on Control Computer Technology, University of Karlsruhe, Karlsruhe, West Germany, 1980.
7. J. Encarnacao, *Computer Graphics*, R. Oldenbourg Verlag, Munich, West Germany, 1975.
8. R. A. Shubert, NC Interactive Graphic Systems, Society of Manufacturing Engineers Technical Paper MS77-971, Dearborn, Michigan, 1977.
9. E. Hoerbst, et al., GMB-Programming System for Interactive Graphics, *CAD Proceedings*, 1976.
10. Graphics Methods Bank, User-Handbook, Siemens AG, Munich, West Germany.
11. Brochure of Actron Industries, Inc., Monrovia, Calif.
12. Electronic Data Processing for Production Planning and Control, VDI Taschenbücher, VDI-Verlag, Düsseldorf, West Germany, 1972.
13. G. Koch, U. Rembold, and L. Ehlers, *Introduction to Computer Science II*, Karl Hanser Verlag, Munich, West Germany, 1980.
14. Data Processing in Design, Methods and Aids: Automatic Generation of Drawings, VDI Guidelines 2211, Düsseldorf, West Germany.

15. J. Kurth, COMPAC, a Computer Aided System to Design Workpieces, reprinted from *Zeitschrift für wissenschaftliche Fertigung*, Feb.-Mar. 1973.
16. A. A. G. Requicha, Part and Assembly Description Languages: I-Dimensioning and Tolerancing, Production Automation Project, Report TM-19, University of Rochester, Rochester, N.Y., 1977.
17. W. Fitzgerald, F. Gracer, and R. Wolfe, GRIN: Interactive Graphics for Modeling Solids, *IBM Journal of Research and Development*, Vol. 25, No. 4, July 1981.
18. N. N. Samual, A. A. G. Requicha, and S. A. Eling, Methodology and Results of an Industrial Part Survey, Production Automation Project, Report TM-21, University of Rochester, Rochester, N.Y., 1976.
19. U. Baatz, Interactive Design with the Help of a CRT, Ph.D. dissertation, Rheinisch-Westfälische-Hochschule, West Germany, 1971.
20. A. A. G. Requicha and H. B. Voelker, *An Introduction to Geometric Modeling and Its Application in Mechanical Design and Production*, Advances in Information Systems Science, Vol. 8, Plenum, New York, 1981.
21. F. L. Krause, Method to Conceive CAD Systems, Ph.D. dissertation, Technische Universität, West Berlin, 1976.
22. J. P. Lacoste, Representation of Workpieces for CAD, Ph.D. dissertation, Rheinisch-Westfälische-Hochschule, Aachen, West Germany, 1972.
23. U. Baatz, Computer Aided Design, Ph.D. dissertation, Rheinisch-Westfälische-Hochschule, Aachen, West Germany, 1971.
24. F. O. Vogel, Interactive Order Processing, Ph.D. dissertation, Rheinisch-Westfälische-Hochschule, Aachen, West Germany, 1976.
25. Computer Aided Design, *Markt + Technik*, No. 6, Feb. 12, 1982.
26. R. Schuster, Experiences Obtained with a CAD System by Automobile Manufacturers, VDI-Berichte 413, VDI-Verlag, Düsseldorf, West Germany, 1981.
27. I. Barda, Experience Obtained with a CAD System by an Aircraft Manufacturer, VDI-Berichte 413, VDI-Verlag, Düsseldorf, West Germany, 1981.
28. J. Encarnacao, *Computer-Aided Design, Modeling, and System Engineering*, Springer-Verlag, West Berlin, 1980.
29. J. Kurth, COMPAC—A Computer Aided System to Design Workpieces, *Zeitschrift für wirtschaftliche Fertigung*, Vol. 2, No. 3, 1973.
30. H. Depler, F. Fricke, and A. Greindl, Automation of Manufacturing Planning Processes, *Zeitschrift für wirtschaftliche Fertigung*, Dec. 1972.

31. W. Fitzgerald, F. Gracer, and R. Wolfe, GRIN: Interactive Graphics for Modeling Solids, *IBM Journal of Research and Development*, Vol. 25, No. 4, July 1981.
32. H. B. Voelcker and A. G. Requicha, Geometric Modeling of Mechanical Parts and Processes, *Computer*, Dec. 1977.
33. An Introduction to PADL, Production Automation Project Report TM-22, University of Rochester, Rochester, N.Y., 1974.
34. H. B. Voelcker, Geomertic Modeling of Mechanical Parts and Processes, *Computer Research Review* (University of Rochester), 1976–1977.
35. C. M. Brown, PADL-2: Core Software for Solid Modeling Systems, *Computer Research Review* (University of Rochester), 1981–1982.
36. H. Kubo and N. Okino, Practical Approaches for an Integrated CAD/CAM-System, Society of Manufacturing Engineers Technical Paper MS73-943, Dearborn, Mich., 1973.
37. N. Okino, TIPS-1/TIPS-2: Report, Relationally Structured Processor-Base CAD/CAM, Fourth NSF/RANN Grantee's Conference on Production Research, IIT Research Institute, Chicago, Dec. 1–2, 1976.

6
Manufacturing Systems

6.1 INTRODUCTION

In this chapter group technology, process planning, and production scheduling are discussed. All three activities play an ever-increasing role in manufacturing. The reader who is a novice in manufacturing technology will find it difficult to understand how these intricate and rather involved problems can be solved reasonably well in a plant.
In Chapter 2 it was mentioned that a medium-sized company often uses as many as 50,000 different components. Each of these components may consists of five to seven parts. To complicate the problem even further, each part may have to be processed by 1 to 10 different processes. To be able to handle this maze of problems, the manufacturing engineer must be able to work in a very systematic and orderly fashion. He or she must be familiar with data processing practices available to control and schedule a plant. In addition, the engineer should be familiar with coding, group technology, and the tools of operations research. It is also important to realize the limitations of all these technologies.

When planning manufacturing systems, the degree of automation that can economically be justified must be considered. There are several software packages available to support the manufacturing activities discussed in this chapter. Experience has shown that the most successful ones are those which are not fully automated. It is very expensive to include special features or exceptions in a universal system. Derivations from conventional practices should be handled by conventional methods.

6.2 GROUP TECHNOLOGY

6.2.1 Part Family Concept

A part that has been designed for manufacturing usually has to be produced by several succeeding manufacturing operations. If there is a large spectrum of parts to be produced, it will be necessary for workpieces to share common processing equipment. For this reason it is of advantage to group parts together to families either according to their geometric similarities or to similar fabrication methods. Figure 6.1 shows different rotational parts that can be machined on the same lathe. A change of parts would only require a new part program to generate a new contour. A tool change in this case probably would not be necessary. The parts form a design family, they are similar in design, and in this case can also be produced by a similar manufacturing process. Figure 6.2 shows cubical parts which are not very similar any more; however, they also form a production family and can be made on the same multiaxis machining center, requiring the same tools. Figure 6.3 shows dissimilar parts requiring at least one common process, which is to drill four holes. In this case, the other processes needed to shape the part would have to be done with different machine tools. These parts are typical for companies producing a wide spectrum of products. Figure 6.4 shows two completely identical designed parts, one made from plastic and the other from steel. The manufacturing processes would be injection molding for the plastic reel and turning for the metal reel. In this case we have a common design family; however, the production processes are unrelated.

In the first two cases discussed, it was possible to group parts together according to similar design attributes and common manufacturing processes. In the last case only certain design and manufacturing characteristics were identical; however, this group would also have to share machine tools. In group technology an attempt is made to find common manufacturing processes and to schedule parts through a manufacturing facility to maximize utilization of available machining recourses and to emulate a flow line production process. Figure 6.5 shows a typical conventional layout of a manufacturing facility for similar

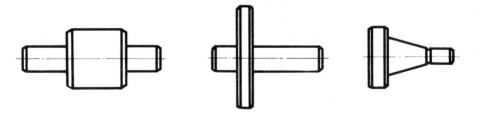

FIG. 6.1 Rotational part family requiring similar turning operations.

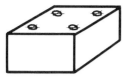 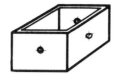 

FIG. 6.2 Similar cubical parts requiring similar milling operations.

 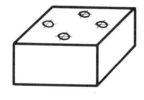 

FIG. 6.3 Dissimilar parts requiring similar machining operations (hole drilling, surface milling).

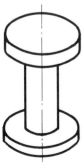 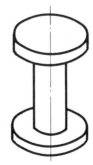

Matl. : Plastic   Matl. : Steel

FIG. 6.4 Identically designed parts requiring completely different manufacturing processes.

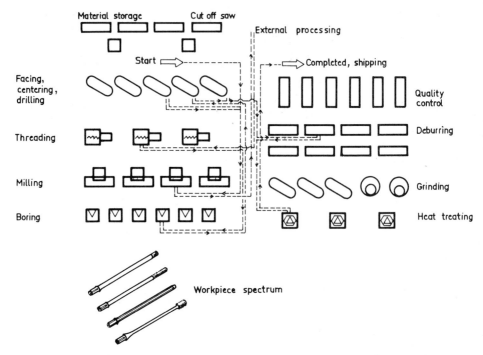

FIG. 6.5 Layout of conventional machine shop by type of machining operation. (From Ref. 1, Courtesy of Society of Manufacturing Engineers, Dearborn, Michigan.)

rotational parts [1]. The workpiece spectrum is shown in the lower part of the figure. The factory is divided into 10 different work centers. The raw material arriving in form of rods will be stored in a warehouse (1). When the part is released for production, the raw material will be retrieved and cut to slabs (2). The first finishing machining operations are facing, centering, and gun drilling (3). The next process is milling (4). Thereafter the part is sent to an external source; for example, to carborize the part surface. Upon return to the factory a thread is cut (5) and a hole bored (6). From here the part travels back to center (3) to be turned again, and then it is sent for deburring (7). The last operations are heat treating (8) and grinding (9). Before leaving the plant the part is inspected (10) and packed for shipping (11). These operations require that the part is being sent back and forth several times through the plant, making the material movement very involved. In addition, there would be several machine tool setup operations and tool changes.

If the parts were to be manufactured according to group technology considerations, the plant would have to be realigned (Fig. 6.6). Now, the production process assumes a flow line operation with machine tools located in the flow line where they are needed. It can readily be seen that intraplant transportation is minimized. Setup operations and tool changes are also reduced.

Group technology is of particular interest for plants having low and medium-volume production lots. It will, however, not be very useful for large volume production runs common in mass production. Here the plants are already laid out by the flow line concept. A good definition of group technology is as follows [2]: "Group Technology is a technique for manufacturing small to medium lot sized batches of parts of similar process, of somewhat dissimilar material, geometry and size, which are produced in a committed small cell of machines which have been grouped together physically, specifically tooled and scheduled as a unit."

6.2.2 Classification Procedures

With group technology the workpieces and machining operations have to be classified. This implies that a suitable method of coding must be found which can easily be used for manual or computer-aided classification procedures. With the manual method the description of parts and processes are cataloged. When a workpiece is scheduled for production, a catalog search is made to find a suitable manufacturing process and sequence. When the computer is used, the information about the part and the fabrication process is stored in a memory

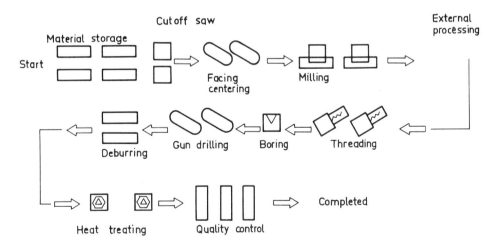

FIG. 6.6 Layout of machine shop according to group technology considerations. (From Ref. 1. Courtesy of Society of Manufacturing Engineers, Dearborn, Michigan.)

peripheral and the manufacturing data are retrieved automatically. Usually, the part spectrum and machining operations within a plant are of such a magnitude that manual classification methods become prohibitive, whereas the computer can easily be programmed to search quickly for common part features, part similarity, suitable machine tools, manufacturing sequences, and so on. One of the main difficulties for coding is to decide which parameters are important for classification. No rigid rule can be given since the parameters may vary, depending on the part spectrum. A common practice is to separate rotational from nonrotational parts. Since the rotational parts are easier to handle, most of the hitherto designed classification systems are for these part families. Figure 6.7 shows a classification system designed by the Boeing Company. The primary distinction is done by shape and by function of the workpiece. These groups are then partitioned into smaller subgroups. Classification by shape usually determines the manufacturing process, whereas the function is of interest to the designer. The designer may want information about the function of a part. If a part that has similar functions is already in existence, the designer does not have to duplicate it. Figure 6.8 shows the result of an industry survey where different parameters for typical workpieces were ranked in order of importance. The most important ones are candidates to be included in a classification system.

This study contains functional, dimensional, material, and machining parameters. Dimensional parameters, including accuracy, were considered to be the most important. There is different emphasis placed on different parameters in design and manufacturing. Figure 6.9 shows the results of a similar study with sheet metal parts. Here the major shape, the material, and the material specification had the highest ranking.

The formation of families is a compromise between many factors. If a part spectrum is being classified, an attempt is usually made first to perform a rough classification and then successively divide the parts into smaller and smaller classes of approximately equal size. Of course, there will be a limit to the size of a family that can economically be justified. For classification there are also many processing parameters that must be taken into consideration. This means that the types of processes within the plant and their limitations must be known. An attempt should be made to use one machine tool or one group of machine tools to produce the entire part without tool changes or with only few tool changes and with a minimum of reclamping of the workpiece. Tool changes and setup times usually add considerably to processing cost. Similarly, the number of machining sequences should be kept to a minimum. The entire machining operation should be controllable by one operator; otherwise, difficulties may be encountered during rescheduling of production runs.

Manufacturing Systems 293

In some cases it may be possible to perform only selected machining functions on a part family, whereas in other cases, special processing equipment cannot be placed in the flow pattern of the plant (Fig. 6.6). These situations are not very desirable since they lead to unnecessary material handling and part routing.

One of the most difficult task is to balance the load of available machine tools. In adverse conditions it may even happen that most operations are done on only a small number of machine tools and that the rest of the machines are underutilized. This may require an unduly large effort to balance machine tools. The solution to this problem may even be part of the benefits of group technology. It will help to locate bottlenecks and underutilized equipment.

Upon initial investigation of the part spectrum it will become apparent that there are many parts that cannot be handled by any known classification method. For this reason it is important to perform economic studies when group technology is introduced and to limit its application to parts that can be grouped together in a family reasonably well. Figure 6.10 shows several products described by a structural bill of materials. In this example group technology can be applied to only about two-thirds of the parts at the lowest level. This figure shows the importance of investigating the entire product spectrum. In the example each product contains parts that can be grouped together with parts of other products.

6.2.3 Benefits of Classification

A well-conceived classification system offers numerous benefits to the entire manufacturing organization. A code applied to a part by design will be used for equipment specification, facility planning, process planning, production control, quality control, tool design, purchasing, and service. The main benefits obtained in these activities are enumerated below.

Engineering
   Reduction of number of similar parts
   Elimination of duplicate parts
   Identification of expensive parts
   Reduction of drafting effort
   Easy retrieval of similar functional parts
   Identification of substitute parts
Equipment specification and facility planning
   Flow line layout of production equipment
   Location of bottlenecks
   Location of underutilized machine tools
   Reduction of part transportation times
   Improvement of facility planning

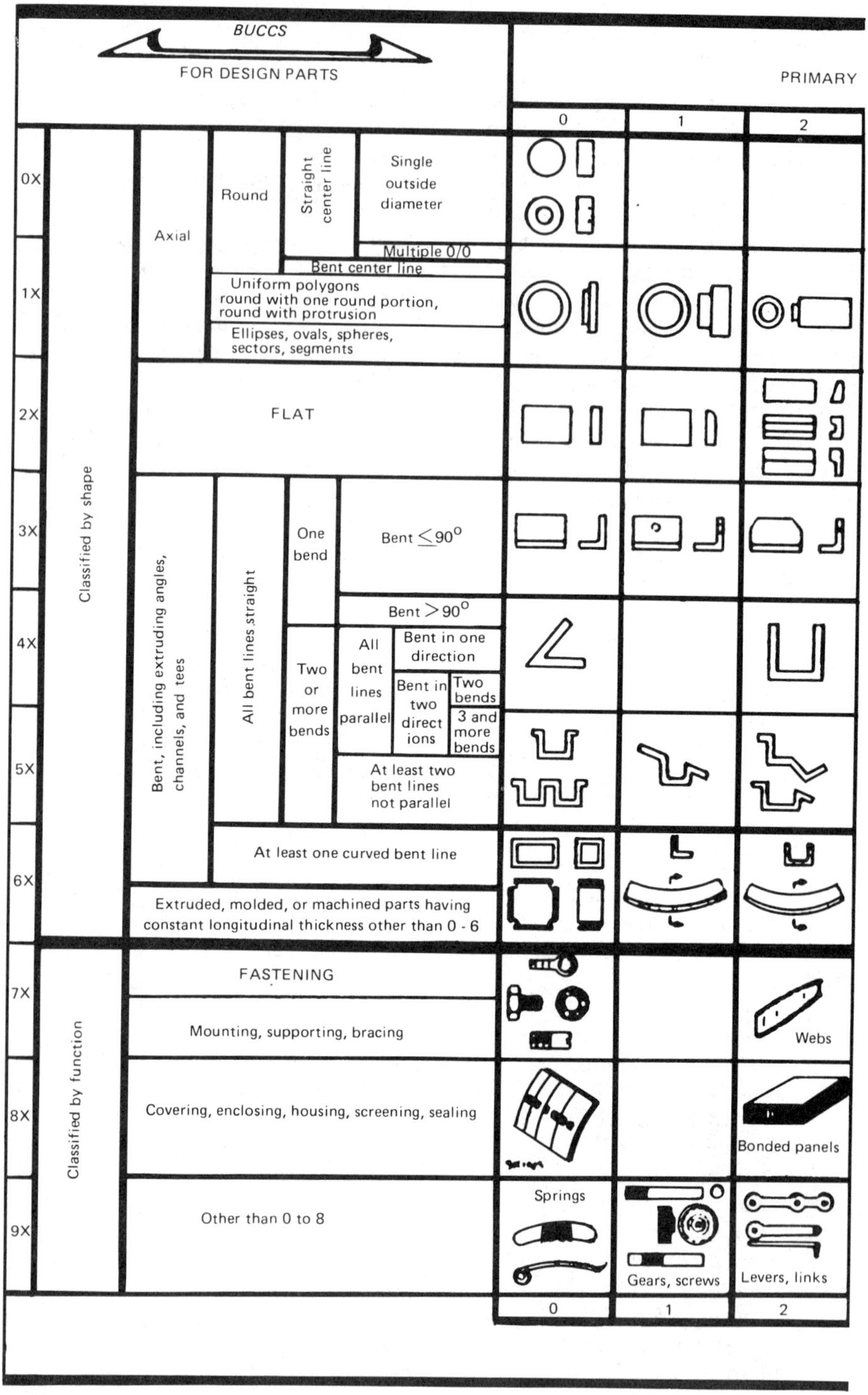

FIG. 6.7 The BUCCS coding system of the Boeing Company. (From Ref. 3. Courtesy of Society of Manufacturing Engineers, Dearborn, Michigan.)

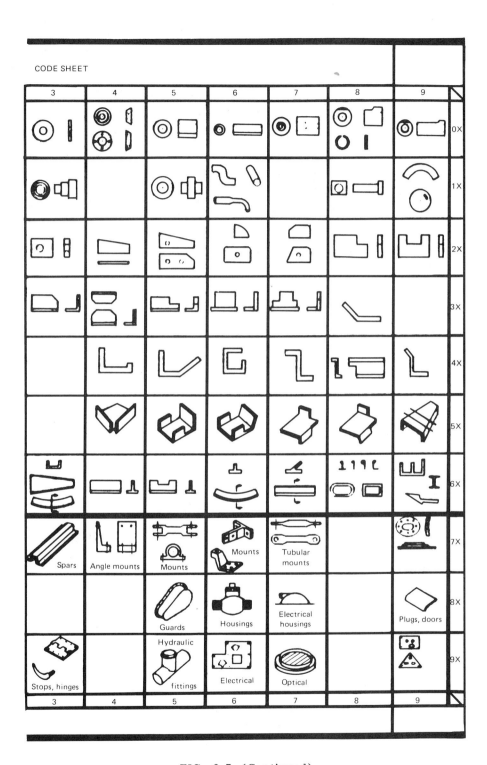

FIG. 6.7 (Continued)

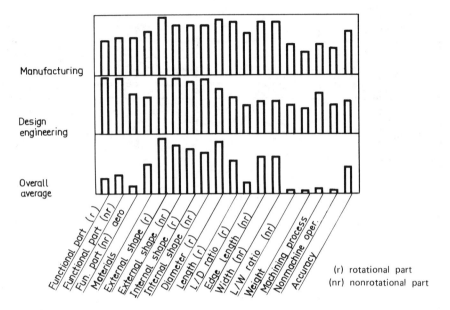

FIG. 6.8 Comparative rankings for major parameters. (From Ref. 4. Courtesy of Society of Manufacturing Engineers, Dearborn, Michigan.)

    Use of universal production equipment
    Use of manufacturing cells
Process planning
    Reduction of number of machining operations
    Reduction of NC programming time
    Shortening of production cycles
    Improvement of machine loading operation
    Reduction of setup and processing times
    Reduction of tools and fixtures to be used
    Easier prediction of tool wear and tool changes
    Improvement of part routing
Production control
    Better control of manufacturing processes
    Reduction of in-process inventory
    Better equipment monitoring
    Easier location of production difficulties
    Reduced warehousing and material movement
Quality control
    Reduced sampling and inspection times
    Improved utilization of measuring instruments
    Fewer and more universal guages
    Reduced time to locate defects
    Easier to install quality inspection procedures

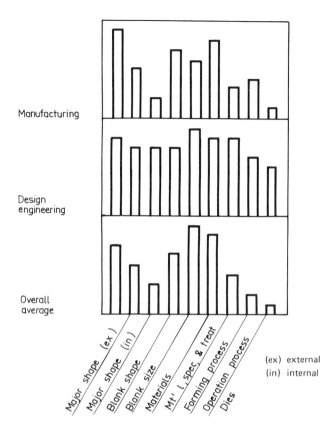

FIG. 6.9 Comparative rankings for major parameters. (From Ref. 4. Courtesy of Society of Manufacturing Engineers, Dearborn, Michigan.)

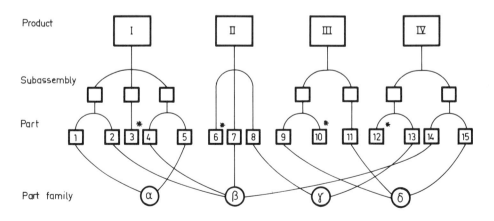

FIG. 6.10 A structured bill of materials and common part families *, no part family found. (From Ref. 5. Courtesy of Society of Manufacturing Engineers, Dearborn, Michigan.)

Tool design
    Easier scheduling of tools
    Reduced number of tools and fixtures
    Less design work for new tools and fixtures
    More universal tools
Purchasing
    Reduction of parts and types of raw material
    Easier ordering of material in lot sizes
    Better evaluation of vendors
    Use of more standardized material
    Easier to perform make-or-buy decisions
Service
    Reduction of inventory
    Easier to find replacement parts
    Better customer service

It can easily be recognized that the benefits obtained through the use of group technology extend far beyond the limited scope of machining parts. When this technology is being introduced in a factory, an extensive study must be conducted to include all functions of the firm that will be effected by it. Otherwise, there will be only a limited use for group technology.

5.2.4 Code Number System

The task of a code number system is to identify a part and to classify it. Each workpiece used in manufacturing needs a unique identification number to identify it positively. When the part progressively changes its shape during processing, it may lose its identity and new unique numbers may be assigned to it to be able to identify it in several semifinished stages. The identification number can exist independently of a classification number. All documents associated with the part must carry the part identification number. For group technology the part must have a classification number assigned to the part, and it may be changed independently of the identification number (Fig. 6.11). Classification may be performed according to common design or production attributes or both. From Fig. 6.1 we have learned that parts that have common design characteristics often will be processed by similar fabrication methods. For this reason, design-oriented

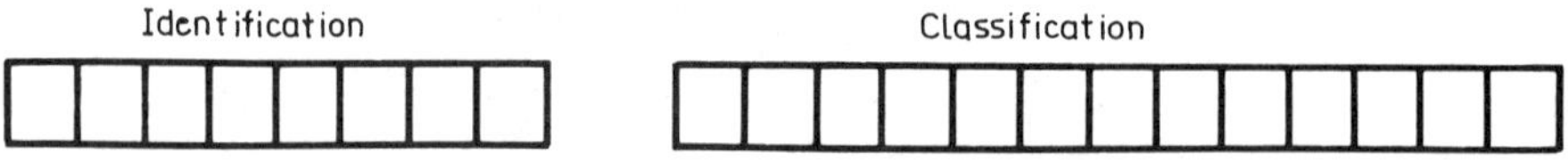

FIG. 6.11 Typical structure of a part number.

classification may suffice in the majority of applications. From Fig. 6.4 it can be seen that different processes may generate parts of similar or even identical design. Thus using the fabrication method as the underlying concept for coding may wrongly separate similar parts. If design and production features are of importance, a coding system incorporating both has to be conceived.

A code number may be of infinite length. This, however, would make it impractical for use. The maximum number of positions of coding systems used by industry presently is approximately 36. The number of values that can be placed in each position depends on the alphabet selected. With the decimal system there are 10, with the hexadecimal 16, and with the Latin alphabet 26 possible position descriptors. Most coding systems use 10 or fewer descriptors for each position. Upon conception of a coding system the alternative of maximizing information contents against clarity has to be weighed.

Coding systems should be designed such that the most significant parameter will be presented by the most significant position. Furthermore, similar items should have similar codes. This enhances identification of families and the possibility of sorting workpieces in ascending code number order. Thus families can easily be brought together. This feature cannot, however, always be achieved, and there is the possibility that upon sorting, only the most significant parameters are brought together, whereas those identified by the less significant position will be dispersed.

Another prerequisite for group technology which we already had stated is that each number should be unique. This means that it must not be possible to identify more than one workpiece with one number. This requirement is very difficult to achieve since it is almost impossible to check a completed coding system for all possible workpiece and machining combinations. A typical problem is shown in Fig. 6.12, where one code number describes two parts. There also may be problems with defining the parameter range to be presented by a code value in a specific location of the number. Since there are only discrete values allowed, a part whose parameter slightly exceeds a borderline may have a different number than a similar part that stays within the borderline.

When a numbering system is selected for coding there are two possibilities to conceive a code. They are the chain type and the hierarchical structures. They are discussed next in more detail.

Chain-Type Code Structure

With the chain-type code, each digit in a specific location of the code describes a unique property of the workpiece. An example of a chain-type code is shown in Fig. 6.13. This type of code is very easy to learn and to use because of the unique attribute of each position

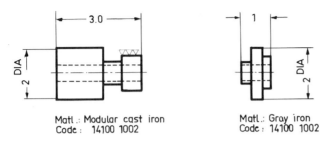

FIG. 6.12 Possible ambiguity with a coding system (Opitz system). (From Ref. 6.)

and its value. It is of particular advantage where the part function or the manufacturing process have to be described. Due to its almost unlimited combinational features, its length often becomes excessive. For this reason, there are two types of chain codes used. One has a fixed length, where the code values are always of the same length and at the same position. The other type has a variable open-ended length. It needs a position identifier. One of the features of this code is that each position can easily be searched without going through the entire number. Most of the present coding systems use this code structure.

Hierarchical Code Structure

With the hierarchical code structure each code character depends on the preceding one, and it in turn qualifies the succeeding character. This implies that parts are compared by the sequence in which they are represented in the code (Fig. 6.14). The code may be much shorter than the chain-type code. It usually has between four and

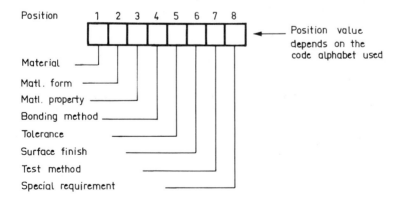

FIG. 6.13 Example of a chain-type code.

Manufacturing Systems 301

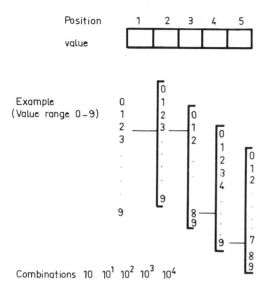

FIG. 6.14 Example of a hierarchical code structure.

eight positions in practical applications. Because of its hierarchical nature the code is suited to represent design features of the workpiece. However, it leads to problems when functions and manufacturing processes have to be described. A rather severe disadvantage is that it requires expertise to conceive such a coding system for a part spectrum. It is also very difficult to learn. When sorting is done to search for specific parameters, this structure may lead to long computer times. In practical applications the chain-type and hierarchical code structures are often combined. The former structure facilitates vertical search and the latter assists in horizontal search.

6.2.5 Classification Methods

There are three classification methods in use:

1. Visual inspection of the part spectrum
2. Classification by design and processing features
3. Classification by process flow

Classification by Visual Inpsection

This is the simplest method, but it requires a thorough knowledge of the entire part spectrum and excellent manufacturing experience. It should be used for initial studies to investigate the feasibility of introducing group technology into a plant or where only a small part spectrum is considered. Because it is very labor intensive, it usually leads to limited savings. For a larger part spectrum it even becomes very difficult to specify families progressively. There will be many parts left over that cannot be accommodated by this method.

Classification by Design and Processing Features

The underlying concept of this method is common design and processing features. In general, similar parts require similar fabrication processes (Fig. 6.1). However, there may also be exceptions, as shown in Fig. 6.4. Here identically designed parts were made by entirely different processes. There are many different classification methods of this type used by industry. Before the method is applied on a large scale, the product spectrum should be sampled and investigated with the help of the visual method.

Properly done, this method can lead to a very systematic approach. Coding is done by design and entered into the computer via the CAD system. This information is used by the computer together with tooling, machine, cost, and time data to generate the process plan. Classification results are comprehensive and will contain all part families that can be formed economically. Its greatest disadvantage is the time required to set up the classification systems and to maintain it.

Possible output information from the computer includes:

1. Groups of part families
2. Automatic selection of standard parts
3. Machine tool, tool, and fixture selection
4. Automatic generation of cutting parameters, cost details, and lead, processing, and completion times
5. Processing sequence for each part
6. Material order information and material allocation

The main features of several chain-type classification systems will be discussed in more detail. This discussion must be very brief, as a detailed description of each of these systems would fill an entire book. The Brisch system is a good example of a hierarchical coding system. It is widely used by industry. The reader who is interested in it may consult the literature.

The Opitz Classification System

This chain-type classification system was developed by the Technical University of Aachen under the auspice of the German Machine Tool Association. It represents the first comprehensive system and is extensively covered in the literature [7]. The basic structure of the system is shown in Fig. 6.15. The code number has a maximum of

Manufacturing Systems 303

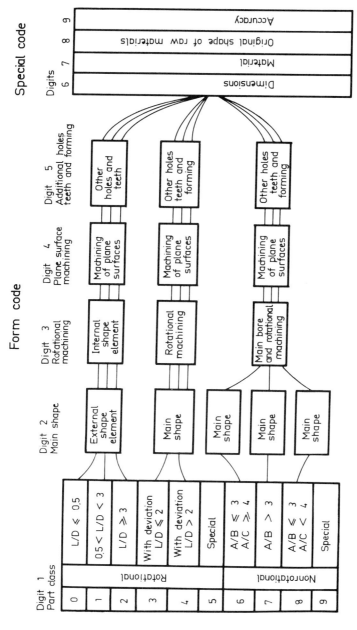

FIG. 6.15 The Opitz parts classification and coding system.

13 positions. Each position may assume 10 different values (attributes). The digits 1 to 5 describe the design of the workpiece and the surface contour. The manufacturing process is described by the digits 6 to 9. This includes the shape of the blank and the surface finish. Digits 10 to 13 identify the production processes and their sequence. A description of the attributes defined by the first six digits for rotational parts is shown in Fig. 6.16. The meaning of the last four digits identifying the production process can be defined by the user of the coding system. For rotational parts, for example, they may be defined as follows [8]:

10 Length of workpiece before machining
11 Gross weight of workpiece
12 Complexity of the part defined by the number of external and internal diameters
13 Gives information on the lot size to be produced

Figure 6.17 shows an example of a shaft and Fig. 6.18 an example of a square flange, both coded by the Opitz system.

CODE System

This system was developed by industry and is marketed by Manufacturing Data System, Inc., Ann Arbor, Michigan. The chain-type code has eight positions, each of which can describe 16 different attributes. A sample page from the CODE book is shown in Fig. 6.19. This pictorial, together with Fig. 6.20, easily explain the use of the system.

When the code number is entered into the computer, the specific dimension of the part, its number, the material, and the surface finish also have to be provided. A typical computer printout is shown in Table 6.1. The reader who scans this table carefully may extract important information. The contents of the second column suggest the use of common material to facilitate standardization of material purchased. The information in the columns specifying the maximum outer diameter and length is used to select machine tools and fixtures. It also helps to detect production bottlenecks. The contents of the lot size and standard cost columns contains economic information. It can be determined why different parts are very expensive compared to others which are almost identical. It also suggests that some low-lot-size parts should be eliminated and substituted for with standard parts. This may need a slight redesign of the product. The CODE system is design oriented and is not as comprehensive as the Opitz system; however, it serves its purpose for many applications.

Manufacturing Systems

Digit 1			Digit 2			Digit 3		
Part class			External shape, external shape elements			Internal shape, internal shape elements		
0		$L/D \leq 0.5$	0	Smooth, no shape elements		0	No hole, no breakthrough	
1	Rotational parts	$0.5 < L/D < 3$	1	Stepped to one end or smooth	No shape elements	1	Smooth or stepped to one end	No shape elements
2		$L/D \geq 3$	2		Thread	2		Thread
3			3		Functional groove	3		Functional groove
4			4	Stepped to both ends	No shape elements	4	Stepped to both ends	No shape elements
5			5		Thread	5		Thread
6			6		Functional groove	6		Functional groove
7	Nonrotational parts		7	Functional cone		7	Functional cone	
8			8	Operating thread		8	Operating thread	
9			9	All others		9	All others	

FIG. 6.16 Form code for rotational parts of the Opitz system.

Digit 4

	Plane surface machining
0	No surface machining
1	Surface plane and/or curved in one direction, external
2	External plane surface related by graduation around a circle
3	External groove and/or slot
4	External spline (polygon)
5	External plane surface and/or slot, external spline
6	Internal plane surface and/or slot
7	Internal spline (polygon)
8	Internal and external polygon, groove and or slot
9	All others

Digit 5

		Auxiliary holes and gear teeth
0		No auxiliary hole
1	No gear teeth	Axial, not on pitch circle diameter
2		Axial on pitch circle diameter
3		Radial, not on pitch circle diameter
4		Axial and/or radial and/or other direction
5		Axial and/or radial on PCD and/or other directions
6	With gear teeth	Spur gear teeth
7		Bevel gear teeth
8		Other gear teeth
9		All others

FIG. 6.16 (Continued)

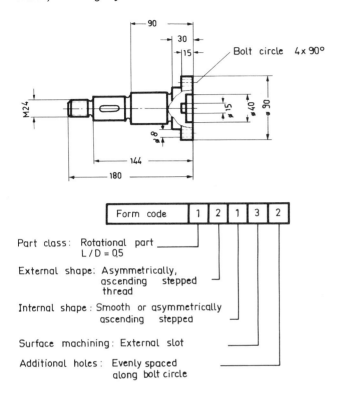

FIG. 6.17 Classification of a rotational part. (From Ref. 9.)

MICLASS System

The system was developed in the Netherlands. The abbreviation is derived from the name Metal Institute Classification System. The developer had in mind a classification system to standardize drawings, to facilitate the retrieval of parts, and to standardize process routing. It was conceived for interactive communication with the computer. The designer can enter a new code for a part under computer guidance.

In this chain-type coding system, a minimum of 12 and a maximum of 30 positions are provided (Fig. 6.21). The position descriptors can range from 0 to 9. The first 12 positions are standard code used for any part and describe the following attributes:

1	Basic shape
2 and 3	Shape element
4	Location of the shape element

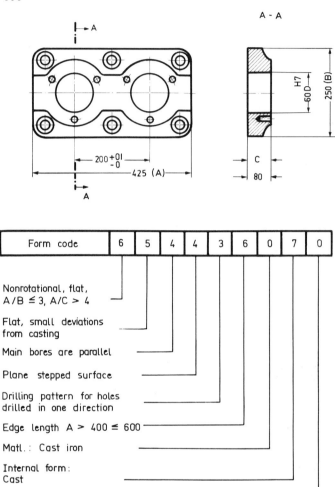

FIG. 6.18 Square cast-iron flange classified by the Opitz system. (From Refs. 7 and 8.)

Manufacturing Systems

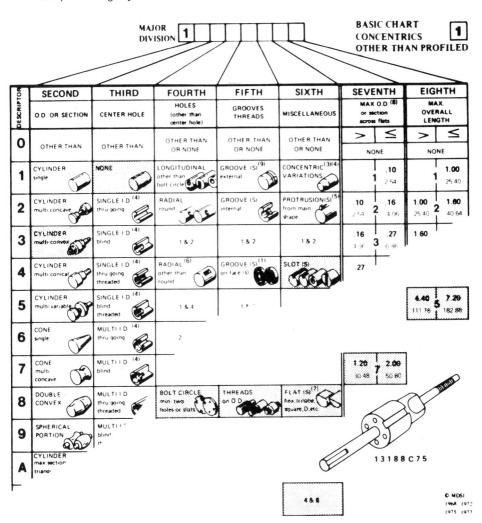

FIG. 6.19 Sample page of the CODE system. (From Ref. 10. Courtesy of Society of Manufacturing Engineers, Dearborn, Michigan.)

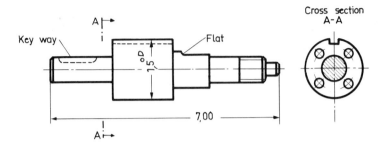

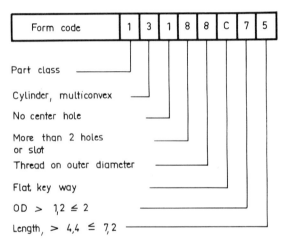

FIG. 6.20 Example of a workpiece coded by the CODE system.

    5 and 6   Primary dimension
    7   Ratio of dimensions
    8   Supplementory dimensions
    9 and 10   Tolerances and finishes
   11 and 12   Material

The user is able to specify the attributes of the additional 18 positions. A typical example of the definition of locations 19 to 27 for rotational parts is as follows [11]:

    19   Number of outside diameters
    20   Number of inside diameters or specific shape
    21   Rotational grooves or knurls

TABLE 6.1 Sample Printout of the CODE System[a]

CODE NO.	MATL.	FIN	HT		MAXIMUM O.D	SMALLEST O.D	SMALLEST I.D.	MAXIMUM LENGTH		LOT SIZE	STANDARD COST	PART NO.
13D0956C					812	438	313	10.578		329	8.2341	9876523
13D0957E	1050			A	1.600	.750	438	14.358		61	11.0078	6956333
13D0957E	1045				1.600	.750	438	14.358		44	12.3363	30163859
13D0957E	1050			B 1	1.600	.750	438	14.358		24	15.5471	10034876
13D0957E	1050				1.600	.750	438	14.358		5	7.6463	21012756
13D0957E	1050				1.600	.750	438	15.848		6	9.8730	45628019
13D0957E	1050				1.600	.750	438	15.848		92	9.6871	40628737
13D0957E	1050			B 2	1.600	.750	438	15.848		73	14.6165	5003131
13D0957E	1050				1.600	.750	438	15.848		5	19.7045	7396211
13D0957E	4140				1.625	.997	438	15.865		14	8.7327	21035687
13D0957E	4140				1.625	.997	438	15.865		27	15.5839	11286598
13D0957F	1045				1.625	.985	438	16.015		5	52.7780	27913621
13D0957F	1045				1.625	.985	438	16.015		20	17.8950	47603001
13D0957F	1045			D	1.625	.985	438	16.015		6	38.7534	39368598
13D0957F	1045				1.625	.985	438	15.015		115	5.7089	13507765
13D0957F	4140				1.625	.997	438	17.365		120	7.5679	36975106
13D0957F	1045				1.625	.997	438	17.365		34	5.2199	7834590
13D0957F	4140				1.625	.997	438	17.365		101	8.2131	9234501
13D0957F	1045				1.625	.985	438	17.515		75	8.3187	13706948
13D0957F	1045				1.625	.985	438	17.515		40	13.2586	14713526
13D0957F	1045			E 1	1.625	.985	438	17.515		10	32.4476	47329873
13D0957F	1045				1.625	.985	438	17.515		26	17.1325	50016713
13D0957F	1045				1.625	.985	438	17.515		14	28.7932	50173721
13D0957F	1045				1.625	.985	438	18.015		19	21.4583	30659875
13D0957F	1045				1.625	.985	438	18.015		20	16.6572	30264781
13D0957F	1045			E 2	1.625	.985	438	18.015		12	31.9997	26145697
13D0957F	1045				1.625	.985	438	18.015		6	39.0012	10301002
13D0957F	4140				1.625	.997	438	19.365		87	15.1214	21203697
13D0957F	1045				1.625	.985	438	19.515		120	5.5562	8340021
13D0957F	1045			F	1.625	.985	438	19.515		115	5.4136	9569572
13D0957F	1045				1.625	.985	438	19.515		5	28.7397	40078561
13D0957F	1045				1.625	.985	438	19.515		10	13.7008	84182111
13D0958F	4120	00			2.125	.969	438	18.797		2000	9.6878	4078336
13D0958F	1140	00		G 1	2.125	.969	438	18.797		25	27.7252	10102593
13D0958F	1140	00			2.125	.969	438	18.797		100	23.6501	73965111
13D0958F	1140	00		G 2	2.125	.969	438	20.859		1000	15.4715	6474372
13D0958F	1140	00			2.125	.969	438	20.859		1100	13.2603	72964344

[a](From Ref. 10. Courtesy of Society of Manufacturing Engineers, Dearborn, Michigan.)

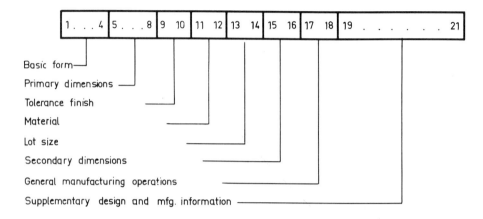

FIG. 6.21 The structure of the MICLASS coding system.

22 Close tolerance diameters
23 Splines
24 Gears
25 Sprockets
26 Pitch diameter/diameter pitch
27 Number of teeth

6.2.6 Composite Component Concept

When group technology is applied to part manufacturing and a plant is laid out according to this concept, many processes will be grouped together. Grouping can be done by several different methods, depending on the part spectrum, the available machine tools, or the required flexibility. Typical criteria are the following:

1. The use of universal machining centers
2. Grouping together of different machine tools to one production unit
3. The use of a flexible manufacturing system or a suitable flow line machine layout

In this concept the part spectrum is analyzed for related machining operations that can be performed by one of the aforementioned machine tool groups. After all possible operations have been identified, a composite part which may be existing or nonexisting is formed. This part will contain all machining operations of a combined family. The principle of the concept is shown in Fig. 6.22. Once a composite has been

Manufacturing Systems

External shapes

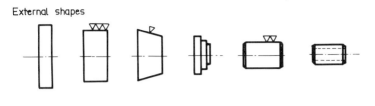

Internal shapes

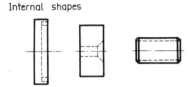

Composite (real or hypothetical part)

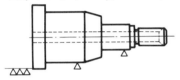

FIG. 6.22 The composite concept.

found, a proper machine tool or a group of machine tools can be set up which are capable to machine all parts from which the composite was derived.

There are certain rules to be observed when this technique is used:

1. Machining operations should be similar.
2. A tool change should only be made with an automatic tool changer, if not avoided altogether.
3. If possible, one tool should be used for all cutting operations.
4. Dissimilar operations such as cutting and grinding cannot be done on one machine tool.
5. The bar stock used as the raw material should have similar metallurgical and machining characteristics.
6. Clamping and chucking devices should be similar.

Figure 6.23 shows a composite parts sheet of the Boeing Company. In this case the composites were made rather small, which is an accepted method when a wide part spectrum and large production runs are used.

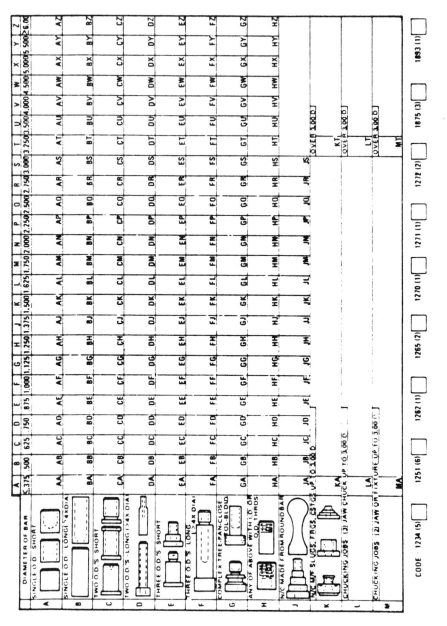

FIG. 6.23 Composite parts code sheet of the Boeing Company. (From Ref. 3. Courtesy of Society of Manufacturing Engineers, Dearborn, Michigan.)

6.2.7 Factory Flow Analysis

Until now we have mainly considered similar design attributes of a part spectrum. Generally, the design attributes also require similar processes. In factory flow analysis we look for part families that have similar manufacturing sequences. Here parts are assigned to groups that require the same routing through the machine shop. This implies that parts may be of different design; however, they must be made by the same sequence of processes. The flow of parts through the plant is contained in the routing sheet. It determines to which department a part is brought and which machine tools are used. Since, in general, the layout of the plant was done prior to the formation of a part family, this method does not guarantee optimal routes. However, if group technology and factory flow analysis are combined during the plant layout, efficient manufacturing routes can be established and maintained, as long as there is no change in the part spectrum or the design of the part. Since only the operation sequence is of interest there is no classification system necessary to perform factory flow analyses.

This method also requires a very thorough knowledge of the part spectrum which can be grouped together, and the available processes. Grouping starts with an analysis of the departmental structure of the factory and the machining resources in these departments. Thereafter a group analysis is conducted to form part groups and their associated machining groups. For each part a record is made indicating the part number and the machining sequences. This record is entered in a factory flow matrix (Fig. 6.24). The parts are presented by the row numbers and the machines by the column numbers. An entry in this matrix denoted by "×" indicates that the part will be processed by this machine.

PART NO.	MACHINE NO.															m
	1	2	3	4	5	6	7	8	9	10	11	12	13	14	15	
1		x								x	x	x				4
2			x		x			x					x		x	5
3	x				x			x						x		4
4	x			x				x						x		4
5			x		x			x					x		x	5
6	x			x		x		x						x		5
7		x					x			x	x	x				5
8			x		x			x					x		x	5
9				x		x		x						x		4
10		x					x			x	x	x				5
n	3	3	3	3	3	2	3	4	3	3	3	3	3	4	3	

FIG. 6.24 Unsorted factory flow matrix. (From Ref. 12.)

```
              MACHINE NO
PART       14  9  15 13 12 11 10  8  6  5  4  3  2  1  7   m
NO.
 9          x  x                      x     x              4
 4          x  x                            x        x     4
 3          x  x                      x              x     4
 1                      x  x  x                   x        4
10                      x  x  x                   x     x  5
 8                x  x              x     x     x          5
 7                      x  x  x                   x     x  5
 6          x  x                      x     x     x        5
 5                x  x              x  x  x                5
 2                x  x              x  x  x                5
 n          4  4  3  3  3  3  3  3  3  3  3  3  3  3  2
```

FIG. 6.25 Result of the first grouping.

Numerous algorithms have been developed for sorting the contents of the matrix to machine clusters. A recently published method will be discussed in detail [12]. When this algorithm is applied, it probably will have to be done with the help of a computer. Since for a large part spectrum and many machine tools, it will be necessary to go through many iterations to reach an acceptable solution. The use of the algorithm will be shown with the help of Figs. 6.24 through 6.27.

    1. Count the number of entries × in each column. Record these numbers. (In Figs. 6.24 and 6.25 they are recorded in the auxiliary column.) Count the number of entries × in each row (row n). Rearrange the matrix with columns in decreasing order of ×, and rows in increasing order of ×, respectively.

```
              MACHINE NO.
PART       14  9  15 13 12 11 10  8  6  5  4  3  8  1  7
NO.
 9          x  x                      x     x
 4          x  x                            x        x
 3          x  x                      x              x
 6          x  x                      x     x        x
 8                x  x              x     x     x
 5                x  x              x  x  x
 2                x  x              x  x  x
 1                      x  x  x                   x
10                      x  x  x                   x     x
 7                      x  x  x                   x     x
```

FIG. 6.26 Result of the second grouping.

Manufacturing Systems 317

	MACHINE NO.													
PART NO.	14	9	6	4	1	15	13	8	5	3	12	11	10	2	7
9	x	x	x	x											
4	x	x		x	x										
3	x	x	x		x										
6	x	x	x	x	x										
8						x	x	x	x	x					
5						x	x	x	x	x					
2						x	x	x	x	x					
1											x	x	x	x	
10											x	x	x	x	x
7											x	x	x	x	x

FIG. 6.27 Result of the final grouping.

2. Take the first column of the new matrix and move all rows that have an entry × in this column to the top. Repeat this procedure with succeeding columns until all rows are rearranged.
3. If the current matrix and the one that preceded it are identical, go to 6; otherwise, go to 4.
4. Take the first row of the matrix and move all columns that have an entry × in this row to the leftmost position. Repeat this procedure with all succeeding rows until all columns are rearranged.
5. If the current matrix and the one that preceded it are identical, go to 6; otherwise, go to 2.
6. Stop.

If we follow this algorithm carefully, the results in Fig. 6.25 are obtained after the first grouping. Now the algorithm is applied iteratively. In column 1 under machine 14 the parts 9, 4, 3, and 6 are moved to the top. The next column is already in proper order; thus the algorithm goes to column 3 of machine 15. Rows 8, 5, and 2 are moved up. Upon completion of this procedure, the results of Fig. 6.26 are obtained. Now with the help of the algorithm, the columns are rearranged. The final grouping is given in Fig. 6.27. By applying the algorithm again to the contents of this figure, no change can be obtained. Thus the result shows the final grouping.

When one machine is used for processing of many parts, complete clustering may not be obtained. As a matter of fact, this machine will block the formation of clusters. The machine can be deactivated for further grouping until clusters have been obtained. Thereafter, it will be reactivated and assigned to those groups where it is needed. This method is also shown in Ref. 12.

6.3 PROCESS PLANNING

The task of process planning is to translate design data to work instructions to produce a part or a product. This implies that process planning should interact directly with the information presented on the workpiece drawing and the bill of materials. This interface is possible with the aid of the computer and a common data base used in an integrated manufacturing system. The information rendered by process planning leads the workpiece through the individual manufacturing stages. They start with the selection of raw material and end with the completion of the part.

The form of the raw material used depends on the part and the manufacturing process. Typical materials are metal sheets, rods, bar stock, forgings, and blanks or slabs of metal. The raw part may have almost any shape. It is usually designed such that the metal cutting or forming operations to produce the part are kept to a minimum. The most common forming processes are turning, boring, drilling, milling, grinding, broaching, punching, bending, forging, sintering, electrodischarge machining, and chemical milling. With computerized process planning an attempt is made to automate the planning function and to reduce the nonproductive residence time for a part in a factory (Fig. 2.1). With a conventional manual system the planner relies heavily on personal manufacturing experience. He must know possible machining operations and sequences to produce a part and the available cutting tools, machine tools, and measuring instruments. In addition, he needs tables and handbooks to determine obtainable tolerances and surface finishes as well as the depth of cut, optimal feeds, and speeds for machining. A planner usually tries to minimize the manufacturing cost or time. However, this may not be possible in every case, since there are many parts competing for the same machining resources. Thus the planner has the additional burden of investigating manufacturing alternatives to utilize idle machine tools and to search for short material flow routes. With a wide part spectrum or intricate parts, the planning process is very tedious and demands much skill and endurance. It is usually very difficult to find good manufacturing experts who are willing to do this type of repetitive work. For this reason selected processes and machining sequences are often impractical, time consuming, and expensive.

With an automated planning system the computer can be given decision rules to generate the process plan. For this purpose it must have access to the central manufacturing data base, containing information on customer orders, engineering specification, available machine tools, and manufacturing processes. It may also contain optimization rules to utilize the plant resources fully. The particular asset of the computer is its high speed, which allows investigation of many different manufacturing alternatives.

An automated process planning system should have the following features:

1. It should operate as an integrated planning aid that obtains input data automatically from engineering and sales to generate a complete set of process plans to be used by production planning as well as production, material, and quality control.
2. It should render basic data for work order routing, fabrication schedules, payroll accounting, and material release.
3. It should be of generalized design to accommodate different types of parts.
4. It should have an interactive on-line system to utilize the potential of the computer fully.
5. It should be user friendly and provide operator guidance.
6. The system should be of modular design for easy expansion, modification, and maintenance.
7. Like all other production equipment, it should be cost justified.

Before we discuss process planning in more detail, the structure of the bill of materials will be explained because planners have to interact directly with this record.

6.3.1 Bill of Materials

The bill of materials and the drawing are the principal documents needed to produce a part. Each part contained in the final product is represented in the bill of materials. There are several prerequisites to be observed when it is assembled. First, each part is to be identified uniquely, including raw material and subassemblies. Second, the contents of an item must be defined uniquely. For example, the same subassembly numbers may not be maintained in case components are changed. Third, the state of completion of a product should be reflected by the structure of the bill of materials. Thus this document describes not only the product but also its different states of creation.

Figure 6.28 shows a typical product structure. At the highest level, the product consists of the subassemblies A and B and the part 1. Items A and B are broken down further, whereas item 1 is a final part. Subassembly A contains subassembly B and part 2. Finally, subassembly B contains parts 3, 4, and 5. Figure 6.29 shows a single-level bill of materials for these parts. This document describes the product, its components, and the quantity needed. It is well suited for determining the types and number of components required. The structure of the product cannot be depicted.

It is easy to design from Figs. 6.28 and 6.29 an indented bill of materials (Fig. 6.30). From this the structure of the product can be identified. However, it becomes difficult to identify the quantity of

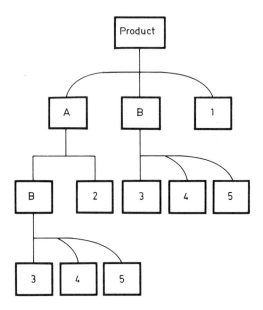

FIG. 6.28 Product structure. (From Ref. 9.)

each item needed. In addition, if the product is very complex, any changes may be very difficult to perform.

Another presentation of the product structure of Fig. 6.28 is shown in Fig. 6.31. In this block-type bill of materials the information on each subgroup of the product has to be stored only once. The required memory space in the computer is minimized. Additions or deletions can easily be performed since they have to be done only once. This

Product	
Part	Quantity
1	1
2	1
3	2
4	2
5	2

FIG. 6.29 Bill of materials for parts.

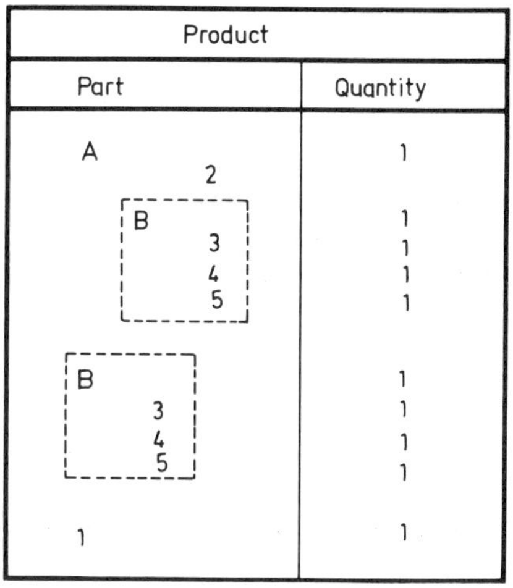

Manufacturing Systems

Product	
Part	Quantity
A	1
2	1
B	1
3	1
4	1
5	1
B	1
3	1
4	1
5	1
1	1

FIG. 6.30 Indented bill of materials.

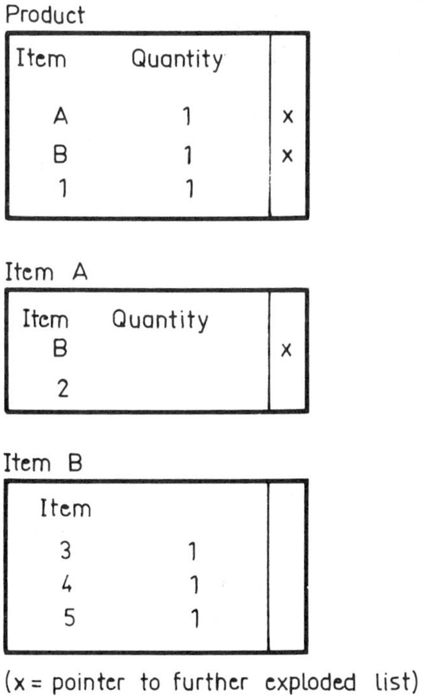

Product

Item	Quantity	
A	1	x
B	1	x
1	1	

Item A

Item	Quantity	
B		x
2		

Item B

Item		
3	1	
4	1	
5	1	

(x = pointer to further exploded list)

FIG. 6.31 Block-type bill of materials.

structure, however, makes it difficult to calculate total quantities required. Also, the product structure cannot readily be seen. Despite these difficulties, many computer users prefer this structure for implementing a bill of materials on the computer.

A typical contents of the bill of materials is shown in Table 6.2. In practical applications not all of these items may be needed, or others have to be supplemented. This table shows that a real bill of materials is rather complicated in comparison with those shown in Fig. 6.29 through 6.31.

6.3.2 Requirements for Process Planning

Figure 6.32 shows a simplified planning sheet used to produce a workpiece. It contains different types of information. First, the heading identifies the planning sheet, important dates, its origin, the number of pieces to be manufactured, and any significant signatures to verify its contents. Second, there is workpiece-related information which identifies the part, the drawing, the classification, the part family, and its physical parameters. The third part of information describes the operation number and description, the machine tools to be used, tools, and setup and operation times. An automated planning system must be capable of generating these different types of information to the manufacturing department.

Table 6.3 shows the functions necessary to generate a process plan. The first step is to select the raw part, which may be a slab, a blank,

TABLE 6.2 Typical Contents of Bill of Materials

Number of current line	Change status
Identification number	Process plan number
Quantity needed	Type of material used
Measurement unit used	Number of raw material
Name of part	Quantity of raw parts
Validity status of information	Measuring unit for raw part
Existing identification number: yes/no	Number of die or template
	Where was part obtained from
Are basic data on file? yes/no	Further explosion of bill of materials: yes/no
Classification number	Finished weight
Type of standard used	Production status
Standard number	Status of part usage: active/passive
Drawing number	
Drawing format	Value of part

Source: Ref. 9.

Manufacturing Systems

Operation Sheet No.			Date			
Part No.		Part Name			Drawing No.	
Orig.	Checked		Changes		Approved	
Pieces		Matl.		Weight		
Op. No	Operation	Mach. Tool	Tools	Fixtures	Set-up time, Hr	Operation time, Hr
5	Rough turning	Lathe 1	T1	Chuck	0.1	0.15
10	Fine turning	Lathe 2	T2	Chuck	0.05	0.15
15	Drilling	D Press 1	D1	Drill jig	0.15	0.1
20	Chamfer	D Press 1	Ch1	Drill jig	0.05	0.05
25	Counterboring	D Press 1	D2	Drill jig	0.05	0.08
30	Heat-treat	Furnace			0.1	0.5
35	Grinding	Grind 5			0.1	0.05

FIG. 6.32 Simplified manual operation sheet.

a rough forging, or a rod. It is necessary to determine the required oversize of the part to assure a flawless surface after machining. The weight of the raw part also has to be calculated. In the next phase, the individual operations to produce the part are determined. For this purpose, in most cases, group technology methods are used for coding and classification. The principles of these methods were explained earlier. The part spectrum is coded as family groups, whereas the class criterion may be either common design features or common manufacturing methods. The system has access to information on the manufacturing resources via a classification code matrix. The possible operational sequences are maintained in a sequence file. It is important to distinguish between a global sequence and a local sequence. The global sequence determines all manufacturing steps in sequential order. The local sequence relates to operations at a given workplace done at one setup.

TABLE 6.3 Functions of the Process Plan

Selection of raw material or blank
   Shape
   Dimension
   Weight
   Material

Selection of process and sequence of machining operations
   Global operations
   Local operations at a given workplace

Machine tool selection

Auxiliary functions
   Fixtures
   Tools
   Manufacturing specifications
   Measuring instruments

Manufacturing times
   Feeds and speeds
   Setup time
   Lead time
   Processing time

Text generation

Process plan output
   Process plan header
   Process plan parameters

The next activity is the selection of the machine tools. For this purpose, the system must contain a machine tool matrix file. It holds information on the functions of the machine tools and their physical restrictions. To determine auxiliary functions, the system has to contain fixture, tool, and measuring instrument matrix files. In addition, a file containing manufacturing specifications must be provided. In a succeeding step the manufacturing times are calculated. This includes determination of feeds and speeds and setup, lead, and processing times. Information on these activities may be obtained from look-up tables or with the help of formulas. The text generator produces the tabulated process plan. It may be displayed on a CRT or printed as a hard copy. Manufacturers that have implemented an automatic planning system have found that its design, implementation, and maintenance can be a time-consuming and rather horrendous task. The data base must contain the following files:

Manufacturing equipment of the plant
Available manufacturing processes
Part spectrum to be manufactured
Tool description
Jig and fixture description
Manufacturing times
Manufacturing unit costs

Most of the process planning systems that have been developed to date can handle only rather simple tasks, such as planning of drilling and turning operations. In addition, there are some systems available for milling and grinding. The aircraft industry is trying to automate process planning of complex sheet metal part families.

Before the conception of a process planning system, a feasibility study must be performed in which the entire part spectrum of the plant is to be investigated. A similar feasibility study to automate the design process was discussed in Chap. 5. The investigation must include possible automation of all functions of the process plan (Table 6.2). In many cases it may not be economically feasible to perform a complete vertical integration. Some planning functions may have to be done manually or semimanually. Similarly, horizontal integration of the entire workpiece family must be carefully investigated. It may not be economical to include seldom-used or odd-shaped parts. For these situations a manual interface to the process planning system must be provided.

Figure 6.33 shows the result of an investigation where the planning effort of different process planning phases for various industries was investigated. The part spectrum consisted of rotational workpieces such as shafts and bell-shaped workpieces and simple bent sheet metal stampings. The machining operations to produce the parts were turning, boring, reaming, thread cutting, stamping, and bending. There are two different results entered in this figure, one obtained from the manufacturer's own investigations and the other through a questionnaire. Figure 6.34 shows a pictorial in which the results obtained for different industries were averaged. The phases "determination of machining sequence" and "determination of manufacturing lead times" consume the major share of the planning effort. This implies that most of the emphasis in conceiving a process planning system should be directed toward these areas. The phases "text planning" and "process plan output" are not considered, since they are needed for any process planning system.

Similar investigations were made to determine the frequency distribution of part families within different industries (Fig. 6.35). This pictorial indicates that rotational and cubical workpieces are the predominant parts. It also suggests that a generalized process planning

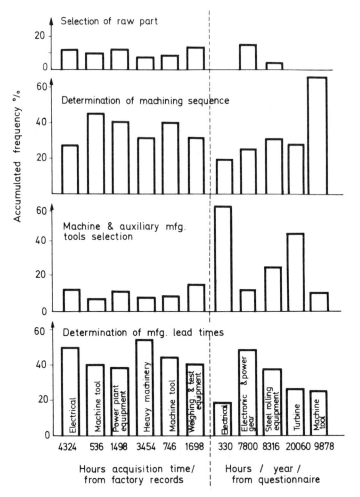

FIG. 6.33 Process planning effort by different industries. (From Ref. 13.)

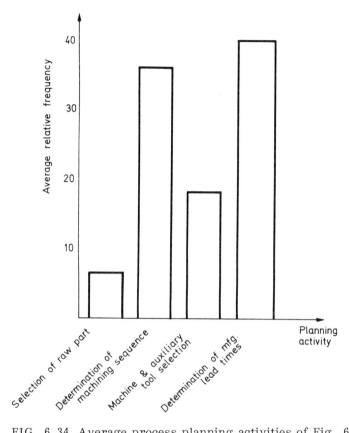

FIG. 6.34 Average process planning activities of Fig. 6.33. (From Ref. 13.)

system for rotational and cubical parts should be possible. Figure 6.36 contains a pictorial in which the results of Fig. 6.35 are averaged.

To automate a planning system with the help of a computer, the activities to generate a process plan should be described by algorithms. This is very easy to do for standard rotational parts. In most cases they can be represented by two-dimensional descriptive methods similar to those discussed in Chap. 5. The machining operations for these parts are also easy to describe. This method is not adequate when the part has more than one axis, asymmetrical protrusions, or is machined from other than round raw material. In this case a three-dimensional descriptive method must be used.

Cubical parts such as formed stampings, housings, and workpieces confined by spatial surfaces should also be represented to the computer

	0	1	2	3	4	5	6	7	8	9
Electrical & electromotor	35,5	2	40,7							
Electronic & power gear	36	11	24	1,5	2,5		21	3	2	
Heavy machine tools	21	20	17	5	2	2	12	12	10,5	1
Machine tools	25	16	12	5	1,5		14	12	14	
Lathes	33	23	20	3	1	2	5	6	8	2
Presses & Lathes	23	17	14	3	1,5		16	14	16	
Transfer machines	30	23	14	7	4		7	5	21	1
Turbines	20	29	21	9	1,5		9	5,5	7	
Turbines	18	21	6	30	9		9	12,5	6	
Pumps	31	8	23	29	1			1,5	9,5	
Textile machines	16	16,5	14	4	10	2	7,5	14	17	3
Steel rolling & pipe welding	18	9	10				54	11	1	
Weapons	20	26	16	7	3		11	9	12	

FIG. 6.35 Relative frequency (%) of part classes produced by different types of firms. (From Ref. 13.)

Type of parts	0	1	2	3	4	5	6	7	8	9
Relative frequency %										
Maximum	35.5	28.2	40.7	29.5	9.7	2.2	54	13.7	16.7	3
Median	24.2	16.9	17.4	7.6	2.7	0.5	13.2	8	9.6	0.5
Minimum	16.3	2	5.5	0	0	0	0	0	1	0

FIG. 6.36 Relative frequency (%) of part classes produced by sampled firms of Fig. 6.35. (From Ref. 13.)

by a three-dimensional descriptive method. Here the description of spatial curves or randomly oriented planes usually leads to difficulties. These parts generally have to be machined in different positions, requiring refixturing or reclamping. Such operations are difficult to present with algorithms. For many machining operations feeds and speeds can be represented with empirical formulas that are easily translated into algorithms. However, severe difficulties may be encountered when cutting tool and fixture geometries are to be described by an algorithm.

6.3.3 The Process Plan

In the preceding section general process planning considerations were discussed. The selection of a process planning method depends on the part spectrum and on the internal structure of the manufacturing organization. The process plan contains a great number of different parameters, the most important of which are shown in Fig. 6.37. One has to distinguish among three different types of parameters:

1. Parameters specific to a part
2. Parameters entered periodically, such as material cost, manufacturing cost, and so on
3. Parameters automatically generated by the system

With an automatic process planning system the number of parameters specific to a part should be kept to a minimum. Otherwise, the system may be burdened with too many input errors. The majority of the source data for the process plan are recorded on the drawing and the bill of materials. To automate the system, it should be designed in such a way that these data can be readily accessed.

The process planning system must be conceived of as an integrated part of manufacturing (Fig. 6.38). The flow of orders will first be processed by production planning. Therefore, information about the preliminary manufacturing sequence and the process description enters the process planning activity. Here a current process planning file is generated which is the input to process planning monitoring. Furthermore, from process planning the detailed operation instructions are entered into manufacturing monitoring and control. The last two activities are needed to supervise proper execution of the process plan and manufacturing. This rather involved interaction between all these activities has to be done with high precision and full understanding of the entire system.

There are three different types of process planning systems in use: (1) generative process planning, (2) variant process planning, and (3) a hybrid system consisting of generative and variant process planning. The two first systems will be discussed in more detail. Coding

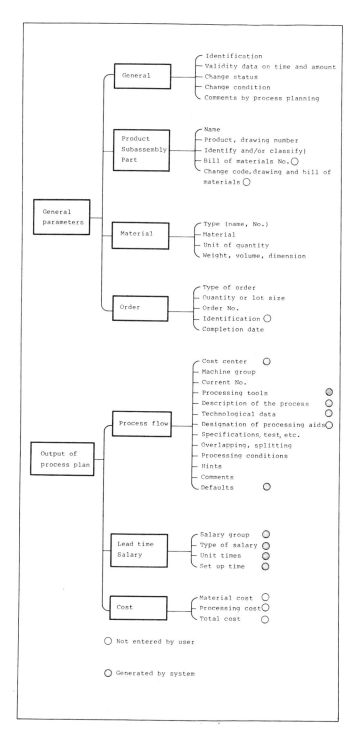

FIG. 6.37 Information related to the process plan. (From Ref. 14.)

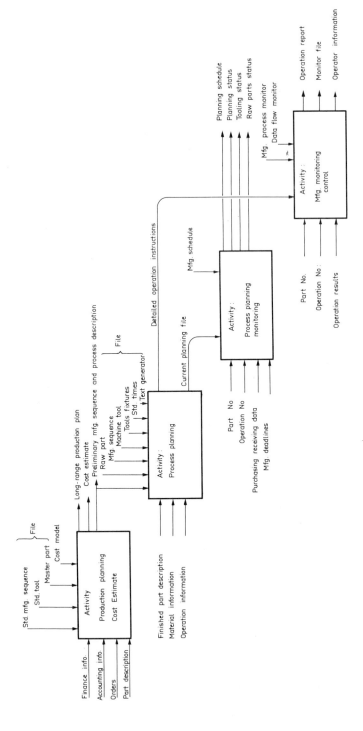

FIG. 6.38 Process planning and its manufacturing environment. (From Ref. 15.)

Manufacturing Systems

and group technology concepts play an important part in process planning. They were presented in an earlier section and it is assumed that the reader is familiar with them. Since both the generative and variant methods have their limitations, they are often combined in a hybrid system. However, this method will not be covered here.

6.3.4 Generative Process Planning

For the development of an automatic generative process planning system the part surface to be created and described by the drawing has to be related to a manufacturing method. The planning algorithm must contain modular machining primitives, which are capable of selecting from the dimensional parameters of the workpiece and other important physical properties the proper manufacturing process or sequence of processes. The designer may wish, for example, to provide a hole in a sheet metal bracket. In this case there could be two machining primitives available. The first one would select a drill press and the desired drill. The workpiece would be fixtured and the hole drilled. The second one would select a punch press and a punch. Again the workpiece would be fixtured, probably with a different type of fixture, and the hole would be punched. There may, however, be certain requirements as to the dimensional accuracy of the finish of the hole surfaces. In this case the system will be more selective and may find only one primitive that fulfills this requirement. In general, there are many more parameters that have to be considered, such as number of pieces produced, depth of hole, its location, hardness of the bracket, and many others.

A machining primitive may contain one or several operations. If, for example, a workpiece has to be provided with a threaded hole, the operations centering, drilling, and chamfering that proceed tapping may be implied by a machining primitive tapping (Fig. 6.39). It may also be convenient to combine a series of operations to produce a complex workpiece to a "super" primitive or a machining module. Such a module may even contain sequential machining operations to be done on different machine tools.

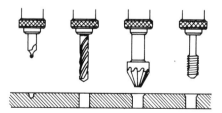

FIG. 6.39 Machining sequences to produce a thread.

A generative process planning system needs a rather complex decision logic (Fig. 6.40). It has to have manufacturing know-how of all processes used in a plant. It also needs information on possible succeeding processes, manufacturing alternatives, and conflicting or mutually excluding processes. For this reason most of these systems are still in the experimental stage. In the future, however, they will play an important role in manufacturing. With the advent of low-cost supercomputers and results from research work in the field of artificial intelligence, it will be possible to conceive generative process planning systems with limited human intelligence. With these systems it will be possible to obtain optimal information on machining and machine tool sequences, cutting tools, fixtures, and machining parameters.

At the present time generative automatic process planning is of particular interest in low-volume production and for non-part-family manufacturing or in the case where a variant process planning system has to be built up from process primitives. This is similar to design, where variant design modules have to be constructed from simple point, line, and curved elements (Chap. 5).

Several conditions must be met when an automatic planning system is being implemented. First, the part has to be broken down into shape primitives with the help of which machining primitives can be defined. Dimensions of the part, dimensional accuracy, surface finish, and physical properties are major factors in determining the machining operation. Second, rules have to be devised by which certain primitives can be combined to machining modules. Third, a method must be found to formalize the description of machining primitives and modules. It would also be of advantage to conceive generalized machining modules which can be used universally through parameterization. Finally, decision rules have to be found to determine the sequence of machining operations and alternatives.

The basic information on the process plan can be entered into the computer by textual programming, coding, or by a graphic technique. If a user guidance method is used, the latter techniques should be preferred over the first. It is easy to learn, is less sensitive to input errors, and in general is user friendly, requiring no programming expertise. Since the whole process plan is to be generated by the computer, the system should be capable of printing the entire operation sheet (Fig. 6.32). As discussed earlier, it contains heading data and data describing the process sequence. This implies that a syntax or instruction format has to be conceived to allow the user to enter the data. The instruction formats will differ in different organizations.

It may even be necessary to reformat the heading of the hitherto used manual operation sheet. Typical instruction formats to describe parts were shown in Chap. 5. If an integrated CAD/CAM system is being conceived, it is of advantage to use a similar syntax for process planning as for CAD. Furthermore, the system designer should follow

Manufacturing Systems

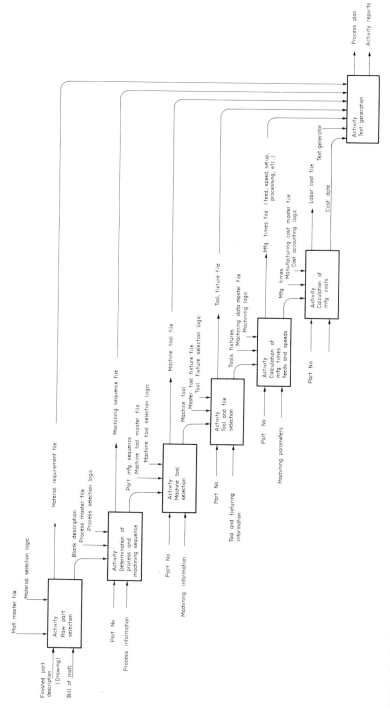

FIG. 6.40 A simplified generative process planning activity.

modern software engineering methods. The planning activities shown in Table 6.2 and Fig. 6.40 require the conception of seven main modules which will be linked together by a planning processor. First, the user codes the application and then a syntax check is made. After correction of possible errors, the program modules are compiled and the planning processor links them together in an executable program. In the following section the individual modules are discussed in more detail. Because of the many different alternatives to be considered in process planning, decision and look-up tables are widely used.

Activity: Raw Part or Blank Selection

In this activity the raw part or blank from which the finished workpiece will be made is selected. Here one of the main functions is to determine the machining allowance. For this the system must know the permissible machining allowance for each material-machine combination. Since a part can be fabricated by different manufacturing methods and from different materials, there must be decision rules that also consider economical aspects. For example, with a small production run the lowest cost to make a bracket may be to machine it from a solid blank. However, in mass production a sheet metal operation may have to be considered. Actually, for raw part selection all succeeding activities have to be considered (Fig. 6.40), since the material selected also determines the machining process. This activity gets its basic information from the material master file. In case there is no information on the part in this file, it has to be provided with the help of a service program. The output of part selection is a material requirement file used for material management. This information also enters the next activity.

Activity: Determination of Process and Machining Sequence

In this activity the surface to be machined is described with equivalent machining elements. These determine the type of process to be used. There are usually several different machining elements needed to generate the final contour of the part. It may also be possible to have different processes available to fabricate the part. Knowledge of the processes and possible manufacturing sequences is obtained from the process master file (Fig. 6.40). Data on part number and part description are entered into the computer and with the help of these a suitable process and manufacturing sequence are determined. Figure 6.41 shows alternative process selection methods for forming metal duct halves. In this case there are many parameters to be considered when selecting a suitable process. For example, a 36-in.-deep aluminum duct needing a blank size of 194 in. will be manufactured most economically on a stretch press. However, this method does not allow for return flanges, compound planes, and so on. For other

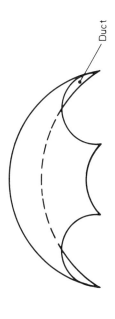

				Complexity of part					
Equipment	Material type	Blank size (in.)	Draw depth (in.)	Simple contour	Reverse contour	Return flanges	Compound planes	Compressed or stretch flanges	Bend radius
Hammer	Aluminum Stainless steel Titanium Hastalloy	60 × 96	20	Yes	Yes	No	Yes	Severe	Severe
Bag press	Aluminum	42 × 150	10	Yes	Yes	Yes	Limitations	Limitations	Severe
Hydropress	Aluminum	54 × 120	4	Yes	Yes	Yes	Limitations	Limitations	Severe
Hydroform Press	Aluminum Stainless steel Titanium	24 Diameter	12	Limitations	Limitations	No	No	Severe	Severe

FIG. 6.41 Typical process selection matrix for forming duct halves. (From Ref. 3. Courtesy of Society of Manufacturing Engineers, Dearborn, Michigan.)

Equipment	Material type	Blank size (in.)	Draw depth (in.)	Complexity of part					
				Simple contour	Reverse contour	Return flanges	Compound planes	Compressed or stretch flanges	Bend radius
Electroshape press	Aluminum Stainless steel Titanium Hastalloy	20 Diameter	6	Yes	Limitations	No	Yes	Severe	Severe
Stretch press	Aluminum	48 × 194	36	Yes	No	No	No	No	Limitations
Stretch draw	Aluminum	82 × 84	10	Yes	Yes	No	Yes	No	Limitations

FIG. 6.41 (Continued)

Manufacturing Systems

flanges there are several alternatives. In this case the computer should also be supplied with manufacturing cost data and possible equipment to be preferred. For determining possible machining sequences it is necessary to compile a list of all available fabrication processes to generate the part and to construct a part family process grid (Fig. 6.42).

With the help of this information it is also possible to investigate manufacturing alternatives, material flow routes, and to do part grouping to use common processes. The subject of group technology was discussed earlier in the chapter. The final output of this activity is the machining sequence file used in the process plan and by production scheduling. Input data are also generated for the next activity.

Activity: Machine Tool Selection

This activity is actually very close related to the activity that determines the process sequence. For this reason they are very difficult to separate. The main source of data is the machine tool master file describing all available machine tools in the plant. Typically for selecting a machining process the workpiece is regarded as a composition of unit machining surface.

For each process or machine tool there should be a process capability file. It will contain several or all of the following parameters:

Types of surfaces and dimensions that can be machined
Workpiece material and initial condition
Types of machining operations and machinability data
Types of tools and tool approach directions
Obtainable tolerances, surface finish and form geometry
Setup and machining times and cost data

Figure 6.43 shows a simplified process grid for selecting lathes for rotational parts. The process boundary is the length/diameter ratio. All machine tools in this figure will have approximately the same accuracy and produce the same surface finish.

When there is a choice of different machine tools, accuracy and surface finish become very important parameters which determine the process boundary. A typical accuracy analysis to produce holes is shown in Table 6.4. The parameters recorded are listed in the left-hand column. They are straightness, parallelism, roundness, true position, and surface finish. A computer program that has to check all these parameters will get quite involved. For economic reasons the number of process boundaries needed should be carefully investigated. They should actually be kept to a minimum.

FIG. 6.42 Part family process sequence grid. (From Ref. 17. Courtesy of Society of Manufacturing Engineers, Dearborn, Michigan.)

				PART NUMBER	8Q236	8C2009	57R2152	2102025	2103244	2102068	2102OFA	57R2244	75R212	57R2313	65R2211	65R2185	
				DIAMETER	438	438	375	312	375	312	375	312	500	312	437	437	
				APPROX. DEV. LENGTH	18.42	19.62	20.00	20.70	12.15	8.06	22.50	7.75	15.50	16.87	18.06	37.50	
				MATERIAL	1018	1018	1018	1018	1018	1018	1018	1018	1018	1018	1018	1018	
SEQ	BC	MACH.	OPERATION	PART NAME	CROSS SHAFT	CROSS SHAFT	ACCELERATION ROD	ACCELERATION ROD	CONTROL ROD	CONTROL ROD	CONTROL ROD	BRAKE ROD	ACCELERATION ROD	ACCELERATION SHAFT	ACCELERATION SHAFT	ACCELERATION SHAFT	MANUFACTURING RULES
010	51	MAN	CLEAN AND DELIVER		X	X	X	X	X	X	X	X	X	X	X	X	ALL COLD ROLLED PARTS START AT 51 MANUAL CLEAN AND DELIVER
020	51	4389	SAW TO LENGTH WELLS		X	X	X	X				X		X	X	X	
030	51	4607	CHAMFER BOTH ENDS		X	X	X	X				X		X	X	X	IF QUANTITY IS LESS THAN 25 AND IF PART IS A CROSS SHAFT AND IF DIAMETER IS LESS THAN 3/4"
040	27	4811	CHAMFER THREAD CUTOFF														THEN 51 4389 SAW TO LENGTH WELLS
050	74	4897	CHAMFER TURN CUT THREADS CUTOFF						X	X		X					IF QUANTITY IS GREATER THAN OR EQUAL TO 300 AND IF LENGTH IS LESS THAN 10" AND IF DIAMETER IS LESS THAN OR EQUAL TO 3/4"
060	40	M	HOLLOW MILL THREAD (1) END					X			X						THEN MAKE ON 74 4896 AUTOMATICS
070	40	M	CUT THREADS ON BOTH ENDS						X				X				
080	40	M	HOLLOW MILL (1) END					X	X			X					IF ANY THREADS THEN GO TO 40 M
090	94	W	WASH									X					IF DONE ON AUTOMATICS AND IF THERE ARE ANY THREADS
100	94	0427	CUTOFF BURR						X	X		X					THEN 94 W WASH
380	12	INSP	INSPECT BEFORE VENDOR		X	X	X	X	X	X	X		X	X	X	X	ALL PARTS DONE ON AUTOMATICS 94 0427 CUT OFF BURRS
390	39	VHT	OUTSIDE VENDOR HEAT TREAT		X	X	X		X	X	X					X	IF ANY HEAT TREATING IS SPECIFIED THEN VENDOR HEAT TREAT
400	39	VPL	OUTSIDE VENDOR PLATE		X	X	X	X	X	X	X		X	X	X	X	
420	12	INSP	FINAL INSPECT		X	X	X	X	X	X	X	X	X	X	X	X	IF ANY PLATING IS SPECIFIED THEN VENDOR PLATE

Manufacturing Systems 341

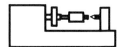

		Lathe 1	Lathe 2	Lathe 3	Lathe 4
Length	L = 50 cm	x	x		
	L = 100 cm		x	x	
	L = 200 cm			x	x
	L = 400 cm				x
Diameter	D = 20 cm	x	x		
	D = 30 cm		x	x	
	D = 40 cm			x	x
	D = 50 cm				x

FIG. 6.42 Part family process sequence grid. (From Ref. 17. Courtesy of Society of Manufacturing Engineers, Dearborn, Michigan.)

Figures 6.44 through 6.46 show various methods used to describe the machining of slots. In the first example (Fig. 6.44 and Table 6.5) there are 32 different attributes by which a slot can be presented to the computer to select the appropriate machining process. A set of notations can be used to describe a machined surface [17].

$$S_i = \{k, t, z, -, \square, //, L, /, \oplus, \odot, \div, \frown, \cap\}$$

The set characters are

   k = surface to be machined (slot, plane, hole, etc.)

   t = dimensional tolerance

   z = dimensionality

The other variables are form geometric presentations FG(n) of the workpiece characteristics. With this connotation and the selected attributes, process boundaries can be described as follows:

$$P_k = \{K, T, Z, FG(1), \ldots, FG(10)\}$$

If $S_i \in P_k$, then the surface i can be machined by process k. Another possibility for describing slots and their equivalent computer instructions is shown in Fig. 6.45. The instruction to perform the machining operation is

TABLE 6.4 Hole Process Boundary Matrix: PPLIM(ITYPE, IBNDRY)[a]

IBNDRY	ITYPE			
	Twist drill	Spade drill	Boring	Spot drill
Smallest tool diameter	.0625	.75	.375	.125
Largest tool diameter	2.0	4.0	10.0	.75
Negative tolerance (undersize)	$.007(DIA)^{.5}$	$.004(DIA)^{.5} + .0025$	.0003	.001
Negative tolerance (oversize)	$.007(DIA)^{.5} + .003$	$.005(DIA)^{.5} + .003$	.0003	.005
Straightness	$.0005(l/d)^{3} + .002$	$.0003(l/d)^{3} + .002$	.0003	.002
Roundness	.004	.004	.0003	.003
Parallelism	$.0010(l/d)^{3} + .003$	$.0006(l/d)^{3} + .003$	.0005	.002
Depth limit, l/d ratio	12.0	4.0	9.	1.0
True position	±.008	±.008	±.0001	±.002
Surface finish, CLA	100	100	8	60

[a]All dimensions in inches. Surface finish in microinches.
(From Ref. 17. Courtesy of Purdue University, Lafayette, Indiana.)

TABLE 6.5 Attributes Used to Describe Slots (a slot is a cavity or slit in a workpiece characterized by symmetry about at least one axis)

ATTRIB (1) = slot number
ATTRIB (2) = surface type
            = 3 - slots

              ATTRIB LIBRARY FOR SLOTS

ATTRIB (3) = SLOT TYPE
            = 1; SIMPLE
            = 2; DOVETAIL
            = 3; RADIUS (WIDTH)
            = 4; RECTANGULAR
            = 5; RADIUS (LENGTH)
            = 6; RECTANGULAR W/RADIUS (WIDTH)
            = 7; T-SLOT
            = 8; Y-SLOT
            = 9; VEE BOTTOM
            = 10; COMPOUND
            = 11; OTHER

ATTRIB (4) = SURFACE FINISH (BOTTOM) -CLA
ATTRIB (5) = PRECAST INDICATOR
            = 0; WITH CASTING SKIN
            = 1; W/O CASTING SKIN
ATTRIB (6) = MAX STOCK REMOVAL
ATTRIB (7) = MATERIAL TYPE
            = 1 - CAST IRON
            = 2 - MALLEABLE IRON
            = 3 - STEEL
            = 4 - STEEL FORGING
ATTRIB (8) = BHN
ATTRIB (9) = DIRECTIONAL VECTOR
            = 1; LINEAR
            = 2; CIRCULAR
            = 3; ELLIPTICAL
            = 4; HELICAL
            = 5; OTHER
ATTRIB (10) = THRU OR BLIND INDICATOR
            = 1; THRU BOTH ENDS
            = 2; THRU ONE END
            = 3; BLIND BOTH ENDS
ATTRIB (11) = BOTTOM SLOT INDICATOR
            = 1; FLAT
            = 2; RADIUSED WIDTH
            = 3; RADIUSED LENGTH
            = 4; BOTTOM ANGLE
            = 5; RADIUSED BOTH LENGTH AND WIDTH

---

[a]From Ref. 17. Courtesy of Purdue University, Lafayette, Indiana.

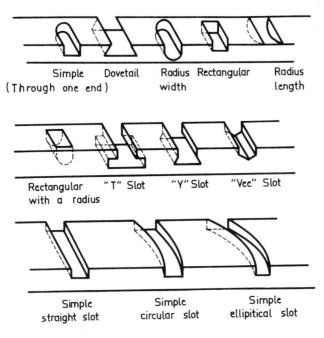

FIG. 6.44 Different slot configurations and their definitions. (From Ref. 17. Courtesy of Purdue University, Lafayette, Indiana.)

$$\left.\begin{array}{c} \text{MCOARSE} \\ \\ \text{MFINE} \end{array}\right\}, \text{MTOL}, +\text{K}, -\text{K}, \left\{\begin{array}{c} \text{TA} \\ \text{TB} \\ \text{TC} \\ \text{TD} \end{array}\right\}, \text{No.}$$

The modifier is explained in Fig. 6.46. The output of the machine tool selection is recorded in the process plan and in the next activity, which selects tools and fixtures.

Activity: Tool and Fixture Selection

The task of this activity is to select the correct tools and fixtures. To optimize manufacturing productivity, as few tools and fixtures should be used as possible; otherwise, tool change and approach times may slow down the production considerably. The tool and fixture activity gets its input from the master tool and fixture file and the machine tool selection activity.

No.	Slot shape	Instruction
1	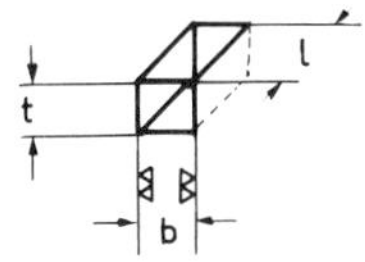	SLOT / a, b, l, t a = 1
2	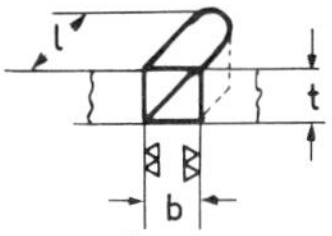	" a = 2
3	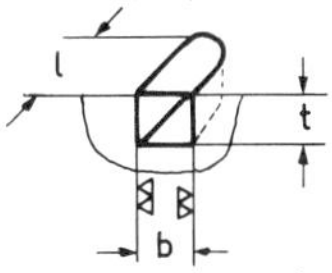	" a = 3
4	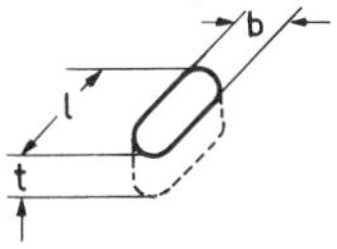	" a = 4
5	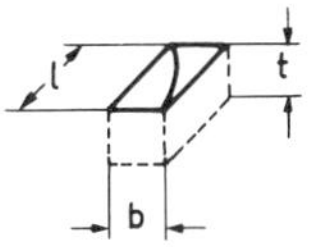	" a = 5
6	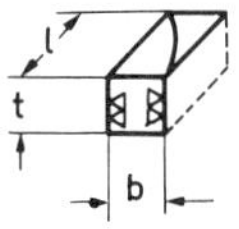	" a = 6
7	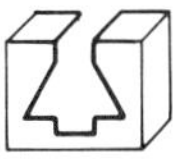	SLOT / NSTD Nr, Type, b, l FSTD
		NSTD National Standard FSTD Factory Standard

FIG. 6.45 Description of slots and their equivalent computer instructions. (From Ref. 14.)

Modifier		Value
Surface finish	MCOARSE MFINE	K = 0.063 mm RMS K = 0.061 mm RMS
Machining tolerance	MTOL	+K, −K 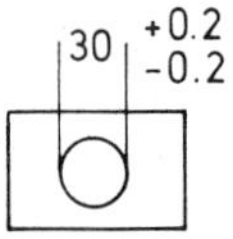 30 +0.2/−0.2
Technology	TA, TB, No. TC, TD	Additional information given by the user E.g. TB 0.05 Roundness TD 0.03 Parallelism

FIG. 6.46   Modifiers to describe machining variables.

For each machining element needed to generate the workpiece, a cutting tool is selected. Several alternatives may have to be considered in order to accommodate as many machining elements as possible with one tool. For this purpose the tool file will contain information on the type of tool, the geometry, its approach directions, and the cutter bit. Each of the tools capable of producing a machine element will have a priority attached to it. It may be based on chip removal rate, tool geometry, or other manufacturing considerations. Figure 6.47 shows four tools that would be able to generate a cylindrical surface. Similar tables would have to be set up for fixturing devices. The output of this activity is the tool and fixture file needed to set up the machine tools. Information on the tools is also supplied to the next activity.

Activity: Calculation of Manufacturing Times, Feeds, and Speeds

When the tools have been selected feeds, speeds, setup times, cut times, tool change times, and tool life have to be determined. These parameters will be affected by the material to be cut, the cutting tool, machining parameters, and the design of the machine tool. There are ample machining data available in various handbooks [18]. The main problem is to decide what data are essential and to limit the file to an acceptable size. The most common cutting tools are high-speed and carbide-tipped steels. The cutting speed to be used can be calculated as

$$V = V_0 \times F_f \times F_d$$

where

V_0 = reference cutting speed, m/min
F_f = feed factor
F_d = depth of cut factor

The preceding parameters can be obtained from tables [18] and can be implemented in the process planning program. For turning and boring machines the machining data can easily be calculated.

A suitable file structure containing machining data is of hierarchical design (Fig. 6.48). Calculation of manufacturing times is one of the most important tasks of process planning. This information is needed to determine labor cost, lead times, to locate manufacturing bottlenecks, and to perform investment planning. Investigations have shown that time calculations need about 30% of the entire planning activity. This is a rather high share, in particular if the low accuracy associated with this activity is considered. A reasonable accuracy can be obtained only when all subjective influences have been eliminated.

Tool	Priority	Tool		Main cut direction	Cutter bit type
		Family	Type		
	1	6	2	1	4
	2	6	5	1	1
	3	6	3	2	4
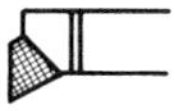	4	6	3	2	4

FIG. 6.47 Different tool alternatives to generate a given surface.

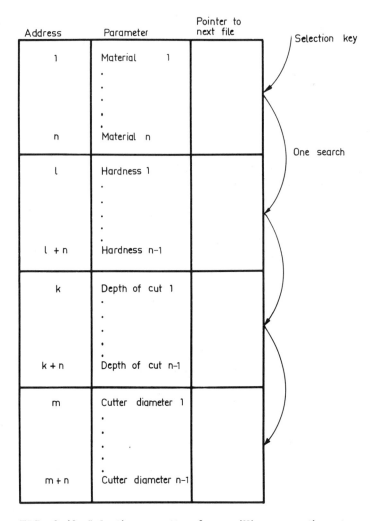

FIG. 6.48 Selecting a cutter for a milling operation at constant cutting speed. (From Ref. 17.)

Good time data may be gathered with computer-controlled data acquisition equipment. The total machine time is [17]

$$T = T_c + T_{lu} + T_{ar} + T_m + T_d \frac{T_c}{T_l}$$

where

T_c = cutting time = $\pi \cdot$ diameter $\cdot$ length$/V \cdot f$
T_{lu} = load and unload time
T_{ar} = tool approach and return time
T_m = measuring time
T_d = tool change time
T_l = tool life

The times T_c and T_{ar} can be calculated with the help of operating parameters. The tool life T_l is obtained from a modified Taylor equation:

$$T_l = \frac{C}{V^\alpha f^\beta d^\gamma}$$

where

V = cutting speed

f = feed rate

d = depth of cut

C = machinability factor

α = speed exponent

β = feed rate exponent

γ = depth-of-cut exponent

The factors C, α, β, and γ are obtained from experiments and are published in handbooks. They depend on the material to be cut and tool hardness, surface structure, cutting geometry, tool stability, number of interrupted cuts, and other factors. The times T_{lu}, T_m, and T_d should be taken from operational data or time estimates. The output of this activity will be placed in the manufacturing time file and is also used to calculate manufacturing costs.

Activity: Calculation of Manufacturing Costs

The manufacturing cost can be calculated with the help of the time data obtained from previous operations and the manufacturing cost master file. As these calculations are straightforward accounting functions, they will not be discussed in detail. Cost data are the basis for optimization calculation. In order to perform such calculations, optimization functions have to be established and described with the help of algorithms. Usually, such functions are very complex and nonlinear and are difficult to solve.

6.3.5 Variant Process Planning

The variant process planning method assumes that for a workpiece to be manufactured, there must be available a process planning model of a nonparameterized variant. During process planning the designer queries a variant catalog and searches for a part having design and manufacturing similarities. When the part has been found, the parameters of the new part are entered and inserted by the computer into the process model. With this information the computer automatically generates the process planning sheet. This method is much simpler to use than the generative method; however, it assumes that variant macros exist and that a similar workpiece can be manufactured by similar processes (Fig. 6.49). The two spools shown in Fig. 6.4 could not be planned by the variant method since they have only design and no processing similarities. Most of the process planning systems presently available are of the variant type and use some kind of group technology principles.

Prior to setting up such a variant process planning system, a study of the entire part spectrum must be conducted. Similar parts of similar design and manufacturing characteristics will be grouped together and non parameterized variants will be designed. The variant may be constructed either with the help of the generative process planning method or from simple primitives describing individual processes, machine tools, cutting tools, and so on. At present, the latter method is preferred, since for generative process planning many basic problems have yet to be solved. In many cases it may be uneconomical or even impossible to conceive variants for a part spectrum. In this case the workpiece may have to be broken down into form elements as discussed in Chap. 5. A separate process plan will be drafted for each of the form elements. It is the designer's function to combine the form elements to one workpiece and to configure one process plan. This method may not be very easy to use since form elements can be machined much more easier and with more universal tools than variants. There also can be the possibility that when the computer selects a machining methods for one form element, the cutting tool will interfere with an adjacent form element.

Manufacturing Systems

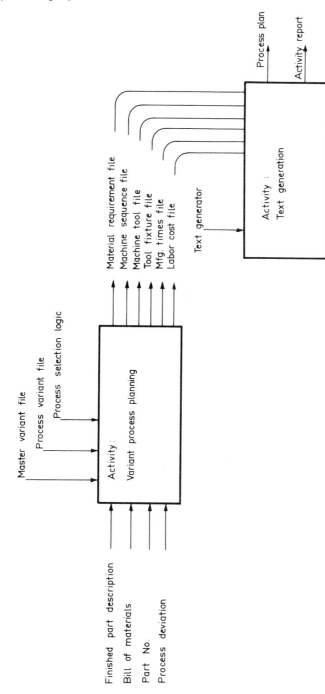

FIG. 6.49 Simplified variant process planning activity.

With proper application of group technology and machining know-how, this problem may be resolved.

Modeling process variants from design variants is not always possible. Figure 5.46 shows different gears that can be described by one nonparameterized variant. In practical applications there may be a size problem if the part spectrum contains small and large gears which for economic or dimensional considerations have to be machined on completely different types of machine tools or even made by different processes. The same problem may arise with gears having the same size but different shapes.

CAPP System

With variant process planning there are considerably fewer data to be entered into the computer than with generative process planning. For this reason this method is well suited to be used on-line in an interactive mode. Such an on-line planning system, CAPP (CAM-I Automated Process Planning System), developed by Computer Aided Manufacturing-International, Inc., will be discussed briefly. This system is being developed under the CAM-I program and funded by the supporting member organization from industry, government, and education from Europe, North America, and Japan.

The goal of the process planning project is to automate this activity completely. The following objectives are to be achieved:

1. To develop a software system for process planning
2. To normalize manufacturing information and data structure, define a common external interface, improve the human-machine interaction, and spur development of application programs.
3. To solve immediate requirements of process planning for a wide industry spectrum, centralize master planning files for source data, provide file maintenance and integrity features, reduce manual work, and thereby improve the skill of the planner
4. To create a common manufacturing technology data base

There are different versions of the CAPP system; they can be operated on large and small computers. The system is based on group technology as it was discussed in this chapter. The classification and coding method are determined by the user. The code length can be selected, up to a maximum of 36 digits. The data structure of CAPP has six files (Fig. 6.50). The user must provide the family matrix, standard sequence and operation plan files, work elements (machine elements), and work element parameters into the data base.

The user loads the part family matrix into the matrix file (Fig. 6.51). A pointer from each part family points directly to its standard manufacturing plan, also supplied by the user. There is a standard plan information structure provided by CAPP (Fig. 6.52). It contains

Manufacturing Systems 353

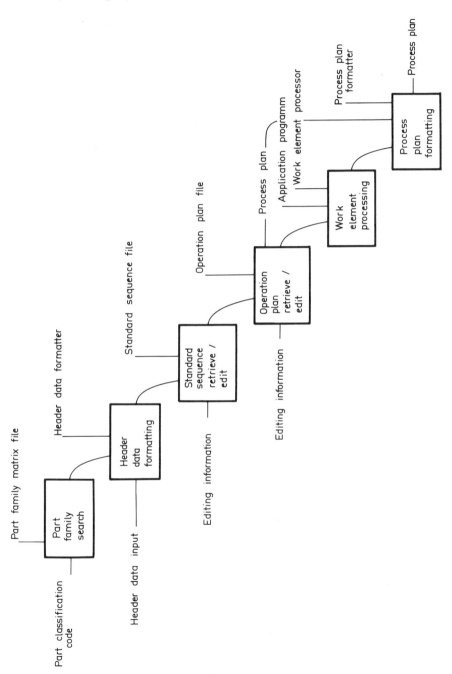

FIG. 6.50 Flow diagram of the CAPP process planning system. (From Ref. 19. Courtesy of Manufacturing Engineers, Dearborn, Michigan.)

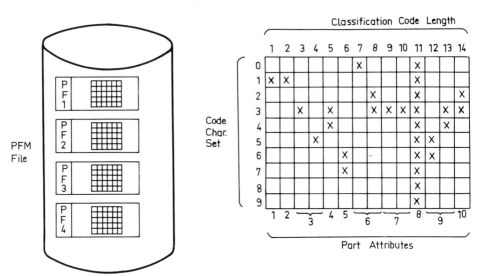

FIG. 6.51 Part family matrix file. Each PFM points directly to a unique standard manufacturing plan. (From Ref. 17. Courtesy of Society of Manufacturing Engineers, Dearborn, Michigan.)

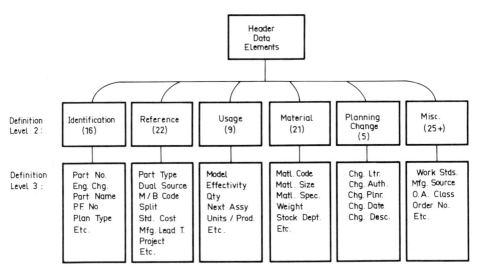

FIG. 6.52 Standard plan information structure. (From Ref. 19. Courtesy of Society of Manufacturing Engineers, Dearborn, Michigan.)

```
                    HEADER DATA

     OS  OP  PP  SI  SC  FS  DS  MM  PD  PU  FI  LD


    PF NUMBER 20                  MOD 30
    CLASS CODE 21385597           US 1
    PART NUMBER 63A480638-2005    EFFECTIVITY F15A/TF15A
    PART NAME ARM, ACTUATOR       S/N 26
    MATL D6AC                     THRU UP
    SPEC NIL-S-8949               NEXTR ASSY 68A480638-5003
    SIZE 6BA480638 FORGING        BANNER INTERCHANGEABLE
    ORIG PLNR RVK                 SPECIAL INFO SEND PART TO
    DATE 2/3/76                   VENDOR FOR SHOT PEEN-
    AUTH DCN A                    CORROSION PROTECT PER PS2001
    SPCDS                         PROTECT PART FROM NICKS
    CODE 1/N/M                    SURFACE DAMAGE IN TRANSIT
    JOB 668                       AND STORAGE BETWEEN OPERATIONS
    COST/CODE 165
```

FIG. 6.53 Input and edit of header data. (From Ref. 19. Courtesy of Society of Manufacturing Engineers, Dearborn, Michigan.)

```
                     OPCODE SEQUENCE
     IS  IN  DL  RV  RT  RN  OP  DS  FS  MM  SI  SC  LD

            PART NUMBER 68A480638-2005
            PART NAME ARM, ACTUATOR
            PF NUMBER 26
            CLASS CODE 21385597
            OPCODE SEQ
         10 FURNMAIL        140 VENDSHOTP
         20 DEGREASE        150 INSP
         30 HTREAT          160 CADPLT
         40 VMILL           170 PRICE
         50 DRILREAM        180 INSP
         60 VHVDRO          190 PKG
         70 HMILL           200 ASSEMBLE
         80 SGRIND
         90 DEBURR
        100 ID
        110 INSP
        120 ALKCLEAN
        130 STRESSLV

        DL/80 -- DL/170 -- IS/70, VMILL2 -- RN
```

FIG. 6.54 Display and edit of operation code sequence. (From Ref. 19. Courtesy of Society of Manufacturing Engineers, Dearborn, Michigan.)

TABLE 6.6 Typical Instructions Used by the CAPP System

Instruction	Action
FS	Part family search
FP	Format plan
DP	Delete plan
DL/(Operation No.)	Delete operation No.
HD/(Part No., Plan type, Status)	Retrieve header
OS/(Part No., Plan type, Status)	Retrieve operation code sequence
OP/(Part No., Plan type, Status)	Retrieve operation plan
PP/(Part No., Plan type, Status)	Process plan
MS/(Classification code)	Matrix search
SP/(Part family No.)	Retrieve standard plan
DS/(Part family No.)	Display part No.
SA/(Attribute, Value)	Search attribute
CS/	Continue search
MM	Return to main menu
FI	Position cursor
WX/(Operation No.)	Expand work element
LO	Log off
.	.
.	.
.	.

approximately 100 items which have to be selected by the user for the header data upon initialization. The planner cannot change the header's format. A typical header is shown in Fig. 6.53. There are provisions to define an operation code menu for the various manufacturing operations. Figure 6.54 shows a sequence of operation codes for a part family. The operator has a user language of approximately 33 instructions available to work interactively with the system. Some instructions are shown on the top line of Figs. 6.53 and 6.54. Several instructions and their meaning are shown in Table 6.6. Upon careful review of these instructions, it can be seen that the system is user friendly and very powerful. For example, a part family search can be made with the help of the instruction FS. The system may show that this family is already retrieved or it will interactively assist the operator to enter a new standard process plan. There are good editing functions available. For example, in case a process or process sequence has to be changed, this can be done easily by pointing to the operation sequence number at which the change has to occur. Deletion or addition of single or several operations is possible.

6.4 PRODUCTION PLANNING AND CONTROL

6.4.1 Master Production Schedule

Manufacturing resources comprise labor, energy, material, machine tools, and material movement equipment. When the production schedule is being prepared for a line of products, there often is an unlimited combination of different types of machining operation, labor skills, material flow routes, and inspection procedures to be investigated in order to optimize productivity or to maximize profit. Within the last decades, many mathematical methods have been developed to improve the allocation of factory resources, including heuristic approaches, queueing theory, linear and dynamic programming, and many others. Allocation and scheduling algorithms based on these methods usually have to be solved with the aid of the digital computer. Despite the enormous speed of the computer, it often may take many hours of computing time to arrive at an acceptable result. This holds true in particular in situations where hundreds of products are to be manufactured, involving hundreds of machine tools with thousands of machining operations and sequences, and where a high number of workers are employed. It may also become a very difficult task to build a model that is able to accommodate the entire scope of variables and combinations needed to build a mathematical image of the production system. For this reason optimization is usually done with smaller computing models which may render only suboptimal solutions. The results may, however, offer a substantial improvement over conventional planning and scheduling methods.

Resource planning starts with a long-range plan where a projection is made into the future of the company. This is done by observing the market, the resource supply, and new products and demographic developments. With this knowledge, elements that are needed for the manufacture of a new product are extrapolated into the future (Sec. 2.5.2). Once the long-range goals are agreed upon, manufacturing scheduling for the near future is activated (Sec. 2.5.6). For this it is necessary that orders have been entered and that the design of the product and its manufacturing methods and sequences are known. From the customer orders and from the forecast of future demand, a master production schedule is conceived, representing a list of end products to be manufactured, the quantities ordered, and the due dates. An end product may be a completed assembly, a subassembly, or a workpiece. A typical master production schedule is shown in Fig. 6.55. The plan contains a matrix listing the number of products to be completed in various time periods. The planning horizon is divided into a firm and a tentative planning section. The firm section contains the end items for which committed orders have been received. The master production plan serves as input to material requirement and manufacturing resource planning (Fig. 6.56).

6.4.2 Material Requirement Planning

A product to be manufactured usually consists of numerous components that may be produced from different engineering materials. To

Time period (month) \ Product	1	2	3	4	5	6	7	8	9	10	11	12
Vertical lathe*	10		12	13		8	3	10		20		15
Horizontal lathe*	1	3		2	2			3	2		5	1
NC lathe*	5		4	8	1	3	4	2		1		5

Planning horizon (Time periods): Firm (periods 1–7) / Tentative (periods 8–12)

FIG. 6.55 Master production schedule of a machine tool manufacturer.
* Quantities ordered.

meet the completion date for a product, the materials have to be ordered with sufficient lead time (Fig. 6.57). This requires accurate calculations to meet the due dates set by marketing. The due dates are obtained through a back-scheduling procedure. A considerable amount of experience is needed to calculate the lead time for raw material and components. The planner must know the company's production capabilities and those of the material and component suppliers. Standard materials and parts may be ordered periodically, since they are used in many products, assemblies, and subassemblies, whereas special or expensive materials may be delivered only on specific dates. In modern manufacturing organizations, material requirement planning is usually highly computerized.

The planning algorithms are, in general, not very complex; however, the number of parts and materials needed to complete an entire product spectrum may be so numerous that conventional planning procedures are too cumbersome and expensive. Material requirement planning produces information on the right quality, right parts, and the correct timing for production, raw material, and components. This information is derived from the following documents (Fig. 5.56): (1) bill of materials and related engineering documents, (2) master production schedule, and (3) inventory master file. The function of the master production schedule was discussed in Sec. 6.3.1 and that of the bill of materials in Sec. 6.2.1.

The inventory master file contains a record of all items that are momentarily in the inventory and which have been ordered (Fig. 6.58). With a properly designed material requirement plan, it is possible to order parts in the exact quantities needed to minimize inventory cost (Fig. 6.59). There are numerous cost items that have to be considered when parts are ordered in lot sizes. They are:

Manufacturing Systems

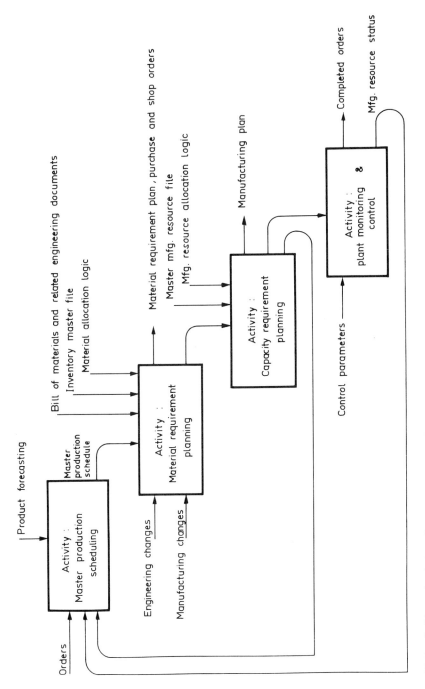

FIG. 6.56 Principle of master production planning activity.

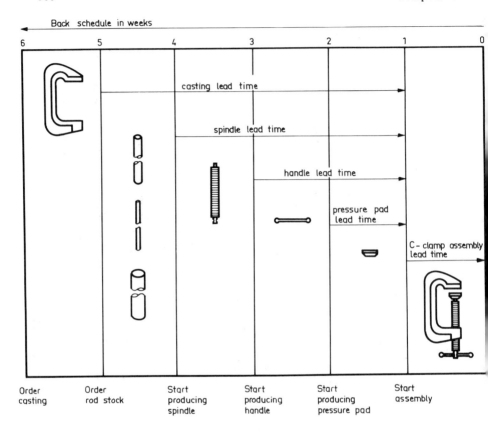

FIG. 6.57 Scheduling manufacture of a C-clamp.

    N = number of parts sold annually
    S = setup or ordering costs
    C = cost to carry inventory
    U = unit cost of part

With this information, an economic order quantity (EOQ) can be calculated under the assumption that parts are used at a constant rate. The economic order size is

$$EOQ = \sqrt{\frac{2NS}{CU}}$$

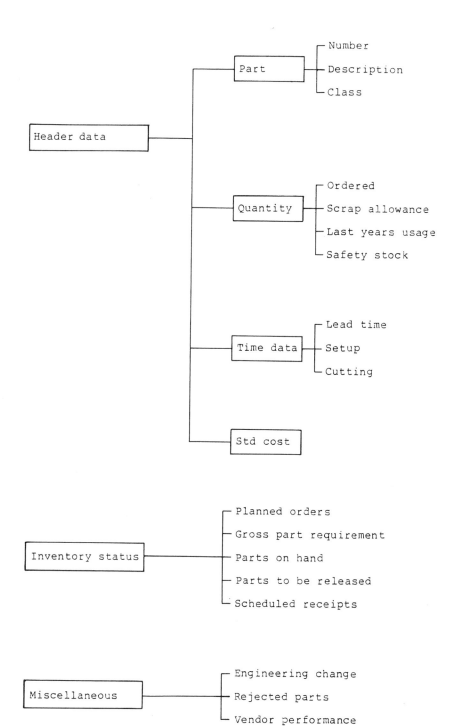

FIG. 6.58 Typical information contained in the inventory master file.

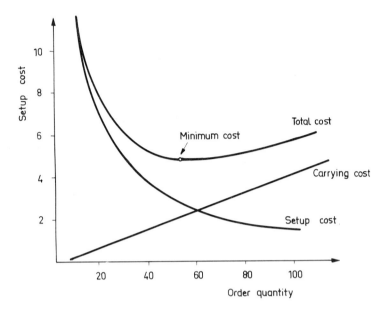

FIG. 6.59 Cost relationship for carrying inventory.

Calculation of the EOQ is not always easy because the demand for a product is often not continuous throughout a year. There are usually difficulties in determining the exact inventory carrying cost. For this reason different lot sizing techniques are being used. The reader may consult the literature on material requirement planning and operations research to become familiar with economic ordering techniques.

When calculating material demand, a distinction has to be made between independent and dependent parts demand. Independent demand refers to most finished goods and spare parts. It must be forecast by statistical methods since it is usually random in nature. The material requirement planning process concerns itself with dependent demand, which is derived from the information contained in the bills of materials. Typically, the structure of a product is of a hierarchical nature (Fig. 6.28) and there are several hierarchical levels in the bill of materials. Thus the number and type of items needed to build a product can be depicted directly from the bill of materials; they depend on the structure of the product.

6.4.3 Manufacturing Resource Planning

The functions of manufacturing resource planning are shown in Fig. 6.60. They consist of (1) capacity requirements planning, (2) order

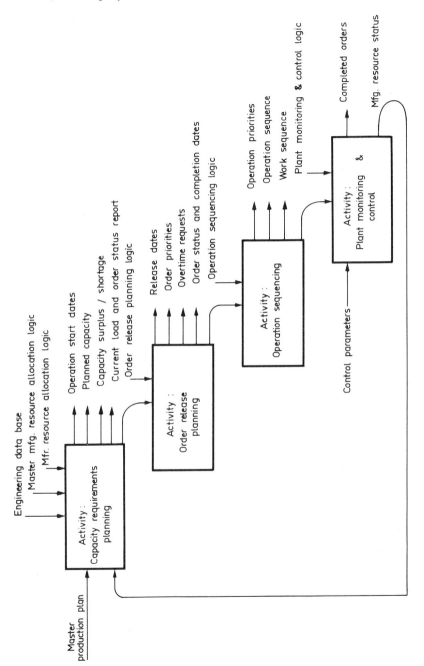

FIG. 6.60 Capacity requirement planning. (From Ref. 20.)

release planning, and (3) operation sequencing. Basically, these activities look at the delivery date of the finished product. They alot to every component of the product, labor, material, and machining resources in order to meet the customer delivery date. All manufacturing times are extrapolated backward from the due date of the time of the completed item. The basic data that go into manufacturing activity planning are the orders and their due dates, the manufacturing resource file, and the manufacturing resource status. Figure 6.61

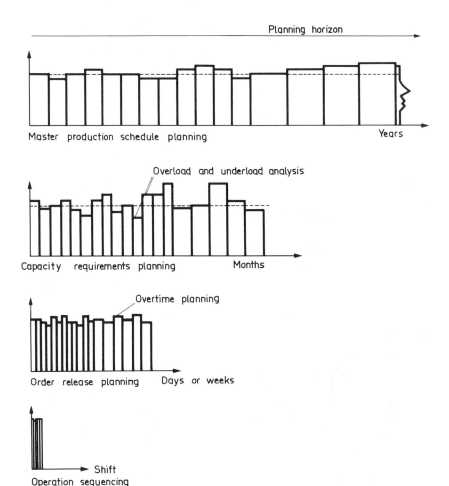

FIG. 6.61 Different stages of manufacturing resource planning. (From Ref. 20. Courtesy of International Business Machine Corporation, White Plains, New York.)

shows how the different phases of manufacturing resource planning are refined from one stage to the next.

The planning process starts with the conception of a master production schedule (Sec. 6.3.1). With this schedule a gross overall balance for the manufacturing facility is made to assure that the planned products can be produced with the available plant resources. The planning horizon is one or several years. The time increments are 1, 3, or 6 months. This schedule is used to perform capacity decisions for the plant, equipment, and labor.

Capacity planning is a refinement procedure wherein the long-range schedule is adjusted to the manufacturing needs of the nearer future. The planning horizon is one or several months and the time increments are weeks. In this stage, capacity adjustments are made involving:

Providing for production equipment
Subcontracting of work
Planning of labor resources

Order release planning renders details for planning of subcontractual work and overtime. Adjustments are made in order release dates if the capacity cannot be scheduled economically or if problems are encountered with production equipment. Normally, the capacity has been planned adequately in the previous stage and only some fine tuning has to be done. The planning horizon is in days for several weeks ahead.

Operation sequencing controls closely queueing of work at every work center. It tries to load equipment to its full capacity by issuing accurate sequencing orders to the manufacturing floor. At the same time, an attempt is made to meet order deadlines. The planning horizon is the shift or the workday, and the time increments are minutes or hours.

An example of how manufacturing resource planning works is shown in Fig. 6.57. The product to be manufactured is a C-clamp. Manufacturing resource planning obtains the information about the product and its components from the engineering drawing and the bill of materials. It is assumed that the order is to be shipped on January 1. From the lead times of the individual components a back schedule is calculated. For example, for the machined casting to be available for assembly, it must be ordered with a lead time of 5 weeks. The order of the rod stock for the spindle is released 4 weeks prior to assembly. The orders for the rod stock for the handle and pressure pad are issued at the same time, to simplify purchasing procedures. Manufacture of the handle starts 2 weeks before assembly and that of the pressure pad 1 week before assembly. With this schedule all components are available at the time when assembly commences. When each part is released for manufacturing, the proper machine tools have to be available and the routing has to be determined. The resource planning activity must know purchasing, lead, tool change, setup, machining, moving, and quality control times. The proper machine tools, machining sequences and a routing plan must also be available. In addition, it may be necessary to have on file an alternative manufacturing schedule for each part, in case other components with higher priority compete for the same resources or in case of an equipment failure. From this example it can be seen that the planning logic may get very involved and that there must be accurate information available on the manufacturing processes and its time constituents. Manufacturing resource planning tries to solve the following problems:

1. To meet the immediate and future master production schedule; to allocate for each processing center the required machining and operator times; in case of a bottleneck, to make capacity adjustment by subcontracting parts, allowing overtime, or adding an extra shift.
2. To determine for each order the completion time and to schedule orders according to priority rules; in case of foreseeable problems, to release orders earlier, to prevent idle times or bottlenecks, thereby leveling the load for each machine to reduce subcontracting, idle time, overtime, and personnel and part movements.
3. To plan queueing, to minimize lead times for machines and operators. In-process inventory is minimized if a workpiece is processed immediately after it has arrived at a workstation, thus making working capital available for other purposes.
4. To plan the sequence of operations needed for each work center; to issue a work sequence plan for the foreman and for manufacturing control.
5. To issue reports on start dates, capacity surplus or shortage, order release dates, order priorities, overtime, operation priorities, and sequences as well as on completed orders. Information will also be made available for new facility and equipment planning.

The algorithms to direct scheduling of the manufacture of a workpiece must contain information about the available plant resources, the equipment status, the composition of the product, and its desired completion dates. For this purpose they combine data from past performances of the plant, with parameters of its present status and with information about future deadlines. When scheduling is automated, the flow of this information into the scheduling algorithm must be guaranteed.

Information about the past and the present can conveniently be accumulated by factory data acquisition devices (Fig. 3.15). From the data

Manufacturing Systems

gathered, knowledge is obtained about machining and inoperative times, queue sizes and machine utilization, and operator performance. The inoperative time elements typically consist of:

Workpiece preparation time (cleaning, marking out, etc.)
Setup times (tools, workpiece)
Queue times (waiting for transportation and machining)
Postoperation times (deburring, inspection, classification)
Transportation time

Processing of Work Queues

The queue time is a decisive factor that determines the utilization of manufacturing equipment. In an ideal plant no queues are needed. Workpieces flow unobstructed from one processing station to another. In the real world, however, equipment cannot be loaded exactly to its capacity and does not always render the desired performance. This is due to the statistical distribution of processing times, problems with equipment breakdown, and operator fatigue. Thus queues are required to ensure against idle time. From the distribution of the queues, valuable information for the manufacturing schedule is obtained (Fig. 6.62).

Figure 6.62a shows a well-balanced manufacturing cell with a normal distribution. The average queue length is determined by equipment cost, type of manufacture, and management philosophy. In case the queue is too long, there may not be enough manufacturing capacity available, or other problems may prevail.

A persistent backlog of work may be caused by an unnecessary workpiece buffer (Fig. 6.62b). In this case, the average waiting time could be shortened substantially by a reduction of the contents of the buffer; otherwise, the distribution of the queue portrays a normal pattern.

A system that is frequently overloaded is shown in Fig. 6.62c. It is the task of scheduling to minimize this problem. A situation where too much capacity is available is shown in Fig. 6.62d. In this case the queues are very short or may hardly exist.

Optimal shop schedules can be obtained only when queuing problems are fully understood and when priority rules are given that render an acceptable job tardiness. The following parameters have a distinct influence on an efficient shop schedule:

Time distribution
    Arrival of the workpiece
    Setup of the tools
    Setup of the workpiece
    Processing of the workpiece
    Transportation of the workpiece
    Equipment breakdown and due dates

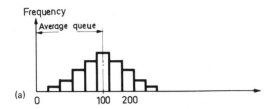

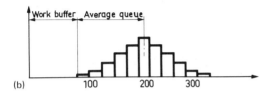

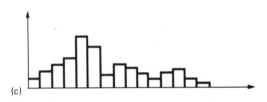

FIG. 6.62 Different distribution of queues at work centers: (a) controlled queue; (b) controlled queue with unnecessary buffer; (c) uncontrolled queue; (d) underloaded work center. (From Ref. 20. Courtesy of International Business Machines Corporation, White Plains, New York.)

Shop configuration
  Type of equipment used
  Equipment vintage
  Transportation system
  Warehouse system
  Plant layout

Shop practices
  Priority rules
  Machining methods and sequences
  Alternative processes and machining sequences
  Processing by single or multiple machines
  Single versus mixed model flow
  Overtime
  Job splitting
  Performance measures
  Performance evaluation methods

Shop load level

The time distribution cannot be easily predicted. In general, they are obtained from measurement of like or similar operations. In case of a new manufacturing process, they have to be estimated for the start-up phase of the operation. As soon as the process has come to a steady-state operation, the various distributions can be calculated from time measurements. The gathering of date may, however, be cumbersome.

The layout of a manufacturing system does not follow any strict rules. Often, the arrangement of the plant has simply grown historically to its present state, and old and new equipment have been merged. Even plants that manufacture an identical product may have completely different configurations. For this reason scheduling rules that pertain to one plant may be of no use to another.

Shop practices may also have historical roots or may have been formed by an individual or a team of experts. The desire of any manufacturing organization to maintain stable employment will greatly influence job practices. In some cases this may lead to situations where some processes are temporarily operated inefficiently. Priority rules are usually selected such that the job tardiness is minimized. Typical rules are [21]:

1. The workpiece with the earliest due date will be placed first in the machining queue.
2. The workpiece with the least processing time in the following queue will be placed first into the present queue.
3. The workpiece with the shortest processing time will be machined first.

4. The workpiece placed first into a queue will be machined first.
5. The workpiece placed last into a queue will be machined last.
6. The workpiece with the smallest ratio of slack time to the number of remaining operations is placed first in the machine queue.

The shop load level is determined by the number of products ordered by the customer. If the plant can meet all due dates timely, the facility works either below or to full capacity. When delivery delays are experienced, the capacity of the plant has to be reviewed. It may be necessary to build a new facility or to add capacity to critical processes.

6.4.4 Load Balancing

In order to utilize manufacturing equipment to its full capacity, it is necessary to perform a thorough load-balancing operation. The basic problem is depicted in Fig. 6.63. There are four machine tools shown which produce three parts. Part 1 needs a turning and milling operation. Part 2 is first routed through the first milling machine, and then to the second one, and thereafter it is completed on the second lathe. The path for part 3 leads from the first milling machine to the second lathe. The reader may already realize that even for this

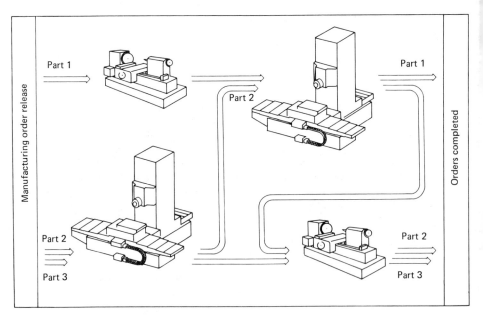

FIG. 6.63 Scheduling of workpieces through a plant.

Manufacturing Systems

simple problem, load balancing can get quite involved. When there are many workpieces to be machined on many machine tools, balancing cannot be done efficiently by hand and the computer should be used. There are many balancing algorithms available which try to optimize machining utilization under the condition that part delivery dates be met [22]. The more efficient algorithms will produce only a suboptimum solution, to reduce computing times. Some methods are also limited to the size of the problem they can handle. The techniques are usually based on the following methods:

Heuristic models
Lexicographic search
Permutation search
Branch and bound

Figure 6.64 shows an example where three workpieces are to be scheduled through a lathe and a milling machine. The processing times are shown in the upper part. There is a precedence relationship given in this problem which states that turning has to be done before milling. Six different permutations are possible for the solution of the problem. Two of them are shown in the pictorial. Eight minutes are needed to complete the workpiece machining sequence c-a-b and 10 minutes for the sequence a-b-c. In the first case the idle time for the milling machine is 2 minutes and in the second case 4 minutes.

There are different balancing techniques that can be followed. In principle it is possible to balance over time (Fig. 6.65a) or to balance over several machine tools (Fig. 6.65b). Both of these methods are linear. Figure 6.65c shows balancing over time and over several machine tools. If there is no preferred balancing direction, the method is nonlinear. In most cases production problems are so complex that nonlinear balancing is necessary.

The frequency of balancing is also of importance (Fig. 6.66). If balancing is done on a weekly basis, equipment problems or material shortages will result in decreasing machine utilization as production proceeds from one day to another. When loads are balanced more frequently, these types of problems are reduced. For example, in the case of a material shortage another product for which material is available can be scheduled and built, thus maintaining a high throughput. Information on defective machine tools or material shortage can quickly be brought back to the scheduling computer by means of a factory data acquisition system.

6.4.5 Scheduling Alternative Manufacturing Equipment

In manufacturing operations parts often have to be schedules through alternative manufacturing routes. This becomes necessary when

Machine Workpiece	Lathe	Milling
a	3 min	1 min
b	2 min	2 min
c	1 min	3 min

1. Machine sequences and times

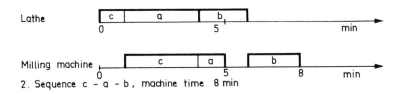

2. Sequence c - a - b, machine time 8 min

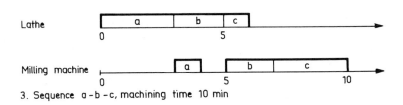

3. Sequence a - b - c, machining time 10 min

FIG. 6.64 Gnatt presentation of two possible work sequences.

when either a part with a higher priority is planned for production or when an equipment failure blocks a process. For this reason the planner must provide the scheduling program with knowledge of alternative manufacturing processes for parts. There may be options for one or several production routes. Often, the alternative will not render a cost-effective solution but will assure that the deadline for the part is met.

The input to the scheduling program is a matrix list in which for each part the latest time period at which production can start is entered. The different manufacturing alternatives are also given. The scheduling algorithm first tries to find an available machine for each part (Fig. 6.67). If all machines that can produce the part are occupied, the algorithm attempts to reschedule to other

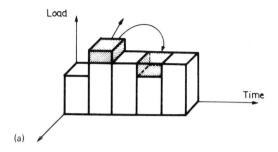

(a)

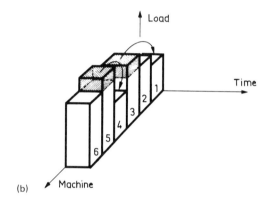

(b)

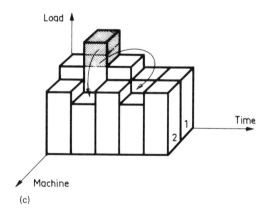

(c)

FIG. 6.65 Different load-balancing strategies; (a) balancing over time; (b) balancing over machine tools; (c) balancing over time and machining tools.

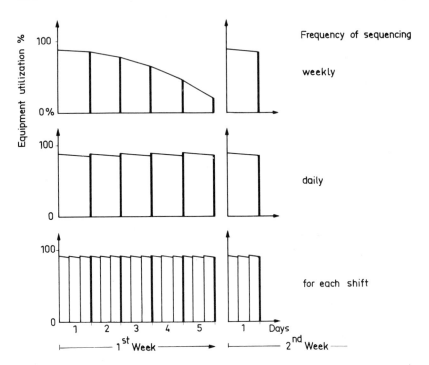

FIG. 6.66 Effect of frequency of sequencing on equipment utilization. (From Ref. 20. Courtesy of International Business Machines Corporation, White Plains, New York.)

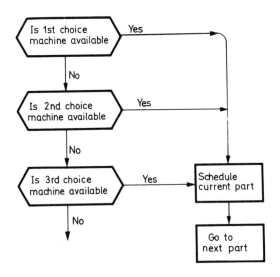

FIG. 6.67 First pass of scheduling algorithm.

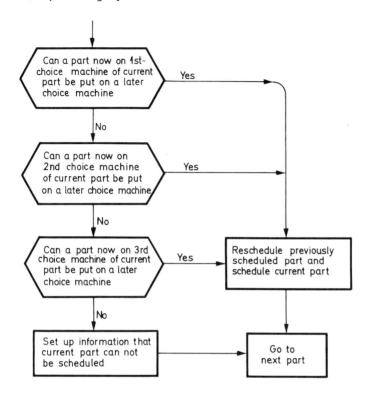

FIG. 6.68 Second pass of scheduling algorithm.

machines parts that are currently in production (Fig. 6.68). The computer output comprises one listing in which for each part the inventory is given for different time periods, and a second listing in which for each machine tool the number and quantity of parts to be produced are stated.

Such a computerized schedule allows the planner to look at a long time frame and to detect bottlenecks. Information on the unavailability of machine tools can help the planner to decide if overtime is warranted or new equipment is needed. It is also possible to minimize in-process inventory or to detect slack periods.

When scheduling equipment for production, the setup cost for tools must be weighed against the cost of carrying inventory. Two different practices are commonly used. With one the machine tool is rescheduled as soon as the part inventory reaches the requirements for a time period. The equipment is then retooled for another production run. Thus for small lot sizes the setup cost will overproportionally burden the part. With the second method parts are produced until the tool has to be resharpened. With this method the inventory carrying cost may be very high for the part. For this reason the scheduling algorithm should contain a routine that minimizes the sum of these two cost components.

6.4.6 Assembly Line Balancing

Principles of Line Balancing. Modern consumer goods are usually assembled according to the line production principle. The product is launched at the beginning of an assembly line and components are added as this product proceeds down the line. Assembly may be done automatically or by hand, whereby the manual method prevails (Fig. 3.37). The total work content W is subdivided into numerous individual assembly tasks which we call work elements WE. To each work element an assembly time named element time T_e is assigned. The total minimum assembly time is

$$T_{min} = \sum_{i=1}^{n} T_{e,i}$$

where n is the total number of work elements.

The element times for the individual work elements may differ considerably. This creates the problem that different assemblers may have different workloads. Assembly line balancing tries to equalize the workload, to give each assembler the same amount of work, and to optimize utilization of assembly effort. As the product moves down the assembly line, each assembler must be given enough station time T_s to complete the work at his or her station. This time is smaller or equal to the actual cycle time T_c of the assembly line:

$$T_s \leq T_c$$

T_c is the cycle time at which a product is launched or at which it leaves the assembly line. It is equal to the production rate. The maximum possible cycle time is

$$T_{cmax} \geq T_{smax}$$

In other words, T_{smax} determines the output of the assembly line.

Since work elements have different element times T_e, it may be necessary to assign the assembly for several parts to one station. Thus T_s for each station is

$$T_s = \sum_{i=1}^{m} T_e$$

where m is the number of work elements assigned to a station. The task of the planner is to create station times T_s of equal length. Since in general this is not exactly possible, the quality of a line balance can be expressed as the balance delay:

$$BD = \frac{pT_c - T_{min}}{pT_c} \times 100\%$$

where p is the number of workstations on the assembly line. Here T_{smax} could be selected as T_c.

Restrictions to line balancing. The distribution of work elements across the entire assembly line is usually not permissible. There are several restrictions that must be considered.

1. *Precedence restriction*: A precedence is necessary when one operation has to be performed before another. For example, the body of a car has to be assembled before the doors. Precedence restrictions are quite numerous and can be presented by a precedence diagram (Fig. 6.69). This picture shows a very simplified assembly of a car and the precedence relationships of the subassemblies. In the upper part the elements to be assembled and the precedence relationships are listed. The lower part shows the precedence graph. This graph can be plotted automatically by the computer. It shows the planner quickly whether proper precedence relationships have been selected. The plot is also a good means of documentation.

2. *Additional restrictions*: In the last column of the upper part of Fig. 6.69, code numbers for restrictions are entered. The numbers are explained in the pictorial. There are usually many restrictions on an assembly line. Typically, they are:

   a. The assembly has to be done at a specific station because of an available tool.
   b. The part must be assembled from the top or the right or left side of the line.
   c. The unit must be turned around for assembly.
   d. There are two assemblers required for the task.

Mathematical models for line balancing. Line balancing has been done since the conception of the assembly line. Only with the introduction of the computer a tool became available which was capable to

/Name /	Element description	/Predecessor	/Restriction/
A1	ASSEMBLE DOORS	-	-
A2	ASSEMBLE CHASSIS	-	-
A3	ASSEMBLE MOTOR	-	-
A4	ASSEMBLE WHEEL RIMS	-	-
A5	PAINT DOORS	A1	8
A6	PAINT CHASSIS	A2	8
A7	TEST MOTOR	A3	4
A8	MOUNT TIRES	A4	-
A9	MOUNT DOORS	A5, A6	3
A10	MOUNT ENGINE	A6, A7	2
A11	MOUNT WHEEL	A6, A8	-
A12	TEST CAR	A9, A10, A11	9

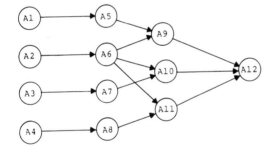

Restrictions: 2 Must be done by 2 assemblers
              3 Right or left side of assembly line
              4 At motor test station
              8 At painting facility
              9 At car test station

FIG. 6.69 Precedence relationship for a simplified assembly of a car.

perform line-balancing calculations within an acceptable time frame. The numerous line-balancing algorithms are based on the following principles:

Heuristic methods
Linear programming
Dynamic programming
Network methods

Heuristic methods have become the most important. They require the least amount of computation time but may render only a suboptimal solution. The result is calculated in successive steps. With the help of heuristic rules at each step, an additional elementary operation is included into the partial solution.

There are several strategies available for assigning work elements to stations. The easiest method is the largest-candidate rule. The work elements with the highest T_e value are assigned to the station. The sums of the T_e values for this station must stay below cycle time T_c of the line. The elements with the next-highest value are successively assigned to stations until no work element is left over.

Another technique assigns work elements to stations according to their position in the precedence diagram (Fig. 6.69). The first elements are selected first, and so on.

A third technique combines the algorithms of the two previous methods. It takes advantage of the T_e value of an element as well as of its position in the precedence diagram. With this information a ranked positional weight value is computed for each element. The elements are then placed into workstations according to the magnitude of this value.

Type of assembly line operations. Assembly lines may have different modes of operations. They are determined by the product spectrum and the piece rate. For this reason several different types of balancing models have been designed. The most commonly used mode of operation is the single model assembly line. This line produces one product for a given time period. After this period another model may be produced on this line, requiring another line balance. Appliance manufacturers typically use this type of operation.

More modern assembly systems allow mixed-model operations. This is a common situation in the automobile industry, where almost every car is custom ordered. The models are intermixed and assembled simultaneously on the same line. Balancing for this type of operation is usually very complex. There are several computerized line-balancing methods available which can be bought as a programming package. They are accessed either by batch type or by interactive operations. A typical computer printout is shown in Fig. 6.70 [23].

Element definition input

/Name/	Element Description	/Time	/Predecessors	/Restrictions/
1HF	GET PALLET TO LINE	100		HF
2HF	CABINET TO PALLET	114	1HF	HF
3HF	REHANG CABINET HOOK	040	2HF	HF
4HF	PUSH CABINET DOWN LINE	055	2HF	HF
5HF	ASSEM (2) TOP LOCKS	150	3HF, 4HF	HF
6	ASSEM MODEL TICKET	094	3HF, 4HF	
7H	ASSEM (3) GUSSETS	224	3HF, 4HF	H
8H	PLACE BOLT THRU GUSSET	040	7H	H
9H	START (2) NUTS & (2) ROL	220	7H	H
10H	GET & ASSEM (1) SIDE GUJ	087	7H	H
11H	GET & ASSEM (1) SIDE GUI	087	7H	H

Sample output for one station

/Element names/Times/

POSITION CHANGE FROM TOP TO REAR		0
REMOVE HARNESS FROM SUMP	28HR	70
PUSH EMPTY HOIST TO BASE	29H	30
TAPE HARNESS TO SUSP ROD	62H	96
OIL TO CENTER POST	30H	55
ASSEM OIL SEAL	31H	33
SEAT OIL SEAL	32H	58
BRUSH GREASE TO AGITATOR	51H	70
Total time		412

(a)

FIG. 6.70 (a) Element description; (b) printout of a line-balancing run for one station; (c) summary output at end of balance.

Summary output at end of a balance

Elapsed time for balancing only is 160 centiseconds
(b)

Maximum station time	414
Number of stations	51
Balance delay time	14.59%
Idle time per cycle	3081 time units

/Endbal/
(c)

FIG. 6.70 (Continued)

Some of the drawbacks of line-balancing methods are the fact that the first stations of the assembly line are balanced the best, whereas the last stations are often underloaded. Another difficulty that usually arises involves the inability of the computer to understand the limitations of assembly line workers. For example, one person may be left-handed and another may have a particular skill for a given operation.

6.4.7 Order Dispatching and Control

When the planning and scheduling tasks have been completed and all factory resources are available, the order can be released for production. Several documents are needed to initiate and monitor the production process (Fig. 6.71). They indicate the materials and tools to be used, the manufacturing operations and their sequences, the machines or work centers involved, and give detailed instructions to process and monitor the orders properly. The principal activities are:

Order release
Order-in-progress monitoring

With the order release the fabrication process is started. The documents issued by this activity are:

Route sheet
Part list
Fabrication release card
Move ticket

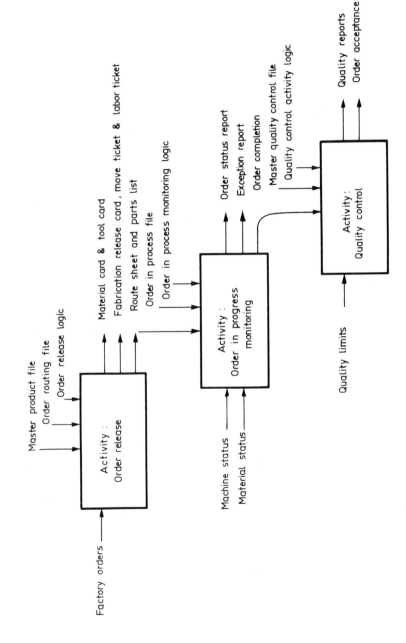

FIG. 6.71 Order release, order monitoring, and quality control.

Labor ticket
Material card
Tool card

Usually all of these documents accompany the workpiece as a work order package through all stages of its fabrication process.

The routes sheet indicates the manufacturing route through the plant. It contains information on the type and sequence of workstations through which the part has to progress. The part list contains information about all the components from which the part is assembled. The fabrication release card is placed on the machines or workstations after the setup of the production equipment has been approved. The move ticket identifies and records material that is being moved from one department to another. The labor ticket records the direct labor hours needed to manufacture the part. From it, labor information for payroll calculation is extracted. It also renders information for accounting and quotation purposes. The material and tool card initiates the release of materials or tools needed to fabricate the order. Order-in-progress monitoring is needed to record the progression of the fabrication process. The appropriate entries are made in these documents at each machine or work setup. For this purpose data acquisition devices may be used which are connected to the central control computer. Thus at any instant plant personnel are able to receive from the computer an up-to-date report on the status of the work in progress.

Typical information entered through data acquisition devices are order and work center identification, start time, completion time, wait time, material shortage, piece count, scrap, and machine breakdown. Order progress monitoring is an important input to almost all manufacturing functions. For example, customer order servicing is able to give the customer information about the status of an order at any time. (Fig. 2.14). The customer can also be informed if an order is on time or late.

When the product has been completed, it undergoes a thorough quality control check and from there it is sent to the customer. In principal, this completes the manufacturing process. A considerable number of data are gathered during this process. The important data must be evaluated and kept and are used as feedback information so that manufacturing can take corrective actions to maintain or to improve service.

REFERENCES

1. D. Schultz, A Cost Perspective for Group Technology, Society of Manufacturing Engineers Technical Paper MS73-512, Dearborn, Mich., 1973.

2. W. F. Hyde, *Classification, Coding and Data Base Standardization*, Marcel Dekker, New York, 1981.
3. A. R. Thompson, Improving Productivity through Classification and Coding, Society of Manufacturing Engineers Technical Paper MS76-727, Dearborn, Mich., 1976.
4. I. Ham and O. Shunk, Group Technology Survey Results Related to I-CAM Program, Society of Manufacturing Engineers Technical Paper MS79-348, Dearborn, Mich., 1979.
5. J. P. Hsu, Putting GT and MRG Together, Society of Manufacturing Engineers Technical Paper MS79-956, Dearborn, Mich., 1979.
6. Y. I. ElGomayel, Group Technology and Computer Aided Programming for Manufacturing, Society of Manufacturing Engineers Technical Paper MS73-980, Dearborn, Mich., 1973.
7. H. Opitz, *A Classification System to Describe Workpieces*, Pergamon, Oxford, England.
8. R. H. Phillips and Y. T. ElGomayel, Group Technology Applied to Product Design, The Manufacturing Productivity Education Committee MAPEC, Purdue University, West Lafayette, Inc., 1977.
9. VDI Taschenbücher T28, VDI-Verlag, Düsseldorf, West Germany, 1972.
10. R. W. Conway, Classification and Coding—A Prerequisite for CAD/CAM, Society of Manufacturing Engineers Technical Paper MS79-977, Dearborn, Mich., 1979.
12. H. M. Chan and D. A. Miler, Direct Clustering Algorithm for Group Formation in Cellular Manufacturing, *Journal for Manufacturing Systems*, Vol. 1, No. 1, 1982.
13. W. Eversheim, W. Fischer, and M. Stendel, Modular System Variants to Automatically Generate Process Plans, Westdeutscher Verlag, Opladen, West Germany, 1980.
14. G. Bachmann, Computer Aided Generation of Process Plans with the Help of Manufacturing Oriented Programming Systems, Dissertation, Technische Hochschule, Aachen, West Germany, 1973.
15. A. V. Tipuis, S. A. Vogel, and C. E. Lamb, Computer Aided Process Planning System for Aircraft Engine Rotating Parts, Society of Manufacturing Engineers Technical Paper MS79-155, Dearborn, Mich., 1979.
16. R. M. Smith, Computer-Aided Fully Generative Process Planning, *Manufacturing Engineering*, May 1981.
17. R. A. Wysk, et al., Analytical Techniques in Automated Process Planning, The Manufacturing Productivity Education Committee, Purdue University, West Lafayette, Ind., 1980.
18. E. Oberg, D. J. Franklin, and H. L. Holton, *Machinery's Handbook*, Industrial Press, Inc., New York, 1979.

19. C. H. Link, CAM I Automated Process Planning, Society of Manufacturing Engineers Technical Paper MS77-260, Dearborn, Mich., 1977.
20. *Communication Oriented Production Information and Control System*, Vol. V, IBM Corporation, White Plains, N.Y., 1972.
21. E. M. Dar-El and R. A. Wysk, Job Shop Scheduling—A Systematic Approach, *Journal of Manufacturing Systems*, Vol. 1, No. 1, 1982.
22. K. R. Baker, *Introduction to Sequencing and Scheduling*, Wiley, New York, 1974.
23. R. L. Bollenbacher, Line Balancing Made Easier, *Industrial Engineering*, Nov. 1971.

7

Control of Manufacturing Equipment

7.1 INTRODUCTION

Today there is a large number of different types of numerically controlled (NC) machine tools available. The NC technology is of great importance for automated manufacturing. NC machine tools are interconnected by material flow systems and assembly cells. This technology has been implemented successfully for a variety of applications in metalworking, such as milling, grinding, drilling, boring, and punching. The most important groups of metal processes are:

   Basic forming (casting, forging, rolling, etc.)
   Re-forming (bending, coining, drawing, etc.)
   Cutting (turning, milling, boring, broaching, grinding)
   Joining (bonding, joining, assembling, etc.)
   Coating (deburring, finishing, annealing, hardening, etc.)
   Material treating

Metal-cutting operations are mainly done by machines, which control the exact relative movement between the workpiece and the tool. This requires control units that are able to generate and follow trajectories and perform logical control functions. The task of a control unit is to generate and execute an efficient control program with the aid of geometrical and technological manufacturing data. Other basic functions are axis control and process adaptation and optimization.

If hierarchical control structures are considered in computer-integrated manufacturing systems, the geometrical and technological manufacturing data are produced on a supervising level. The data are then transferred to the corresponding control unit and are transformed into an machine-dependent part program Fig. 7.1. All operations of

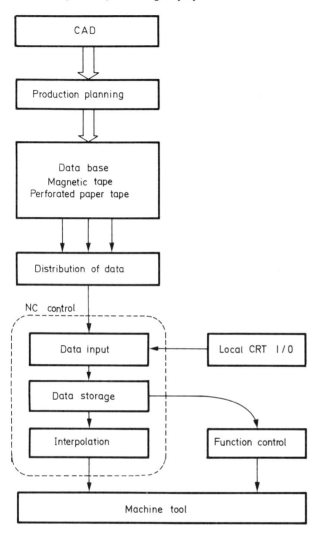

FIG. 7.1 Data flow to operate an NC machine tool.

of the manufacturing unit are accurately defined by this program.
The most important functions can be summarized as follows:

1. Processing of numerically coded information
2. Processing of logic functions
3. Processing of geometric (kinematic) functions
4. Acquisition of the actual tool positions and the status of the machine tool
5. Tool error correction and tool change

First the historical development of the evolution of conventional control (Sec. 7.2) for machine tools will be described. Then the principles of NC control (Sec. 7.2.5) with emphasis on hardware and software are explained. Next, a description is given of the information flow in machine tools and technological and geometrical processing of data. The most important control methods and sensors (Sec. 7.5) are summarized. Section 7.6 introduces NC part programming using the programming languages APT and EXAPT. CNC and DNC systems are discussed in Secs. 7.7 and 7.8. The principles of adaptive control to realize optimal operation of an NC machine tool are introduced in Sec. 7.4. Application of the computer and the programming methods used in the aforementioned area are given priority throughout these discussions.

The following sections cover the control of metal-cutting machines tools that use a geometrically defined cutter. They are of particular importance in piece part manufacturing.

7.2 NUMERICAL CONTROL

During this century, the development of manufacturing was characterized by an increasing efforts to automate machine tools (Fig. 7.2). At the same time, a decreased emphasis on control by human beings can be noted (Table 7.1). The advantage of obtaining higher productivity is accompanied by the disadvantage that the machine tools lose flexibility. New products frequently make it necessary to modify an installation, and new, expensive machine tools often have to be installed. Usually, high production rates imply low flexibility. Therefore, in the early state of manufacturing, high-quality parts with low production rates were manufactured only by hand-operated or semi-automated machine tools. This is generally an uneconomical endeavor.

The development of manually operated and semiautomated machine tools reached its maturity in the 1940s. However, there were increasing requirements for more accurate workpieces and new, hard-to-fabricate materials. These facts and the requirement for more flexibility resulted in a new control concept for machine tools, generally known as

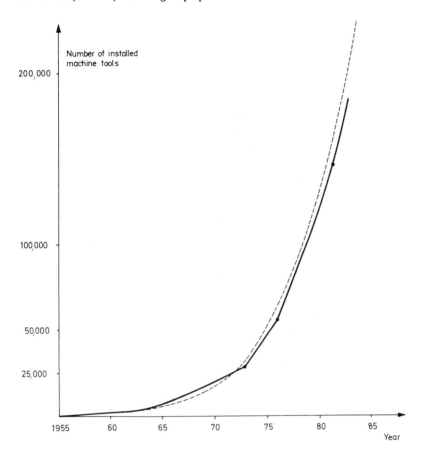

FIG. 7.2 Time history of number of installations of NC machine tools in the United States.

numerical control (NC). In 1948, the United States Air Force, together with the MIT Servomechanical Laboratory, started a project to develop a new control concept to drive the cutter of a milling machine along a continuous trajectory. A control scheme was realized which moves the tool along the defined path in small incremental steps.

The numerical information describing the tool movement was punched on a perforated paper tape and could be transferred to the machine controller by a paper tape reader. Within a relative short period of time, NC machine technology was developed as a flexible manufacturing tool which could be economically applied for milling, boring, and turning. The first successful applications were reported in the mid-1950s.

TABLE 7.1 Degrees of Automation

Degrees of automation	Characteristics
Zero order	Includes all hand tools (knife, hammer, pliers, etc.) which increase workman efficiency but replace no human energy or control.
First order	Includes portable or stationary machines or tools (pneumatic drills, portable electric tools, wood lathes, etc.) which are mechanically actuated, but must be fed and controlled by the operator.
Second order	Includes 100% mechanically powered machines (radial drills, bench lathes, etc.) but the operator must do complete setup and provide on–off control.
Third order	Includes completely self-acting machines (turret lathes, screw machines, grinders, transfer machines, etc.), with open-loop performance.
Fourth order	Includes machines having monitored performance (turning engines, honing machines 5-axis milling, etc.). Machine motions are corrected as necessary (closed loop feedback control).
Fifth order	Includes machines that use small special purpose computers, the control being based on the automatic solution of control equations.
Sixth order	Includes machines or devices wherein control is based on the automatic solution of complex formal logic equations.
Seventh order	Includes machines that learn from mistakes and attempt different modes of operation as necessary.
Eighth order	Machines that can extrapolate from experience (inductive reasoning) and form modes of operation beyond actual experience (could be an automatic operations research machine that sets up much of its own local programming.
Ninth order	Machines having the creativity or originality to devise new concepts and can work beyond their programming.
Tenth order	Machines having all the above potentials, plus the characteristics of dominance (able to give orders to designers and operators).

Source: Ref. 1.

At that time MIT also developed the programming language APT (Automatically Programmed Tools), which allows the description of parts and tools using mathematical relations. Geometrical, technological, and executive instructions are used to define the manufacturing task textually. APT is a general-purpose language and independent of the NC machine. Numerous APT dialects and new NC programming languages, which all use the basic APT concept, have been developed since.

The rapid progress in VLSI technology will accelerate the development of high-order software tools for the efficient control of manufacturing equipment. Integrated CAD/CAM systems, which generate part programs via a hierarchy of control modules, will be soon available for NC, PC, and CNC.

7.2.1 Justification of NC Machine Tools

For introduction of NC machines for multimachine manufacturing systems, economic and technological aspects of the manufacturing process have to be taken into consideration. The critical areas for possible cost saving have to be identified to aid a justification study. The most important aspects are as follows:

Direct labor
Machine setup
Material handling
Production control
Part inspection
Equipment maintenance
Manufacturing engineering
Prototype and new parts (products)

NC technology is finding wide acceptance for the following reasons:

1. The prevailing trend of increasing labor cost and the further reduction of labor hours
2. The high efficiency and productivity of NC machine tools (a factor of 4 compared to conventional machine tools)
3. The higher precision and uniformity needed for parts
4. The trend to higher flexibility because of smaller production lots and shorter product life cycles
5. Optimization of the production flow
6. Decreasing costs of computer hardware
7. Efforts to increase the quality of work life
8. The trend to integrate CAD with CAM

Furthermore, it is expected that with decreasing hardware cost, improved control systems can be obtained. This is due to the increasing integration density (VLSI) of electronic circuits. The modularity of control circuits, high fault tolerance, and available fault diagnostic programs will help to save inspection costs. The benefits of NC technology will not be realized just by the use of individual NC machine tools. However, it will be obtained by combining different types of machines to a manufacturing cell and by operating it as a DNC system (Sec. 7.8). This fact must be observed when economic aspects are considered. Present research and development efforts to improve NC technology include:

1. Automatic adjustment of the cutting edge of the tool
2. Correction for tool wear
3. Automatic prediction of operating times
4. Automatic determination of tool reference coordinates
5. Automatic correction for possible machine distortion by heat
6. Correction of spindle errors
7. Automatic workpiece measurement by the machine tool
8. Automatic correction for measurement errors
9. Verification of the part program via a CRT
10. Display of the machining statistics on a CRT
11. Improvement of operational and planning capabilities of the supervisory control computer
12. Optimization of tool trajectories
13. Simplification of part programming
14. Reduction of cutting cost

New trends that can be expected include the following:

1. *Modularity*: Modular design of the machine tool and its control. The control unit will be located away from the machine to make both units more accessible.
2. *Improved storage facilities for programs*: The floppy disk, dynamic RAM, and the magnetic bubble memory will replace punched tapes.
3. *User comfort*:
   a. Simplified machine operation by dialogue
   b. Operator guidance
   c. Reduction of operator functions
   d. Multiple functions
   e. Error detection and analysis
4. *NC programming via dialogue*:
   a. Direct programming at machine tool
   b. Simple application specific programming languages
   c. Comfortable editor functions (smart editors)
   d. Automatic detection of programming errors

5. *Human-machine communication:*
   a. Numerical display for tool position, velocity, and so on
   b. Automatic conversion to different measurement units
   c. Conversion from relative to absolute positions

In the next section the design and operation of NC systems is discussed, including simple classical logic controls, sequence NC control systems, and the principle of more advanced DNC concepts.

7.2.2 Basics of Logic Control

Hardwired logic circuits are basic modules for the control of automatic machine tools. The logic operations are carried out by logic networks which combine several input signals and produce output signals. This operation is carried out either by relays or by logic semiconductor circuits. The values of the input or output parameters are either true (logical 1) or false (logical 0). They represent the following logic process states:

Switch on/switch off
Voltage on/voltage off
Motor on/motor off
Threshold value/not threshold value etc.

Modern machine tools may contain dedicated semiconductor circuits or more universal microcomputers. The functions that the microcomputer can perform are as follows:

On and off control
AND, OR, NOT operations
Arithmetic operations
Memory functions
Counting and delay functions
Timekeeping

The design of a logic network is done either with the help of truth tables or with logic equations using Boolean algebra. The truth tables contain those combinations of logic inputs that generate an output of a logic 1. Logic equations are a mathematical tool to describe the functional logic relations of input and output.

The signal flow diagram which uses logic symbols for basic logic operations is another, but more technical method that can be used to describe a logic network. The AND, NAND, OR, or NOR circuits, which often have more than two input signals, make use of solid-state electronic components. Various technologies (DTL, RTTL, TTL) are available in which logic operations are realized with integrated circuits

(Fig. 7.3). Advanced technologies for integrated circuits make possible the manufacture of logic chips with powerfull switching capacities. They can replace classical electromechanical relay switches. Higher logic operations such as counting and information storing can be realized with two interconnected NOR gates called flip-flops. A sequence of flip-flops can be combined in a shift register and used as a storage cell for a computer word. Its advantage is a very short storage time for a write cycle. Another important component for logic control is the binary up/down counter, which is usually configured from flip-flop circuits and to which a clock element is added to generate the clock pulses needed for counting.

Architecture of Hardwired Logic Controllers

For logic control, logic operations are applied to binary input signals to generate binary output signals (states). Binary signals may be obtained from the output of binary transducers or from analog transducers, which have a defined threshold. Here the logic switching state stays stable until the analog variable crosses the threshold. A signal converter is necessary between the process, the user console, and the logic control circuit. It converts the input signal to the level of the logic controller. Transistor-transistor logic (TTL) circuits need binary signals in the range 0 to 5 V. Output signals of logic controllers have to be amplified to appropriate signal levels to control, for example, a motor, a valve, or a switching gear. The structure of a hardwired logic control system is illustrated in Fig. 7.4.

An example of a logic control circuit for a feedforward drive control for a machine tool is shown in Fig. 7.5. Three operation modes are implemented: (1) fast-forward feed, (2) fast-reverse feed, and (3) normal forward feed. The operator can activate each mode by pressing a switch. The desired feed mode is valid if it is assured that no other switch is depressed at the same time (interlock function). The feed drive is also interrupted when the spindle reaches a limit switch at the end of the movement axis or if another machine function (hydraulic, lubrication, etc.) is not working. The structure of the logic controller for the interlock functions is shown in Fig. 7.5.

Logic controllers are easy to realize using integrated circuits. They are hardwired and can be built from modular components on a standardized circuit board. Compared with electromechanical technologies such as relays, they have the following advantages:

Short switching cycles
Long service life (no wear)
Low input current
Very compact design

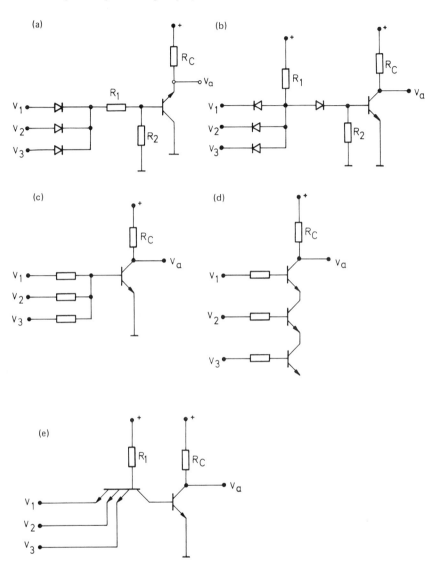

FIG. 7.3 Realization of logic functions: (a) NOR gate (diode-transistor logic, DTL); (b) NAND gate (diode-transistor logic, DTL); (c) NOR gate (resistor-transistor logic, RTL); (d) NAND gate (resistor-transistor logic, RTL); (e) NAND gate (transistor-transistor logic, TTL).

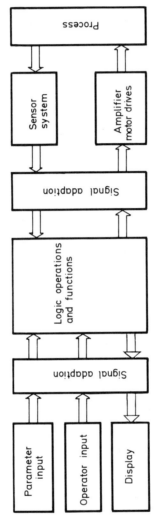

FIG. 7.4 Schematic diagram of a hardwired logic control system.

Control of Manufacturing Equipment 397

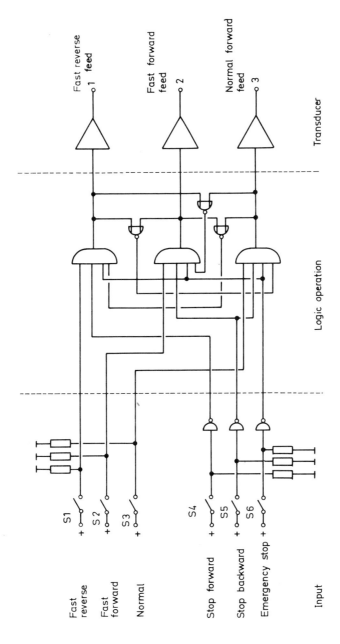

FIG. 7.5 Circuit of a feedforward control using logic operations.

With the introduction of the microprocessor, a programmable tool became available which has a powerful instruction set, including the most important logic operations. Logic control with the aid of a microprocessor is very flexible and can easily be implemented.

7.2.3 Programmable Controllers

Programmable controllers are made from basic digital modules such as counters, registers, arithmetic units (addition, subtraction), and memory. The information to be processed is contained in machine internal binary code. Two types of control modes are in use (Fig. 7.6): (1) Hardwired logic control (Parallel logic control) and (2) programmable sequential control (Sec. 7.2.4).

A hardwired digital controller is very difficult to modify. The logic functions are well defined but cannot be modified without redesign of the corresponding circuits. Contrary to this hardwired concept, the flexible programmable controller offers many economical and technological advantages. The structure of the software for such a device depends on its instruction set and the requirements of the technical process. Three programming modes are in use (Fig. 7.7): (1) hardwired programming, (2) textual programming, and (3) graphical programming.

Hardwired controllers are characterized by their logic hardware elements and their connection wires. They are realized from relays and integrated circuits which are interconnected by wires, diode matrices, crossbar switches, or printed circuit cards. Some logic elements may be interchanged. Thus a limited flexibility can be obtained.

Textually programmed controllers are divided into RAM-oriented and PROM-oriented devices. They are the subject of interest in this discussion. In the case of the RAM-oriented controller, the program

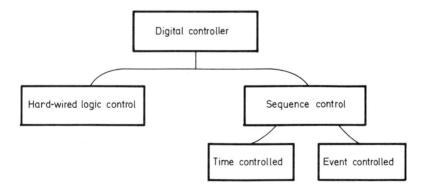

FIG. 7.6 Different digital control modes.

Control of Manufacturing Equipment

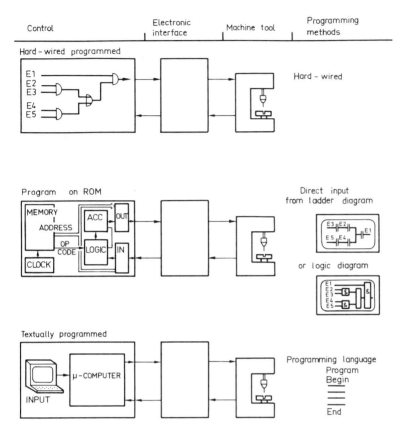

FIG. 7.7 Technologies to realize digital controllers.

can be modified with the aid of a programming system. When the program is changed, mechanical replacing of the memory chips is not necessary. A microprocessor is the heart of the controller. In a PROM-oriented controller, the program can be modified with the help of a PROM programmer. The memory chip containing the control program has to be exchanged by hand.

A programmable controller is driven by clock pulses and operates sequentially. In other words, the instructions are executed one after the other (linear program structure). The execution time of a software program is longer than that of a pure hardware wired program. In particular when complex parallel logic operations are to be performed, critical real-time delays may result. For this reason the controllers are equipped with 10-MHz-driven 8- and 16-bit microprocessors with an optimized instruction set and very short machine cycles. The essential

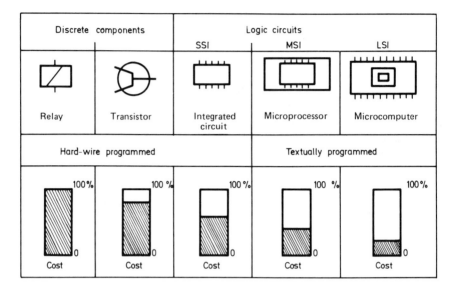

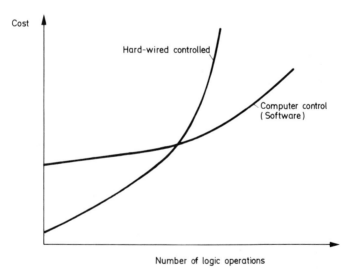

FIG. 7.8 Cost comparison of hardwired and memory-programmed logic controllers.

advantage of a software-programmed controller compared with a hardwired controller is its flexibility and low price. Figure 7.8 shows a cost comparison for these two controllers. With an increasing number of functions, the controller using software renders the more economical solution.

Structure of Programmable Controllers

Programmable controllers perform logic operations using process signals. The input signals are transferred by the I/O interface to the accumulator, where the logic operations are done using Boolean algebra. Additional operations are signal delay, signal time variation, and memory buffering. With this functions simple automatic assembly systems, conveyor belts, metal-cutting machines, and so on, can be controlled.

The internal structure of a programmable logic controller is similar to that of a digital computer or microprocessor (Fig. 7.9). The following components are present:

1. *The input interface*: Digital process signals are transferred via this unit into the processor. Each input line is addressable.
2. *The arithmetic and logic unit*: This component performs the logic operations specified by the program instructions.
3. *The accumulator*: The results of the logic operations are stored in this unit and are then transferred to the output interface.

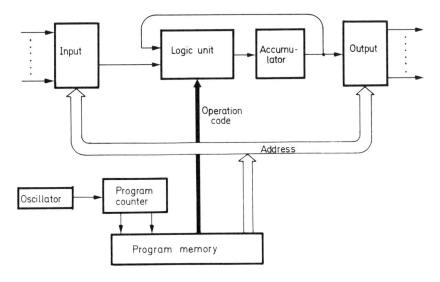

FIG. 7.9 Structure of a programmable controller.

4. *The output interface*: This unit transfers the logic output signals to the process.
5. *The program memory*: This component contains the process control program. The instructions are retrieved from memory under direction of the program counter and executed one after the other. Via the address field of the instruction, the desired I/O lines are selected. The operation field defines the logic operation to be performed.
6. *The program counter*: This unit generates addresses for program execution. The counter is driven by clock pulses.

The programmable controller executes the program sequentially. After the program has gone through one complete cycle, it is restarted and begins a new cycle. Since there is no interrupt possible, process signals do not alter the operation of the program. A reaction to a critical process signal is possible only when the particular process line is being addressed. For this reason the program cycles have to be very short, which means that the program must be of limited length.

In Fig. 7.10 the structure of a linear control program is shown. The program consists of instructions and blocks. Instructions are the smallest elements of the program but can be combined with other instructions to blocks (modules). The format of an instruction consists of an operator field and an operand field. The operation field defines the logic operation to be performed, whereas the operand field defines parameters and addresses of I/O lines. An instruction may only contain an operator. In this case the operand field is empty. Jump instructions are possible to change the execution path of the program.

Programming

Programming a PC is supported either with the aid of a circuit diagram, a logic diagram, a ladder diagram, or logic equations in textual form. The implementation of the logic control task is done with the help of a ladder diagram and an instruction list which are specific to the programmable controller (Fig. 7.11). The most important instructions for processing the I/O signals and the internal logic variables are AND, OR, and NOT operations. Additional instructions are SET and RESET and I/O operations. If a logic diagram is available, conversion to the contact plan or the instruction sequence can be performed directly.

A programming system consists of a keyboard with control keys, alphanumeric input keys, a CRT terminal, and a floppy disk unit (Fig. 7.12). Editor functions, program translation, debugging, and file handling are controlled by an operating system. Programming can be done in two basic modes: (1) on-line programming and (2) off-line programming.

Control of Manufacturing Equipment

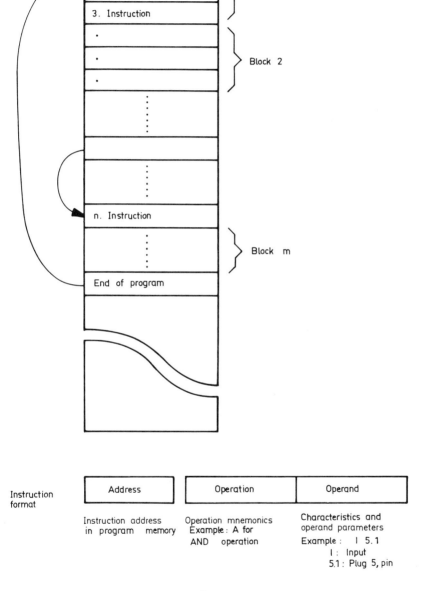

FIG. 7.10 Program structure of a textually programmed logic controller.

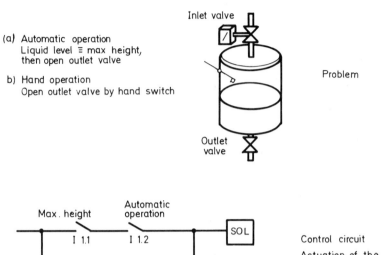
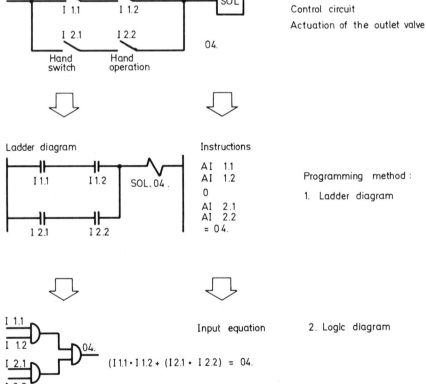

FIG. 7.11 Transformation of a control problem into a contact plan and an instruction list. (Courtesy of Siemens AG, Munich, West Germany.)

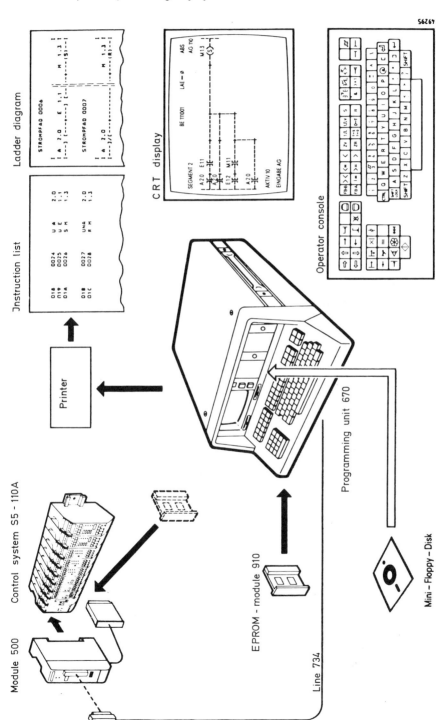

FIG. 7.12 Structure of a modern comfortable programming system for PCs (System 670 of Siemens). (Courtesy of Siemens AG, Munich, West Germany.)

In the case of on-line programming, the technical process is directly interfaced to the programming system. With this method the program is written and tested interactively. Debugging and test operations can be done stepwise or via break points. Program verification, file deletion, and insertion operations can be performed directly.

Off-line programming is done independently of the process with the help of an EPROM. For program testing the EPROM is inserted into the controller and execution of the program is observed. In case of an error, the EPROM is placed back into the programmer and a correction is done. When the program is correct, the memory chip is burned in. The program can be erased with ultraviolet light.

Several user aids for program development are available, such as the following:

1. Graphical output of signal flow
2. Graphical editor
3. Output of a ladder diagram, a logic diagram, or a program listing
4. Interactive user guidance
5. Listing of an error and its analysis
6. File handling
7. Program storage on a floppy disk

Most of the PC manufacturers design their programming system for medium- or high-order languages. A compiler is needed to generate machine-specific operation code.

To control execution of a user program, the operating system has to support the following functions:

1. Input of process signals and their storage in memory
2. Execution of the user program; processing logic variables
3. Output of logic control signals to the process
4. Recovery routines for machine and program errors
5. Synchronization of the data transfer to computers on the next hierarchical control level (DNC level)

For many applications it is advantageous to perform logic counting and delay operations externally to the PC and to process only the results of these with the controller. This will relieve the processor of time-consuming routine operations. The most important modules of PCs are as follows:

AND modules
OR modules
Exclusive OR
Holding circuits

Latch relay
Shift register
Delay timers; short and long intervals
Up/down counters
Equivalence and compare operations

Advanced PCs can be interfaced to hierarchical process computers, for example, to a DNC system via a standard bus (Sec. 4.4). Typical tasks of the process computer are to transfer control programs to the PC and to sample production data, for example, to synchronize the material flow. It also serves as an input device for production data acquisition and quality control.

7.2.4 Sequential Controller

The sequential controller is similar to the logic controller. Program execution is linear but is synchronized by events or status signals. Synchronization is done by time (clock) or event signals of the machine tool, such as threshold or handshake signals or process variables. Two types of sequential controller are available (see Fig. 7.13): (1) sequence control with time synchonization, and (2) sequence control with event synchronization.

Compared with the logic controller, which performs its logic operations on-line, the execution of a sequential program is dependent on a

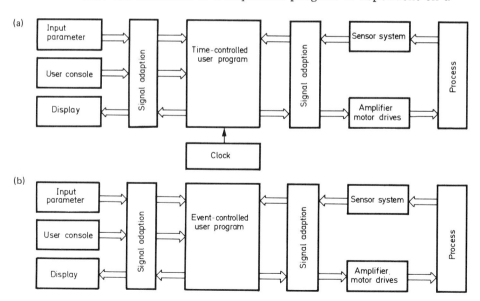

FIG. 7.13 Schematic diagram of sequence controllers: (a) time synchronization; (b) event synchronization.

switching event. Whenever a program instruction has been executed, the program pointer be incremented only when the switching condition is satisfied. These conditions can include:

1. *A waiting-time cycle*: The next program instruction can be executed only after a defined time delay.
2. *An elapsed-time cycle*: An error is assumed when the next instruction is not executed after a defined time.
3. *A handshake condition*: The next program step is dependent on the execution of the previous instruction. This can be a message or a handshake signal (e.g., motor on, valve closed, threshold value reached).
4. *A process state*: The next program instruction can be executed only if defined process states have been reached (e.g., speed, position, temperature, optimal parameters).

In Fig. 7.14 the individual steps of a sequential control program for an automatic boring operation are shown. First the drill approaches the object surface at rapid speed, and stops before it hits the workpiece. Next, a coolant is turned on. Then the drill operates at normal feed until the desired boring depth is reached. After this event the drill is retracted at rapid speed until a limit switch indicates the original start position. The start signal for this operation can be generated by hand or by an external device such as a manipulator.

To realize sequence control operations, logic controller or PCs are used. Programming of the individual program steps is done with (1) crossbar switches, (2) programmable controllers, and (3) microcomputers. Sequence controllers render satisfactory and economic solutions for manufacturing machines that operate in defined modes. Such modes are constant feed (forward and backward) and operations of limit switches. This type of control cannot be used to perform contouring operations (e.g., for lathes, milling machines, etc.). Contouring means to move the tool along a continuous path at a defined speed. In this case the logic control has to be done by numerical control, which allows continuous operation.

7.2.5 Principle of Numerical Control

Machining of workpieces requires logic and sequential control. Thus attention must be paid to the relative movement between the tool and the workpiece. A device capable of generating this movement from defined trajectory parameters is called a numerical controller. It is operated with the help of an NC program, which contains the geometrical and technological information for machining the workpiece. The NC program is processed by the machine control unit (MCU). It also performs machine-specific control functions. The classical NC machine

Control of Manufacturing Equipment

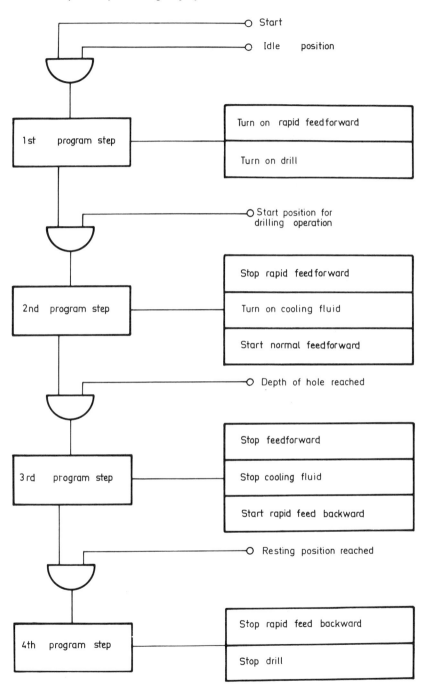

FIG. 7.14 Sequence control of a drilling operation.

used perforated paper tapes to store the part program (Fig. 7.15). Modern systems are equipped with magnetic tapes, disks, punched cards, or semiconductor memories. The possibility of changing a program quickly makes the machine tool very versatile. Standard NC machines can be integrated into CNC and DNC systems. For this reason they may be considered as the basic building block for computer-integrated manufacturing. An NC controller contains the following functional blocks (see Fig. 7.16):

The data input module and address processor
The geometric processing unit
The technological processing unit
The central machine control unit (MCU)
The I/O unit for process and control signals

In the following sections the functions of these blocks are discussed.

7.2.6 Data I/O Module

The part programs and manufacturing data are entered into the machine tool via the data input module. In most cases punched paper tapes contain the NC program. The use of these tapes has a historical and practical background. For example, in the rough manufacturing environment a robust storage medium for the part program is needed. In addition, the tape can be read in at a synchronous speed (30 to 300 characters/sec) at which the machine tool processes the data. One paper tape typically may contain one part program with a size of 10^3 to 10^5 characters. Read-in and processing of the program is done sequentially block by block. First the NC control data are checked for errors (ISO code, 7 bits, 1 parity bit). Then the data are decoded and converted into machine-specific code. In addition, the address calculation is done to store the data at their appropriate address in their own memory. This memory is reserved for geometric and technological data. With the aid of the memory addresses, proper processing and interpolation of the NC information are possible.

In modern controllers the circuits of the I/O modules for NC machines are integrated on a VLSI chip. They allow data transfer over either serial or parallel lines from a host computer, a floppy disk, or tape. The structure of an I/O data module is illustrated in Fig. 7.17.

7.2.7 Processing of Geometric Data

The NC machine is the positioning device for the tool that machines the workpiece. The tool follows the instructions of the part program and moves from point to point along a defined path. The number of degrees of freedom for the movement is determined by the number of axes of

Control of Manufacturing Equipment 411

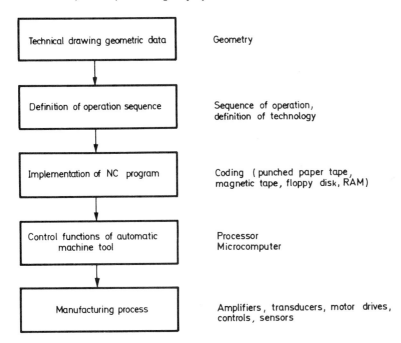

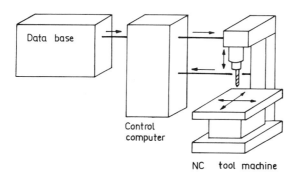

FIG. 7.15 Information flow to generate a NC program for a workpiece.

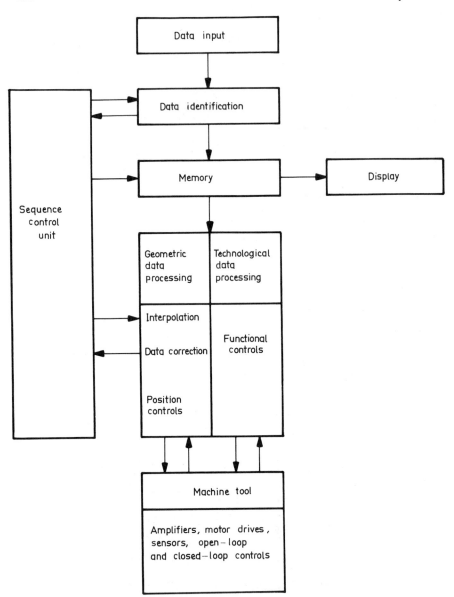

FIG. 7.16 Function blocks of a machine control unit (MCU) for a NC tool machine.

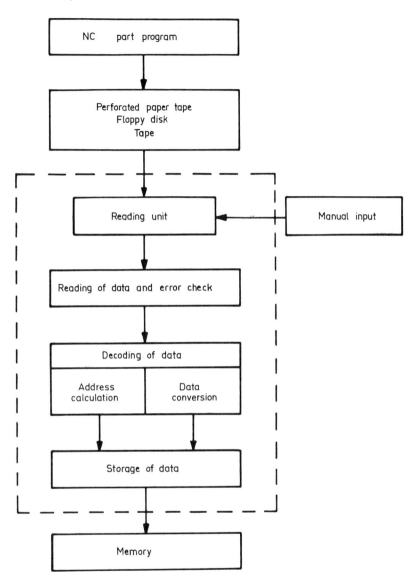

FIG. 7.17 Structure of a NC data input unit.

the machine tool. Modern machine tools have very high resolution; they can position a tool in increments of $\Delta \varepsilon < 10^{-5}$ m. Such an accuracy specification, however, requires the storage of a large number of data.

The tool follows a path in cartesian coordinates to generate the workpiece contour. It is calculated by interpolation. The input to a cartesian interpolator are trajectory parameters, time conditions, the desired incremental resolution, and the sample data rate. The output of the interpolator is the incremental trajectory information $\Delta r = \Delta x + \Delta y + \Delta z$, which defines the cartesian position of the tool. For the general case of six degrees of freedom, additional incremental information is necessary to define the tool's orientation. Decoding is followed by an operation that performs the transformation of the cartesian trajectory into machine-tool-specific coordinates (Fig. 7.18). A characteristic function is the interpolation frequency, which can be specified by the user. In the following section two widely used interpolation algorithms are discussed.

Interpolation Methods

The interpolation method depends basically on the machining application. Machine tools that perform their operation in a stationary position need only point-to-point information without trajectory interpolation (e.g., boring or spot welding). For machining of cylindrical, axis-parallel or axis-vertical surfaces, simple one-axis controls without interpolation can be used (e.g., for milling and turning machines). For complicated piece part contour surfaces, interpolation algorithms that calculate the relative movement between the tool and the workpiece are applied (Fig. 7.19).

The requirements for interpolation algorithms can be summarized as follows:

1. The interpolated curve has to approximate the desired piece part contour as close as possible.
2. Lines and circles must to be interpolated very accurately.
3. The cutter velocity has to be variable and curve independent.
4. The number of selected trajectory parameters should be as low as possible. Typical parameters are (a) start and finish positions, (b) circle center and radius, (c) path velocity, and (d) sample data rate.
5. Each final position must be reached accurately to avoid divergence from the desired cutter path.
6. The interpolation algorithms should be as simple as possible to allow high interpolation frequencies.

The task of interpolation consists of two parts. First, the continuous cartesian time history of the desired trajectory is calculated. The

Control of Manufacturing Equipment

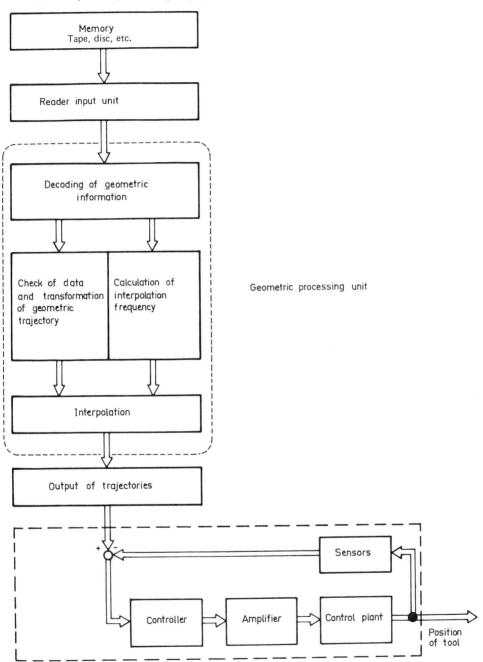

FIG. 7.18 Data flow through the geometric data processing unit of a MCU for NC machines.

Control	Problem	Tool	Application
Point-to-point control	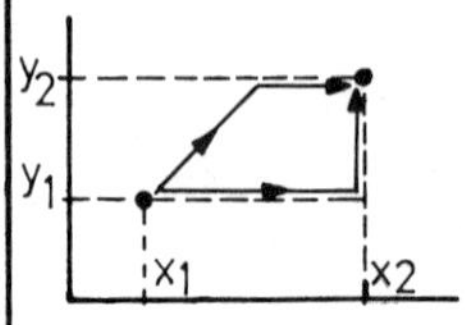 No interpolation	Not active during positioning	Boring, spot welding
Primitive linear path control	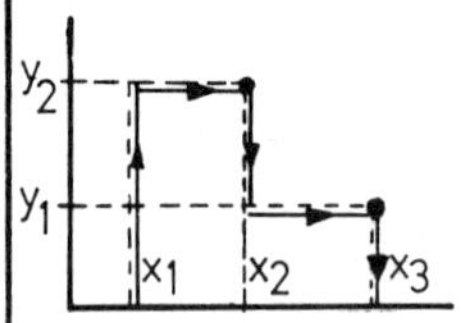 No interpolation	Active during positioning	Cylindrical contouring, milling (parallel to axes)
Linear path control	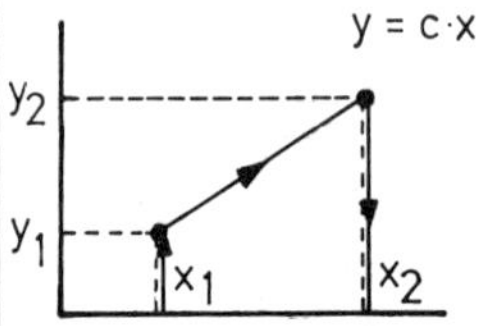 $y = c \cdot x$ Linear interpolation	Active during positioning	Conic contouring turning, milling along streight lines
General path control	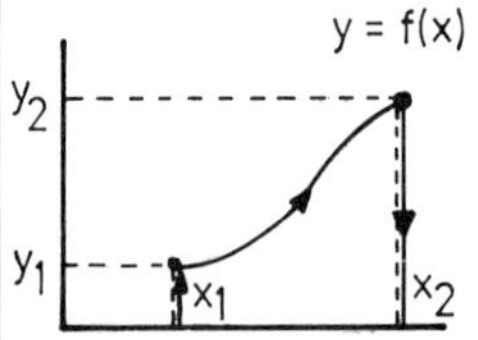 $y = f(x)$ Higher order interpolation	Active during positioning	Contouring, turning, milling along general paths

FIG. 7.19 Trajectory planning methods for different machine tool applications. (From Ref. 2. Courtesy of VDI-Verlag, Düsseldorf, West Germany.)

inputs to this part are parameters that are needed by the analytical algorithm for calculation. The second part is the coordinate transformation of the trajetory from the cartesian coordinate space into the machine-specific coordinate space. This produces the time history for movement of the axes. In some cases a second interpolation algorithm is applied to calculate the final machine coordinates to filter and smooth the axis movements. For NC machines the following interpolation algorithms are uses:

Linear (line) interpolation
Circular (circle) interpolation
Quadratic (parabolic) interpolation
Cubic interpolation
Interpolation with high-order polynomials

High-order polynomials are used to generate complex contours and surfaces. Manipulator devices and industrial robots need sophisticated interpolation algorithms to perform finishing and assembly operations (Chap. 8).

Classical NC machines have hardwired interpolators made from logic circuits. For each interpolation method a dedicated algorithm is designed and implemented. Usually, only a limited number of algorithms (i.e., linear and circular), in the form of hardwired logic circuits, are implemented. More flexible interpolators can be realized with fast 16-bit microcomputers, bit-slice computers, or arithmetic processors. The type of interpolation, the sample data rate, and the desired accuracy are variable parameters. Software interpolators offer a high flexibility. Their use is of particular interest for the computer numerical control (CNC) concept, which was introduced in the early 1970s (Sec. 7.7).

Interpolation by Integration (DDA Method)

Most classical NC machines are equipped with linear, circular, or combined linear-circular interpolators. They are suitable for those applications where the contour of the workpiece is bounded by straight lines and circular segments. Many higher-degree curves can be approximated with a series of circular arcs. In the case of complex free-form designs, parabolic and cubic interpolators are applied. For special-purpose applications requiring extremely high accuracy, high-order general interpolation algorithms are necessary (e.g., for machining the rotor blades of turbines).

In general the approximation of a curve can be done using a series of polynomial arcs. Polynomials have good approximation properties, but in the case of more complicated curves they need excessive computation time. Complex curves are approximated by a series of linear segments or circular, parabolic, and cubic arcs. These curves are transformed by the interpolator into a sequence of elementary cutter increments.

Another approach to solving the interpolator problem is through the use of differential equations. Here a solution is obtained with an interconnected set of integrators. Such an integrator network is called a digital differential analyzer (DDA) [3]. This method allows implementation of simple and efficient linear and circular interpolation algorithms. As an additional benefit, they permit control of the feed rate of the tool.

Linear DDA interpolation. Let us consider a linear curve that the cutter tool has to follow at a constant velocity v from a point P_i to P_e. In the case of two-dimensional movement (x-y plane), the motion is defined by the following set of equations:

$$\left. \begin{array}{l} x(t) = v_x t + x_i \\ y(t) = v_y t + y_i \end{array} \right\} \quad t > t_0 \qquad (1)$$

With $v_x = dx/dt$ and $v_y = dy/dt$, these equations can be written as

$$x(t) = \int_{t_0}^{t} v_x \, dt \qquad (2)$$

$$y(t) = \int_{t_0}^{t} v_y \, dt$$

If the time interval T for the movement between P_i and P_e is given, Eq. (2) can be rewritten as

$$x(t) = \int_{t_0}^{t} \frac{x_e - x_i}{T} \, dt \qquad (3)$$

$$y(t) = \int_{t_0}^{t} \frac{y_e - y_i}{T} \, dt$$

Now let the interpolator operate N steps during the time interval $T = n \, \Delta t$ ($n = 0, 1, \ldots, N$); then we can write

$$x(t) = x_i + \sum_{1}^{n} \frac{x_e - x_i}{T} \qquad (4)$$

$$y(t) = y_i + \sum_{1}^{n} \frac{y_e - y_i}{T}$$

Equation (4) defines the linear interpolation of a straight line between two points using N steps. At the time instant $t_n = n \Delta t$ ($n = 1, 2, \ldots, N$) the x and y components of the line are incremented by Δx and Δy (Fig. 7.20). The implementation of Eq. (4) with hardwired logic requires the following components (Fig. 7.21): (1) a register, (2) an adder, (3) an accumulator, and (4) a counter.

The input to this DDA interpolator is the absolute distance between the initial and final positions of the line ($x_e - x_i$, $y_e - y_i$). Also, the number of interpolation steps N and the interpolation frequency (strobe) have to be specified. The steps Δx and Δy at each time interval Δt have to be smaller than the smallest possible positional increment ΔP of the axis motor drive. To adapt an interpolation algorithm to an axis-drive system, the following conditions have to be satisfied:

$$\text{abs} \frac{x_e - x_i}{N} \leq \Delta P_x \tag{5}$$

$$\text{abs} \frac{y_e - y_i}{N} \leq \Delta P_y$$

$$f_0 = \frac{N}{T}$$

where

f_0 = interpolation frequency
ΔP_x = smallest incremental step of x-axis drive
ΔP_y = smallest incremental step of y-axis drive
T = transition time to process an entire curve section
N = number of interpolation steps

EXAMPLE 7.1 Let us consider a point $P_1(2,2)$ and a point $P_2(6,4)$. It is the task of a cutter to produce a straight line while it moves from P_1 to P_2. There are 10 interpolation steps which have to be performed at time increments of 2 sec. It is assumed that the displacement increments along the x and y axes coincide with the interpolated line increments. With Eq. (4) the following equations can be written:

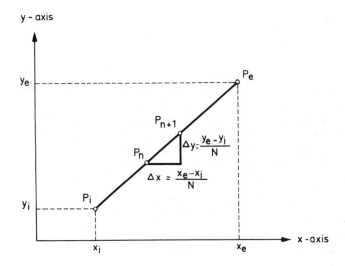
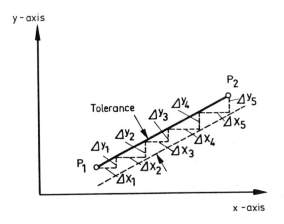

FIG. 7.20 Principle of linear interpolation.

Control of Manufacturing Equipment   421

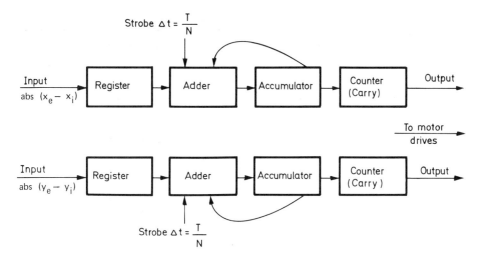

FIG. 21   Logic network of a linear DDA interpolator.

$$x(t) = 2 + \sum_{1}^{10} \frac{4}{T}$$

$$y(t) = 2 + \sum_{1}^{10} \frac{2}{T}$$

The results of these equations are shown in Table 7.2.

In case the interpolation increments are smaller than the increments of the drive axis, a 1 is carried over into the counter when there is an overflow in the accumulator.

EXAMPLE 7.2   The trajectory of Example 7.1 is calculated by using the identical interpolation parameters. However, the increments of the x and y drive axes are assumed to be 0.5. The result is shown in Table 7.3 and Fig. 7.22.

Circular DDA interpolation.   The circular interpolation is the second important method used in NC technology. It generates incremental information for the tool to follow a circular path. The DDA method allows the calculation of incremental steps Δx and Δy along circles or arcs at constant tool velocity v. A circle can be defined by its center

TABLE 7.2 Interpolation Results of Example 7.1

Steps	Increments		Traveled distances	
	Δx	Δy	x	y
0	0	0	2	2
1	0.4	0.2	2.4	2.2
2	0.4	0.2	2.8	2.4
3	0.4	0.2	3.2	2.6
4	0.4	0.2	3.6	2.8
⋮	⋮	⋮	⋮	⋮
10	0.4	0.2	6.0	4.0

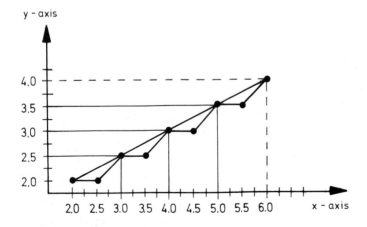

FIG. 7.22 Linear interpolation between two points according to Example 7.2.

Control of Manufacturing Equipment

TABLE 7.3 Interpolation Results of Example 7.2

Steps all 200 msec	Interpolated increments		Carryover		Traveled distances	
	Δx	Δy	Δx'	Δy'	x	y
0	0	0	0	0	2	2
1	0.4	0.2	0	0	2	2
2	0.4	0.2	0.5	0	2.5	2
3	0.4	0.2	1.0	0.5	3.0	2.5
4	0.4	0.2	1.5	0.5	3.5	2.5
5	0.4	0.2	2.0	1.0	4.0	3.0
6	0.4	0.2	2.0	1.0	4.0	3.0
7	0.4	0.2	2.5	1.0	4.5	3.0
8	0.4	0.2	3.0	1.5	5.0	3.5
9	0.4	0.2	3.5	1.5	5.5	3.5
10	0.4	0.2	4.0	2.0	6.0	4.0

$P_c(x_c, y_c)$ and its radius r. The angular velocity of the cutting tool along the circular arc is $\omega = d\varphi/dt$. The x and y components of a circular arc are defined by (Fig. 7.23)

$$x(t) = x_c + r \cos \varphi(t) \tag{6}$$

$$y(t) = y_c + r \sin \varphi(t)$$

With $\varphi(t) = 2\pi t/T$ (T = $2\pi/\omega$, ω is the angular velocity) we obtain

$$x(t) = x_c + r \cos \frac{2\pi}{T} t \tag{7}$$

$$y(t) = y_c + r \sin \frac{2\pi}{T} t$$

The components of the circular velocity along the arc are obtained by differentiating x(t) and y(t) with respect to the time t:

$$\dot{x}(t) = \frac{dx}{dt} = v_x = -\frac{2\pi}{T} r \sin \frac{2\pi}{T} t \tag{8}$$

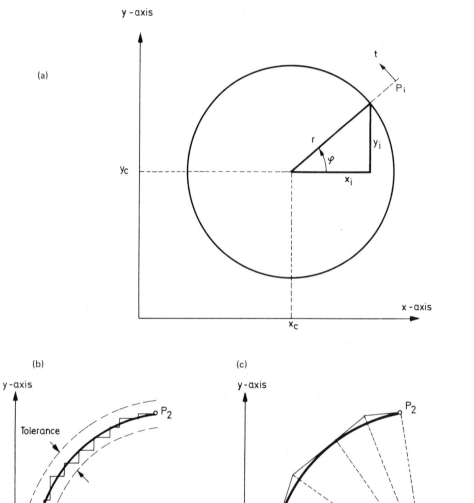

FIG. 7.23 Principle of circular interpolation: (a) definition of a circle; (b) incremental steps of a circular interpolation; (c) approximation of a circle by linear interpolation.

Control of Manufacturing Equipment

$$\dot{y}(t) = \frac{dx}{dt} = v_y = +\frac{2\pi}{T} r \cos \frac{2\pi}{T} t$$

Substituting Eq. (7) into (8) yields

$$\dot{x}(t) = -\frac{2\pi}{T} [y(t) - y_c] \qquad (9)$$

$$\dot{y}(t) = +\frac{2\pi}{T} [x(t) - x_c]$$

By taking x_i and y_i as initial parameters $x(t)$ and $y(t)$ can be obtained through integration:

$$x(t) = x_i + \int_0^t \dot{x}(t) \, dt = x_i - \frac{2\pi}{T} \int_0^t [y(t) - y_c] \, dt \qquad (10)$$

$$y(t) = y_i + \int_0^t \dot{y}(t) \, dt = y_i + \frac{2\pi}{T} \int_0^t [x(t) - x_c] \, dt$$

With the aid of the discrete notation $t = n \, \Delta t$ ($n = 0, 1, \ldots, N$), Eqs. (10) can be rewritten

$$x(t) = x(n \, \Delta t) = x_i - 2\pi \sum_1^n \frac{y(n \, \Delta t) - y_c}{N} \qquad (11)$$

$$y(t) = y(n \, \Delta t) = y_i + 2\pi \sum_1^n \frac{x(n \, \Delta t) - x_c}{N}$$

The DDA algorithm represented by this equation consists of two integrators whose registers are initially loaded with the value of the absolute distances $y_i - y_c$ and $x_i - x_c$. The integrators are connected to each other. The output of the y integrator is fed back to the input register of the x integrator, while the output of the x integrator is reversed in sign and fed back to the input register of the y integrator. Additional interpolation methods are described in Chap. 8.

7.2.8 Functional Control of NC Machines

The technological processing unit of an NC machine controls in parallel to the geometric processing unit the logic states of the machine tool and the sequential steps of the control program (Fig. 7.24). Technological processing includes logic operations, sequence control, event

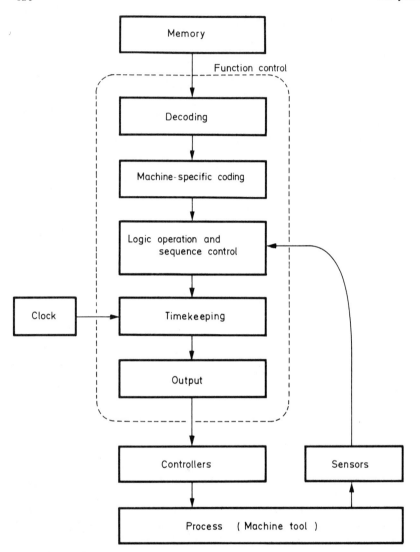

FIG. 7.24 Schematic diagram of the functional control of an NC machine tool.

Control of Manufacturing Equipment   427

control, timekeeping, and the I/O of sensor and control signals. The time- and event-controlled sequences are necessary to perform the following basic machine tool functions:

1. Setup of the feed rate through (a) an approach mode (fast), (b) a work mode (slow), and (c) a reverse mode (rapid) direction of the spindle
2. Setup of the rotation, clockwise/counterclockwise
3. Activation of the tool and of tool changes using an identification number
4. Special-purpose functions (e.g., turn on/off cooling fluid)

The signal flow through the technical processing unit of a functional machine tool control is shown in Fig. 7.24.

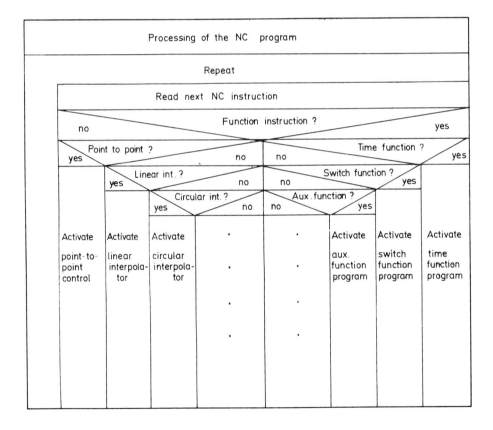

FIG. 7.25 Interpretation of NC blocks.

After interpretation of an NC block, the control signals are separated into geometric and technological elements. Then the technological processing unit can start with execution of the logic program sequence. Sensor and time signals are used to synchronize this event-controlled process. The I/O unit of the NC system transfers the control signals and sensor signals between the process and the machine control unit (MCU). A structogram of an interpretation of an NC block is illustrated in Fig. 7.25.

7.3 CONTROL SYSTEM OF THE NC MACHINE TOOL

The most important task of the control unit of the NC machine tool is to direct the material removal of the tool and to generate the desired surface of the workpiece. A typical accuracy at a feed rate of 10 cm/sec is approximate 0.001 mm. Depending on the size of the workpiece, the amount of material to be removed may vary from a few grams to several tons. In addition, the machining conditions may have to be altered along a given trajectory. Thus the control loop has to be laid out for very stringent requirements. It has to direct the tool to produce the workpiece with a high surface finish and with a minimum dimensional error. In practice, the control system is tailored toward the machine tool. Two principal methods are used to command a machine tool (Fig. 7.26): (1) open-loop control and (2) closed-loop control.

In open-loop control the machine is positioned by a command of a controller. Under normal operating conditions this method may render adequate service. However, since there is no response signal sent back to the controller, a sudden overload or another disburbance may make the machining operation unstable. For example, with a stepping motor a high torque may cause skipping of a step. Despite this fact, stepping motors are frequently used for the control of NC machine tools. For each input pulse the motor produces a defined incremental angular step. The angular velocity can be controlled by varying the frequency of a pulse train.

In case of nondeterministic disturbances, the closed-loop control principle must be applied. Here the controller continuously compensates for deviations occurring between the desired position and the actual position of the tool. Position control is achieved with the help of a sensor that records the position of the tool during the machining operation and sends this information back to the controller. The position signal returned is compared with the input signal obtained from the tape, and any deviation is automatically compensated for. When a control system is selected, the dynamic response of the control loop and that of the machine tool has to be known. In the next section a short overview is given about different types of NC control systems.

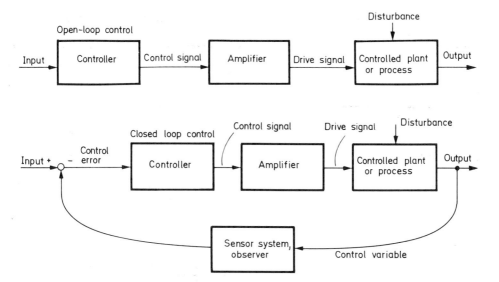

FIG. 7.26 Open- and closed-loop control methods. Control variables are distance, position, velocity, revolution, acceleration, torque, and force. Open-loop control allows only minimal disturbance. Arrows indicate that feedback of process state variables and sensors are necessary (observers).

7.3.1 Elements of a Feedback Control System

The hardware configuration of a typical NC servomechanism consists of the following components: (1) a transducer, (2) a sensor, (3) a servomotor, (4) a control amplifier, and (5) the machine to be controlled. These system elements are integrated into a closed loop which can be illustrated by a block diagram. It consists of graphical symbols that define the information flow and functional relation between the elements of the system. The symbols show the characteristic response of an element to a rectangular input signal (Table 7.4). The block diagram can be interpreted as a signal flow graph. The edges of the graph represent the operands and the nodes represent the operators. The nodes can be of four different types:

1. The signal source, whereby the node generates response.
2. The signal sink; here the node is affected by the system, but it generates no feedback signal.
3. The combination of signals; this node is effected by the system and generates response.
4. The node with memory; this node memorizes the state of past system signals.

TABLE 7.4 Basic Block Diagram Elements to Describe Servo Control Mechanisms

Name	Functional relation	Transfer function	Time history of unit step response	Symbol
Amplifier	$y = k\, u$	k		
Integrator	$y = k \int_0^t u(\tau)\, d\tau$	$\dfrac{k}{s}$		
Differentiator	$y = k\, \dot{u}$ $\left(\dot{u} = \dfrac{du}{dt}\right)$	$k\, s$	Area = k	
Dead time	$y = k\, u(t - T_t)$	$k\, e^{-T_t s}$		
1st order dead time element	$T\dot{y} + y = k\, u$	$k\, (1 - e^{-t/T})$		
Dead time element with differentiator	$T\dot{y} + y = kT\dot{u}$	$k\, e^{-t/T}$		
1st order dead time element with amplifier	$T_N \dot{y} + y = k\, u + kT_Z \dot{u}$	$k\, \dfrac{1 + T_Z s}{1 + T_N s}$	$T_Z < T_N$	
2nd order dead time element	$T^2 \ddot{y} + 2dT\dot{y} + y = k\, u$	$\dfrac{k}{1 + 2dTs + T^2 s^2}$	$d < 1$	

Control of Manufacturing Equipment

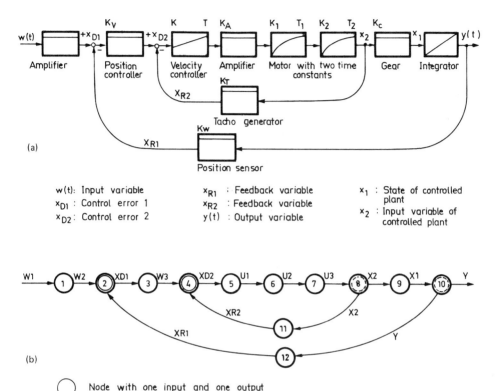

FIG. 7.27 Multiloop control system for NC servo control: (a) block diagram of the system; (b) signal flow graph.

Figure 7.27 illustrates a typical multiloop NC control system together with its block diagram (a) and signal flow graph (b). In such a feedback loop the input signal is combined with the feedback signal x_R, and the difference x_D is entered as the control signal to the system drive via an amplifier. The output signal $y(t)$ actuates a servomotor, which provides the rotary or linear motion required by the axis-drive system.

The analytical description of the dynamic response of the control loop can be described with the aid of the (1) functional relationship, (2) transfer function, and (3) step response. The functional relationship describes the dynamic response of the controller as a function of time (e.g., one consisting of a servo motor, a transducer, and a gear).

Let w(t) be the input function and y(t) the output response of a dynamic system. Then the functional relationship between w(t) and y(t) is defined by an operator f, which transforms w(t) into y(t), y(t) = f[w(t)]. In the case of an integrating system, the functional relation is

$$y(t) = f[w(t)] = k \int_0^t w(t)\, dt \qquad (12)$$

The application of the operator f to w(t) generates y(t). If multiple forcing functions $w_1(t)$, $w_2(t)$, ..., $w_n(t)$ are applied to the input, the functional relationship is written as

$$y(t) = f[w_1(t), w_2(t), \ldots, w_n(t)]$$

In control theory the relationship between the input forcing function and the output response is called the transfer behavior. Thus the behavior of control system is defined by the transfer function ϕ:

$$\phi = \frac{\text{output response}}{\text{input function}} = \frac{y}{w}$$

The transfer function can easily be formulated if the dynamic system is linear. Linearity is defined by the relationship

$$y(t) = f[cw_1(t) + cw_2(t)] = cf[w_1(t) + w_2(t)] \qquad (13)$$

This equation states that the transfer behavior is linear if the input forcing function w(t) yields y(t) and cw(t) yields cy(t) the amplification principle). Furthermore, the superposition of $w_1(t) + w_2(t)$ must enforce an output of $y_1(t) + y_2(t)$.

An important attribute of the dynamic linear transfer element is that the combination of linear transfer elements (e.g., cascade, parallel, or loop blocks) again yields a linear transfer element (Fig. 2.29). To facilitate the analytical and stability analyses and to design the controller, the transfer function F(s) is used.

7.3.2 Transfer Function

The time history of the dynamic response of the NC servo look is often defined by a second-order differential equation. A system with a multiple-axis configuration yields a complex highly coupled mathematical model. The dynamic response of such a systems is defined by a differential equation of the form

Control of Manufacturing Equipment

$$b_m y^{(m)} + \cdots + b_0 = a_n x^{(n)} + a_{n-1} x^{(n-1)} + \cdots + a_1 x^{(1)} + a_0 x \quad (14)$$

where y is the input forcing function [$x^{(n)} = d^n/dt^n$, $y^{(m)} = d^m y/dt^m$] and x is the output response variable. In mechanical systems the relationship n > m usually prevails. For the analysis of such a dynamic system, classical solution methods can be used. In control engineering the complex differential equations are transformed into simpler equations by using the Laplace operator. This operator, which transforms the time function f(t) into the function F(s), is a complex variable of the form s = σ + jω. The Laplace operator is defined in the following manner:

$$F(s) = E\{f(t)\} = \int_0^\infty f(t) e^{-st}\, dt \quad (15)$$

This equation is very useful when a differential equation defining the response of an NC servo controller is investigated. To use the Laplace transform, tables are available that facilitate the analytical work of the control engineer (Table 7.5). An example will illustrate the use of the Laplace operator.

EXAMPLE 7.3 Let us consider a dynamic mechanical system which is defined by the following second-order differential equation:

$$a_2 x^{(2)} + a_1 x^{(1)} + a_0 x = y \quad (16)$$

where $x^{(1)}(t = 0) = 0$ and $x(t = 0) = 0$. Application of the Laplace operator yields

$$a_2 s^2 X(s) + a_1 s X(s) + a_0 X(s) = Y(s) \quad (17)$$

solving for X(s), we obtain

$$X(s) = \frac{1}{a_2 s^2 + a_1 s + a_0} Y(s) = F(s) Y(s)$$

or

$$X(s) = \frac{1}{(s - a)(s - b)} Y(s)$$

TABLE 7.5 Often-Used Laplace Transformations

$\dfrac{1}{s}$	1
$\dfrac{1}{s-a}$	e^{at}
$\dfrac{1}{s^2}$	t
$\dfrac{a}{s^2+a^2}$	$\sin at$
$\dfrac{1}{s(s-a)}$	$\dfrac{1}{a}(e^{at}-1)$
$\dfrac{1}{(s-a)^2}$	te^{at}
$\dfrac{1}{(s-a)(s-b)}$	$\dfrac{e^{at}-e^{bt}}{a-b}$
$\dfrac{s}{s^2+a^2}$	$\cos at$
$\dfrac{s}{(1+as)(1+bs)}$	$\dfrac{ae^{at}-be^{bt}}{a-b}$
$\dfrac{1}{s(s^2+a^2)}$	$\dfrac{1}{a^2}(1-\cos at)$
$\dfrac{1}{s(s-a)(s-b)}$	$\dfrac{1}{ab}\dfrac{ae^{t/a}-be^{t/b}}{b-a}$
$sF(s)-f(+0)$	$\dfrac{df(t)}{dt}=\dot{f}(t)$
$s^2F(s)-sf(0)-f(0)$	$\dfrac{d^2f(t)}{dt^2}=\ddot{f}(t)$
$\dfrac{1}{s}f(s)$	$\int_0^t f(\tau)\,d\tau = f(t)\times 1$
$F_1(s)F_2(s)$	$f_1(t)f_2(t)=\int_0^t f_1(\tau)f_2(t-\tau)\,d\tau$
$\dfrac{1}{2\pi j}\int_{x-j\infty}^{x+j\infty} F_1(\sigma)f_2(s-\sigma)\,d\sigma$	$f_1(t)f_2(t)$

Control of Manufacturing Equipment

If y(t) is the unit step input function $\sigma(t)$, the use of s yields $Y(s) = E\{\sigma(t)\} = 1/s$. The result is

$$X(s) = \frac{K}{(s-a)(s-b)s} = \mp(s) \cdot \frac{1}{s}$$

By applying the inverse Laplace operator, we obtain with the help of the table the response of the differential equation:

$$x(t) = \frac{K}{a-b} + \frac{K}{b-a}(ae^{-t/a} - be^{-t/b}) \tag{19}$$

where x(t) is the dynamic unit step response and F(s) the transfer function of the mechanical system.

The transfer function of an NC servo control loop is composed of the individual transfer functions of the control elements, such as the motor, transducers, gear, sensor, control amplifier, and control plant. Figure 7.28 shows the transfer functions of multiloop servo control elements of an NC machine. They can be reduced to one function representing the comprehensive overall transfer functions of the entire control loop. The rules for combining transfer elements are illustrated in Fig. 7.29. It is shown how the parallel, serial, feedback, and feedforward transfer elements can be reduced to simpler transfer functions.

7.3.3 Implementation of NC Machine Controls

The transfer function is a standard tool in the field of servo engineering to analyze and to design control systems. The most important methods to analyze the system stability have been discussed thoroughly by Bode, Nichols, Popov, Nyquist, and others [4-6]. The interested reader will find many standard books on this subject in this field of system theory. The control engineer uses many different tools to analyze stability criteria of control systems (e.g., the phase, root locus, Nyquist, and Nichols methods. Once the analysis of the system and the theoretical controller design is done, the controller can be implemented. The basic control technologies are (1) analog, (2) hybrid, and (3) digital (DDC).

Figure 7.30 shows an axis control system for a machine tool with feedback control loops for velocity and position. A pulse input is sent to a comparator counter, which calculates the positional error. The output is converted into an analog signal, which is passed to two analog control amplifiers. They both have a compensation loop. The output of the first amplifier is modified by a compensation feedback signal. Both the compensation and velocity feedback improve the performance of the system. Other position controls operate in open-loop or closed-loop mode with only position feedback (Fig. 7.31).

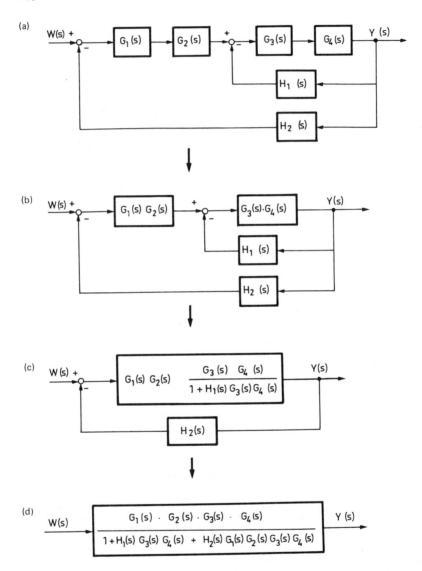

FIG. 7.28 Reduction of a multiloop control system to a single transfer block: (a) first step; (b) second step; (c) third step; (d) final step.

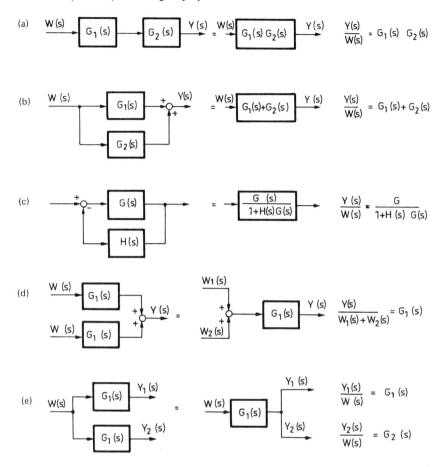

FIG. 7.29 Basic combinations of transfer elements: (a) blocks in cascade; (b) blocks in feedforward; (c) blocks in feedback; (d) blocks with summing point; (e) branching of blocks.

In practice, analog and hybrid servo controls use LSI chips for signal converters, counters, comparators, amplifiers, and logic elements. Complex control algorithms are usually implemented on a microcomputer (8 or 16 bits). It operates as a sampled data device (Fig. 7.32) and is characterized by the sampled data rate and the DDC algorithm. The implementation of the DDC algorithm usually makes a discrete process model necessary. A differential equation that describes the dynamics of the control system has to be transformed into a difference equation, whereas an integral equation can be transformed into, for example, a recursive summing equation [7]. A systematic approach is discussed

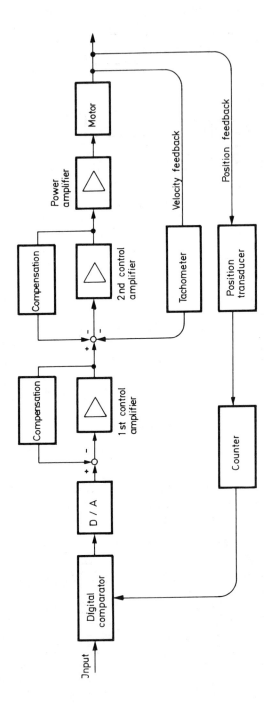

FIG. 7.30 Machine tool axis control with position and velocity feedback.

Control of Manufacturing Equipment

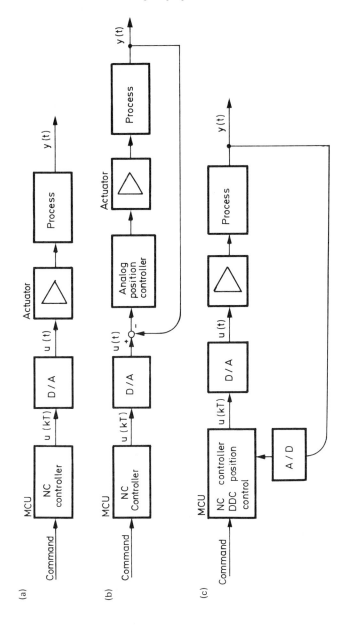

MCU = Machine control unit; $u(kt)$ = Discrete controller variable; $u(t)$ = Continuous controller variable

FIG. 7.31 Various position control methods: (a) feedforward position control; (b) analog feedback position control; (c) digital feedback position control. MCU, machine control unit.

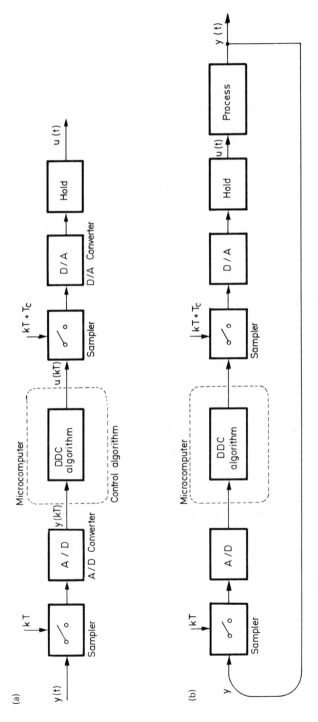

FIG. 7.32 The digital computer as a sample data controller: (a) open-loop mode; (b) closed-loop mode. T_c, computation time; T, sample interval; $k = 0, 1, 2, \ldots, n$.

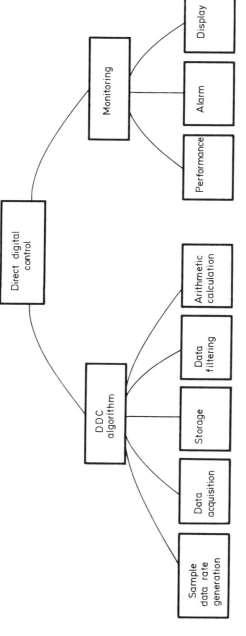

FIG. 7.33 Typical tasks of a DDC system.

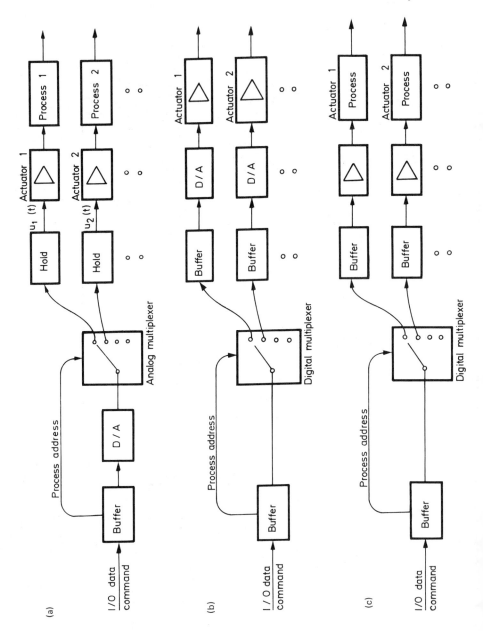

by Jury [8], who explains the analysis and design of discrete control algorithms using the z transform. The z transform uses transfer functions in a similar manner to the Laplace method (Sec. 7.3.2). A DDC system calculates the DDC algorithm and that of the utility functions, such as data acquisition and storage, data filtering, and monitoring and output of control data (Fig. 7.33). If the cycle time for calculating the DDC algorithm is short compared to the sample data rate, multiple processes can be controlled. In this case a real-time multitasking operating system supervises the execution of each DDC algorithm. The I/O of control data can be performed with high reliability using a multiplexer (Fig. 7.34). Higher-order functions typically used for optimal and adaptive controls can also be implemented on a microcomputer-based NC system.

7.4 ADAPTIVE CONTROL

An adaptive controller is defined as a system that operates in an environment where the parameters change stochastically over time. To obtain the desired performance of the plant, the adaptive control system must compensate for these unpredictable environmental changes. Adaptive control strategies consist of three basic tasks (Fig. 7.35):

1. To monitor and identify stochastic changes in the environment which disturb the desired performance of the process
2. To decide how to modify the control strategy and/or the controller or a part of it to obtain optimal performance
3. To modify the control strategy to realize the desired decision

Adaptive controllers are often called self-optimizing systems. "Self-optimizing" includes both feedback strategies and optimizing of the performance.

7.4.1 Adaptive Control for NC Machines

The production time for NC tools consists of the following components:

Loading the NC program
Handling the workpiece
Tool setup

FIG. 7.34 Various I/O configurations for control of multiple processes: (a) analog multiplexed output for analog actuator; (b) digital multiplexed output for analog actuator; (c) digital multiplexed output for digital actuator.

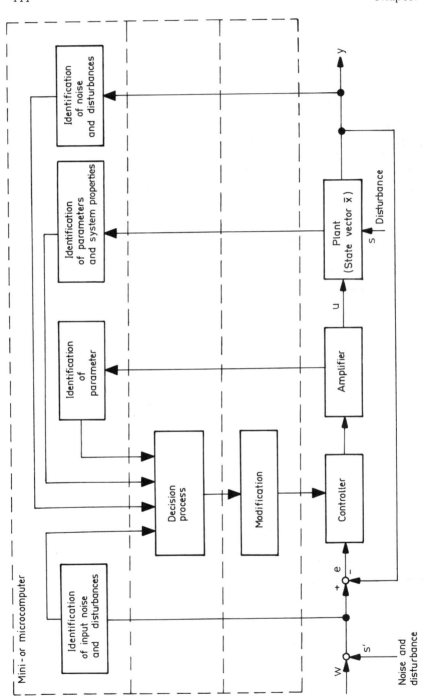

FIG. 7.35 Schematic diagram of an adaptive multiple-task control system.

Machining the workpiece (in-process time)
Visual control by the operator
Inspection of tool wear
Quality control

Adaptive control normally is justified for those NC machine tool applications where the in-process time consumes a significant part of the total production time. The in-process time can be optimized by the modification of feed and speed during machining. This may be necessary due to changes in the hardness of the workpiece, its width or depth of cut, tool wear, or operating temperature. Contrary to NC systems, which direct the path of the tool only during machining, adaptive control can respond to process changes and compensate for them on-line. This control is necessary for all manufacturing processes where changes in raw material, wear of machine components, or tools cannot be predicted. The control adapts the machine to prevailing cutting conditions, time variables, and system changes. Process variations can be classified as follows (Fig. 7.36):

1. Changes due to environmental influences
2. Changes in the raw material (e.g., its microstructure, hardness, strength, and thermal properties)
3. Changes and wear of components (e.g., cutter profile and temperature)
4. Failure of components
5. Changes of the process model involving the (a) parameter, (b) structure, (c) constraints, and (d) signals.

It is the basic task of adaptive control to identify these changes and to use them to evaluate the performance index. This index is needed as input for the decision strategy, which optimizes the process by modifying the control strategy. For NC machine tools geometric and

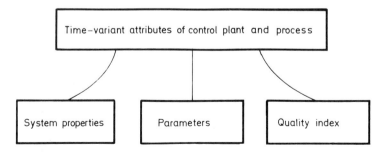

FIG. 7.36 Stochastic process variables that make adaptive control necessary.

technological changes are parameters for adaptive control strategies. For example, geometric adaptive control ensures that the predefined surface accuracy and surface quality is realized. Technological adaptive control optimizes the controller with respect to tool wear or new machining parameters required after an automatic tool change. For example, a cutter with different diameters needs appropriate constraints. Other adaptive methods for machine tool control are ACC (adaptive control constraint) and ACO (adaptive control optimization).

The ACC strategy tries to provide a constant machining operation during cutting. One or several process variables are altered to assure the constant cutting condition within predefined constraints. This strategy tries to utilize the machine at its maximum capacity. It may also prevent machine components and tools from overloading. The control parameters of the system are power, torque, and force.

The ACO strategy tries to maintain a constant quality index by controlling one or several input variables or control parameters. The quality index is a parameter that designates the economic operation of the machine tool. The index has to be calculated with the help of a mathematical model that uses as input sensor data of the process (e.g., the wear of the tool). ACO is frequently used for the automatic calculation of the depth of cut.

7.4.2 The Three Basic Tasks of Adaptive Control

To evaluate the performance of a machine tool and to respond to unpredicted environmental changes, the adaptive control system has to do three basic tasks: identify, decide, and modify. It is difficult to generalize the tasks and to distinguish them from each other. In the following the three tasks are discussed briefly.

The Identification Task

The identification task has to determine the current performance of the process or of the system. To find the current value of the process index, it is necessary to acquire process feedback information and sensor data and to evaluate these. In the case of continuous environmental changes (e.g., tool wear) that influence system performance, the identification has to be done continuously. It involves a number of measurement activities. Often, a mathematical model of the process has to be referenced to estimate the performance index. It includes the computation of the performance index from measurements of process variables. For example, the difference between the actual and the desired optimal performance has to be computed. To measure the actual performance index with a high accuracy, appropriate sensors are necessary. The technological requirements of the sensors depend on the following:

The desired accuracy
The noise of the process and its environment
The attributes of the process
A priori knowledge about the process

The sensors and the sensor processing methodology are divided into two groups:

1. The direct or indirect measurement of the parameters or of the performance index using dedicated sensors (observers).
2. The identification of the parameters or of the performance index with the aid of measured input and output signals, assuming that:
   a. The input signal is measurable.
   b. The output signal and its noise component are measurable.
   c. The calculation of process and environmental changes with the aid of measured input and output variables is precise.

Process and environmental changes can be caused by a flaw in the raw material, the wear of a component, or the failure of a component. The activities for the identification phase are shown in Fig. 7.37. If the process is described by a mathematical model, the identification problem is related to changes inside of the model. The model describes the current state of the process and allows calculation of the performance index or function. For adaptive control of a machine tool the cutting cost often (cost per inch of metal) is used as a performance

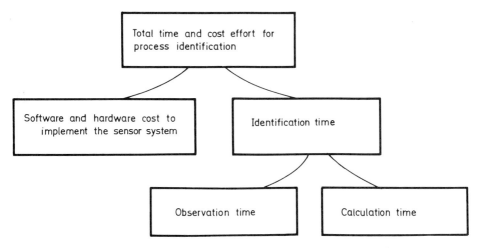

FIG. 7.37 Aspects of the evaluation of technological and computational efforts of the identification task.

index, and it is described by a mathematical equation. There are three methods used to change the mathematical model:

1. *Changes in the model parameters*: It is assumed that the structure of the mathematical model is defined but that its parameters are altered with time to reflect process changes (e.g., tool wear).
2. *Changes in the model structure*: It is assumed that the structure of the mathematical model is known a priori but that it will alter with time to reflect process changes. Usually, limited changes of the model structure are allowed to decrease the computational effort.
3. *Changes in the constraints*: It is assumed that the structure of the mathematical process model is known but that there are different constraints within which the process variables have to operate (e.g., minimum or maximum speed, feed, torque and force).

The mathematical procedure for identification by the model can be classified into three basic types:

1. Identification of the process with parametric models using mathematical relations
2. Identification of the process with parametric models using model tuning techniques
3. Identification of the process with a nonparametric model

If enough information about the process and its mathematical model is known, an on-line identification function can be defined. On-line measurements of input and output signals of the process are used as input for the identification function. Approximation, regression, correlation, and search methods have been developed for this purpose. In some cases the mathematical model has to be selected during identification. If the process is poorly defined or the model is too complex, it is more practical to measure or to estimate the performance quality index directly. In this case heuristic search strategies are applied to find the optimal operational mode of the process.

The Decision Strategy

When the process, its structure, its parametric model, or its performance quality have been identified, the optimization procedure can commence. It is the task of the decision process to determine the optimal control set point or strategy that guarantees the desired performance quality. The decision process has as its input the identified process changes and the desired performance index. The performance

Control of Manufacturing Equipment

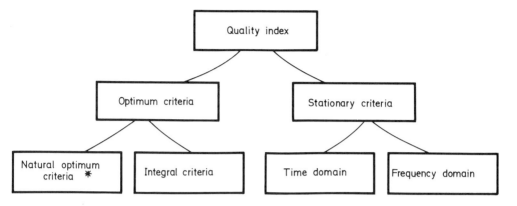

FIG. 7.38 Various types of performance indices or functions. *, Efficiency, Risk, profit.

index or function can be related to constant or optimal criteria (e.g., cost, efficiency, risk, and wear) (Fig. 7.38). If the controller is parametrized, the task is to determine the optimal parameters to achieve the desired process performance. The parameters are calculated with the aid of an optimization strategy whereby the controller coefficients are a function of the identified process parameters and of the performance index. If the structure of the controller is not predefined, the strategy is to determine an optimal control algorithm which achieves the desired process performance. The control mechanism is a function of the process and of the performance index. In the case of parameter optimization, three decision processes can be defined:

1. The regular algorithmic decision process
2. The search algorithmic decision process in open-loop mode
3. The search algorithmic decision process is closed-loop mode

Regular algorithmic decisions. For many adaptive control systems a quality index Q can be defined as a function of the process parameters a_i ($i = 1, 2, 3, \ldots, m$) and controller parameters r_j ($j = 1, 2, 3, \ldots, n$). To optimize system performance the quality index must have a minimum. Thus the conditions

$$\frac{\partial Q}{\partial r_j} = \phi \quad j = 1, 2, 3, \ldots, n$$

have to be fulfilled. If n control parameters have to be optimized, n equations must be solved. For many applications the index Q has the form of an integral. Its general form is

$$Q = \int_{T_i}^{T_e} F(w, y, \bar{x}, s, u, t) \, dt \tag{21}$$

where

    w = input variable

    y = output variable

    $\bar{x}$ = process state vector

    s = noise

    u = drive signal

    t = time

    T_i = start time

    T_e = end time

For the design of an adaptive controller, an efficient quality index must be defined. It is also of importance that the minimum can be found with low computational effort. The adaptive process in Fig. 7.39 requires no calculation of the index Q. Here Q can be controlled directly in an open-loop mode by the equation

$$\bar{r}_{opt} = f(\bar{a})$$

Algorithmic decision process in open-loop search mode. For this strategy the quality index Q has to be calculated by observing dynamic changes in process parameters (Fig. 7.40). The search algorithmic strategy tunes the control parameter r until r_{opt} of Q_{min} is found. The optimum can be obtained by using maximizing strategies such as the trial-and-error, gradient, and univariate methods. A mathematical model is needed for the calculation of Q and for the performance of the search procedure. When r_{opt} is found, the controller is modified accordingly (open-loop adaption).

Algorithmic decision process in closed-loop search mode using a direct measured index. If Q is measured directly, the search strategy to find the optimum is performed with the help of the controller and the controller plant (Fig. 7.41). Here a model is not used. The parameters r are tuned until the index Q is optimal. With NC machine tools, feed and speed are often tuned at defined increments until the index is at its minimum. This is called closed-loop adaption.

Control of Manufacturing Equipment   451

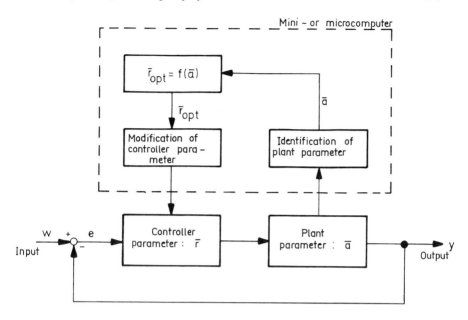

FIG. 7.39 Adaptive control system with regular algorithmic optimization strategy.

7.4.3 Application of Adaptive Control in Machining

Typical parameters that make adaptive control necessary in metal machining processes are:

1. Variations due to different depth or width of cut. This necessitates an adjustment in the feed rate. Adaptive controllers that compensate for these variable geometric conditions are typically applied in contouring and profile milling operations.
2. Variations due to different workpiece hardness and variable cutting conditions. A workpiece of different hardness necessitates an adjustment in feed and speed rates to avoid extensive tool wear.
3. Tool wear. The increasing dullness of the tool edge during cutting increases the cutting force. An adjustment for tool wear is a reduction of the feed rate.
4. Air gaps during cutting. If the cutter is passing through an air gap, the feed rate may be increased to speed up machining.

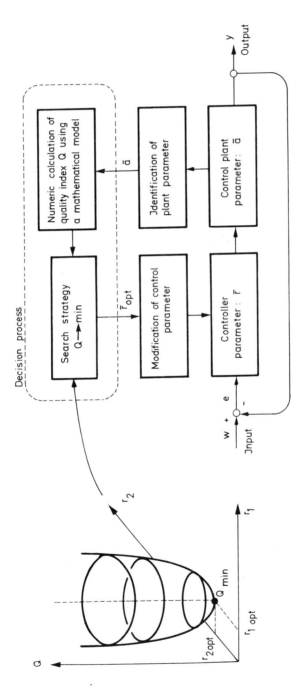

FIG. 7.40 Adaptive control system with a search algorithmic strategy to optimize performance index.

Control of Manufacturing Equipment   453

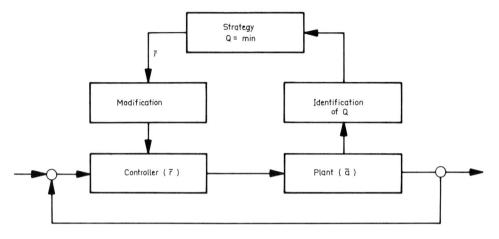

FIG. 7.41 Adaptive control system with direct measurement of performance index.

5. Variations in the workpiece stiffness. In case the stiffness of the workpieces changes during machining, the feed rate may be reduced to maintain stability and accuracy and to avoid chatter.

Figures 7.42 and 7.43 show typical applications of ACC and ACO adaptive controls for metal cutting. Other applications are to determine the depth of cut automatically, to save machining time and to prevent the tool and the drive system from overload and the workpiece from damage (Fig. 7.44). For this purpose several process variable usually have to be measured, such as:

Cutting force and torque
Feet rate
Cutter temperature
Vibration
Machine motor load
Spindle force

Most adaptive control systems for metal cutting optimize with respect to metal removal and tool wear. Because of the absence of accurate, fast, and reliable sensors optimal performance is obtained from indirect index measurements rather than direct measurements. Presently, many attempts are made to develop control systems that calculate the performance index on-line during machining.

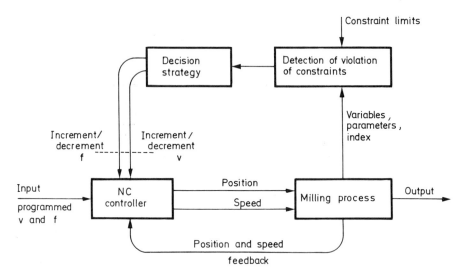

FIG. 7.42 Adaptive control constraint (ACC) structure of a milling process.

7.4.4 Advantages of Adaptive Control in Metal Cutting

Adaptive control offers the following advantages for machining:

Increased production rates
Increased tool life
Improved protection of workpiece from damage
Less operator intervention
Easier part programming

With classical NC control, the part programmer must plan and calculate the feed and speed of the cutter path, taking into considerations the worst-case cutting condition. Usually, a time-consuming procedure of trial and error for program optimization has to be performed until satisfactory cutting conditions are realized and the program has been optimized. In adaptive control the calculation of feed and speed and of constraint limits for forces, torques, horsepower, and for cutter geometry is done by the controller. The programmer only has to define the part geometry and some basic cutting parameters. The time for part programming and program testing will be reduced. However, for the design of adaptive control systems, complex mathematical and technological problems must be solved. The problem is summarized as follows:

1. The original system analysis is very different and involved. Adaptive control loops are highly nonlinear. In some cases the complexity of describing the machining process prevents a reliable implementation.
2. The identification task requires the availability of complex sensors and sensor processing algorithms.
3. The index of performance is highly dependent on the performance of the sensor system. It must be possible to optimize the index by tuning selected control parameters or by designing specific control algorithms.

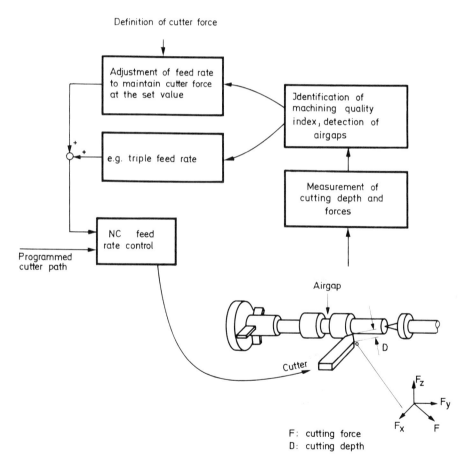

FIG. 7.43 Configuration of an adaptive control optimization (ACO) machining system using cutting depth and force to calculate the performance index.

Control of Manufacturing Equipment

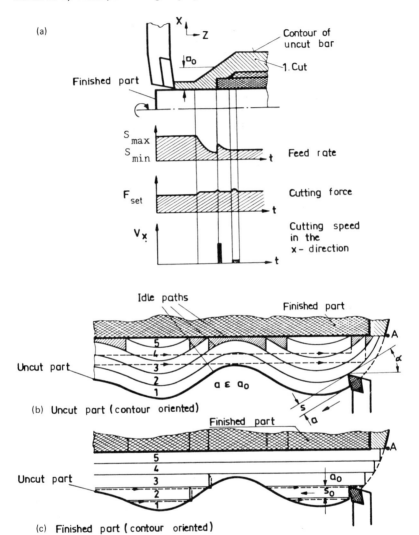

FIG. 7.44 Principle of automatic determination of depth of cut in adaptive control and the two basic strategies to find the tool path: (a) time history of feed rate, cutting force, and cutting speed; (b) uncut part contour-oriented cutter path; (c) finished part contour-oriented cutter path.

4. There are no reliable, fast on-line sensor systems available.
5. The optimal decision strategy has to be calculated on-line and should not disturb the stability of the system.
6. There are economical limits for the hardware costs and the implementation effort. The advantages of the improved performance must warrant the added cost.

Future adaptive control systems will perform intelligent functions which today are done by the human operator. They will have learning and self-optimizing capabilities and will be integrated in CIM systems.

7.5 SENSORS FOR COMPUTER-CONTROLLED MACHINE TOOLS

Important functional elements of an NC system are the signal transducer and the sensor system. They are needed for the acquisition of measurement data and the observation of the state variables of the cutting process, including quality parameters of the tool and of the product. The accuracy of a machine tool depends to a high degree on the resolution of the sensor system. Example of measurement parameters are:

1. Travel distance and the rotational angle of the tool table and the workpiece
2. The speed of the power drive
3. The cutting force and torque
4. The approach of the tool to the surface to be cut
5. The tool temperature
6. The depth of cut

These parameters are determined with the following types of sensors:

1. Position and velocity sensors
2. Temperature sensors
3. Force and torque sensors (a) at the tool and (b) at the workpiece
4. Tactile sensors
5. Proximity sensors
6. Optical sensors
7. Safety sensors
8. Sensors to detect the physical characteristics of the material of the workpiece
9. Vision
10. Acoustic and vibration sensors

Depending on the sensor application, the measurement data may have different formats. Binary signals describe threshold values, end points, critical limits, and logic operating states. The physical states of a machine tool are described by continuous analog signals that have to be digitized. In the case of an optical sensor (e.g., a vision system) the geometric features of the workpiece have to be extracted from gray-scale information and binary patterns. To process the sensor signals, a suitable interface to the machine tool has to be provided.

7.5.1 Principles of Measuring Displacement and Angles

To control the relative movement between the workpiece and the tool, the machine displacement and angular orientation have to be measured. The displacement transducers can be of the following type:

Analog
Absolute digital
Relative digital (incremental)

Analog measurements are made with the help of linear and rotational potentiometers. The measured value is represented by a continuous analog voltage signal. Because of their poor resolution and of their sensivity to noise generated by the contact impedance of the contact wiper, these devices are not often used for machine tools.

The characteristic feature of a digital measuring system is that it records the measured value as a displacement or as an angular increment. The signal is recorded with the aid of a coded disk or a coded linear scale using a photoelectric recording principle (Fig. 7.45).
The displacement information is available in either coded or incremental form. It is recorded by a counter. Measuring systems of this type are universally used for machine tools. A rotational speed is measured either with a tachometer or an incremental transducer. The tachometer produces an analog signal of low resolution. The incremental transducer generates for each angular step a rectangular pulse. The distance between two succeeding pulses is recorded and transmitted to the controller. Processing of the incremental signals is done with the help of an integrated circuit counter and a buffer (Fig. 7.46).

7.5.2 Cut Force and Torque Sensor

To determine the interaction between the workpiece and the tool, the cutting force has to be recorded at the tool holder and the torque at the drive shaft. These parameters are needed for adaptive control and automatic contour generation to optimize machining. Similarly, for assembly operations by a robot the applied and reaction forces and

Control of Manufacturing Equipment

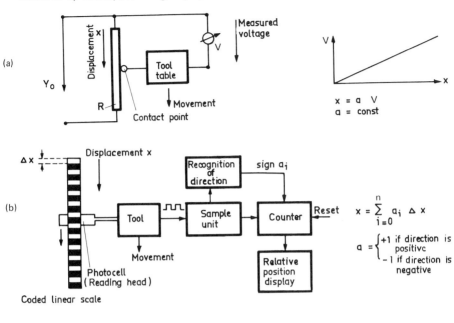

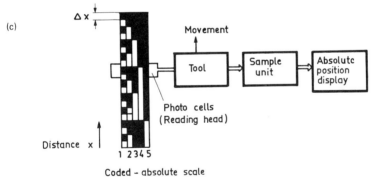

FIG. 7.45 Various methods to measure displacement: (a) using potentiometer; (b) relative-coded linear scale; (c) absolute-coded scale; (d) binary-coded disk.

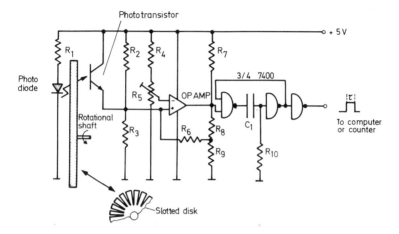

FIG. 7.46 Principle of measuring rotational speed and angular displacement.

torques have to be recorded, including the gripper forces. In general, a force cannot be measured directly. However, it is determined via a motor current or the deformation of an elastic measuring device. Typical measurement sensors are:

1. *The strain gauge*: Here the elongation of a measurement wire is recorded as a resistance change.
2. *The piezoelectric crystal*: The deformation of a piezoelectric crystal generates a voltage that is proportional to the load.
3. *Magnetoelastic transducers*:
   a. The deformation of an elastic element is determined by changes in its magnetic properties (inductivity).
   b. The deformation of an elastic element is determined by changes in the magnetic field between a primary and secondary winding (transformer principle).

A torque can be measured by the following principles (Fig. 7.47):

1. *Current measurement*: The torque is proportional to the armature current of a motor, which is energized by a constant stator current.
2. *Strain gauge*.
3. *Magnetoelastic transducer*: A shaft subjected to torsion changes the magnetic coupling between a primary and a secondary magnetic coil.
4. *Photoelectric transducer*: Two slotted disks record the torsion of a shaft with the help of a photoelectric transducer.

Control of Manufacturing Equipment

The requirement for sensors used in machine tools are:

High linearity
Good stability
Low hysteresis
Defined frequency response
Good working characteristics
Wide measuring range
Low measuring errors

7.5.3 Wear Measurement

An important parameter for a machine tool is the tool wear. The economy of a machining process depends on the tool wear and can be improved by adaptive optimization methods. The most important measurement methods are (1) continuous measurement directly during the operation of the machine tool, (2) intermittent measurement directly at the tool, (3) indirect measurement during operation of the machine tool, (4) indirect measurement at the workpiece during machining, and (5) indirect measurement at the chip. In case the wear cannot be measured directly, it has be determined indirectly by a correlation method. The principle used depends on the application. The measuring methods shown in Fig. 7.48 will be described briefly.

Direct continuous wear measurement. In this case an electric resistor is embedded into the tool tip. The change in resistivity is proportional to the wear. This method can be refined with the help of a resistor network where the resistivity changes at quantified steps during wear.

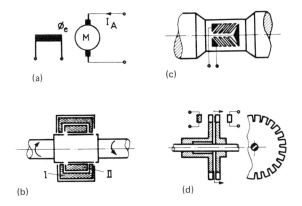

FIG. 7.47 Principle of measuring torques: (a) armature current; (b) magnetoelastic; (c) strain gauge; (d) photoelectrical.

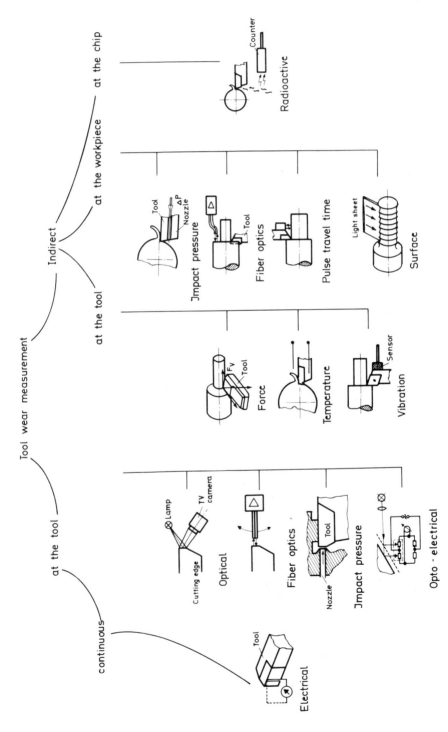

FIG. 7.48 Principle for tool wear measurement.

Direct intermittent wear measurement. With this method the machining process is interrupted periodically and the tool is positioned for measurement. The profile of the cutter edge is recorded. Some methods use optical sensors, such as a TV camera or a photodiode, to measure the width of the worn edge. Other methods record the increasing dullness of the tool by measuring the cutting force with a pneumatic sensor. The force is proportional to the amount of wear.

Indirect measurement of wear during machining. Here the wear is determined with the help of the following process parameters:

Cut force
Cut temperature
Vibration of the tool edge (tool chatter)

The correlation with wear depends on the material and the machining method. It must be determined experimentally.

Indirect wear measurement at the workpiece during machining. With this principle the wear is determined from the workpiece properties. Examples are:

1. *Impact tube principle*: Wear is measured by recording over the machining time the decreasing head pressure between an impact tube and the worn tool.
2. *Fiber-optic wear measurement*: Here the intensity of reflected light between a sensor and the workpiece surface is correlated with wear.
3. *Ultrasonic method*: The wear of the tool is detected by recording the travel time of a sound wave between an emitting and a receiving transducer.
4. *Light sheet method*: Here the optical appearance of the workpiece surface is evaluated.

Indirect wear measurement at the chip. Here two principles are employed. With the first method the tool tip is made radioactive. During machining, radioactive parts of the tool tip are worn off the cutter edge and detected with the scintillation counter. With the other method the shape of a chip is investigated.

7.5.4 Proximity Sensors to Detect the Tool Approach

To optimize machining time the tool usually approaches the workpiece rapidly. At close proximity to the surface to be machined, the tool is slowed down to its cut speed. The most important proximity sensors use the inductive or capacitive principle. With the inductive principle the tool and the workpiece are part of a magnetic loop which changes

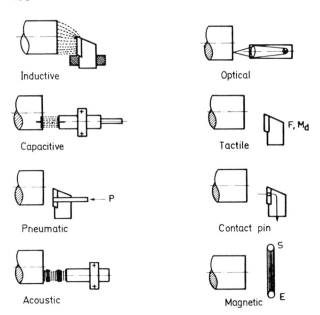

FIG. 7.49 Measuring principles for tool approach.

its properties as a function of distance. Capacitive proximity sensors use a similar principle, where the capacitance between the tool and the workpiece changes with the approach distance.

Other methods use mechanical, acoustic, and photoelectric proximity sensors. In case the exact dimensions of the raw part are known, these parameters can be used to optimize the tool approach. Figure 7.49 shows various principles of tool approach sensors. A vision system can also be used to supervise the approach. The principle of machine vision is explained in Chap. 8.

7.6 NC PART PROGRAMMING

The NC machine tool concept is based on textual programming methods to describe a workpiece with the help of control surfaces. The description of the part contour is taken from the drawing, converted into a code, and entered on a code carrier such as a punched tape. This information is read into the machine tool and is used by its controller to instruct the cutter to machine the part surface. For this the format of the control data and the machine tool commands must be defined in detail. The control program consists of a sequence of commands that have a standardized symbolic format. The instructions are

Control of Manufacturing Equipment 465

usually coded on perforated paper tape, computer cards, magnetic tape, or a floppy disk. The control program is transferred to the machine control unit (MCU), which translates the instruction to control information (Sec. 7.2). The operational instructions are divided into tasks such as to control the geometric path the cutter has to follow, to operate the feed rate of the tool, and to activate the coolant flow. Three types of basic information can be distinguished:

Geometric information
Technological information
Executive instructions

Geometric information is used for positioning of the tool and the workpiece. The technological information consists of logic switching functions, tool specification, physical machining parameters, and miscellaneous instructions. Each task is represented by a sequence of commands called a block. Each block begins with its number and ends with a carriage return sign.

The following sections deal with the basics of part programming. The discussion includes simple manual programming and sophisticated automatic programming with the aid of NC languages.

7.6.1 Structure of the NC Program and the NC Punched Tape

An NC part program is defined as follows: The NC program consists of blocks of words, which represent basic control instructions for the NC machine control unit. A programming word consists of a functional operator field which is coded by a character and followed by a parameter (operand) coded by a number. Each block begins with a word containing the block number. The meanings of the most important functional symbols, often called addressing characters, are listed in Table 7.6.

The numerically coded parameter that follows the functional instruction field defines either a geometric position or a key number describing technological information. The code entered on perforated paper tape is defined by the ISO. The tape has seven tracks used for information and one for a parity bit. In addition to the ISO code, the EIA code (EIA 244) is often used.

7.6.2 Manual Part Programming

For better understanding of computerized automatic NC programming, the basic manual NC method is discussed. First, manual NC part programming is done on the lowest level, usually called the machine level. It can be compared to assembly programming. The manual preparation of the NC program is done by the programmer, who enters the program

TABLE 7.6 Basic Characters Used for NC Tapes and Their Control Functions

Character	Address
A	Angular dimension around x-axis
B	Angular dimension around y-axis
C	Angular dimension around z-axis
D	Angular dimension around special axis or third feed function
E	Angular motion around special axis or second feed function
F	Feed function
G	Go, condition for motion, feed function
H	Unassigned
I	Interpolation parameter or feed rate parallel to the x-axis
K	Interpolation parameter or feed rate parallel to y-axis
L	Unassigned (can be used e.g. for line feed)
M	Miscellaneous functions
N	Sequence number
O	Do not use
P	3rd rapid-traverse dimension parallel to x-axis
Q	2nd rapid-traverse dimension parallel to y-axis
R	1st rapid-traverse dimension parallel to z-axis
S	Spindle speed
T	Tool function, tool number
U	Secondary motion dimension parallel to x-axis
V	Secondary motion dimension parallel to y-axis
W	Secondary motion dimension parallel to z-axis
X	Primary x-motion dimension
Y	Primary y-motion dimension
Z	Primary y-motion dimension

Control of Manufacturing Equipment

by direct coding on perforated paper tape. The programmer has to consider all operations and interactions between the machine tool and its MCU. Three basic steps are performed:

1. *Preparation of the machining operation*: The drawing of the workpiece is used as the principal document to prepare the part program. First the surface of the workpiece is described with the help of points, lines, circles, ellipses, and so on. This is done with regard to a machine-independent coordinate system. The next phase is to determine the individual steps of the tool path and to correct for the workpiece stiffness and its fixturing requirements.
2. *Programming of operation sequence*: With the help of the geometric and technological data of the operation plan, the individual machining sequences are entered into a standard programming form. The entries into a row represent a record (block) and describe an individual machining step. A series of sequential records make up a program.
3. *Preparation of the NC tape*: In the next step, the NC program is punched onto a paper or plastic tape, which is used as a storage device. The NC machine reads the tape and translates the coded program into machine tool instructions.

The phases of manual part programming are illustrated in Fig. 7.50. The reader can see that NC part programming requires expertise and high-level skills. This qualifications must include the following:

1. Skills to read a drawing and to translate the workpiece pictorial into machine instructions
2. General knowledge of production methods
3. Background in analytical geometry to calculate the cutter path and to correct for its offset
4. The ability to think in three dimensions
5. The ability to combine the information of the part drawing with those of mathematical tables and to conceive a part program
6. Debugging and testing of the part program
7. Preparation of punched tapes or cards
8. Determination of tooling, fixture, cutter, and feed rate parameters

This list of responsibilities could be extented. To simplify the cumbersome manual programming method, efficient software tools and programming aids were developed. They will be discussed next.

7.6.3 Computer-Aided Part Programming

The manual programming method presented in the preceding section shows that even for the description of a simple workpiece, a series of

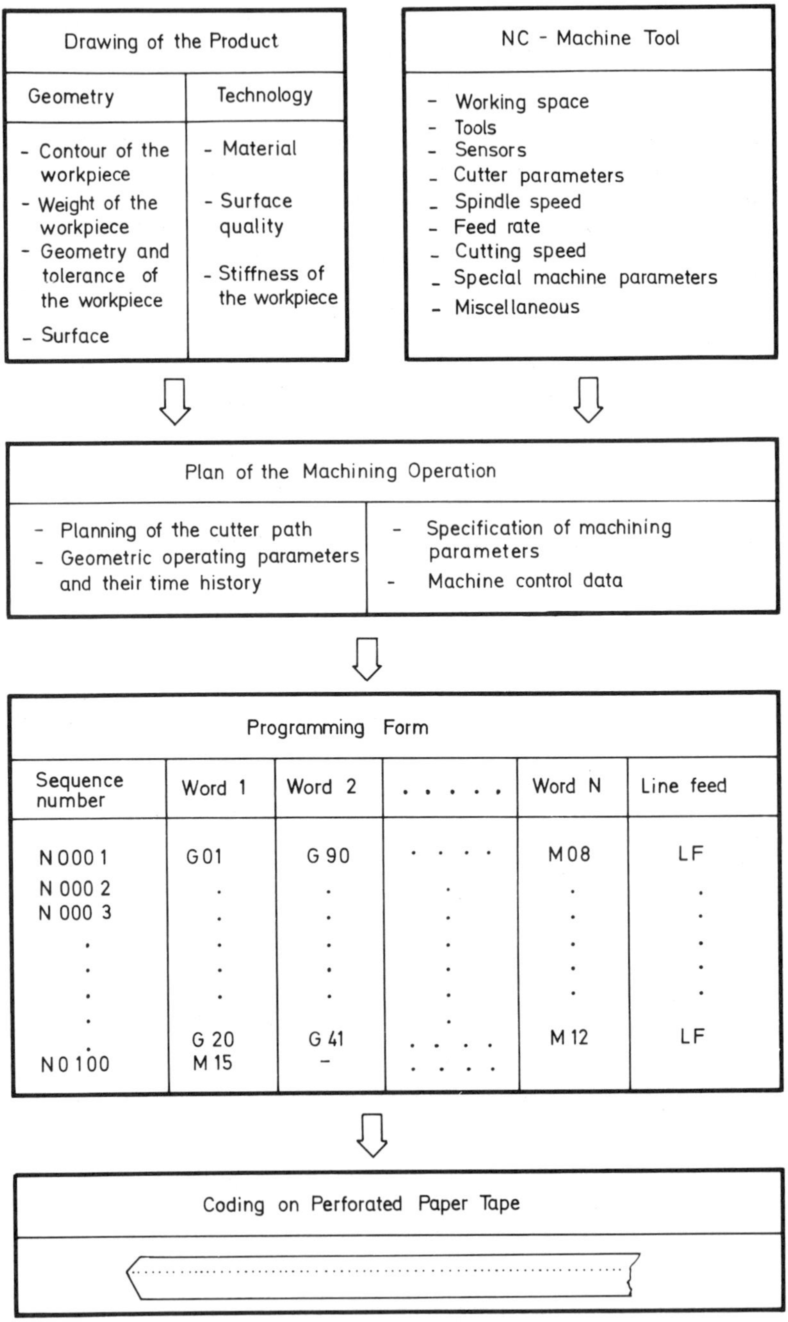

Drawing of the Product	
Geometry	Technology
- Contour of the workpiece - Weight of the workpiece - Geometry and tolerance of the workpiece - Surface	- Material - Surface quality - Stiffness of the workpiece

NC - Machine Tool
- Working space - Tools - Sensors - Cutter parameters - Spindle speed - Feed rate - Cutting speed - Special machine parameters - Miscellaneous

Plan of the Machining Operation	
- Planning of the cutter path - Geometric operating parameters and their time history	- Specification of machining parameters - Machine control data

Programming Form					
Sequence number	Word 1	Word 2	. . . . .	Word N	Line feed
N 0001	G 01	G 90	. . . .	M 08	LF
N 000 2	.	.	.	.	.
N 000 3	.	.	.	.	.
.	.	.	.	.	.
.	.	.	.	.	.
.	.	.	.	.	.
.	G 20	G 41	. . . .	M 12	LF
N 0 100	M 15	—	. . . .		

Coding on Perforated Paper Tape

FIG. 7.50 Phases of manual programming of machine tools.

Control of Manufacturing Equipment

involved calculations to generate geometric and technological data is necessary. Programming of complex components is very difficult and the number of calculations increases substantially. Writing NC blocks is a time-consuming and repetitive task. Frequent references are needed to tables and diagrams to calculate the geometric and technological parameters.

To overcome these problems, computer-aided automatic part programming methods were developed. The use of the modern fast computer provides many advantages and enables economic NC part programming. Time saving, accuracy, and elimination of manual tasks are some of the benefits. The high-level NC programming languages offer the following capabilities:

1. The definition of the geometry of the product (Fig. 7.51)
2. The definition of the technology (machine tool information)
3. The specification of executive parameters (tool path)

All these features are necessary to generate NC blocks. The principle of computer-aided part programming is outlined in Fig. 7.53 and 7.54. Three basic steps have to be performed by the computer to generate NC blocks:

1. Compilation (translation) of the part program written in a defined language format
2. Calculation of the cutter location (CLDATA) and the geometric and technological data with the help of tool, material, and machining information. The output is a general-purpose program that can be used to control similar NC machine tools of different manufacturers.
3. Conversion of the general-purpose CLDATA to the NC code, which is needed to operate the selected machine tool configuration. This operation is performed by a postprocessor.

A wide variety of programming systems have been developed which can be classified according to the underlying language concepts into three basic types. Type I are special-purpose languages developed for the use of a specific machine tool. They offer a limited but powerful programming comfort and have fixed and mostly numeric formats for the definition of the cutter path. Two-dimensional cutting operations (lines and curves) can be defined. To describe the cutter path the system uses special routines which are stored in a library. Type I programming systems are implemented on computers of different sizes. Design of the MCU and the special-purpose language is often done by the same machine tool manufacturer. Thus it is assured that the programming system is optimized for a specific NC machine tool.

The type II programming systems are of the construct of general-purpose languages which allow programming of an entire class of related

Chapter 7

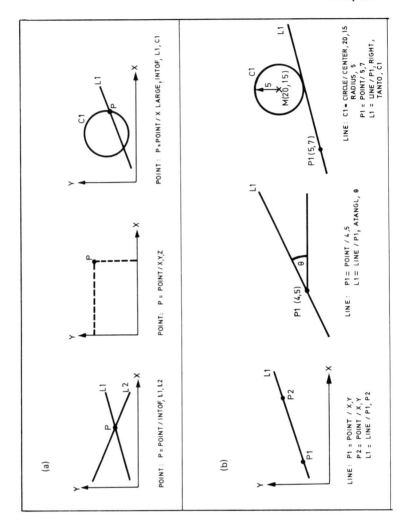

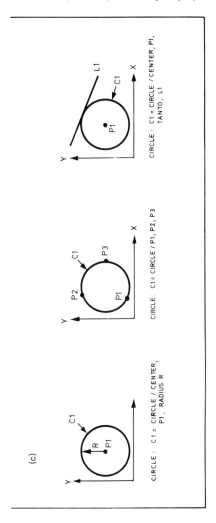

FIG. 7.51 Some possibilities in APT to define: (a) points; (b) lines; and (c) circles.

machine tools. The input language is easy to learn and makes extensive use of symbolic notations with English-like mnemonics.

General-purpose languages allow the description of two- and three-dimensional machining surfaces and that of the tool path. There are many different methods to define surfaces and the tool path. The computer is very versatile; it calculates incremental motions for many different curves without use of a curve routine library. This makes the system suitable for various machine tools. The most widely used programming languages are APT and EXAPT. They will be introduced later. A type II programming system can be implemented on medium-sized and large computers. The translation task is more complex then a single-purpose language that uses library functions. Another helpful feature is the capability of defining technological machine tool information by symbolic notations. Thus all necessary parameters for the MCU can be defined without manual calculations.

The type III programming system accepts programs that define the shape of the product to be machined. The output of the system is the cutter path for machining the contour shape of the workpiece. A part program consists of symbolic statements describing the analytic surface by three-dimensional geometric formulas (plane, cylinder, cone, ellipsoid, hyperboloid, paraboloid, torus, etc.). Machine instructions, tolerances, and tools can also be defined by symbolic statements. The format of type III languages is free; it uses analytic symbolic notations. The computer procedure for generating the tool path is of generalized form. The output of the system consists of general coordinate information which can be adapted by a postprocessor to different machine tools. The type III system is implemented on medium-size or large computers.

General-Purpose NC Programming Languages

Both the type II and III NC programming languages are of general-purpose design. They allow symbolic description of machine tool tasks and calculate the exact cutter path. General-purpose NC part program languages consist of statements which allow description of the (1) product geometry, (2) tool motion (path), and (3) machine tool information (feed, speed, miscellaneous functions).

The most widely used language by industry is APT (Automatically Programmed Tools). There is a continuous effort to provide the language with very sophisticated capabilities. Many dialects and extensions of APT have been developed (ADAPT, UNIAPT, EXAPT, IFAPT, 2CL). They cannot be discussed in detail. However, they have a common kernel which is defined by ISO.

The APT Language

APT (Automatically Programmed Tools) is the first high-order language to program NC machine tools. It can be recognized as the origin

Control of Manufacturing Equipment 473

of many other languages and dialects. The first attempt to conceive this language dates back to the late 1950s. An organized development effort started in 1961, when the APT long-range program was created and contracted to the Illinois Institute of Technology Research Institute (IITRI). Over 130 companies supported the creation of this language through joint funding.

APT uses English-like instructions to describe the geometry of a part. The input statements are of four types:

1. Geometric statements are used to define the part configuration (Fig. 7.51). With these the programmer is able to describe geometric elements such as points, lines, circles, ellipses, planes, cylinders, cones, and general conics and quadrics with different surfaces. The contour surfaces of the workpiece can be described with the help of similar definitions, as shown in Table 5.2.
2. Motion commands that control the path of the cutter along the surface of a workpiece. The repertoire includes startup and point-to-point instructions, modifiers to change cutter movement direction, and methods to describe the cutter.
3. Postprocessor commands that control different machine functions such as spindle speed, feed rate, acceleration, deceleration, and coolant supply.
4. Special control instructions that generate translations, rotations, and output listings. There are also possibilities for program loops, jump instructions, and subroutines.

The contour of the part is described by the programmer with the aid of a sequence of instructions. Figure 7.52 shows how the tool is directed to generate the contour of the workpiece with the help of control surfaces. The tool travels along the part and drive surfaces until it encounters the check surface. Here its path is changed and the tool follows a newly defined part surface. This process is repeated until the entire part contour has been generated. A processor is used to transform the part program into a tool path description (Fig. 7.53). The translator compiles the program language into computer executable instructions which are processed by the arithmetic unit. The mathematical calculations needed to generate cutter location coordinates are done by this unit. It allows the inspection of the actual path the cutter will follow. The output of the processor is a machine-tool-independent program. Since the control unit of a machine tool is of machine-specific design, the cutter location data are adapted by a postprocessor to machine-specific code. The program to generate the part contour consists of dimensional and technological data which are necessary to operate the machine tool.

APT was conceived as a generalized program development tool to handle many different machining processes. For this reason the

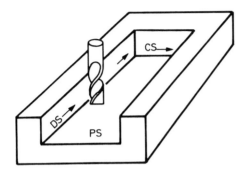

FIG. 7.52. Principle of programming with APT: DS, drive surface; CS, check surface; PS, part surface.

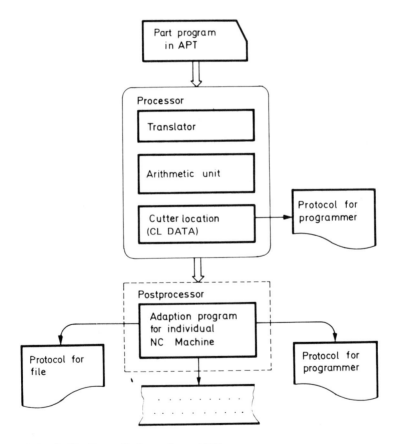

FIG. 7.53 Translation of an APT program.

the processor is very large and needs a considerable amount of memory space. This fact, however, limits the use of APT to manufacturers who have access to large computers. During the 1960s and 1970s attempts were made to simplify the programming system so that it would be available to smaller computers. This has led to the conception of APT-like languages and subsets of APT. As a matter of fact, several minicomputer manufacturers now offer APT program development systems for their equipment. In the following section, the German development EXAPT (EXtended APT) is discussed.

The EXAPT Programming System

Syntactically, EXAPT is identical to APT. In addition to the geometrical description of the workpiece, technological data of the machine tool and workpiece can be considered. The EXAPT system is of modular design and can be operated on smaller computers. The universities of Aachen, Berlin, and Stuttgart developed this system during the late 1960s. Now it can be purchased from the EXAPT Verein located at Aachen, West Germany. The fundamental concept of the EXAPT system is shown in Fig. 7.54. With this system the most important two and three axes machining operations can be handled. The concept of EXAPT is as follows.

BASIC EXAPT. This module is the basis of all other modules. It can be used similarly to APT, to program a tool path to generate a workpiece surface. In addition, feeds, speeds, and the geometry of tools can be described. The degree of automation that can be obtained is limited.

EXAPT 1. This programming module was devised for boring operations. The following capabilities were added to the BASIC EXAPT features:

Test for tool collision
Automatic calculation of number of required cutting paths
Automatic selection of machining parameters
Selection of tools
Selection of machining cycles

EXAPT 1.1. With EXAPT 1.1, boring and milling operations can be programmed. For boring, all features of EXAPT 1 are available. For milling, the additional capabilities are testing for tool collision and calculating automatic cutter path segmentation.

EXAPT 2. This module was designed to handle turning and boring operations on lathes. It is possible to describe the contour of the raw workpiece and that of the finished workpiece. Additions to EXAPT 2 are the following:

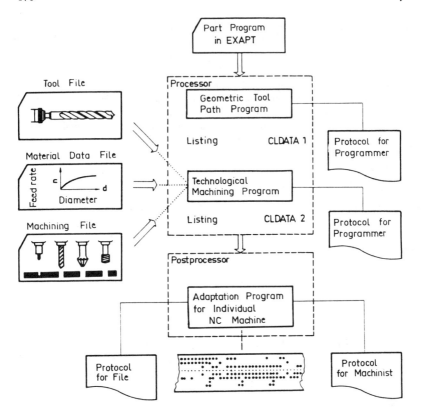

FIG. 7.54 Sequence of programming an NC machine tool with EXAPT. (From Ref. 9. Courtesy of EXAPT Verein, Aachen, West Germany.)

Test for tool collision
Automatic cutting path segmentation
Selection of machining parameters

Figure 7.55 shows how with an increasing degree of automation the EXAPT system is being integrated into design. At the present time a strong development effort is being made to close the gap between CAD and CAM and to add the following features for example for boring:

Selection of machining sequences
Selection of machine tools
Selection of fixtures and jigs

The automation of the programming process for milling and turning is not that easy. However, in the long term, solutions will also be found to tie CAD together with CAM for these machining operations.

Control of Manufacturing Equipment 477

The process of preparing the NC tape with an EXAPT module is shown in Fig. 7.54. The part program is entered into the computer and interpreted by the processor. In the first phase the geometrical data are processed as with APT. In the second phase the technological data are incorporated into the program with the aid of information obtained from the tool, material data, and machining files. The output of the processor is a machine-independent part program. The following postprocessor serves the same function as that used with APT.

A workpiece to be machined is represented in Fig. 7.56a. The corresponding part program for EXAPT is shown in Fig. 7.56b. The program consists of the following parts:

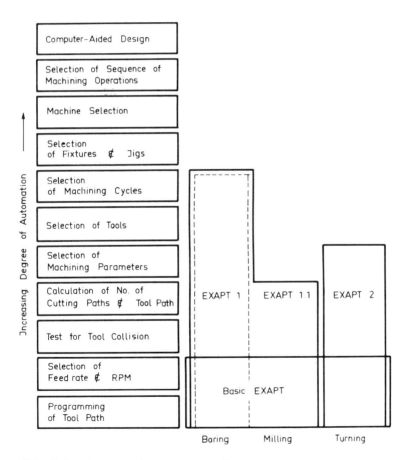

FIG. 7.55 Degree of automation of EXAPT. (From Ref. 9. Courtesy of EXAPT Verein, Aachen, West Germany.)

1. *Header data*: defines the part, machine tool, and the coordinate system
2. *Geometrical data*: describes the workpiece
3. *Technological data*: describes the machining operations
4. *Executive instructions*: directs the machining operation
5. *End instruction*: completes the machining cycle

Interactive Symbolic Programming

When workpieces are complex, program development by the APT and EXAPT principle may be quite time consuming and cumbersome. For this reason, several machine tool manufacturers have developed symbolic programming facilities. In the simplest case, programming is done via a keyboard directly at the machine tool. Each key contains an instruction: for example, "Drill, Countersink, Grid pattern,

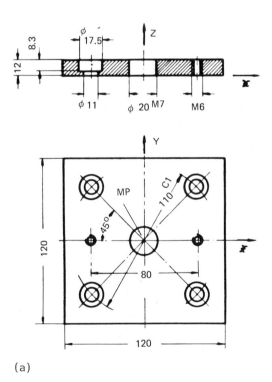

(a)

FIG. 7.56 (a) Workpiece to be machined; (b) EXAPT 1 program of a workpiece. (Courtesy of VDI-Verlag, Düsseldorf, West Germany.)

Control of Manufacturing Equipment

1	PARTNO/PLATE	⎫
2	MACHIN/BORE	⎬ HEADER
3	TRANS/200,100,0	⎭
4	ZSURF/12	⎫
5	MP = POINT/0,0,12	⎪ GEOMETRICAL
6	C1 = CIRCLE/CENTER,MP,RADIUS,55	⎬ DEFINITION
7	PAT = PATTERN/ARC.C1,45,CCLW,4	⎭
8	PART/MATERL,12	⎫
9	CLDIST/0.8	⎪
10	BORE1 = DRILL/SQ,DIAMET,5,DEPTH,	⎪
	12,TOOL,315	⎪ TECHNOLOGICAL
11	THREAD = TAP/SQ,DIAMET,6,DEPTH,	⎬ DEFINITION
	12,TOOL,416	⎪
12	BORE2 = DRILL/SQ,DIAMET,11,DEPTH,12	⎪
13	SINK = ·SINK/SQ,DIAMET,17.5,DEPTH,8.3	⎪
14	REM = REAM/DIAMET,20,DEPTH,12	⎭
15	COLLNT/ON	⎫
16	FROM/−100,0,100	⎪
		⎪
17	WORK/WORK/BORE1,THREAD	⎪
18	GOTO/−40,0,12	⎪
19	GOTO/40,0,12	⎬ EXECUTIVE
		⎪ DEFINITION
20	WORK/REM	⎪
21	GOTO/MP	⎪
		⎪
22	WORK/BORE2,SINK	⎪
23	GOTO/PAT	⎭
24	FINI	} END

(b)

FIG. 7.56 (Continued)

Keyway etc." With the help of the keyboard the programmer enters the part program step by step into the computer, which converts the instructions into machine code. Visually, this programming method is designed for specific applications such as turning.

More advanced symbolic programming systems use a keyboard in connection with an interactive terminal. The FAPT TURN system, developed by the Siemens Corporation, will be discussed in detail. The program development system consists of a computer, a CRT, a keyboard, a bubble memory, and a tape reader/punch combination. The rotational part shown in Fig. 7.57 is taken as an example to show how programming works. The various steps, which are greatly simplified, are explained below.

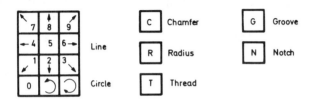

Symbol keys to describe a workpiece contour

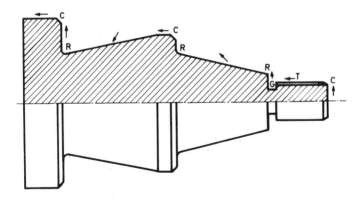

Rotational workpiece to be machined

Keys to be depressed to program workpiece contour

FIG. 7.57 Programming a rotational workpiece with the EXAPT system.

Control of Manufacturing Equipment   481

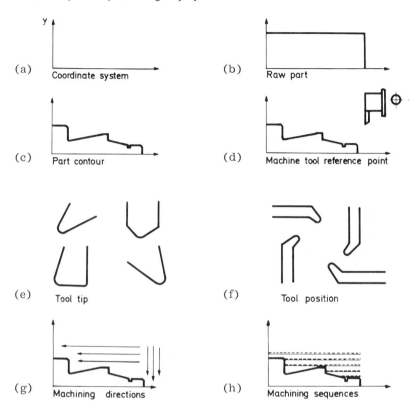

FIG. 7.58 Interactive CRT display of different programming phases.

1. Initiation of the programming system.
2. Selection of the raw material; there are 17 different choices.
3. Selection of the surface finish.
4. Positioning of the coordinate system (Fig. 7.58a).
4. Raw part selection, cylinder, hollow cylinder, or a special shape (Fig. 7.57b).
6. Generation of the finished part contour (Figs. 7.57 and 7.58c). This is done with the help of symbol keys. The sequence of keys to be depressed is shown in the lower part of Fig. 7.57. Basically, lines and circles can be entered. The system will ask for more information when a line or a circle is indicated: for example, dimensions, diameters, and so on.
7. Generation of the thread by indicating length, pitch, and cutting direction.
8. Groove cutting and dimensions.
9. Chamfer and dimensions.

10. Programming of mathematical functions, such as sin, cos, and square root.
11. Specification of machine tool reference point (Fig. 7.58d).
12. Tool tip and position selection (Fig. 7.58e and 7.58f).
13. Determination of toolholder.
14. Determination of cutting parameters, feed, and speed.
15. Selection of machining directions (Fig. 7.58g).
16. Contour segmentation (Fig. 7.58h).

For machining of certain segments it may be necessary to redefine the surface finish. This can be done with the aid of parameters when the surfaces are described.

Special-Purpose Languages

Many factories produce special-purpose parts which have common features. For example, the workpiece shown in Fig. 7.59 has a typical design found in sand-mold castings. Since it is very difficult to program such a part by hand, a computer is normally used. In this particular case, a manufacturer of precision patterns (Anderson Industries, Muskegon, Michigan) introduced a programming system which can easily handle this type of parts. The objectives of the programming system are:

1. Short programming instructions
2. Simple syntax
3. No hand computations
4. The ability to handle many different geometries and machining operations
5. The ability to program a wide spectrum of machine tools (e.g., from point-to-point machines to five-axis continuous-path equipment)

The lower half of Fig. 7.59 shows a section of the program that describes the center cavity of the part. There is a 10-field record into which only numbers are entered. The points that are needed to describe the contour are 1, 2, 3, 5, 6, 7, and 8. The contour points of the radii and parabolic fibers as well as the coordinates of point 4 are calculated by the programming system.

The known points are encircled and designated as code points. They are enumerated in field 7. They may be defined absolutely or incrementally from a code point entered previously. For example, point 8 is defined incrementally from code point 7. The number 500 in field 7 separates the geometrical description of the cavity from the milling instructions. First the compiler calculates the coordinates of point 4 (see entry 19 in field 7). It is located at a distance of 6.48 from code point 6 at a positive angle of 61° from the x axis. The -4

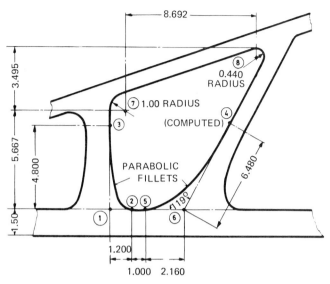

FIG. 7.59 Example of a special-purpose NC programming language.

at the end of line 19 is an instruction to calculate point 4. The 50 in the next line turns on the spindle and coolant. It commands the tool from the home position to its work position at a travel clearance of 1 in. above the part. The tool is lowered by 1 in. to contact point 5. The next line has a 2 in. field 1 and a 1 in. field 2. These numbers define the line at which the parabola is tangent to point 2. Similarly, the 1 in. field 4 and the 3 in. field 5 define the second line to which the parabola is tangent to code point 3. The cutter is moved from point 5 to point 2 and then along the parabola to point 3. The next line entry, 23, instructs the cutter to move from point 3 to the beginning of the circular element, defined by point 1 and the radius originating at point 7. The -1 in this line indicates that the radius of the filler circle is 1 in. and that the cutter approaches the circular element from its inside. The next line indicates that the cutter travels from the circular element whose center is point 7 to the element whose center is point 8. Line 24 is the instruction to bring the cutter from the circular element about point 8 to point 4. Line 28 instructs the cutter to move back to point 5 along the parabola defined by the tangent going through point 6. The instruction 70 brings the cutter back to its home position.

This coding method might look complex to a reader who is not familiar with it. However, it is of great benefit and easy to use when there are always similar parts to be coded.

Generative Programming by the Machine Tool Control

With the EXAPT language it is possible to calculate the number of cutting paths needed to machine a part. For this purpose feed and speed parameters must be available for different material and tool combinations. For future programming it will be possible to optimize the cutting operation with the aid of intelligent controllers built into the machine tool. Therefore, it is necessary to equip the machine tool with sensors (Fig. 7.60). The part program is sent to the control computer of the lathe. It describes the raw part and contour of the workpiece. In addition, preliminary feed and speed conditions are given. The part is chucked and machining is initiated (Fig. 7.61). First the tool rapidly approaches the raw part at its smallest diameter until a proximity sensor detects the workpiece. At this instance, the computer directs the tool drive to assume the normal cutter speed.

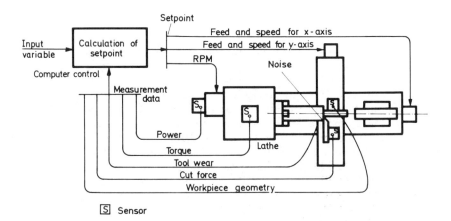

FIG. 7.60 Principle of an adaptive control system for a lathe.

Control of Manufacturing Equipment

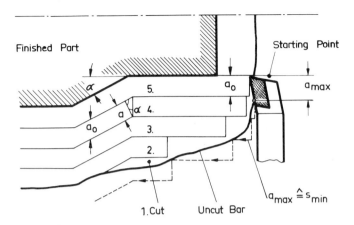

FIG. 7.61 Finished part contour-oriented selection of cuts with contour parallel movements.

The tool now engages the workpiece and tries to start machining. A sensor tells the computer that the cut force is too high. The tool is retracted and a new cut attempt is made at a diameter farther out. This procedure is repeated until the permissible cut force is obtained. From here on, the control commences its machining operation and partitions its own cuts. The operation may be done under the supervision of an optimization model. Optimization criteria may be minimum cost or maximum throughput (Sec. 7.4).

7.7 CNC SYSTEMS

A CNC (Computer Numerical Control) system performs similar control function as an NC system. Modern CNC control systems have a microcomputer or multiprocessor architectures which are highly flexible. Logic control, geometric data processing, and NC program execution are supervised by a CPU. In other words, CNC is a software control system which has no dedicated circuits normally found in a conventional NC system. A CNC system may also be integrated into a DNC (Direct Numerical Control) system. It may be designed for local shop floor programming without the use of a higher control level.

7.7.1 Architecture of CNC Systems

The architecture of a CNC system is shown in Fig. 7.62. The microcomputer performs the following basic tasks:

System management
Data I/O
Data correction
Control of NC program execution
Processing of operator commands
Output of NC process variables to the display

The interpolator is either software or hardware oriented. This depends on the interpolation method used and on the geometric parameters that are needed for the calculation of the cutter path. Presently available CNC systems are of three basic types (Fig. 7.63):

1. CNC with an interpolator implemented completely in software.
2. CNC with woftware interpolator for course interpolation; fine interpolation is performed by hardware.
3. CNC with no software interpolator; interpolation is completely performed by hardware.

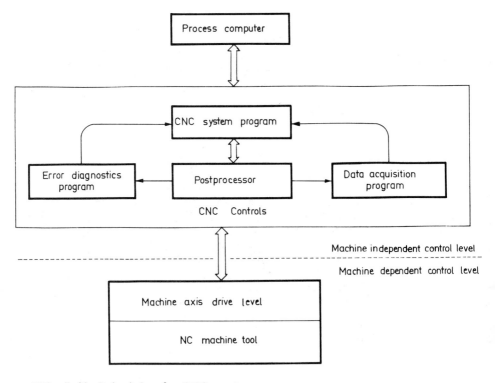

FIG. 7.62 Principle of a CNC system.

Control of Manufacturing Equipment

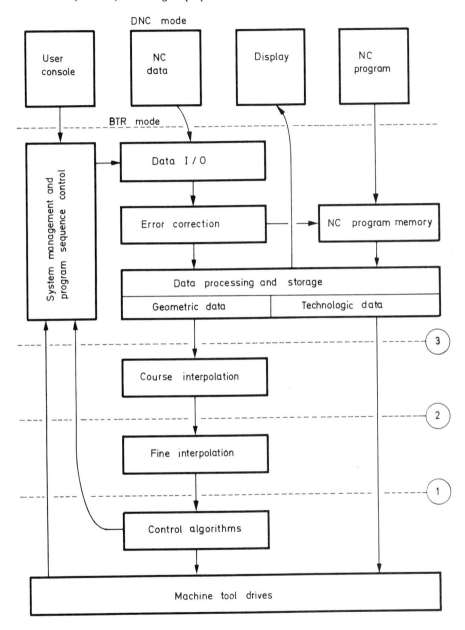

FIG. 7.63 Structure of a CNC controller. BTR, behind tape reader.

The first type needs fast computational support for the calculation of the tool path. Machine tools having a high number of control axes need a multiprocessor architecture with dedicated CPUs to perform the arithmetic calculations. Bit-slice and I/O processors are also employed to speed up the control task.

The second type of CNC controller usually has no computational bottleneck because the fine interpolation is performed by a hardware circuit. The desired interpolation of the tool path is prepared and parametrized by a program. The output of this program are the parameters of the curve segments described, such as starting point, end point, velocity, interpolation frequency, and so on. They are the input to the hardware fine interpolator. The required calculation effort of the interpolator increases with increasing speed and accuracy as well as with decreasing radius of the tool path.

A CNC system with no software interpolator usually has no realtime speed problems. However, its flexibility is relatively low, because the interpolation method is fixed by the design of the hardware. Systems of the third type are realized with a single CPU architecture.

The available memory of a CNC controller is of importance. Early CNCs were 4K and 8K memory versions. The 4K CNC usually operates in two modes. One mode is for the generation the NC program. It occupies about 3 kilobytes and calculates the cutter path for automatic contouring of the workpiece (Fig. 7.64). The second mode is for execution of the NC program. For this, 1 kbyte of memory is reserved for the storage of the geometric data describing the raw part and the end product. The 8-kilobyte CNC allows program generation and program execution at the same time (Fig. 7.64).

The prevailing price decline of high-capacity memory chips has greatly influenced the development of advanced CNC controllers. For

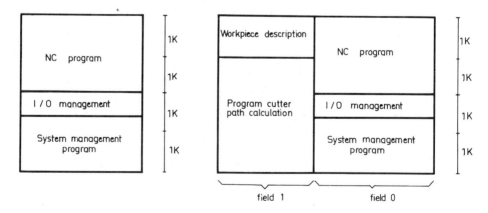

FIG. 7.64 Memory layout for a typical CNC system.

Control of Manufacturing Equipment

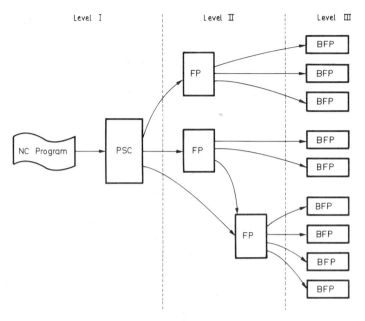

FIG. 7.65 Example of a three-level CNC software structure: PSC, program sequence control; FP, functional program (task); BFP, basic functional program (subtask).

example, to aid program generation on the shop floor, the operator can enter instructions via a terminal and can verify the part program on a display. The contour of the workpiece as well as the calculated cutter path may be shown. All CNC software modules are incorporated into the CNC system program. The input of the piece part program can either be done with the aid of a paper tape via the serial interface of the CNC system or via an operator keyboard.

7.7.2 Program Structure of a CNC System

The software of a CNC system is of modular design. It may have a machine-independant hierarchical structure. The adaption of the NC program to a dedicated machine tool is performed by a low-level postprocessor. An example of a three-level CNC software system is shown in Fig. 7.65.

An interpreter on level I executes stepwise the CNC part program (program sequence control). For each NC block the functional machine program modules are referenced (path control). The programs (tasks) on level II activate basic functional routines (subtasks) in a defined sequence (axis control). The decomposition and the sequencing of the tasks depends on the actual NC block. The software modules on level III are of two basic types. The first type is the postprocessor, which transforms the control data into machine-tool-specific coordinates and which controls the I/O between the CNC and the machine tool. The second type consists of logic and arithmetic programs which perform acquisition of the process data and which generate information for the operator or the DNC computer.

7.7.3 Multiprocessor Architecture for CNC Systems

Figure 7.66 shows a CNC system architecture with three microprocessors. Each microprocessor has its own memory and a local bus. The data transfer between the microprocessors is performed via a common multiport RAM. The individual microprocessor has a von Neumann hardware architecture. The multiprocessors operate assymmetrically according to the master-slave principle. The master controls the communication between the microprocessors, NC program execution, real-time multitasking, and I/O management. The slave processors perform real-time interpolation and other arithmetic calculations as well as the execution of the control algorithms (DDC algorithms). In this example the CNC system operates with two basic software packages, one for lathes and one for milling machines. The adaptation of the basic software to a specific machine tool is performed by the postprocessor. It is machine tool specific and has to adapt the part program to the control system of the machine and has to be specified by the user.

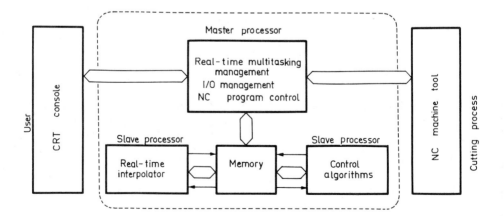

FIG. 7.66 CNC computer control architecture.

Control of Manufacturing Equipment

7.7.4 Use of Graphic Aids for CNC

An efficient tool to aid shop floor NC programming is the graphical terminal. By displaying the contour of the piece parts on the screen it is possible to interactively develop the program and to simulate the machining process. The execution of a NC block (e.g., the movement of a cutter between two points) can be observed by the graphical presentation. In addition, it is possible to display the position of the workpiece and of the fixtures to check for possible collision. The use of such graphical aids for part programming for CNC machines is limited only by the capacity of the image memory. Complex cutting operations require extensive three-dimensional graphic systems which are available only on higher control levels. Some CNC systems allow program development parallel to the operation of the machine tool. This improves efficient use of the CNC systems on the shop floor level.

7.7.5 Conclusion

The control of CNC systems is supported by single or multiple processor architectures. They add flexibility to the machine tool which could not be realized with classical hardwired NC controllers. The software structure allows the conception of functional modules, graphical aids, and the integration of the machine tool into a DNC system. The most important attributes of the CNC system are:

1. The use of microprocessors allows for a reduction in hardware. This increases the reliability of the equipment.
2. The basic hardware is application independent and can be used for different types of machine tools. Programs for the different functions, such as tool control, sequential control, and data acquisition, are implemented as software modules. Production changeover from one part to another can be done easily.
3. Control functions can be changed easily. The addition of new parameters as well as of functional modules can be done on-line without stopping the machine tool.
4. The postprocessor is implemented in software. The logic equations for the machine tool control are defined by the machine tool manufacturer and can be implemented as part of the postprocessor. This means that the controller can interface the equipment at a very low level. The implementation of diagnostic programs is possible. Quality data, production data, and operating times of the machine tool can be acquired automatically. Figure 7.62 shows a CNC system which includes an error diagnostic module and a data aquisition program.
5. The 24-bit addressing capacity of modern microcomputers allows an address space of more than 1 megabyte. The CNC

operating system, the part programs, the programming system, and graphical aids can be stored in one memory.
6. A code transformer allows the use of different coded NC programs (e.g., EIA and ISO code).
7. Interpolation routines facilitate part programming. Higher interpolation routines, such as hyperbolic or elliptic interpolation, are available. It is also possible to interchange axes commands, to perform mirror cutting information, and to use subroutine techniques.
8. CNC systems allow monitoring and correction of tool wear.
9. Smart editors detect on-line syntax errors. It is possible to check for collision and for optimal feed rates.
10. NC program errors can be corrected via the operator console. From this a correct NC paper tape can be punched. Errors are reported back to the DNC system and can be used for program verification.
11. CNC systems are very user friendly. If graphical aids are available, the human-machine interface allows a maximum of information exchange.

Future developments in CNC technology will support ACC and ACO strategies to allow the efficient use of machine tools. The integration of CAD and CNC into CIM systems is currently under study in laboratories and in industry.

7.8 DNC SYSTEMS

Numerical controlled machine tools present the lowest level of the automation hierarchy in a CIM system. To manufacture a workpiece, there must be an individual program available which is machine tool specific. With complex workpieces and the need to change products frequently, the administration efforts in regard to (1) storage of workpiece data and programs, (2) distribution of piece part programs to the machine tool, and (3) the interconnection and synchronization of machine tools can be very extensive. To solve these administrative problems and to support process planning, DNC (Direct numerical control) was conceived. DNC is a system used to schedule and supervise several NC machine tools directly with the aid of a digital computer (Table 7.7).

An additional motivation is to interconnect the administrative computer with the executive system. Thus it becomes possible to integrate material handling, material storage, workpiece processing, and assembly activities. Feedback information from these functions can now be acquired by a data aquisition system (Chap. 4) and processed centrally. With the DNC concept, machine tool controls are interconnected to an

Control of Manufacturing Equipment   493

TABLE 7.7 Basic Function of a DNC System

Administration of part programs for machine tool
  Input, storage, output, protocoling, editing,
    copying, erasing
  Administration of NC part program and NC
    data
  Library functions, storing, updating, retriev-
    ing, controlling memory access

Distribution of part programs
  Activation of NC, CNC, and PC programs,
    acknowledgment of program calls, control-
    ling of start conditions, status reporting
  Activation of communication interface
  Transfer of program or block to the machine
    tool control
  Control of the data transfer, status, and com-
    pletion of transfer

external process computer. The controller of the machine tool may be laid out for NC, PC, or CNC. An interface is provided to tie the controller to the DNC process computer.

In Sec. 7.6 the principle of automatic piece part programming with APT and EXAPT by the process planning activity was discussed. The control data or NC piece part programs generated are administered and distributed by the DNC system. Thus this system represents the interface between process planning at the higher level and machine planning at the lower level. The essential task is to distribute control information among several machine tools so that the machine controllers can perform their operations. Since there are a multitude of different DNC systems available in the United States, Europe, and Japan, it is possible within the context of this book only to discuss their structure, function, and work principle.

7.8.1 Structure of DNC Systems

The principal structure of a DNC system is shown in Fig. 7.67. Here a number of NC machine tools which may have an NC, CNC, and PC controller are connected via an interface to a process computer. The process control computer sends and requests the piece part program or sections of it from its mass memory to the interface and transfers these to the machine tool. The basic function of a DNC system can be divided into (1) administration of the piece part program, and (2) distribution of the piece part program.

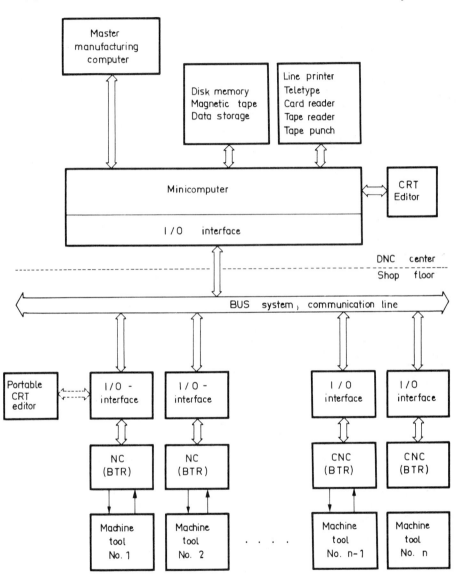

FIG. 7.67 Typical structure of a DNC system.

Control of Manufacturing Equipment 495

If the DNC computer is connected to an autonomous numerical controller, the operating mode is called a BTR (behind tape reader) operation. With this mode the part program is entered directly into the controller without the use of a NC tape. The data formats are identical to those used on the paper tape. In case the tape reader is part of the machine tool, it is possible to use it in conventional mode. In Sec. 7.2 the important functions of the machine tool controller were discussed. They were divided into (1) data input and conversion to machine-specific code, (2) geometric data processing, (3) technologic data processing, and (4) central sequence control. If part of these functions are integrated into a DNC system, the machine tool loses its autonomy; however, its operation may become very flexible. Interpolation routines, data conversion, and program corrections can be performed much more efficiently by the process computer than by the machine tool controller. In this case the DNC system has control capabilities.

7.8.2 DNC System with "Behind Tape Reader Mode"

This is the system used most often. Each machine tool maintains its autonomy and can operate by itself with the help of a tape or a CRT terminal. Thus the machine tool can be operated independently when the DNC system has a failure or the machine tool is needed for a specific application. With this feature the shop floor programming capability is maintained.

The communication between the control computer and the DNC peripherals is shown in Fig. 7.68. First the number of the desired part program and the address of the required machine tool are entered via a keyboard. The computer acknowledges the request and reads the piece part program from its mass memory into its working memory. For the synchronization and possible correction of errors clock and status signals are generated by the transmitter and receiver. Operating protocols about the process, workpieces counts, disturbances, waiting times, setup times, downtimes, and so on, are aquired and processed by a factory data aquisition system. Thus it is possible to generate operating statistics and to determine the efficiency of the machine tool. The most important elements of a DNC system are the following:

A process computer
An I/O interface
A communication interface
A numerical control unit
A supporting software system

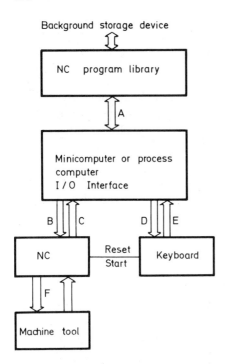

FIG. 7.68 Data flow through a DNC system: A, part program; B, data lines for NC programs and synchronization line; C, acknowledge signal to computer, syntax error, synchronization, status; D, acknowledge signal to user; E, NC number of part program and of machine tool; F, control signal.

The configuration of the components of the DNC system depends on parameters which have to be specified by the user (Fig. 7.69). Of particular significance is the required degree of automation, the information flow, and the plant computer hierarchy. The size of the mass storage peripheral is determined by the amount of user data to be stored for the timely execution of the programs. To store data of n numerically controlled machine tools and m piece part programs, a minimum of

$$K = \sum_{i=1}^{n} m_i$$

Control of Manufacturing Equipment   497

files are needed. The administration of the data is supervised by the operating system of the process computer. The data flow is significantly influenced by the power of this process computer, by the data communication interface, and by the NC, CNC, and PC components. Figure 7.70 shows three principal data paths connecting the background mass memory and the machine tool controller. For the simplest case (A), data are read blockwise for each machine tool and they are temporarily buffered. The blocks are distributed to this buffer. In the case of NC control (B), the programs and NC blocks are read from the background memory into the working storage and distributed blockwise in the BTR mode via a buffer. For case (C), a CNC environment is assumed in which each machine tool has its own

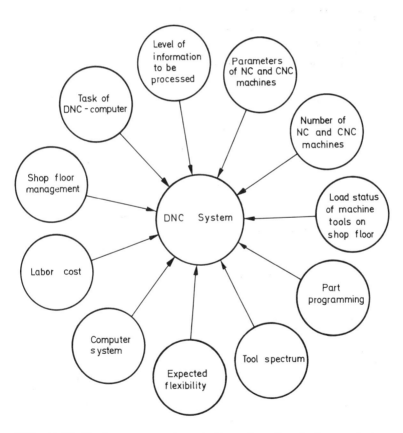

FIG. 7.69 Various parameters influencing the design and the operation of a DNC system.

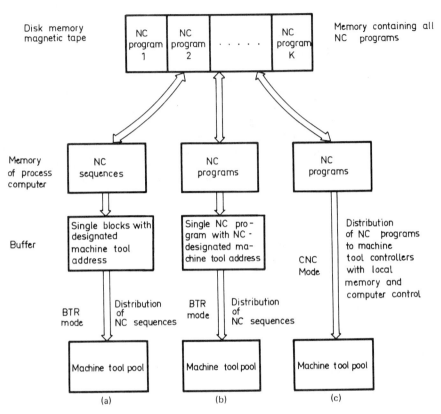

FIG. 7.70 Basic data flows in DNC systems: (a) transfer of NC program blocks from background memory and their distribution via a buffer; (b) transfer of NC programs to the process computer and their distribution via a buffer; (c) transfer of NC programs and their distribution to the CNC machine tool controllers with local memory.

microcomputer and local memory. The data flow consists simply of retrieving the NC program from the NC library and distributing it to the corresponding microcomputers.

In case the DNC system performs additional tasks, such as NC data correction, data aquisition and concentration, control functions to supervise material flow and tool handling, execution of NC functions or adaptive control, and NC programming, a powerful process computer with real-time multitasking capabilities must be used as a DNC superviser.

REFERENCES

1. F. W. Wilson, *Numerical Control in Manufacturing*, McGraw-Hill, New York, 1963.
2. M. Weck, Machine Tools, Vol. 3, VDI-Verlag, Düsseldorf, West Germany, 1978 (in German).
3. T. Sizer, *The Digital Differential Analyzer*, Chapman Hall, London, 1968.
4. N. B. Nichols, Theory of Servomechanics, MIT Radiation Technology Publication, Massachusetts Institute of Technology, Cambridge, Mass., 1946.
5. E. P. Popov, *The Dynamics of Automatic Control Systems*, Addison-Wesley, Reading, Mass., 1962.
6. D. Eckmann, *Automatic Controls*, Wiley, New York, 1958.
7. J. V. Wait, Stata Space Methods for Designing Digital Simulations of Continuous Fixed Linear Systems, *IEEE Transactions on Electronic Computers*, June, 1967, pp. 351–354.
8. E. I. Jury, *Theory and Application of the z-Transform Method*, Wiley, New York, 1964.
9. EXAPT Verein, information on EXAPT, Aachen, West Germany, 1974.

8

Computer-Controlled Parts Handling and Assembly

8.1 INTRODUCTION

Full integration of all production facilities into a computer-controlled manufacturing system requires the automation of parts handling and assembly. In mass production, assembly usually is done by specially designed equipment, for which a computer may not be needed. On the other hand, for small and medium-sized production runs, custom-designed equipment is uneconomical. Here the industrial robot may offer a flexible solution. It consists of a mechanical arm that must have at least one translational or one rotational axis. It is controlled by one or several microprocessors. To the end of the arm, an effector is fastened which can be used as a tool or a toolholder.

When automating material handling and assembly, the aspect of quality of work is also an important consideration. Robots are often used for undesirable and hazardous work (e.g., where human beings are exposed to excessive heat, dust, noise, and monotony).

8.1.1 Historical Development

The development of the modern industrial robot was influenced by two technologies: (1) the teleoperation, which is used, for example, to handle radioactive materials; and (2) the numerically controlled machine tool. At the end of the 1950s, the first robots were designed in the form of a teleoperator device controlled by NC machine servo drives. At the beginning of the 1970s, the company Unimation started to manufacture the first programmable manipulator. More recent manufacturers of robots include Cincinnati Milacron, Volkswagen, Kuka, Asea, Olivetti, Hitachi, and Kawasaki. Until the present most of the fundamental research on robotics was done by universities and research institutes with government funding. Following is a brief historical summary:

Parts Handling Assembly

1945   The teleoperator with a force sensor is developed for nuclear, maritime, and space research.
1950   NC machines are developed, together with the design of simple pick-and-place devices.
1960   Mechanical memories and plugboards are used to program the handling devices.
1970   Electronic memories and digital computers replace mechanical control. Continuous-path control and textual programming are introduced.
1980   The integration of sensors, powerful processors, and communication networks takes place. Robot control becomes part of a larger manufacturing control system.

It can be seen from this historical overview that the use of electronic processors and memories started after 1970. These two devices had a substantional impact on the versatility of the modern robot controller.

Originally, programming of the robot was done with simple methods that had little resemblance of conventional computer programming. Since the 1960s scientists had been working to introduce artificial intelligence into robot technology. So far, this attempt has only reached the testing state. The higher programming languages for robots were developed at the end of the 1970s. Presently, industry is at the threshold of high-level programming languages and programming methods. These tools will substantially help to increase the use of robots.

8.1.2 Areas of Application of Industrial Robots

The task of robots in manufacturing can be divided into (1) machining and processing, (2) handling, and (3) assembly (Fig. 8.1). Since there are presently very few sensors available, most of the robots are used for the first two activities. All of these tasks are of repetitive nature. Once the robot has been programmed, its assignment will very seldom be changed. In most cases the program is conceived and entered into the robot control unit by very primitive programming methods. A representative frequency distribution of present robot application is shown in Fig. 8.2. Seam and spot welding are the most predominant tasks.

In material handling the robots are used for a variety of tasks (Table 8.1). They mostly perform pick-and-place operations. However, they are also employed for separating, distributing, and transporting (Fig. 8.2). A typical assignment of the robot is to transfer workpieces from a machine tool to a material handling system, or vice versa. In this way it is possible to automate tiring work and to obtain constant work cycles which are independent of a human operator. This will also shorten cycle times and facilitate planning of manufacturing runs.

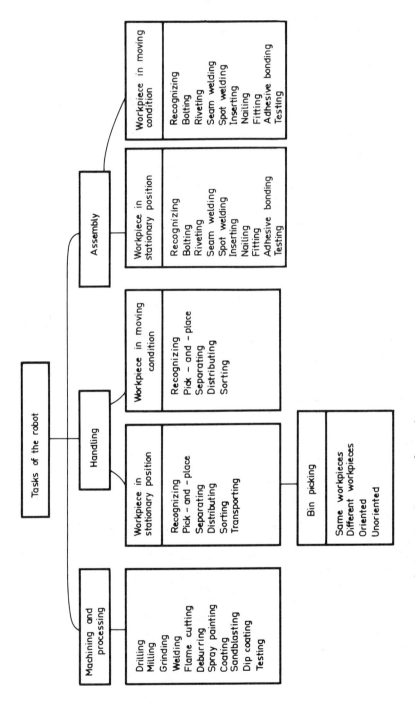

FIG. 8.1 Task of the robot in manufacturing.

Parts Handling Assembly

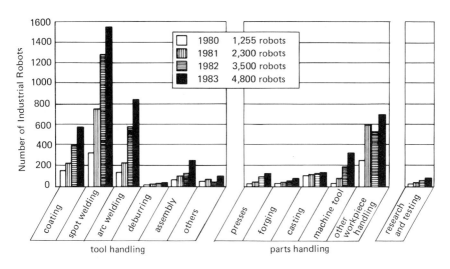

FIG. 8.2 Work assignment to robots in Federal Republic of Germany. (From Ref. 1. Courtesy of Institut für Produktionstechnik und Automatisierung (IPA), Stuttgart, W. Germany.)

TABLE 8.1 Typical Tasks of a Robot

Metal stampings:	Plates, angles, transformer laminations, fan blades
Rotational parts:	Shafts, motor armatures, hydraulic cylinders, screws, pegs, gears, wheels
Cubical parts:	Gear boxes, housings, subassemblies, integrated circuit components
Injection molding; Die casting:	Handling of ladle, removal of hot parts from machine
Painting:	Dip painting, spray painting
Insertion of tools in machine tool:	Drills, screw drivers, pliers, spray guns, welding torches, special tools
Manufacturing operations:	Routing, drilling, riveting, welding, glueing, grinding, polishing
Assembly:	Electrical motors, Solenoids, car bodies, circuit boards

Source: Ref. 2.

With this automation tool machining and conveyance equipment can be controlled by a coordinating supervisory computer.

Bin picking and handling of moving workpieces is very difficult. For these operations a new generation of robots with complex sensors systems has to be developed.

The application of the robot to assembly is still in the experimental stage. Automatic assembly will probably first be done by automotive, electronic, or appliance manufacturers. Industry-wide application of robots in this area depends on the availability of sensors and sensor processing systems, not on the development of the mechanics of the robot. The most important sensors are for vision. With it, "eye-hand coordination" will be possible. Advanced assembly robots will be capable of solving assembly problems automatically and learning from past experience (Table 8.2). This will greatly simplify the solution of assembly work.

The complexity of the tasks described above will require robots of different stages of intelligence. For this reason, robots are often divided into different generations (Table 8.2). The main distinctions between these generations are their sensory and programming systems.

The second generation of robots will be provided with intelligent sensors. A complete manufacturing cell may contain a conveyor, a robot, a vision system, magazines, and machine tools. The cooperation of the individual components requires a supervisor module, which may be part of the robot controller. For example, after the robot has received a start signal, it will take a part from a conveyor and insert it into the machine tool. Then it sends a control signal to start machining. Upon completion of the machining operation, the robot will receive an acknowledge signal. Thereafter it picks up the part again and performs an inspection operation with the help of a sensor. In the last work phase the part is placed on another conveyor.

Usually, several manufacturing cells are combined in a flexible manufacturing system. Such a system may consist of a number of machine tools, part handling devices, and several robots. An operator fills and empties the part handling devices and the robots serve the machine tools.

The selection of a suitable robot system for an automation task is done using the following criteria:

1. Lot size within a defined time frame
2. Frequency of switchover to new manufacturing runs
3. Design and general technical state of the entire manufacturing system
4. Type of handling task
5. Design of the workpiece and the work tools
6. The sequence of fabrication steps
7. The manufacturing machines to be served

TABLE 8.2 Characteristic Features of Robots of Differential Generations

Generation	Characteristics of control algorithms	Characteristics of information processing	Formulation of the handling task by:	Typical sensor information
1. Program controlled robot	– Sequential program flow, simple-move statements, simple-move statements, – Some special functions e.g. automatic program selection, user defined interpolation – Simple work assignment	Control instructions are executed to follow strictly the programming sequence	Simple teach-in and some textual programming	Position and displacement sensor
2. Robot with tactile and vision capabilities	Expanded tasks: – Combination of teach-in and textual programming – Position correction – Recognition of shape and orientation – Identification of the workpiece and its environment – Tracking of moving objects – Self monitoring	Additional to the first generation. – Signals from sensors may be used for logic decisions and change of program sequence. – Program execution according to defined algorithms.	– Explicit programming – High level language – Acoustic programming	In addition to the first generation: Tactile, optical and state sensors. Recognition of objects and their orientation
3. Intelligent industrial robot	World modelling – learning algorithms – during program development and execution the control strategy is optimized.	Generation of actions derived from the identification of the robot world – capability of adaptive learning – the robot can react to unforeseen events.	– Implicit programming – Programming from CAD data base	Optical sensors combined with tactile sensors are used to recognize new states and possible obstacles

8. Fixtures and auxiliary equipment needed
9. Work safety

8.1.3 Design Considerations for Assembly by Robots

During the design of a workpiece, its manufacturing process and sequence are determined. With precision machined products, the assembly effort amounts to 50% of the entire manufacturing cost. An assembly can usually be simplified by the following:

A reduction in number of parts
A reduction of fasteners and pins
The use of new fastening techniques, such as snap connectors

When an industrial robot is used for assembly, the designer must observe that a robot does not have human sensoric and kinematic capabilities. An example is the almost trivial task of bolting together two loose plates with the help of a bolt and nut (Fig. 8.3). For all practical purposes three robot arms with corresponding effectors are needed to perform this work. The assembly requires the availability of force and torque sensors to supervise the insertion and fastening process. In case of an assembly without fixturing, a complex vision system to locate parts and tools is needed. In addition, the movement and cooperation of three arms has to be coordinated. The latter task has not been resolved satisfactorily to date. A robot-specific solution would be to clamp the two plates into a fixture and to fasten the plates with a pop rivet with the aid of one effector (Fig. 8.4).

Our conventional assembly systems assume the availability of two arms with many degrees of freedom, complex sensors, and a huge data base containing assembly experience. A design for assembly by robot is seldom conceived today. An automatic assembly system can succeed only if this is done properly. Most assembly operations that are practiced today are very difficult if not impossible to do with the robot. In addition, they are very expensive.

In case a product is conceived for assembly by robot, the following points have to be observed:

Type of assembly functions
Analysis of the part clearances
Definition of the assembly process
Design of assembly system
Economical justification

8.1.4 Economic Considerations

From many pilot installations the limits of the robots were soon discovered. It was found out that universal robots are very complex and

Parts Handling Assembly 507

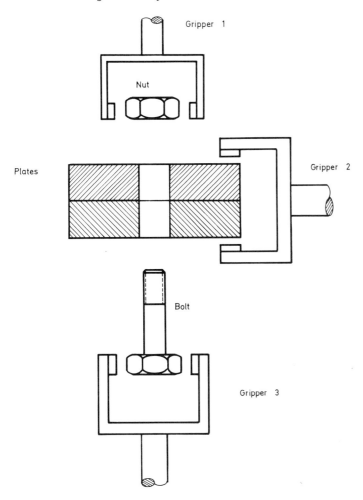

FIG. 8.3 Fastening of two plates with three grippers, poor design for assembly by robots.

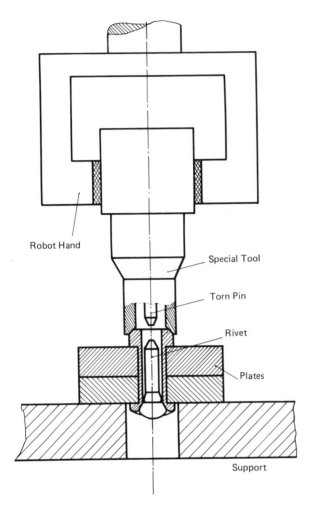

FIG. 8.4 Fastening of two plates with the help of a pop-rivet and one robot gripper, improved design.

Parts Handling Assembly

expensive. New developments are trying to implement problem-oriented solutions. For each class a specific robot system is developed. This trend is supported by the availability of cost-efficient programmable controllers. For the mechanical parts of the system, modular design techniques are gaining popularity.

During the system layout and the selection of the robot, a thorough analysis of the handling process is necessary. For example, it can be shown that for simple mating tasks as they occur in small part assemblies, simple manipulators are performing an adequate service. This development is supported by optimization efforts of the manual assembly process and the conception of new design rules for automation.

An investigation of selected U.S. consumer products showed that the majority of the parts of a product are assembled from one direction only. The analysis included assembly of electrical saws, toasters, electrical motors, and similar devices [3]. For this product spectrum, the following conclusions were drawn. Approximately 80% of the mating operations have a preferred direction. About 35% of all work elements are simple insertions of a pin in a hole. The next important element, with a frequency of 30%, is fastening of screws. From this observation the following rules can be deduced:

1. The number of degrees of freedom should be kept to a minimum.
2. Robots designed to work in a cylindrical coordinate system are better suited for assembly than those with cartesian coordinates.
3. Assembly tasks should be performed from one direction only.
4. The product should be designed in such a manner that only the predominant assembly operations have to be performed.

If these rules are observed, a cost-efficient and highly accurate robot can be used which is easy to program. For specific tasks, alternative concepts may be developed.

8.1.5 Trends in the Development of Assembly Robot System

The majority of assembly tasks are very complex and need robots with advanced sensor and control systems. The discussion above showed that the robot of today is only able to carry out simple assembly tasks. The concepts of assembly lines which are being planned for the near future start with the fact that reliable sensor and control systems are presently not available. Several studies were made to estimate the future development of the robot market. Most of them are not very conclusive because the development for intelligent sensors cannot be foreseen. However, it is clear that control systems for universal assembly robots must be capable to perform coordinate transformations and that they must be equipped with advanced sensors. It is estimated that by 1990 these robots will have a market share of 50%. Also, the programmability of the future robots is of interest. For general assembly tasks, higher programming languages which are capable to handle sensor signals must be available. Here it is estimated that off-line programming will be used by more than 50% of all robots. For general assembly tasks, precise vision systems and force/torque and approach sensors must be developed. Some important parameters of sensors are as follows [4]:

1. A simple vision system must be developed with high priority and it should not cost more than $7000.
2. With 30% of all assembly processes the robot must be able to track the object to be handled.
3. Touch sensors should cost less than $2000.
4. For most applications, including those involving moving conveyor systems, a positional deviation of ±25 mm and an angular deviation of ± 20 degrees are permissible.
5. Inaccuracies should be corrected with the help of a vision system or with sensors.
6. Bin picking requires a complex vision system which is presently not available.

8.2 STRUCTURE OF INDUSTRIAL ROBOTS

The components of a robot are shown in Table 8.3. In general, a robot is conceived as a modular design. Thus it is possible to add or delete motion axes and sensors and to adapt the robot to many manufacturing tasks. The simultaneous movement of several axes and drives are essential features of a robot. They determine the work space, the maximum speed and acceleration, and the allowable weight of the workpiece. Very important for the robot system is its controller, which functionally directs the work of all components. It uses information from sensors and leads the robot through its work cycle.

There are two approaches to obtaining a versatile robot. The first one uses standard components. Here the robot is assembled from modular components and tailored to a particular application. With this approach the robot becomes rather rigid and it is difficult to reprogram it if a new product or a design change makes this necessary. The second approach is to design a versatile robot that can be universally adapted to many applications. This robot needs many degrees of freedom, advanced sensors, and a powerful programming system (Fig. 8.5). Within the context of this book the second robot is of particular interest. The general components of a robot are listed in Table 8.3. They are: (1) kinematics, (2) effector, (3) power drive, (4) control, (5) sensors, (6) programming system, and (7) computer. The discussion in the following sections follows the order indicated. This will be done with the aid of examples.

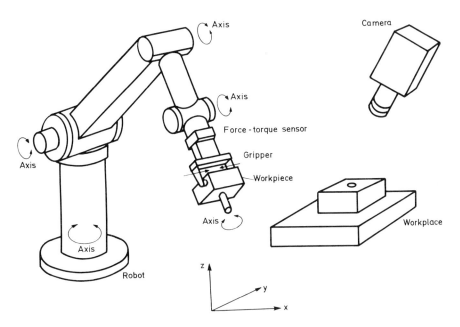

FIG. 8.5 Typical robot system.

TABLE 8.3 Components of an Industrial Robot

Subsystem	Function and task
Kinematics	Design of the manipulator, degrees of freedom, axes, work space Affixation of effector to arm
Effector	Gripping of the object and the tool Performance of assembly
Sensor	Sensing of the internal states of the manipulator Sensing of the states of the object to be assembled and of its environment Measurement of physical parameters Scene analysis, identification, location of the object
Drive	Supplying motion energy
Control	Control and supervision of assembly sequence Synchronization of assembly and supervision of manipulator
Programming	Software for assembly, complier, interpreter, test system, simulator
Computer system	Initiation and supervision of assembly Support of program development

8.3 KINEMATICS

Kinematics is a discipline of theoretical mechanics which describes the motion of a point in space. It is assumed that the point has no mass. When we view a robot we use the laws of kinematics to describe the motion of the joints with respect to their geometry. With the help of the joints the industrial robot is able to move to any position within its work space. The technical realization of a kinematic system is done with joints, linkages, and drives. Each combination of joint-link-drive is called a robot axis. Each robot axis normally represents a degree of freedom.

In general, each point in a three-dimensional space can be reached with three independent motions, which follow the direction of the three main coordinate axis. The axis are perpendicular to each other. It is not necessary that the coordinate system be a cartesian one. It may also be cylindrical or spherical (Fig. 8.6). A robot designed for a cartesian workspace is of simple construction and is simple to program. The positioning accuracy is very high, but it has a limited work space. The versatility of the robot increases with a cylindrical work space and is the highest with the spherical work space. As the versatility goes up, programming becomes more difficult and the accuracy decreases.

So far, nothing has been said about the orientation of the gripper. For example, it may point upward or downward upon arrival at a destination. In case the orientation is to be uniquely identified, three additional degrees of freedom are necessary. In other words, to reach a point in space and to assume any orientation, the robot must have a minimum of six degrees of freedom. The combination of translational and rotational axes determines the work space of the robot, which can be described by spatial surfaces. The three basic kinematic principles, which are most often used for industrial robots, can be realized as follows:

Three translational axes
One rotational and two translational axes
Two rotational and one translational axes
Three rotational axes

To control the arm the angular orientation of the joints has to be specified. The parameters of the rotation angles are θ and those of the translational movement, (Fig. 8.5). Direct programming of complex moves with the aid of these rotations and translations is very cumbersome for a multiple-joint robot. The effect the angular motion of each joint has on the movement of the effector cannot be predicted easily. In general, the cartesian coordinate system, which is more familiar to human beings, is best suited to describe the arm movements. It is used to calculate translations and joint angles with the aid of coordinate transformations.

Parts Handling Assembly

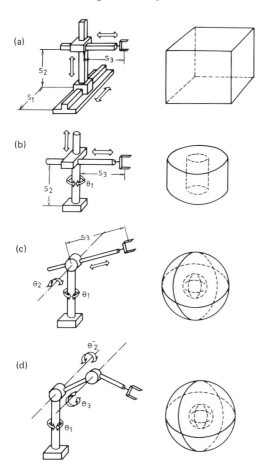

FIG. 8.6 Configuration of manipulators with three degrees of freedom: (a) Cartesian work space; (b) cylindrical work space (hollow); (c) spherical work space; (d) spherical work space.

Since the interaction between the industrial robot and the object to be handled is done with the effector, it is useful just to describe its position and orientation (Fig. 8.7). To calculate points on the three-dimensional trajectory of the effector, two variables are needed:

1. The position $\bar{r}$ of the effector in reference to the coordinate system defined by the orthogonal vectors $\bar{x}_0$, $\bar{y}_0$, $\bar{z}_0$
2. The orientation of the effector defined by the three unique vectors $\bar{x}_e$, $\bar{y}_e$, $\bar{z}_e$

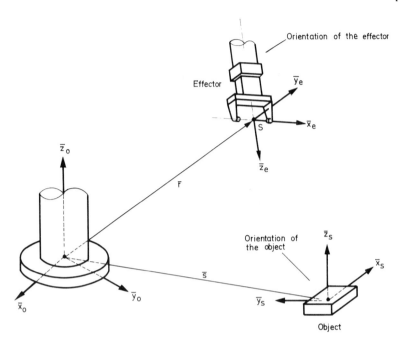

FIG. 8.7 Relationship between the coordinate systems of the robot, effector, and object: $\bar{r}$, location vector of the effector; $\bar{s}$, location vector of the object.

With these two variables the destination point of the effector can be described (position and orientation). The position $\bar{r}$ of the effector is defined by a 3 × 1-vector having the components x_s, y_s, z_s. The orthogonal base vectors $\bar{x}_e$, $\bar{y}_e$, $\bar{z}_e$ describe the orientation of the effector with 3 × 1 vectors. They are represented by a homogeneous 4 × 1 matrix as follows:

$$\underline{E} = \left| \begin{array}{ccc|c} \bar{x}_e & \bar{y}_e & \bar{z}_e & \bar{r} \\ \hline 0 & 0 & 0 & 1 \end{array} \right| \underline{E}(\bar{\theta})$$

The vector $\bar{x}_e$, $\bar{y}_e$, $\bar{z}_e$ shows the orientation and $\bar{r}$ the position of the effector. All four vectors are functions of the joint angles θ (robot coordinates) of the robot. In the following sections the relationship between the robot coordinate system and the absolute coordinates (world coordinates) of the effector is described on the base of homogenous coordinate transformations.

8.3.1 Coordinate Transformation

When a three-dimensional trajectory of the effector is described as a function of time in the cartesian coordinate system, the movement of the individual joints of the robot can also be specified as a function of time. Thus a transformation of cartesian world coordinates to robot joint coordinates must be performed. On the other hand, if the joint angles are given, a transformation of the robot coordinate system to the world coordinates can be done to determine the position and orientation of the effector. In modern industrial robots, both transformations are implemented to plan the trajectory and control the movement along it. The structure of the coordinate transformation is sketched in Fig. 8.8.

Generally, both types of transformations can be described as follows:

1. The transformation of robot coordinates into world coordinates is carried out via 4 × 4 Denavit-Hartenberg transformation matrices. The calculation can be formalized through matrix multiplications and leads to unambiguous results [5].

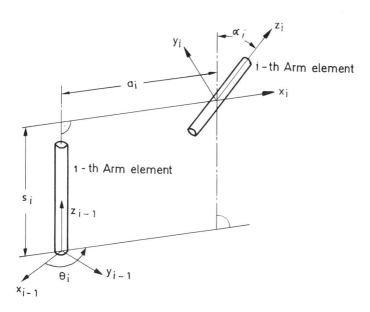

FIG. 8.8 Principle of coordinate transformation for robot joints. (From Ref. 6.)

2. The transformation of cartesian world coordinates into robot coordinates may be ambiguous. The calculation depends on the geometric configuration of the arm and can lead to ambiguous results because of inverse trigonometric terms.

8.3.2 Transformation of Robot into World Coordinates

Transformation of the robot coordinates into world coordinates is done using Denavit-Hartenberg matrices. With this method the position of each motion axis of the robot can be expressed in terms of the cartesian coordinate system of the robot (Fig. 8.8). This method requires the definition of a cartesian coordinate system in each joint. They have to be positioned according to the following orientation rules with respect to the neighboring joints:

1. The z_{i-1} axis of joint points in direction of the motion axis of the i-th arm element.
2. The x_i axis is perpendicular to the z_{i-1} axis and points away from it.

Coordinate systems that are oriented in this manner can be transformed by two translations and two rotations (Fig. 8.8). This is done according to the following rules:

1. A rotation θ_i about the z_{i-1} axis to make x_{i-1} parallel to x_i
2. A translation s_i along z_{i-1} to the point where z_{i-1} and x_i intersect
3. A translation a_i along the x_i axis to bring the origins of the coordinate systems together
4. A rotation α_i about the x_i axis in order to align the z axes with each other

The four operations are done with the help of the 4 × 4 Denavit-Hartenberg matrix.

$$\underline{D}_{i-1,i} = \begin{bmatrix} \cos\theta_i & -\cos\alpha_i \sin\theta_i & \sin\alpha_i \sin\theta_i & a_i \cos\theta_i \\ \sin\theta_i & \cos\alpha_i \cos\theta_i & -\sin\alpha_i \cos\theta_i & a_i \sin\theta_i \\ 0 & \sin\alpha_i & \cos\alpha_i & s_i \\ 0 & 0 & 0 & 1 \end{bmatrix}$$

It transforms a point with homogeneous coordinates of the i-th coordinate system of the robot to the (i-1)-th coordinate system.

$$\begin{bmatrix} x_{i-1} \\ y_{i-1} \\ z_{i-1} \\ 1 \end{bmatrix} = \underline{D}_{i-1,i} \begin{bmatrix} x_i \\ y_i \\ z_i \\ 1 \end{bmatrix}$$

Parts Handling Assembly

By multiplying all Denavit-Hartenberg matrices of the robot joints, one obtains

$$\underline{E} = \underline{D}_{0,1} \underline{D}_{1,2} \underline{D}_{2,3} \cdots \underline{D}_{n-1,n}$$

as a new 4 × 4 matrix which indicates the position and orientation of the effector in world coordinates (x,y,z).

$$E = \begin{bmatrix} x_{ex} & y_{ex} & z_{ex} & x_s \\ x_{ey} & y_{ey} & z_{ey} & y_s \\ x_{ez} & y_{ez} & z_{ez} & z_s \\ 0 & 0 & 0 & 1 \end{bmatrix}$$

The components of this matrix can be interpreted as follows:

1. The (x,y,z) components of the unit vector $\bar{x}_e$ of the coordinate system of the effector are x_{ex}, x_{ey}, x_{ez}.
2. The (x,y,z) components of the unit vector $\bar{y}_e$ of the coordinate system of the effector are y_{ex}, y_{ey}, y_{ez}.
3. The (x,y,z) components of the unit vector $\bar{z}_e$ of the coordinate system of the effector are z_{ex}, z_{ey}, z_{ez}.
4. The (x,y,z) components of the unit vector $\bar{r}$ which points from the origin of the world coordinate system to the origin S of the cartesian coordinate system of the effector are x_s, y_s, z_s.

In case the joint angles are known, it is possible to calculate with the help of the matrix $\underline{E}$ the position $\bar{r}$ and orientation $\bar{x}_e$, $\bar{y}_e$, $\bar{z}_e$ of the effector in world coordinates.

In fig. 8.9 a manipulator with five rotational joints is shown. There are five coordinate systems, one for each joint. The transformation matrices are shown in the upper right corner of Fig. 8.9. By multiplying them,

$$\underline{E} = \underline{D}_{0,1} \underline{D}_{1,2} \underline{D}_{2,3} \cdots \underline{D}_{4,5}$$

the new 4 × 4 matrix $\underline{E}$ describing the position and orientation of the effector is obtained.

8.3.3 Transformation of World into Robot Coordinates

To transform the world coordinates of the effector into arm-specific robot coordinates, the equation for $\underline{E}$ has to be solved with respect to the arm variables θ and s. The orientation and position of the effector is described by the following equations:

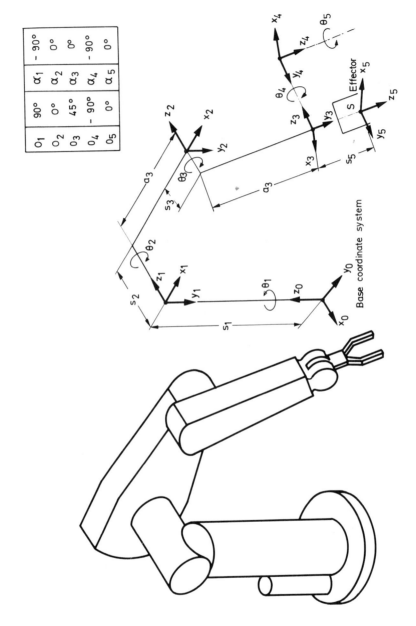

FIG. 8.9 Coordinate systems of the PUMA 500.

Parts Handling Assembly

$$\underline{D}_{0,1} = \begin{bmatrix} \cos\theta_1 & 0 & -\sin\theta_1 & 0 \\ \sin\theta_1 & 0 & \cos\theta_1 & 0 \\ 0 & -1 & 0 & s_1 \\ 0 & 0 & 0 & 1 \end{bmatrix} \quad (\alpha_1 = -90°)$$

$$\underline{D}_{1,2} = \begin{bmatrix} \cos\theta_2 & -\sin\theta_2 & 0 & a_2\cos\theta_2 \\ \sin\theta_2 & \cos\theta_2 & 0 & a_2\sin\theta_2 \\ 0 & 0 & 1 & s_2 \\ 0 & 0 & 0 & 1 \end{bmatrix} \quad (\alpha_2 = 0°)$$

$$\underline{D}_{2,3} = \begin{bmatrix} \cos\theta_3 & -\sin\theta_3 & 0 & a_3\cos\theta_3 \\ \sin\theta_3 & \cos\theta_3 & 0 & a_3\sin\theta_3 \\ 0 & 0 & 1 & -s_3 \\ 0 & 0 & 0 & 1 \end{bmatrix} \quad (\alpha_3 = 0°)$$

$$\underline{D}_{4,5} = \begin{bmatrix} \cos\theta_4 & 0 & -\sin\theta_4 & 0 \\ \sin\theta_4 & 0 & \cos\theta_4 & 0 \\ 0 & -1 & 0 & 0 \\ 0 & 0 & 0 & 1 \end{bmatrix} \quad (\alpha_4 = -90°)$$

$$\underline{D}_{4,5} = \begin{bmatrix} \cos\theta_5 & -\sin\theta_5 & 0 & 0 \\ \sin\theta_5 & \cos\theta_5 & 0 & 0 \\ 0 & 0 & 1 & s_5 \\ 0 & 0 & 0 & 1 \end{bmatrix} \quad (\alpha_5 = 0°)$$

Normally, these equations contain products of sine and cosine functions with different arguments θ. To determine these arguments the equations have to be solved for the individual angles θ_i. In case the robot has n joints, n equations are necessary to determine θ_i.

The configuration of the robot determines whether or not these equations can be solved. The smaller the number of axis, the easier it is to calculate the robot coordinates. The transformation of world into robot coordinates is facilitated if the robot is of a good design.

8.4 THE EFFECTOR

A robot handles a workpiece with the help of its effector. Attempts have been made to develop universal or even programmable effectors. In general, however, the robot user gives preference to a gripper

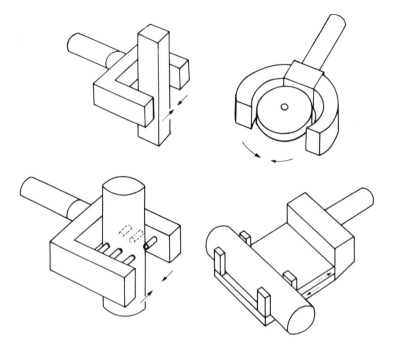

FIG. 8.10 Principles of application-oriented grippers.

designed for the specific application (Fig. 8.10). The following parameters are of interest when a gripper is selected:

The tool to be used
The object to be handled
The surface and geometry of the object
The weight of the object
Position and orientation of the object
Accessibility of the object
Grasping parameters of the object
Friction between finger and object
Reaction forces and moments
Compliant forces

If work tools are used, the energy supply and feeding of parts must be given proper attention. In this section a simple two-finger gripper, and in the following section several sensors for this gripper, will be discussed. The mechanical design of the gripper is shown in Fig. 8.11. Its two parallel fingers are actuated by a dc motor and a gear drive. The fingers are exchangeable to grasp objects from the inside or

Parts Handling Assembly 521

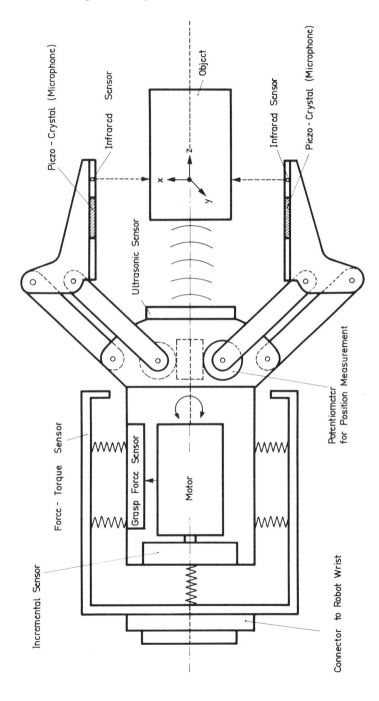

FIG. 8.11 Design of a universal sensor-equipped effector. (From Ref. 7.)

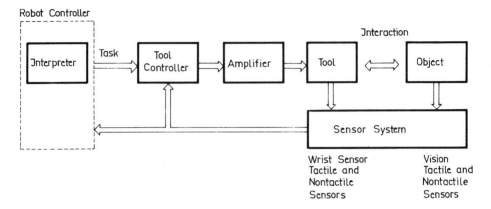

FIG. 8.12 Control loop of the gripper motor.

outside with a defined force. When the fingers open, they follow a radial trajectory. This means that the grasp position of the object changes dynamically. It has to be adjusted with the help of a software routine to guarantee exact positioning. The following operations can be performed by the gripper:

1. To move the fingers with a defined speed
2. To move the fingers with a defined speed to a defined opening position of the jaw
3. To grasp with a given force at defined limits
4. To move the fingers with a defined speed to a defined opening position of the jaw and then to apply a defined force
5. To stop the fingers when a defined position has been reached
6. To measure dimensions of an object
7. To activate other sensors
8. To transfer sensors data to the robot computer

The first six tasks make necessary closed-loop control for the speed and grasp force. This may be done with either software or hardware. Figure 8.12 shows the concept of a control loop for the gripper.

8.5 SENSORS

A robot obtains its versatility with the help of a complex sensor system. There are numerous sensors available to help the robot to perform its work similar to a human operator. Advanced sensors have a limited amount of intelligence to process measuring parameters close to

Parts Handling Assembly 523

the source of their origination. The sensor environment for a robot can be classified as follows:

Internal sensors
  Positioning sensors for the robot axes
  Velocity and acceleration sensors to control the movement of the robot
  Force/torque sensors to control handling of the object, to supervise assembly, and to protect the robot from overloading
  Gripper force and velocity sensors
External sensors
  Approximation sensors
  Touch and slip sensors
  Compliance sensors
  Two- and three-dimensional vision sensors
  Simple sensors to measure length and displacement
  Special-purpose sensors

Internal sensors are part of the closed-loop axis control. They have to sample data with high speed. External sensors work at lower speed and process data to update the robot model. The sensor system consists of a data acquisition and a data processing module. The data acquisition module samples the process data and generates performance pattern. The data processing system identifies the pattern and generates frames for the dynamic world processor.

It is beyond the scope of this book to discuss in detail all available sensors. Only some representative external sensors will be described in connection with the versatile hand shown in Fig. 8.11. In addition, the last part of this chapter is devoted to vision.

8.5.1 Touch and Slip Sensor

This sensor is a microphone made from a piezoelectric crystal (Fig. 8.11). It responds to touch and slippage of an object. To reduce noise originating from the gripper, it is embedded into the finger surface with an elastomeric substance. Noise signals are generated by handling of most nonresilent workpieces. The sensor can be used to center the gripper when it tries to pick up an object. If the object is not in the proper position first, one finger will touch it during grasping. This incident is recorded by the touch sensor. With its internal joint position sensors, the robot records the location of the object. Then it adjusts the hand approach. It is possible to construct a sensor matrix from piezoelectric crystals. With such a device an object and its location between the fingers can be identified.

8.5.2 Ultrasonic Approach Sensor

The vertical approach of an object can be sensed and controlled with an ultrasonic sensor (Fig. 8.11). A simple device consists of a membrane-covered transmitter/receiver to which pulse trains with different ultrasonic frequencies are applied. A columnated ultrasonic sound wave is sent to the object and reflected back to the receiver. The travel time of the sound wave between transmission and reception of the signal is calculated. From this information the distance to the object is determined. A variation of the frequency may be necessary to adjust the sensor for different reflection characteristics of different materials. Such a sensor will have a resolution of 1 mm. An integrated transmitter/receiver system needs a given time to switch from transmitting to receiving. This time interval limits the lower measuring distance of the device. A typical minimum distance that can be measured is 20 cm. If closer distances have to be measured, an independent transmitter and receiver must be used.

8.5.3 Recognizing an Object Between the Gripper Fingers

The microphone discussed above may be used for this purpose. However, it will be activated only upon touch of a workpiece, not by its proximity. For this reason there are usually infrared sensors installed in the fingertips. Figures 8.11 and 8.13 show a sensor system consisting of one transmitter and two receivers. When an object is located between the sensors, the infrared light emitted by the transmitter will be reflected to the receivers. In case the object surface is parallel with the surface of the finger, both receivers will detect the same amount of reflected light. In case the object is skewed with respect to the finger, surface signals of different intensity are received. The robot may now reposition its gripper to be parallel with the object. This sensor will also detect the proximity of a workpiece by evaluating the intensity of the two reflected signals. Both the sensitivity and the approach distance can be adjusted within the limits of the sensor system.

8.5.4 Force-Torque Sensor

Many different force-torque sensor principles have been developed. This type of sensor is needed to control handling of the object, to supervise assembly, and to protect the robot from overload [10]. It is usually installed in the wirst of the effector and should be able to render accurate and repeatable results throughout the entire load range of the robot. The sensor itself has to be protected against possible overload, shock, and vibration. The basic principle of most sensors uses four cantilevers to which eight strain gauges are attached (Fig. 8.14). Careful attention must be paid to a good design so that

Parts Handling Assembly 525

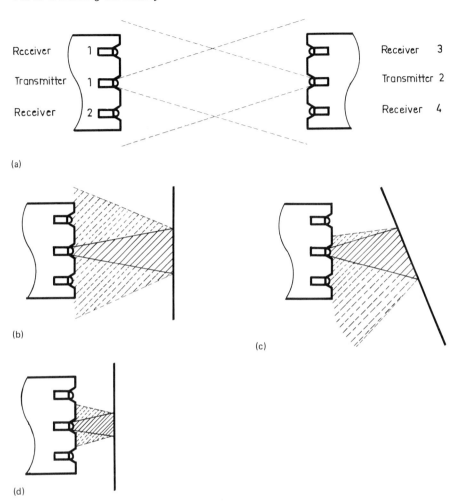

FIG. 8.13 Principle of detecting an object between fingers: (a) location of infrared transmitter and receiver; (b) object parallel with fingers; (c) object in skewed position; (d) measuring approach to object.

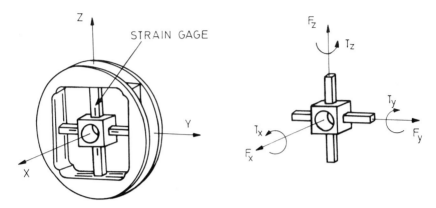

FIG. 8.14 Force-torque sensor. (From Ref. 8.)

the measurement signals are highly linear with the applied force. The analog strain gauge signals are converted to force and torque information via a computer. To obtain the orthogonal components of the measured parameters in direction of the main axes of the effector's coordinate system (frame), a matrix multiplication is necessary. It is assumed that the elongation ε_i (i = 1, 2, . . . , 8) and the force-torque vector $\xi = (F_x, F_y, F_z, M_x, M_y, M_z)$ are linear. Thus one can write

$$\varepsilon = C_{11}F_x + C_{12}F_y + C_{13}F_z + C_{14}M_x + C_{15}M_y + C_{16}M_z$$

Here $\underline{C}$ is a n × 6 matrix. With the eight strain gauge signals, $\bar{\varepsilon}$ represents a 8 × 1 vector and $\underline{C}$ a 8 × 6 matrix. To obtain the 48 components of $\underline{C}$, the sensor is successively loaded with the 6 linear independent force-torque vectors $\bar{\xi}_i$ (i = 1, 2, . . . , 6) (Fig. 8.15). The six measured elongation vectors ε_i can be written with the following matrix presentation:

$$\underline{E} = \underline{C}\ \underline{\xi}$$

Here $\underline{E}$ is a 8 × 6, and $\underline{\xi}$ a 6 × 6 matrix. $\underline{\xi}$ is regular and has the inverse $\underline{\xi}^{-1}$. By multiplying the equation above from right to left, one obtains

$$\underline{C} = \underline{E}\ \underline{\xi}^{-1}$$

Since $\underline{C}$ is a nonquadratix 8 × 6 matrix, it has no inverse. For this reason an inverse pseudo matric $\underline{C}^I$ is found where $\underline{C}$ is regular with regard to its rows and columns:

Parts Handling Assembly

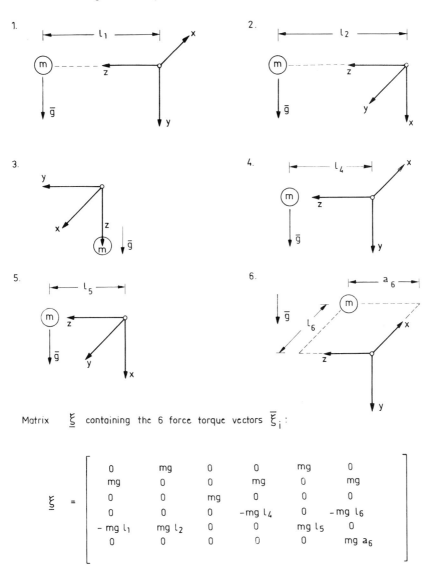

FIG. 8.15 Calibration matrix for a force-torque sensor.

$$\underline{C}^I = (\underline{C}^T\underline{C})^{-1}\underline{C}^T$$

With this the following force-torque matrix can be calculated:

$$\overline{\xi} = \underline{C}^I \overline{\varepsilon}$$

This algorithm can be implemented on a microcomputer which uses a parallel arithmetic processor (Fig. 8.16). The measurement of $\overline{\varepsilon}$ may be done with an integrated data acquisition chip and the coordinate transformation and the correction for the gravity force with an additional processor.

8.5.5 Sensors to Control the Action of the Fingers

To measure the opening distance of the fingers of the sample effector (Fig. 8.11), a potentiometer is used (Fig. 8.17). It is fastened to a pivoting joint of a finger link. The motor operating the finger via a gear drive is suspended on springs inside the wrist. When a force is applied on the object to be handled, the motor moves in direction of the z axis of the wrist. The displacement of the springs is a measure of the force exerted by the fingers. It can be measured with the help

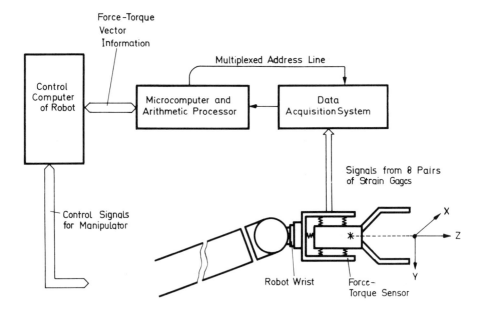

FIG. 8.16 Principle of a force-torque measuring system.

Parts Handling Assembly

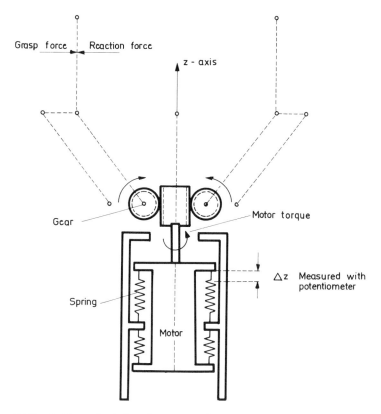

FIG. 8.17 Principle of a grasp force's measuring system.

of a translational potentiometer. The deflection Δz of the spring is proportional to the grasp force. The relation between the motor torque M and the deflection Δz is given by the following motion equation:

$$a\ddot{z}(t) + bz(t) + c \sin \dot{z}(t)[d|z(t)| + r] = n$$

The terms a, b, c, and d are constants and r represents the constant friction factor. The equation can be implemented as a control algorithm on a microprocessor.

8.5.6 Control of a Sensor-Equipped Effector

When the interpreter of the control computer of the robot is initialized, the sensor circuits and the gripper motor are activated (Fig. 8.18).

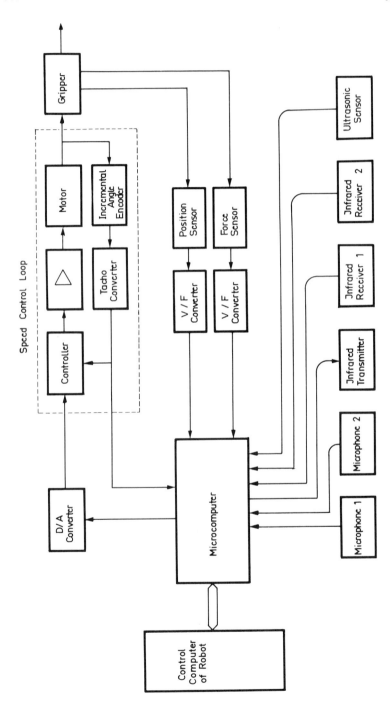

FIG. 8.18 Microcomputer to control operation of a gripper.

Parts Handling Assembly

The software routines of these devices start to acquire and process sensor data and control the gripper and robot operation.

Each data transfer begins with a start character to indicate the beginning of a transfer. A sequence of characters follows to identify the data control word. Then the parameters and values are transferred. The identifier is used to define the activities of the gripper system and to describe the control and sensor data to be exchanged (Tables 8.4 and 8.5).

The control programs are stored in PROMS and are initialized when the power is switched on. After the bootstrapping operation the microprocessors are in a polling mode waiting for a request for input or

TABLE 8.4 Control Functions for a Sensor-Controlled Effector

Primitives	Function
POS	Activates position and velocity control algorithms
FORCE	Activates force control algorithms
VEL	Transfers parameters to velocity control
SENS ON I	Activates microphone and infrared diodes
SENS ON II	Activates force sensing wrist
SENS ON III	Activates ultrasonic sensor
SENS OUT I	Interrupts microphone and infrared diodes
SENS OUT II	Interrupts force-sensing wrist
SENS OUT III	Interrupts ultrasonic sensor
FORCEPS	Activates SENS ON I and POS
STOP	Stops finger in shortest possible time
INF I	Transfers data about the gripper's state from the sensor data table to the robot controller
INF II	Transfers the force-torque vector to the robot controller
INF III	Transfers data from the ultrasonic sensor to the robot controller
POSJU	Adjusts position sensor
FORCEJU	Adjusts force sensor
WRISTJU	Adjusts force-sensing wrist

TABLE 8.5 Types of Data Processed with a Sensor-Controlled Effector

Type of data	Interpretation
POSITION	Desired finger position
FORCE	Desired grasp position
SPEED	Grasp speed
POSLIMIT	Fastening force limit
THRESHOLD	Critical force-torque vector, when an interrupt to the robot controller has to be generated
DISTANCE	Distance to an object, when an interrupt is to be generated
ACTPOS	Readout of actual finger position
ACTFORCE	Readout of actual grasp position
TRAJECTORY Z	Actual Z-coordinate of fingertip for position correction
ACTDISTANCE	Actual distance (vertical) to object
VECTOR WRIST	Actual force-wrist vector
STATUS	Status of sensor system

output data. If control data are received, the corresponding program will be activated. All sensor data are written on output tables, so that a data transfer is possible at any time. From the viewpoint of a central robot controller the peripheral microcomputers are working as I/O modules. They require no additional communication software. The microcomputers have a stripped-down software configuration consisting of a local monitor which allows program modification by the user via an own terminal interface.

8.6 DRIVES TO OPERATE THE ROBOT JOINTS

Robot joints may be moved by pneumatic, hydraulic, or electrical drives. The electrical motor is the predominently used device. It has a good positioning accuracy and its speed and acceleration can be easily controlled. The disadvantage is its high weight. Often, the drive of the gripper is exposed to rough environmental conditions. For example, the gripper of a forging manipulator may have to endure temperatures

Parts Handling Assembly

of 1200°C and that of a spray robot has to stand up to agressive paint constituents. The cost of the drives is 5 to 20% of that of the robot.

8.6.1 Electrical Drives

The following types of electromotors are used: (1) stepping, (2) dc servomotors, and (3) pancake motors. The stepping motor gets its pulses from the controller. There is no feedback loop and positional encoder needed. Each pulse causes a defined angular displacement. The dc servomotors has a position controller and an angular decoder as a feedback device to exactly control the movements of a joint. It can react quickly to acceleration and deceleration commands and has very short reaction times. Its rotational speed can be controlled continuously and it operates very smoothly. Resolvers are used to control speed and torque. The pancake motor has very good control characteristics. Its disadvantage is size and weight.

8.6.2 Hydraulic Drives

These drives permit continuous control of the motion axes and they can move a robot very accurately to defined spacial points. The drives consist of a hydraulic motor and of continuously variable valves which can transmit high forces and torques. Because of changing oil viscosity with changing temperature of the robot environment, the control characteristic of these drives may change. Oil leaks may cause messy operation.

8.6.3 Pneumatic Drives

Pneumatic drives are of simple design, low in cost, and have fast reaction times. The arms are actuated directly by air cylinders. Because the air is highly compressible, it is very difficult to obtain accurate position and speed control. Pneumatic drives are best suited for point-to-point control with fixed start and end positions.

8.7 CONTROL

The task of the control is to give the robot the flexibility it needs to perform its complex handling assignment. It interconnects directly with the application program and translates the instructions into mechanical motion. By doing so it controls and supervises the robot's action and the communication between the peripherals and sensors. The computer of the robot may be used for program development. In general, however, the hardware and software components needed for programming are separate units.

It is not customary to build the robot and the control system by different manufacturers. This is contrary to the NC technology. The control parameters for the robot movements have to be matched exactly with the kinematic and functional properties of the mechanics. In addition, there are no standard interfaces available to interconnect the multitude of existing controllers, drives, and measuring systems. In Germany an effort is being made to define a standard interface called IRDATA between the control and the programming system. The standardization concept is based on CLDATA (cutter location data) code used by NC controls. This concept is enhanced to accommodate robot-specific features, such as addressing of variables, handling of stacks, processing of sensor data, and accommodation of administrative elements.

The control has to interact with the internal components and external environment of the robot. The operation for these assignments can be grouped as follows:

1. Planning the trajectory in world coordinates
2. Calculating motion parameters in robot coordinates for each joint with the aid of coordinate transformation
3. Smoothing the trajectory by interpolation algorithms (adaption to the mechanics)
4. Generating the control signals and the forces and torques needed for the movement
5. Reacting to and correcting deviations from the trajectory with the aid of exact control algorithms
6. Acquiring and processing sensor information
7. Modifying and adapting the control problem to the handling process

Depending on the degree of integration of these components, it is possible to classify robots by characteristic features. Thus we distinguish, for example, between point-to-point and trajectory control. Similarly, there are robots that follow the instruction of the program rigidly, and those which are able to adapt themselves to changing work conditions. The latter must be equipped with sensors. Other control systems may be capable of handling two or several robots simultaneously. In the following sections a control system for a complex assembly robot will be discussed. For simple handling tasks only very basic controllers are used.

8.7.1 Comprehensive Control System for an Assembly Robot

The general structure of a sensor-controlled robot system is shown in Fig. 8.19. It consists of the mechanical system (Sec. 8.2 to 8.4), the programming system (Sec. 8.8), and the real-time controller. In this part the task of the real-time controller will be explained. The sensors

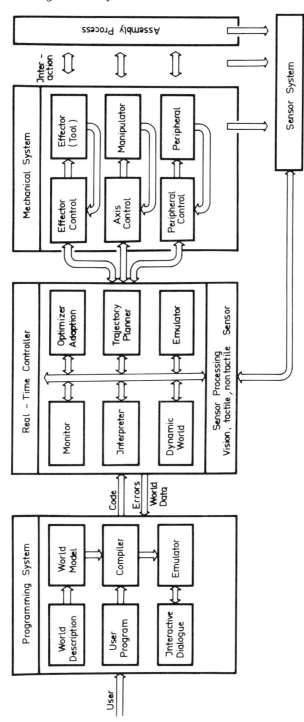

FIG. 8.19 Schematic structure of a robot control system. (From Ref. 11.)

that provide information about the environment of the robot are treated in Sec. 8.5 and 8.10. The controller needs the sensor information to plan the trajectory and to respond to any changes of the robot world during manipulation.

Most modern general-purpose robots have path control by which the effector follows a defined trajectory. The trajectory is planned in real time by the robot controller (trajectory planner) with the aid of sensor information. Disturbances of the control path may be caused by (1) the gravity force, (2) centrifugal force, (3) the Coriolis force, and (4) friction and reaction forces. A large number of control algorithms have been developed to solve these control problems [12–16].

The high number of arithmetic operations to be performed by a control algorithm makes a simplification of axis control necessary. When control algorithms are designed, an attempt is made to reduce the number of additions and multiplications to a minimum. Usually, joint control is done in joint space coordinates (robots coordinates). The joint space trajectory is generated by a postprocessor located between the robot controller and the axis controller (Fig. 8.20). The postprocessor transforms the robot-independant cartesian trajectory into robot-dependent coordinates. Efforts are made to use the resolved acceleration method to control the end effector directly in cartesian world coordinates [17].

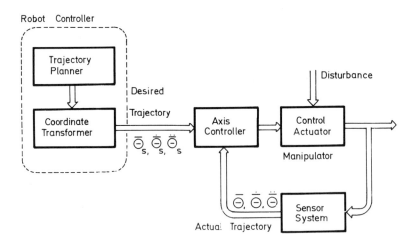

FIG. 8.20 Axis control system.

8.7.2 Central Robot Controller

The robot controller has to perform the proper execution of the manipulation sequence specified by the user program. The individual tasks can be summarized as follows:

- Interpretation and execution of the user program in real time
- Multitask scheduling
- Trajectory planning under sensor guidance
- Coordinate transformation
- Effector control
- Control of peripherals
- Sensor control and data processing
- Bus management and communication with peripherals

For these different functions a modular polyprocessor architecture is necessary. The concept involves the distribution of multiple tasks to a processor environment of autonomous processing nodes. Each node consists of one or more processors with memory, working under a local operationing system. The nodes are of the following types:

- Floating-point arithmetic
- Fixed-point arithmetic
- Sensor
- Logical
- I/O

The communication is done with the aid of a hierarchy of protocols for bus access, transmission of data, conversion between logical and physical links, and communications between processes (active tasks). For multiprocessing the tasks and subtasks are distributed for parallel operation.

8.7.3 Trajectory Planner

Trajectory planning determines the motion of the manipulator under sensor guidance. The parameters of the trajectory are the starting conditions, velocity, end points, and so on. They are defined by the user program and may be modified by the sensor system. They are the input data to the trajectory planner. The different types of motions are:

- Point-to-point
- Straight line
- Defined curves with defined velocity
- Tracking of moving objects (tracking)
- Assembly operations
- Collision avoidance

Depending on the desired effector motion different interpolation algorithms can be applied. Following are the well-known interpolation methods (Fig. 8.21a):

1. Linear interpolation along straight lines
2. Linear interpolation with transition between straight-line segments
3. Trajectory interpolation with algebriac functions (polynominals)
4. Interpolation between time-variant points with time function or gradient strategies

In practice the generation of the trajectory is done in cartesian coordinates at a typical data rate of 100 Hz (Fig. 8.21b). Trajectory planning can be a robot-independent process.

The programmer specifies the points of the effector trajectory with the help of frames (Sec. 8.8). A frame is described by a DH matrix. With the aid of a frame the joint angle and trajectory point can be calculated. For point-to-point control (PTP), only the end points have to be determined. In case the effector has to follow a defined trajectory, interpolation is necessary. The travel times during which the individual joints of a robot traverse along their trajectory may differ considerably, depending on the robot geometry, the operation speed, and the type of trajectory.

With PTP control all robot joints are operated with maximum velocity and acceleration until the desired angle of an axis has been reached. For this reason it is possible that some axes will arrive at their final destination earlier than others. Thus the trajectory of the gripper is undefined. With trajectory control there is a functional relationship between the movements of the individual axes. The trajectory controller assures that all axes start and end their movement at the same time.

The linear interpolation in cartesian coordinates is of particular interest. Here the effector follows a straight line. Other interpolation methods use circular, parabolic, or complex function algorithms. The three control methods PTP, axis control, and linear interpolation will be explained with the help of a simple four axes robot (Fig. 8.22). Only the axes θ_1 and θ_2 will be considered. The coordinate transformation can be calculated as follows (Fig. 8.23):

1. Conversion of robot to cartesian coordinates

    $x = 400 \cos \theta_1 + 250 \cos(\theta_1 + \theta_2) + a$
    $y = 400 \sin \theta_1 + 250 \sin(\theta_1 + \theta_2) + b$

2. Conversion of cartesian to robot coordinates

    $r = \sqrt{(x-a)^2 + (y-b)^2}$

Parts Handling Assembly

(a)

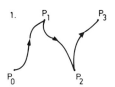

Point to point planner

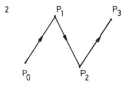

Linear interpolation

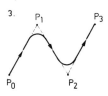

Linear interpolation with segment transition
(Circular, quadratic, cubic etc)

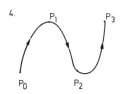

Interpolation with polynomials

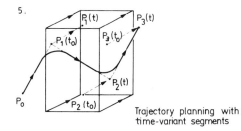

Trajectory planning with time-variant segments

(b)

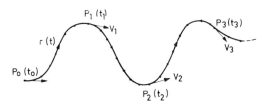

FIG. 8.21 Robot trajectories: (a) different procedures to generate robot trajectories; (b) a sampled curve with equal-time distant points ($\Delta T < 10$ msec).

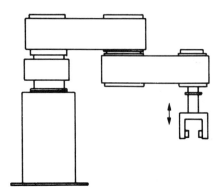

FIG. 8.22 Four-axis robot to be controlled. (From Ref. 17.)

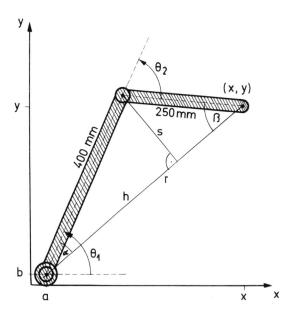

FIG. 8.23 Positioning axes of the robot.

$$h^2 + s^2 = 400^2$$

$$(r - h)^2 + s^2 = 250^2$$

$$h = \frac{48{,}750}{r} + \frac{r}{2}$$

$$\alpha = \arctan \frac{s}{h}$$

$$s = \sqrt{160{,}000 - h^2}$$

$$\beta = \arctan \frac{s}{r - h}$$

$$\theta_1 = \arctan \frac{y - b}{x - a} + \alpha$$

$$\theta_2 = -\alpha - \beta$$

In this example the robot is to be brought from position

$$x = 435, \ y = 510 \quad \text{or} \quad \theta_1 = 60°, \ \theta_2 = -25°$$

to position

$$x = 252, \ y = 161 \quad \text{or} \quad \theta_1 = 70°, \ \theta_2 = -140°$$

With PTP control the axis θ_1 is moved with a maximum angular velocity of 2° per time unit and axis θ_2 with 10° per time unit. The resulting trajectory is shown in Fig. 8.24. The discontinuity of this curve is caused by the fact that axis θ_1 has completed its motion before axis θ_2. With axis interpolation, the movement of θ_1 is slowed down to 0.87° per time unit. Thus both axes come to a standstill at the same time at their destination (Fig. 8.25).

With linear interpolation the points along the trajectory are calculated in cartesian and not in robot coordinates. The points are then transformed into robot coordinates (Fig. 8.26). The coordinate transformations are very time consuming. For this reason it is not possible to perform the interpolation in small increments. The resulting trajectory follows small circular elements and not a straight line. The deviation can be minimized with the aid of an arithmetic processor.

The results of the three methods are shown in Figs. 8.27 and 8.28. The motion diagram for the cartesian interpolation shows that θ_1 and θ_2 have a nonlinear behavior. θ_2 traverses through a temporary position of 80° before it reaches its final destination of 70°. It can also be seen from these examples that the programmer has to be aware of a possible swingover of axes which may cause collisions. If sensor-controlled trajectories are planned, any geometric change of the robot's

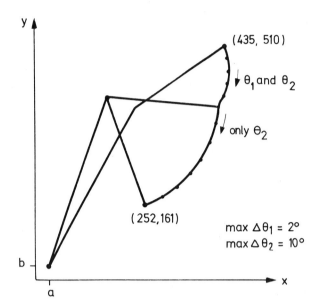

FIG. 8.24 Point-to-point control (PTP).

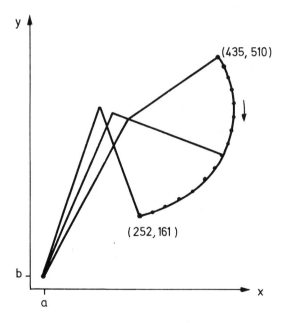

FIG. 8.25 Linear interpolation.

Parts Handling Assembly

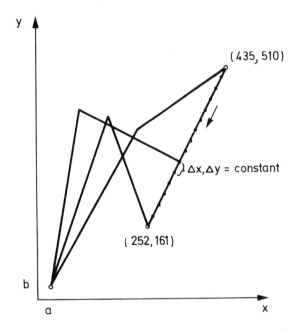

FIG. 8.26 Linear interpolation in the cartesian coordinate space.

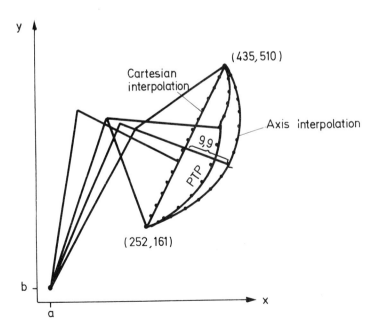

FIG. 8.27 Comparison of three trajectory planning method.

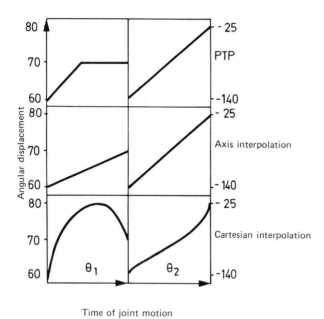

FIG. 8.28 Trajectory control methods for robots.

work space has to be identified, for example, by a vision system. The following parameters must be considered:

Position and orientation of the workpiece
Distances
Diameters
Geometry of the workpiece
Path of the workpiece (position, velocity)
Contours
etc.

These geometric data have to be transformed from sensor coordinates into cartesian world coordinates by a coordinate transformer (Fig. 8.29). For real-time control, the transformation will be done by (1) arithmetic processors, (2) tightly coupled processors, or (3) bit-slice processors.

8.7.4 Dynamic World Processor

An implicit robot programming system needs a world model as a reference. This model can be generated automatically or by hand. It defines the geometric relationship between objects. The frame concept

Parts Handling Assembly

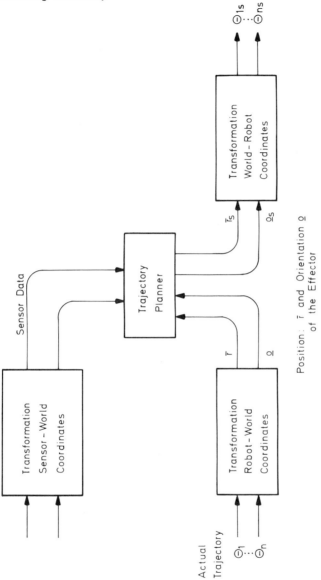

FIG. 8.29 Incorporation of external sensor data for trajectory planning.

(Sec. 8.8) is applied to describe the interaction between the object and the effector. In case of a well-structured assembly operation, the frames have to be defined only once. If an unstructured and time-variant environment prevails, the frames have to be refreshed continuously to present their actual states. This is necessary if:

1. The workpiece has a variable position and/or orientation.
2. Tracking of a moving object (on a conveyor) is necessary.
3. A variable collision scenario prevails.
4. A variable object (contour, weight) is handled or if there is an interaction between multirobot arms, a conveyor, and a manufacturing machine.

The robot world is constantly sampled by the dynamic world processor (Fig. 8.30). It compares actual process data with the world reference model. If a significant event is detected, the reference model will be refreshed and the actual data will be used by the monitorlike interpreter to change the robot's trajectory. An optimizer will perform the program modification with the aid of a predefined strategy.

8.7.5 Modular Structure of the Robot Controller

The control of a complex robot can be done by a hierarchy of control levels, whereby each level has a specific assignment. Beginning with the lowest level, the hierarchy can be defined as follows (Fig. 8.31):

1. *Servo control level and effector control*: At this level the drive signals for the actuators are generated to move the robot joints. Special optimizing control algorithms are used to control the individual mechanical systems. This control level is highly robot dependent. The input signals are the desired joint trajectories in joint coordinates.
2. *Coordinate transformation level*: The transformation of the cartesian trajectory into robot-specific joint coordinates is done at this level by a processor. Its programs are transformation routines which are robot dependent. The cartesian end effector trajectory may be calculated for a different robot configuration if it has the same degree of freedom and the same work space. At this level robot-independent trajectories are transformed for execution.
3. *Trajectory interpolation level*: At this level the robot interpolation routines generate the continuous trajectories in world coordinates which are executed by the robot. The inputs to the control module are sets of parameters (e.g., trajectory

Parts Handling Assembly

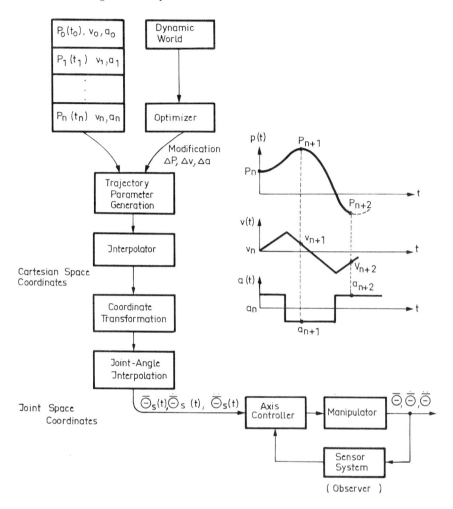

FIG. 8.30 Planning sensor-controlled trajectories.

points, speed, etc.), which define the trajectory. Depending on the application, different interpolation routines (linear, quadratic, cubic, circular, polynomial, special functions) are used. All kinematic calculations for the robot are done in cartesian coordinates.
4. *Trajectory control level*: Data from the user program, the world model, and the sensor system are processed to generate the trajectory parameters. At this level, strategies are

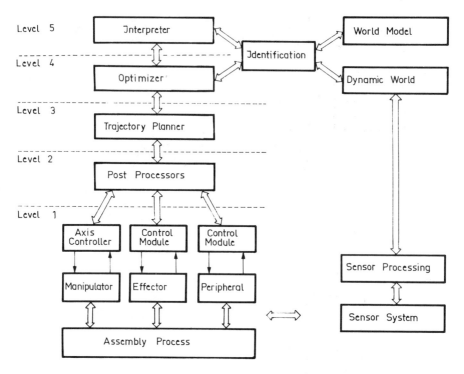

FIG. 8.31 Information flow between different levels of a robot control system.

implemented which optimize the operating parameters of the robot with the help of sensor data.
5. *Central robot controller*: The central controller is realized as an interpreter. Real-time processing, synchronization, and multitasking are done at this level. References are made with the aid of software routines to the world model, to the dynamic world processor, to a program library, and if necessary, to the trajectory control level. Scheduling and dispatching of tasks for the multiprocessor environment are controlled from this level.

8.8 PROGRAMMING LANGUAGES AND PROGRAMMING SYSTEMS FOR ASSEMBLY ROBOTS

Within recent years the industrial robot has matured to a universal tool that can handle efficiently many manufacturing operations. It first was used for simple material handling work and for welding.

Parts Handling Assembly

However, its greatest potential lies in the automation of assembly. For this purpose it must be equipped with sensors and with a comfortable programming system. To describe the work of a robot, the language must have conventional constructs and also those which are robot specific. With these requirements a programming language for robots usually becomes very complex. Researchers first started to develop simple assembly languages. Valuable experience was gathered from their use and lead to various concepts of explicit high-level languages. Presently, there are languages being developed for implicit programming dealing with CAD data. With these it will be possible to solve assembly tasks with the help of expert systems.

8.8.1 General Requirements for Programming Languages for Robots [6]

One of the essential features of a computer language for an industrial robot is the programming of the movement trajectory. For this purpose it is necessary that the programmer has the tool to enter all operating positions of a movement. In general, it is possible to describe textually any trajectory with respect to a given coordinate system with the aid of a sequence of points. However, it is very difficult for a human operator to visualize this series of points in a three-dimensional space and to describe the coordinates of each point via a programming language. To obtain accurate parameters, the exact location of a point has to be found with the aid of a measurement. This, however, is very time consuming and awkward. In practice, the problem is resolved by leading the robot's effector through its desired path and by reading the coordinates of the corresponding robot joints at predetermined points along the trajectory. The parameters of these points are entered into the robot memory. The robot path is then reconstructed by the compiler with the aid of an interpolation algorithm. In addition to the parameters of the trajectory, the control system of the robot must have information about the orientation of the effector. This parameter can also be entered into the program by the teach-in method.

Frame Concept

Modern explicit programming languages for robots use the "frame" concept [18,19]. With this method a spatial point is described by a position vector with respect to a standard coordinate system. The orientation of the gripper is described by a rotation or rotations. Thus a frame consists of a position vector and an orientation (Fig. 8.32). When the effector is described by a position and an orientation, the following convention is useful (Fig. 8.33):

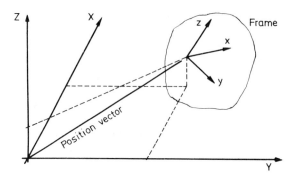

FIG. 8.32 Geometric position of a frame.

1. The end point of the position vector is located in the middle of the centerline which connects the two gripper jaws.
2. The gripper points into the direction of z axis of the frame coordinate system.
3. The y axis of the effector runs through the gripping points of both jaws.
4. The x axis is orthogonal to the y axis and the z axis. The orientation of the axes must follow the right-handed-screw rule.

A geometric relationship can be defined between frames. It consists of a translational and a rotational component. The translation starts at the origin of the first frame and ends at the origin of the second frame (Fig. 8.34). The rotational part is needed to align one frame with the other. The relative definition of frames is of special interest if a change of the value of one frame results in a change of the value of the other frame. In other words, the geometric relation between

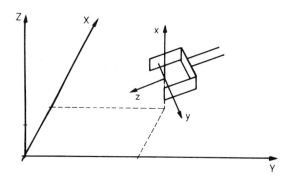

FIG. 8.33 Geometric position and orientation of a gripper.

Parts Handling Assembly

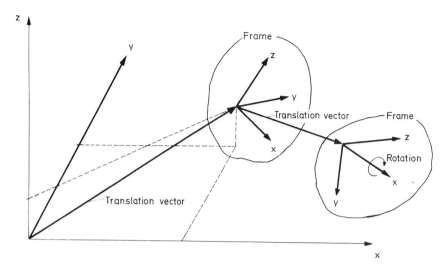

FIG. 8.34 Relations between frames.

the frames will be maintained. Another advantage of the frame concept is that the cartesian coordinate system is a common base for robot and sensor coordinates. As an example, the frame *goal* may be defined by the programmer as follows:

goal: = FRAME(ROT(ZAXIS,180),VECTOR(55,-37,23));

The $\underline{D}$ matrix (Sec. 8.3.2) representing the frame in the computer memory is

$$\underline{D}_{goal} = \begin{bmatrix} -1 & 0 & 0 & 55 \\ 0 & -1 & 0 & -37 \\ 0 & 0 & 1 & 23 \\ 0 & 0 & 0 & 1 \end{bmatrix}$$

This matrix shows that the directions of the x axis and y axis of the frame are inverted.

Another advantage to represent a frame by a $\underline{D}$ matrix is the simplicity of calculating a translation and/or rotation of the frame. For example, the frame *goal* may be rotated by 90 degrees about the z axis of the base coordinate system (Fig. 8.35). The resulting frame *newgoal* is calculated as follows:

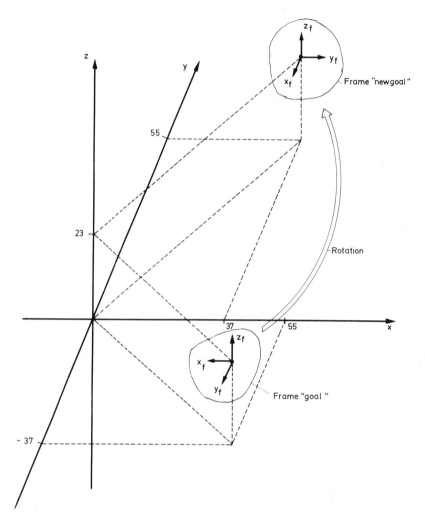

FIG. 8.35 Rotation of the frame goal.

Parts Handling Assembly

$$\underline{D}_{newgoal} = \begin{bmatrix} 0 & -1 & 0 & 0 \\ 1 & 0 & 0 & 0 \\ 0 & 0 & 1 & 0 \\ 0 & 0 & 0 & 1 \end{bmatrix} \begin{bmatrix} -1 & 0 & 0 & 55 \\ 0 & -1 & 0 & -37 \\ 0 & 0 & 1 & 23 \\ 0 & 0 & 0 & 1 \end{bmatrix}$$

ROT(ZAXIS,90)   frame *goal*

$$\underline{D}_{newgoal} = \begin{bmatrix} 0 & 1 & 0 & 37 \\ -1 & 0 & 0 & 55 \\ 0 & 0 & 1 & 23 \\ 0 & 0 & 0 & 1 \end{bmatrix}$$

The zero point of the new frame *newgoal* has the coordinates x = 37, y = 55, and z = 23. Its x axis points in the negative y direction of the base coordinate system, its y axis points into the x direction, and the z axis into the z direction as before (Fig. 8.35).

Special Capabilities of a Programming Language for Robots

In case one compares the assembly instructions for a human operator with programming instructions of a computer language, the following factors are of interest:

1. The transfer of the dimensions from the drawing to the actual assembly object is done by the operator by looking at the drawing and by translating this information directly into an action.
2. Missing information is supplemented automatically by the operator with the help of his or her experience. For example, the instruction "Assemble a Flange" suffices to perform the described operation. The operator searches for the flange, places it on the assembly object, and inserts the fasteners. Exact information about the position of the insertion holes is not necessary.
3. The operator performs automatically a sensor-controlled positioning operation. For example, if the operator enters a screw into a hole, he or she uses several sensors. A positioning is done with the aid of vision and fine positioning under the guidance of the touch sensors of his or her fingers. In case the thread of the screw does not engage with that of the bolt, the operator instinctively takes a correction action.
4. Missing routine tasks may be automatically supplemented for by the operator. For example, the insertion of a screw implies that a screwdriver has to be used and that the fastening process has to be done according to a given sequence of operations.

5. Fixturing needed during assembly may automatically be done by the operator without explicit instructions from the drawing.

It can readily be seen that many of the aforementioned tasks have to be programmed explicitly if a robot would perform the assembly. To obtain a fast and flexible programming system for robots, the following components should be provided:

1. A sensor for vision, force-torque, slip, proximity, and so on, should be installed in the robot.
2. The robot should be controlled by a run-time system which is able to adapt itself to on-line changes during assembly.
3. The compiler that translates the programmed work cycle should have a component which can generate missing information automatically.

These criteria imply that the robot should be controlled by a computer and that a higher programming language is available. To support this statement, several programming methods for industrial robots will be compared:

8.8.2 Programming Methods for Industrial Robots

All programming methods that will be discussed here need a facility to program trajectories. From the user point of view, the methods can be classified as follows:

Manual Programming

With this method the end points of the trajectories are set by hand. They are either hardwired or set through a plug board. Other functions are not provided. For this reason no control computer is needed.

Programming with the Help of the Brake System of the Robot

For programming purpose the brakes of the robot axes are disengaged and the end effector is brought by the operator into the desired work position and orientation. By depressing a function key, the coordinates of the individual axes are entered into the computer memory. To reproduce the desired trajectory, the coordinates are read back from memory and the computer calculates for the axes drives the required motion parameters. For this programming method, the robot must be provided with position decoders.

Sequential Optical or Tactile Programming

With this method the effector is lead by the operator through its trajectory. A guide handle with tactile or optical sensors is attached

Parts Handling Assembly

to the arm. A computer polls at request or continuously these sensors and reads along the traveled effector path the coordinates of all joints into the memory. With the aid of these sensor data the robot is capable to reproduce the required movement along the trajectory.

Master-Slave Programming

For this programming method a second, often small master robot is used to follow the desired work path. The movement is recorded via sensors and stored in memory. The coordinates are then transferred to the slave robot. Thus it is possible to program positions and trajectories.

Teach-in Method

A teach-in pendum (control box) is used to direct the robot along its trajectory (Fig. 8.36). At strategic points the coordinates of the robot joints are polled and entered into the computer memory. This is initiated by the actuation of a pushbutton. The pendum has control buttons for each robot joint. With these it is possible to lead the robot joint through any practical movement and to point the effector into any desired direction. More advanced teach-in systems allow the movement of the effector along the x, y, and z axes of a cartesian coordinate system. Additional features implemented in the pendum are:

Control of the movement velocity
Control of the movement time
Simple program branching
Setting of counters
Miscellaneous functions (control mode, weight compensation, single-step operation)

Textual Programming

These programming methods are similar to those used by data processing. Operations and data are described with the help of character strings. The user enters into the computer the sequence of instructions that describe the movement of the robot. The information is translated by the compiler into machine code. This can be done offline independently of the robot. However, sometimes it is not possible to program positions and orientations. These parameters have to be entered on-line by a teach-in method. For this reason, often no pure textual programming is done.

Acoustic Programming

With this method the program is entered via voice communication into the memory of the robot. Thus the programmer is free from the

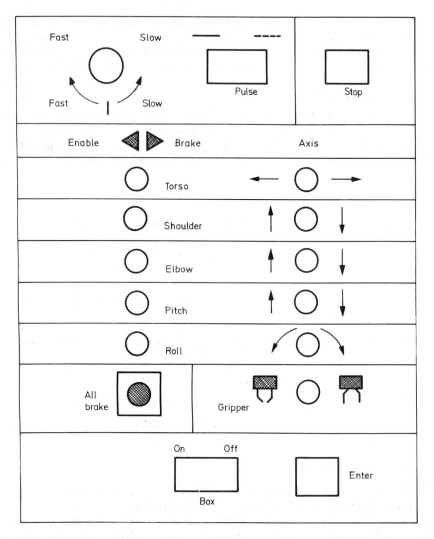

FIG. 8.36 Control box to program robots. (Courtesy of Computer Design, Littleton, Massachusetts.)

cumbersome process of writing down the instructions on paper. However, he still has to learn the vocabulary and the rules of the acoustic programming language. Programming by natural language, including its syntax and semantics, will be discussed later.

Design Considerations for a High-Level Language

The teach-in method and the textual programming language not only allow the description of spatial moves, but also permit branching, looping, and the use of subroutines. Presently, the most frequently used programming method of industry is teach-in. However, modern robots are increasingly being employed to handle complex tasks such as assembly work. Here, teach-in may become very cumbersome and there is an ongoing trend toward the use of higher-level languages. Several robot manufacturers offer a combination of textual programming and teach-in. A spacial point can be entered by defining it and by assigning a name to it. The name is used by the program. It is also possible to lead the effector to the spacial point and to enter in the teach-in mode with the aid of a function key an instruction into the control computer of the robot. This procedure is supported by an editor. The system enters the text of a motion instruction into the program. It is parametrized with a defined speed and the position of the point. With this method it is possible to program a complete sequence of movements and to obtain readable software. The program can also easily be changed. In the future, however, the textual method will predominate for programming of assembly work in the factory. It offers the following advantages:

1. The program is readable to programmers and other users.
2. The program can easily be changed and expanded by other programmers.
3. The program may have variables. During programming only a data type is assigned to it, not its value.
4. The program can be stored in readable form, which is important for documentation.
5. The program can be written off-line without the availability of a robot; no definition of the operating points is necessary.

In principle, the instruction set of a programming language for robots is similar to that of the teach-in method. Thus for each basic symbol of the language, a function key may be provided and installed in a keyboard. For simple assembly tasks, this programming method may render adequate service. However, textual programming in combination with teach-in is needed when the following conditions exists:

1. Complex assembly work is done which requires frequent program branching and subroutine calls.

558   Chapter 8

   2. The sensor signals of the robot have to be processed.
   3. External data have to be accessed (e.g., from a data file of
      the manufacturing system).
   4. An internal world model is used and its data are manipulated.

In addition, textual programming permits the implementation of special
strategies which assure an orderly and safe assembly. The following
operations may be included:

   1. *Measurement of a reference position during program execution*:
      For example, if the exact height of a workpiece is unknown, the
      effector can be lowered slowly until it touches the object surface.
      A force sensor records this event and this position of the work-
      piece is then entered into the computer with the help of an in-
      struction.
   2. *Supervision of an operation*: The success of an assembly is
      monitored by sensors to watch for misalignment, breakage,
      or missing parts. For example, the insertion of a pin into
      a hole can be monitored with a force-torque sensor. During
      insertion the effector may move slightly in the direction of the
      axes perpendicular to the centerline of the pin. A force signal
      in either direction will indicate that the pin is surrounded by
      its mating hole; otherwise, no hole is present. In the latter
      case, the robot will be alerted and a corrective move can be
      done.
   3. *Search strategies*: During bolting or mating operations, the
      workpiece may not be in its exact position. In this case the
      robot may try to locate the center of a hole by a search opera-
      tion. For example, it may try to target the object by a spirally
      shaped search procedure.

For the following reasons, the number of such special programming
instructions should be kept to a minimum:

   To reduce the assembly time of the object
   To save memory space
   To reduce compilation time
   To reduce the number of programming errors

8.8.3 Survey of Existing Programming Languages

There are numerous programming languages under development. A
ranking of different programming languages is shown in Fig. 8.37.
The majority of these are explicit assembly or compiler languages.
Here every move has to be described explicitly by the programmer.
The number of statements to program an assembly task will be quite

Parts Handling Assembly

numerous, even for a simple problem. Presently, there is only one implicit programming language known and that is AUTOPASS [21]. It has limited capabilities to describe simple assembly primitives. A typical instruction would be "PICK UP A BOLT, INSERT IT INTO A HOLE." There are three languages which are based on the NC language concept. They may be used in connection with loading and unloading of NC machine tools. In this case the programmer only has to be familiar with the NC language concept. The NC-type languages, however, become very complex when variables have to be defined and when sensor data derived from moving objects have to be processed. Several of the languages listed are conceived for universal applications; others are designed for a specific type of robot.

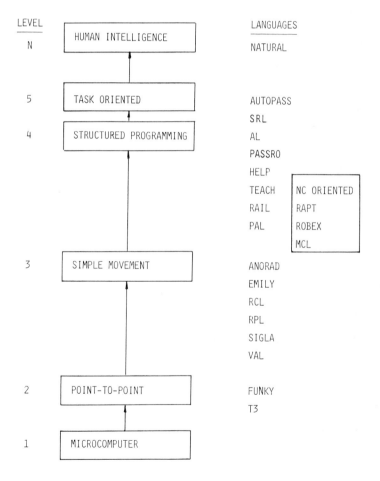

FIG. 8.37 Ranking of different robot languages. (From Ref. 20.)

> Teach-in programming
> Control structure
> Subroutines
> Nested loops
> Data types
> Comments
> Trajectory calculation
> Effector commands
> Tool commands
> Parallel operation
> Process peripherals
> Force-torque sensors
> Touch sensors
> Approach sensors
> Vision systems

FIG. 8.38 Features of programming languages for robots. (From Ref. 20.)

Figure 8.38 shows desired features of programming languages for robots. In addition to the constructs of conventional languages, there should be several specific to robots. For example, typical data types are vector, frame, or rotation. It should also be possible to describe to the robot an effector trajectory and how to handle the synchronization of the work of several arms. The robot must be able to operate the effector and the work tools under program control. In addition, there must be language constructs available which can handle sensor signals to which the robot is capable of reacting.

The many languages presently available suggest that the output of a compiler is a standard intermediate code (Fig. 8.39). In this case the robot manufacturer has to lay out its control system in such a manner that the interface of the controller accepts the intermediate code [22]. Thus it is possible to use different languages for different robots via a standardized interface.

8.8.4 Concepts for New Programming Languages

Presently, there is a considerable amount of research work done to develop new programming languages and systems for robots [23]. Two different development trends can be observed:

Parts Handling Assembling

1. The individual robot is made autonomous. It obtains the capability to adapt itself to the work environment, to make its own decisions, and to take actions to solve unforeseeable situations. In other words, the robot will be provided with an amount of limited intelligence which is needed to perform its assembly task.
2. The entire manufacturing process will be completely automated to eliminate the necessity for human intervention. The robot becomes an integral part of a manufacturing facility which is supervised by a hierarchy of computers. The description of the workpieces and that of the assembly is automatically generated from the design process and transferred from the CAD system to the control computer of the robot. There is no need to program the robot directly and to describe to it the geometrical shape of the object, its surface description, gripping position, and so on. This information is all available from a central CAD data base.

Both of these developments will find their application in different industries:

1. The autonomous and flexible robot, equipped with sensors and a routine system, will be used by small and medium-size companies. This system can be adapted and reprogrammed quickly to new production runs. It may make its own decision with the aid of an expert system (Fig. 8.40). The drawback of this robot is its complexity and the unavoidable high development effort.

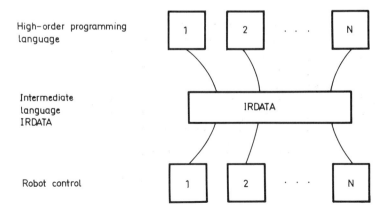

FIG. 8.39 Use of an intermediate language.

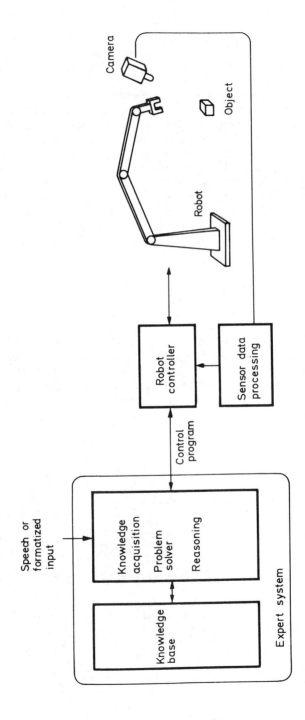

FIG. 8.40 Future autonomous robot.

Parts Handling Assembly

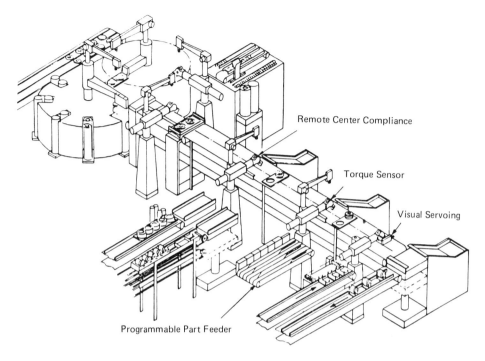

FIG. 8.41 Adaptable programmable assembly system. (From Ref. 24.)

2. Systems with several robots, where each unit performs a specific task, will be used in mass production; they will be the domain of larger companies. Such an integrated manufacturing system will include machine tools, robots, material handling, peripherals, and conveyors (Fig. 8.41). The investment cost will be quite high. With increasing product diversity the intelligent robot may also be installed in flexible manufacturing systems which can handle many product variants.

Independently of these two developments, several high-level programming languages have been developed to solve assembly tasks. Most of them use the frame concept, which was described earlier. However, their use requires that the programmer have the capability to view the assembly object in a three-dimensional space. For example, the programmer must be able to visualize the permissible path of an effector to move from one point in the assembly space to another. Typical features of a high-order programming system will be explained for the SRL language developed by the University of Karlsruhe [25].

TABLE 8.6 Some Robot Language Features

	SRL	AL	VAL
Move by joint specification	DRIVE	DRIVE	DRIVE
Move by frame expression	Yes	Yes	No (only predefined frame values)
Relative move	Frame + frame Frame * frame Frame + vector	Frame + frame Frame + vector	DRAW (without changing the orientation)
Move specification by:			
Velocity	VELOCITY or V	SPEED-FACTOR	SPEED
Duration	DURATION or D	DURATION	
Robot posture	POSTURE	—	RIGHTY, LEFTY, ABOVE, BELOW
Accuracy of control	ROUGH FINE	NO-NULLING NULLING	COARSE FINE
Acceleration	ACCELERATION	—	—
Constant/variable orientation during the move	CONSTORIENT	—	—
Via points of trajectory	VIAFRAMES	VIA	CP
Approach/deproach	APPRO DEP	APPROACH DEPARTURE	APPRO DEPART

Feature	Language 1	Language 2	Language 3
Sensor data can control robot moves	All kinds of sensor data	FORCE TORQUE	—
Reacting to events	ALWAYS WHEN	EVENT-variable in move statement	REACT (IGNORE)
Multitasking	Yes, by SECTIONs	Only parallel blocks	—
Parallel task execution	PARALLEL	COBEGIN	—
Cyclic task execution	EVERY	—	—
Time delayed task execution	AFTER	—	—
Program flow control: Jump	—	—	GOTO
Conditional jump (on interrupt)	IF, IFSIG	IF	IF, IFSIG
Loops	FOR, WHILE, UNTIL	FOR, WHILE, UNTIL	—
Distributor	CASE	CASE	—
Subroutines	PROCEDURE	PROCEDURE	Other programs
Standard data types Numerical	INTEGER, REAL, BOOLEAN	SCALAR	INTEGER
Geometrical	VECTOR, ROTATION, FRAME	VECTOR, ROT, FRAME, TRANS	FRAME (without arithmetic)
Structured data types	ARRAY, RECORD	ARRAY	—

The SRL Language

The new language SRL was designed as a result of an extensive study of AL [27], PASCAL [28], and VAL [26], which was implemented on an industrial hardware configuration and served as the main research tool to conceive the new language. To obtain a very general language for a wide field of applications, several existing languages for industrial robots were also compared. As a result of this research work, the SRL language was designed and implemented with the following features:

1. The general and structured data concept was taken from PASCAL.
2. The geometric data types vector, rotation, and frames are described by predefined RECORDs.
3. Facilities are provided for parallel, cyclic, or time-delayed execution of tasks (called SECTIONs).
4. There is general structure for compound statements to distinguish between:
   a. A syntactical compound statement (BEGIN . . . END).
   b. A block or procedure with its own variables (BLOCK . . . END_BLOCK, PROCEDURE . . . END_PROCEDURE).
   c. A time compound statement for the internal sequential execution of statements, but parallel execution to other program parts (SECTION . . . END_SECTION).
5. Input/output to digital or analog ports and sensors.
6. System components such as robots, sensors, or interrupts are specified.
7. There is a general sensor interface for structured sensor data.
8. There are several move statements for different kinds of interpolation and trajectory calculation.

The robot language SRL was designed hardware independent, which is an improvement over AL. SRL allows very flexible programming of a robot and the adaption to its environment. Therefore, it includes a task concept and elements for reading sensor data. In the following, some features of SRL are explained in more detail. The reader should be familiar with PASCAL and should have some knowledge of AL to understand the subject matter fully.

Part. The language SRL includes a system specification part for adapting a program to different sensors, robots, effectors, absolute addresses, ports, interrupts, and data bases. With the help of the system specification the programmer can write programs which are more or less hardware independent, self-documenting, and portable. If a program does not have a system specification, the programmer has to use the predefined identifiers such as ROBOT or CHANNEL.

Parts Handling Assembly

If the programmer specifies a sensor, the system creates a variable with the symbolic sensor name. The type of the variable has to be declared with respect to the type of the sensor data read in from the run-time system. An example will demonstrate the effect of the SENSOR description. Let the PUMA be connected to a vision system. The input data of this vision system is read in from channel 3 by the operating system with the help of a special communication procedure. That means that the communication protocol is not specified by SRL. The interpreter stores the incoming data as an integer number for the workpiece which the vision system has identified. The positional x and y coordinates are stored as real numbers. The sensor specification is as follows:

```
SENSOR:  visioninfo = CHANNEL (3);
         STRUCTURE visioninfo = RECORD
                         part: INTEGER;
                         x:    REAL;
                         y:    REAL;
                         END;
```

An input from the vision system is initiated by the program as an INPUT statement. Thus the programmer can use the sensor name as a normal variable in an expression or, for example, in a conditional statement:

```
INPUT (visioninfo);
IF visioninfo.part = part3 THEN SMOVE puma TO box3;
ENDIF;
```

Figure 8.42 shows the memory management for processing a sensor variable.

As in AL the language SRL includes powerful arithmetic facilities for explicit computations of geometrical entities or trajectories.

Multitasking. Besides traditional program flow control and blocks, SRL has new and powerful language elements for multitasking. With these language constructs, parallel, cyclic, and time-delayed tasks can be handled easily. The PARALLEL statement effects parallel (or quasi-parallel in a monoprocessor system) execution of all statements or tasks within the PARALLEL statement.

The EVERY statement is for cyclic execution of statements for every specified number of milliseconds within a program or a task. Cyclic execution is initialized when the EVERY statement is executed the first time during the normal program flow and it lasts until the outer task or program is finished.

If execution of statements is not cyclic but time delayed, then in SRL the AFTER statement is used. Time count down is started

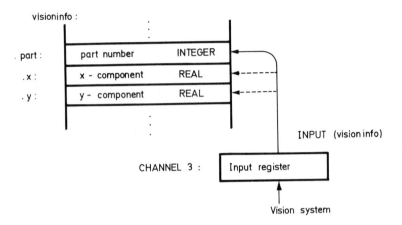

FIG. 8.42 Sensor integration by SRL.

after execution of the AFTER statement. Then the execution of the statements within the AFTER statement will be started after the specified time, except when the outer task or program has been finished. For integration of sensor data or when reacting to external signals the WHEN statement is used. Also, time conditions or logical expressions can be monitored by this statement.

If during the execution of the specified section the given condition becomes true, the statement of the DO part will be executed. The condition can be either an interrupt or a logic expression. When writing a logic expression the programmer can specify the trigger time that the system should use to monitor the condition. A sensor threshold can be specified using the INPUT statement in an EVERY statement before the WHEN statement.

EXAMPLE:

```
    EVERY 100 MS START read_tactilesens WITH PRIO = 5;
    WHEN tactilesens.xaxis > 50 MONITORED EVERY 100 MS
    DURING
            SMOVE puma TO table;
```

Parts Handling Assembly

```
    END DURING
DO WITH PRIO = 1
    STOP puma;
```

Every 100 msec the value of a tactile sensor is read in. During the move of the Puma every 100 msec, the value of coordinate of the x direction of the sensor is compared with the value 50. If it is higher than 50, the puma stops its move.

The specification ALWAYS effects that the statement of the DO part is executed every time when the condition becomes true (after it previously had changed to the value false) or if an interrupt occurs. With ALWAYS WHEN the programmer can install an interrupt service routine.

For optimal time scheduling of the system task the programmer can specify the trigger time; this condition is monitored. After this time delay, the system will check if the condition is true or not.

Move statements. To distinguish between different types of interpolation, SRL includes several move statements. The programmer is able to define off-line exactly a differentiated robot move and can specify many move parameters such as velocity, duration, or via points. Reactions to sensor conditions or data during a move are handled by the WHEN or ALWAYS WHEN statements.

PTPMOVE	Movement without any synchronization between the robot axis. Each axis is moved with maximum acceleration and speed. No general specifications are allowed.
SYNMOVE	Linear interpolation in robot joint coordinates (i.e., all axis will be synchronized). General specifications possible.
SMOVE	Movement on a straight line by linear cartesian interpolation. General specifications allowed.
LANEMOVE	Trajectory calculation by polynomials, similar to the MOVE statement of AL. General specifications allowed.
CIRCLEMOVE	Movement along circular segments. Specifications: center point, end point, angular displacement, or an angle, velocity or duration, fine/course interpolation, and positioning.
MOVE	Move statements with interpreter or controller-dependent parameter specifications. This statement

DRIVE   can be used for all future types of interpolation if the controller includes the control modules.
Movement of one or more robot axis. Specifications: velocity or duration, force, fine/course positioning.

Data and world model. To overcome hardware dependence and to support structured and self-documenting programming, SRL includes the above-stated language constructs. As a new facility, SRL has an interface to a general world model at program run time. The world model can contain data about objects and its attributes, such as workpieces, fixtures, robots, frames, and trajectories.

The goal of the development of SRL is the design of a language that can easily be learned and adapted for further development and application. It will also provide an interface between future planning modules and the "traditional" programming system. A planning module will be used to generate SRL statements from a task (goal)-oriented specification. This will replace explicit programming for every single action (Fig. 8.43). Therefore, SRL is well structured and universal in nature, and it includes all features of robot programming and process control. The standard data types and the RECORDS as structured data types defined by the programmer are from PASCAL. There are new data types added to improve handling of synchronization between the program and external events. Predefined records can be used for geometrical computation needed for robot moves. The standard data types of SRL are:

INTEGER	From PASCAL
REAL	From PASCAL
BOOLEAN	From PASCAL
CHAR	From PASCAL
VECTOR	From AL
ROTATION	From AL
FRAME	From AL
SEMAPHOR	For synchronization of parallel tasks
SYSFLAG	For synchronization of parallel tasks

The programmer describes textually with a frame in SRL notation the position and orientation of the robot effector. It consists of a position vector and a rotation. The rotation can be defined by one or several rotations and it is stored as a 3 × 3 rotation matrix. The user has access to the matrix.

A semaphore is used for synchronization and queueing of tasks within a program. The system flag SYSFLAG is introduced to synchronize programs. The programmer has no direct access to the data type SEMAPHOR or SYSFLAG but can use the statements SIGNAL and WAIT to handle them. Another fundamental feature of SRL is the

Parts Handling Assembly

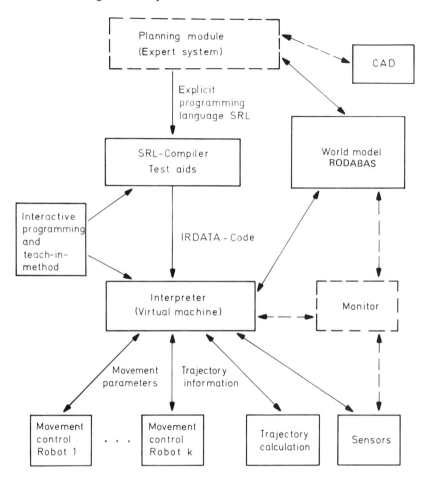

FIG. 8.43 Hierarchical programming structure for industrial robots.

language PASCAL. The data concept and file management are taken from Pascal because it gives the user a very flexible and problem-oriented data structure.

SRL includes the structured data types ARRAY, RECORD, and FILE of PASCAL. Furthermore, the programmer can define his or her own problem-oriented data types as in PASCAL by enumeration and subrange. There are also pointers included in SRL. With respect to data types, the programmer can write records and its components in any expression as in PASCAL. Example: Save_distance: = posvector.x + tolerance;

8.8.5 Programming with a Natural Language

Natural language programming systems are being developed to aid untrained personnel to teach the desired movements of a task to a robot (Fig. 8.44). Because of the complexity of a natural language, its entire vocabulary cannot be used. Usually, simple syntax and semantic are selected which allow to describe the task of the robot by a quasi-natural language. The programming system performs the syntactical and semantical analysis of the speech input, extracts the pertinent information, and makes a plausibility check. Thereafter the corresponding formal robot program is executed. These systems need large memory and require long execution times.

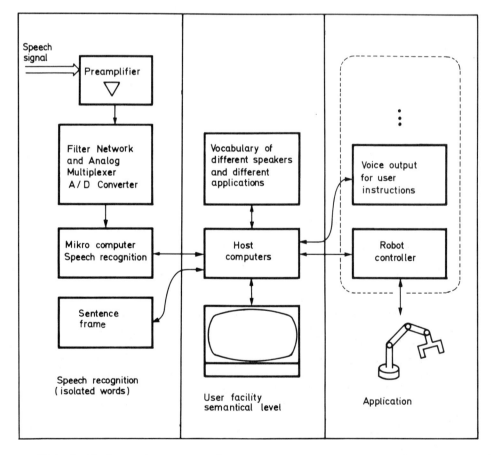

FIG. 8.44 Acoustic programming.

Parts Handling Assembly 573

8.8.6 Implicit Programming Systems

Systems to program robots automatically have been under development since the 1970s. Here the programmer does not need to formulate explicitly every instruction of a task (e.g., "MOVE ARM TO POS. 1"). However, the programmer gives task-oriented instructions (e.g., "FASTEN FLANGE WITH 4 BOLTS"). The system tries to interpret this instruction and plans its execution. The system searches in its library for different operators which will perform the required robot actions. Starting with the initialization state, each succeeding state is planned until the finish state has been reached. The result of this search is a sequence of operators and states which can be visualized as an operation plan. This plan is equivalent to a program obtained from a programming language. A system capable to set up automatically such a plan is called a problem solver [29]. The individual steps to assemble the flange may be as follows:

1. First the flange has to be recognized.
2. Then the effector picks up the flange and places it on the mating part.
3. Now, a check is made with a sensor for proper alignment.
4. The next steps are to fasten a screwdriver to the effector, to locate a bolt, to pick it up, and to insert it into a bolt hole of the flange. To assure proper tightness, a torque sensor supervises the fastening operation.
5. Next, the other three bolts are inserted.
6. In the last step the presence of all four bolts is verified by a vision system.

8.8.7 Programming Aids

In addition to the language there must be a powerful programming system available, consisting of several software packages and of a low-cost program development computer. Figure 8.45 shows a comprehensive programming system for assembly robots. The user describes to the robot the object and the workplace with the help of an application-oriented language. This information is processed by a geometry processor and entered into a world model. Similarly, the movement of the robot is functionally described by implicit instructions and a syntactical analysis is performed. This program is combined with information from the world model. The result is sent to the SRL compiler. It is also possible to communicate interactively with the SRL compiler to enter or edit instructions. The output of the SRL compiler in form of interpretative code is loaded down to the control computer of the robot. Sensor signals from the robot can be brought back to the sensor data processing module. In case an object or a workpiece had

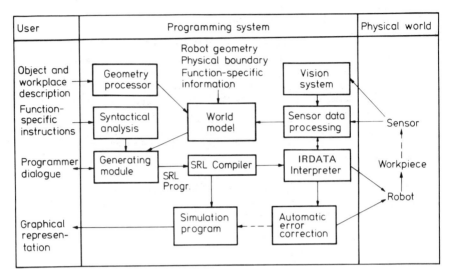

FIG. 8.45 Advanced programming system SRL.

changed its position, this module will send instructions to the world model to update it. The same information is needed by the interpreter to correct the movements of the effector. There is also a simulation program available which allows the programmer to display graphically the work environment of the robot, to check its movements, and to detect possible collisions.

The graphical emulation system is part of the programming system or of the real-time controller (Fig. 8.46). In the early stage of the robot application, the kinematic attributes (joints, links, end effectors), the assembly cell, and its environment can be described on a graphic display. Trajectory planning, the interpolation in cartesian coordinates, and the corresponding coordinate transformation can be tested and optimized. By adding a program for the simulation of the robot's dynamics, the response of the axis motor drives and their control can be traced, to evaluate the dynamics of the robot. For debugging of assembly programs, the emulated robot is interfaced to the programming system, which defines multiple moving tasks. With an off-line program test facility, the workpiece and the robot components can be emulated without the risk of collision. When it is certain that all assembly sequences are performed without conflict, the program can be transferred to the robot control computer for execution in real time to move the mechanical manipulator. Verification of the assembly can be performed in this stage. Figures 8.47 and 8.48 show how a robot is constructed from basic components with the aid of a simulator.

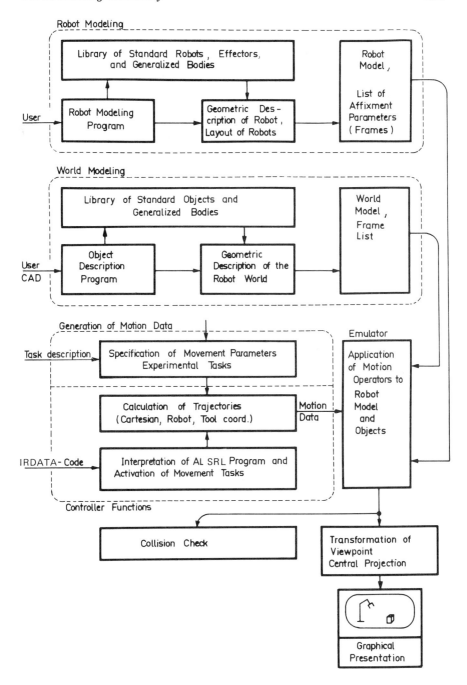

FIG. 8.46 Graphical simulation system for robots. (From Ref. 30.)

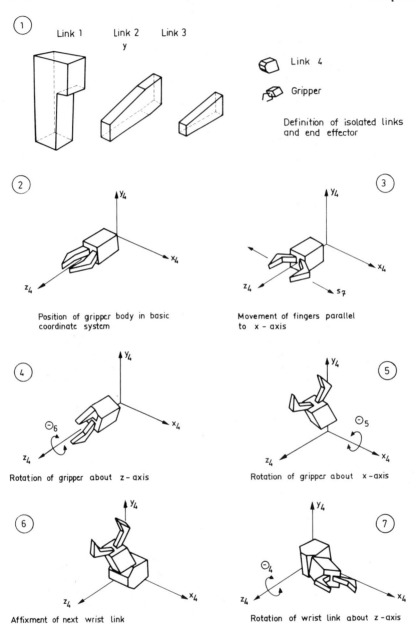

FIG. 8.47 Modeling and composition of basic arm elements.

Parts Handling Assembly 577

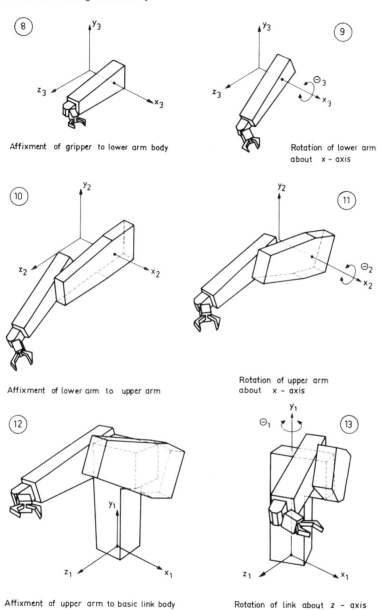

FIG. 8.48 Successive composition of a PUMA 600.

8.8.8 Components of a Programming System for Robots

Nearly all programming methods for industrial robots require software components such as an editor, an interpreter, or a compiler. These usually quite complex software tools form the programming system of the robot. A typical system may include all or only some of the following system programs:

1. Operating system(s) of the robot control computer
2. Interpreter
3. Compiler
4. Editor
5. Frame editor
6. Test aids and simulation programs
7. Geometric processor
8. Interactive components to program special move points by the teach-in method

The specific components needed for the system depend on the underlying programming language or the programming method. The components are described briefly below.

Operating System for a Robot

When a special program development computer is used, for example to run a compiler or a simulation program, the normal operating system of this computer may be used. It is well structured and there is no need to change or extend it for robot programming. The operating system implemented on the control computer of the robot includes additional software components for robot programming. They are:

The programming or positioning module
Modules for coordinate transformations
The robot control module
A module for trajectory calculations
A monitor to observe collisions
An interpreter for user programs

The manual programming method needs no control computer. Programming with the help of the brake system of the robot is a very simple procedure. It only needs components for reading the joint coordinate values and for storing them on an external memory such as a floppy disk. All other programming methods require a positioning module as part of the programming facility (Fig. 8.49). Here the robot is moved to its work positions under control of the program. In smaller

Parts Handling Assembly

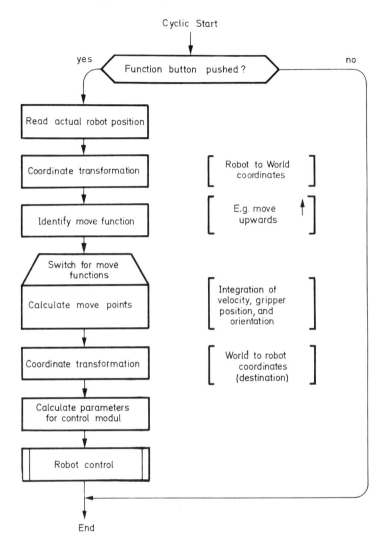

FIG. 8.49 Executing a teach-in-move function.

systems teach-in is done using robot coordinates. More comfortable systems include track instructions in cartesian coordinates.

Move and collision monitoring is performed during programming as well as during the execution of the user program. The controlling of this activity are:

Limits for joint motions with respect to the robot geometry
Emergency stop
Limits for velocity and acceleration
Safety space
Provisions to avoid collision with other objects

If a safety threshold is reached, all robot motors are stopped immediately under the highest priority.

Systems for Pure Move Programming

For controlling the robot's moves, an operating system includes modules for positioning, move interpolation, coordinate transformation, robot control, and perhaps collision monitoring. It operates in two modes:

1. *Positioning mode*: defining the motion points, respectively, the frames
2. *Program mode*: execution of the stored sequence of trajectory points, respectively, the frames

The user can switch between the two modes.

Systems for the Teach-in Method

When using the teach-in method a command mode is necessary to handle the command input via the teach box (e.g., start or stop a user program) (Figs. 8.50 and 8.51). Additionally, many systems allow to define the control mode, such as point-to-point control, path control, and coarse/fine interpolation. Also, numerical parameter values, such as velocity or loop counter, may be entered manually.

Systems for Textual Programming

Programming languages for robots based on BASIC do not need a support computer, but they require an editor for the control computer to perform a syntax check. The operating system contains a module for file input/output to store the edited and coded user program. Figure 8.52 shows the system structure for such a programming language.

If the compiler is implemented on a support computer (e.g., a VAX 750), its editor is used to edit and store the user robot program (Fig. 8.53). In addition, the compiler needs a fast external memory

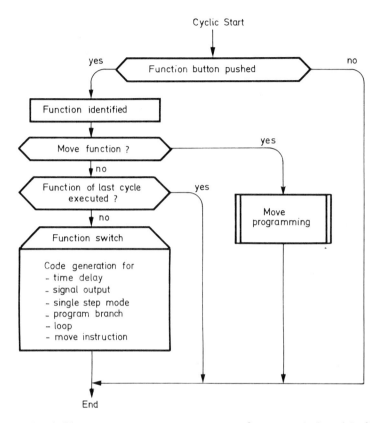

FIG. 8.50 System program structure for a teach-in with function button.

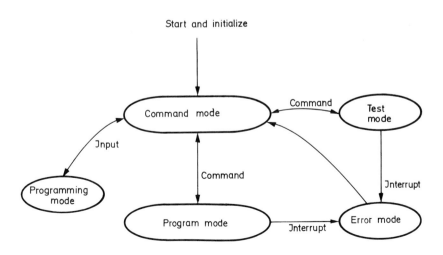

FIG. 8.51 Modes for a programming system with teach-in.

582   Chapter 8

to aid the syntactical analysis and to build up the syntax tree. The operating system can be enhanced with a geometry processor or simulation aids. Both the operating system on the support computer and that on the control computer may include interface modules for a computer network. Thus the code produced by the compiler can be loaded directly into the control computer for execution. Otherwise, a program transfer with the help of an external memory is necessary (e.g., a floppy disk, a magnetic cassette, or a punched tape as used for NC technology).

Interpreter for Robot Programs

An interpreter can be visualized as an abstract machine whose language is the code generated by the compiler. After decoding a code record, the interpreter initiates the performance of the required robot actions. For this it may also use service routines of the operating system. As a matter of fact, in most implementations one cannot distinguish between the operating system and the interpreter. Therefore, the interpreter can be regarded as a stand-alone system,

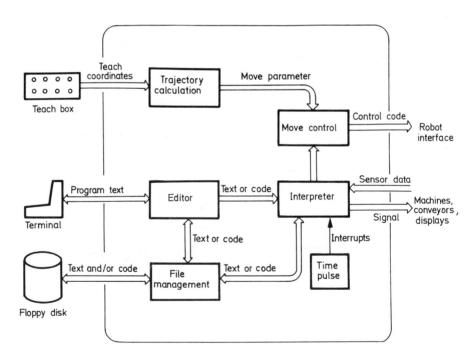

FIG. 8.52 System components for a BASIC-like robot programming language.

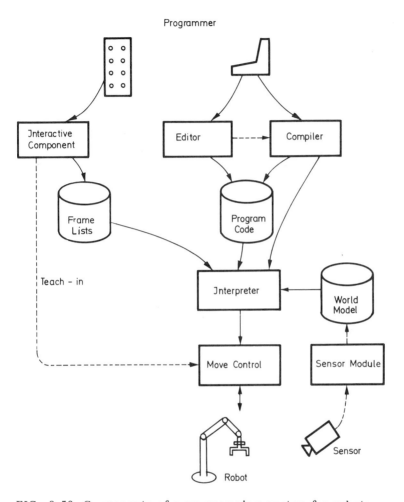

FIG. 8.53 Components of a programming system for robots.

including drivers, coordinate transformation routines, and robot control algorithms. When a high-level robot programming language is used, the interpreter in general handles the assignment of variables, execution of arithmetic expressions, parameter insertion, and block nesting with the help of a stack.

Compiler

A compiler for a robot programming language differs in principle little from a compiler for a "normal" computer language. It has the following basic structure:

Lexicographical analysis (scanner)
Syntactical analysis (parser)
Semantic analysis
Code generation
Code optimizing
Error handling and listing

A compiler used for numerical control can calculate the tool or robot trajectory. The user specifies geometric features or relations. From it the compiler builds up internally a geometric world model to extract the explicit values of the robot movement. At the present there are developments under way to extend the world model and the compiler to generate explicit statements. In this case the compiler is divided into two modules: (1) for normal compilation and (2) for the planning tasks.

Editor

In general, the editors for robot programming languages are similar to those of other languages. In some industrially used languages such as VAL the editor includes an extension for a teach mode. During this mode the programmer can teach the gripper position and orientation (frame) to the robot. After pressing a function button RECORD, the editor reads the robot coordinate values and generates the text for an explicit move instruction for this trajectory position. Thus the user can program long move sequences by the teach-in method and obtains a readable program text which can be edited if necessary.

8.9 INTERFACING OF A VISION SYSTEM WITH A ROBOT

For material handling and assembly work, a vision system may have to be used. The system has to locate the part, identify it, direct the gripper to a suitable grasping position, pick up the part, and bring it to the work area. Frequently, this work has to be done on a moving conveyor. Both, the camera and the robot have a unique coordinate system. The camera identifies the object in regard to its own coordinate system. However, for grasping, the robot must know where the object lies in reference to its own coordinate system. This requires the transfer of the object's location from the camera to the robot coordinate system. The following two mechanisms can be used for coordinate transformations: (1) with the use of a high-level programming language for the robot and/or for the vision system, or (2) without the use of a robot programming language.

Parts Handling Assembly

8.9.1 Robot Calibration with the Use of a High-Level Programming Language

Most vision systems have three features which facilitate the integration with a robot programming system:

1. Parts to be recognized are described by user-defined symbolic names which are represented by ASCII strings.
2. Position and orientation of the workpiece are determined relative to a cartesian coordinate system.
3. Parts to be recognized are taught to the system by "showing." The symbolic name is also entered.

Identification of parts is done with the aid of a feature vector, including area, number of holes, minimum and maximum diameter, and perimeter. Normally, the user does not have to enter these parameters. The system automatically generates the feature vector for a part during teach-in. When the program is executed, the system compares the stored feature vector with that produced from the actual image and tries to identify the part.

The VAL extension VAL 11V includes several commands for a service program which calibrates the camera, learns the identity of the part, and stores the feature vector under a symbolic object identifier. The user program addresses the vision system with the two additional VAL instructions VPICTURE and VLOCATE. The result of VPICTURE is that of taking the picture and storing its pixels in the image buffer. VLOCATE results first in a search for a given object in the image buffer. In case the search is unsuccessful, the program flow branches to a special label or the program execution stops and an error message is printed. In case of a successful search, the vision system stores the position and orientation under a frame identifier which is the same as the name of the object [17].

The coordinate values of the object frame refer to the cartesian coordinate system of the camera and to the base coordinate system of the robot. Figure 8.54 shows schematically the interaction between the vision system and the robot program written in VAL 11V. The user can easily calculate with the aid of relative frames the origin of the camera coordinate system in reference to that of the robot. A frame describes with the help of a vector and a rotation matrix the position and orientation of the gripper or of the tool of the robot. By just looking at the position vector, a relative frame defines that position, which is obtained by adding the position vector of the frame to the vector of the base frame. The calculation of the relative orientation is done similarily by a matrix multiplication.

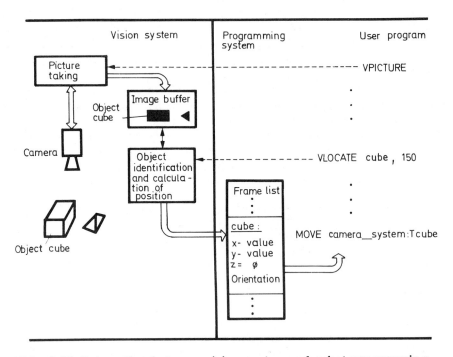

FIG. 8.54 Interaction between vision system and robot programming system (VAL 11V).

For calibration a small disk or ring is placed into the field of view of the camera. Then the gripper is moved under the teach-in mode to the center of the disk. Here the position and orientation are recorded and stored under the frame with the name calib_robot. Now, the robot is moved out of the field of view of the camera. Thereafter the following simplified program is executed to calculate the origin and orientation of the camera coordinate system with respect to the robot coordinate system.

```
HERE  calib_robot
MOVE  home_position (*robot out of field of view*)
VPICTURE
VLOCATE  calib_camera, 150
INV  inv_calib_camera = calib_camera
SET  camera_system = calib_robot : inv_calib_camera
      .
      .
      .

VPICTURE
VLOCATE  object, 150
MOVE  camera_system : object
```

Parts Handling Assembly

Figure 8.55 shows the geometric representation of the relative frames which are necessary for the calculation. For simplification only the position vectors are shown. A similar calculation is made for the orientation of the frame.

The HERE instruction defines the center of the disk as the frame calib_robot with respect to the robot coordinate system. VLOCATE defines the center of the disk as the frame calib_camera with respect to the camera coordinate system, which means that the two vectors point at the same point in space. The frame calib_camera is inverted. The result is the position vector inv_calib_camera, which now points from the center of the disk to the origin of the camera coordinate system. The relative frame calib_robot:inv_calib_camera indicates the position and orientation of the camera coordinate system with respect to that of the robot. The vector of the frame points from the origin of the robot coordinate system to that of the camera. To simplify the notation, the result is assigned to the frame camera_system. After these calculations have been performed, all following object frames that are determined by VLOCATE are referenced by the MOVE statements relative to the frame camera_system. This example emphasizes

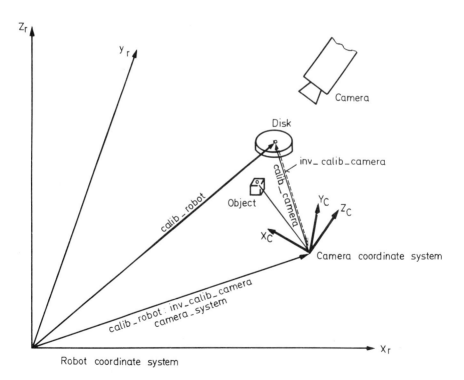

FIG. 8.55 Geometrical representation of relative frames for the transformation of the camera coordinates into robot coordinates.

the necessity of the frame concept and of geometrical operators for vision system applications.

8.9.2 Interfacing of a Vision System with a Robot Without the Use of a Programming Language

To explain the transposition of the workpiece parameters from the coordinate system of the camera to that of the robot, we use Figs. 8.56 and 8.57 [31,32]. A typical vision system is shown in Fig. 8.56. To simplify the pictorial, a photodiode matrix consisting of 64 × 64 pixel units was selected. The length of one pixel is denoted by one PIX. Present cameras have a resolution of 256 × 256 pixels or more. In general, the vision system has its own computer, which determines from the two-dimensional image of the workpiece its identity, location, and orientation. The last two parameters have to be transferred to the control computer of the robot. It in turn will instruct the control circuit to grasp the object and bring it to a predetermined location.

Now we select for our example a simple robot with two translational and two rotational axes (Fig. 8.57). These axes determine the operating range of the robot. The movements in direction of the lift and

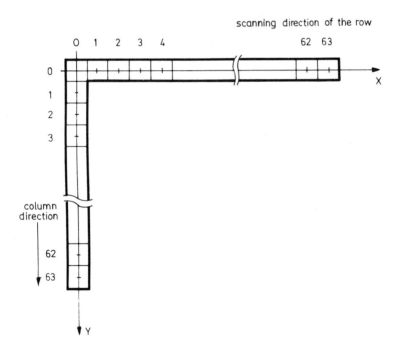

FIG. 8.56 Cartesian coordinate system of a vision sensor.

Parts Handling Assembly

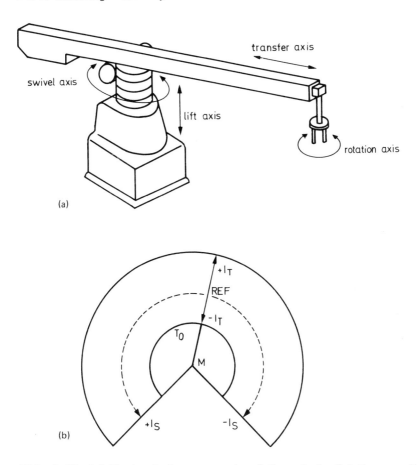

FIG. 8.57 (a) Mechanical components of the robot; (b) the coordinate system of the robot. I_T, increments of the transfer axis; I_S, increments of the swivel axis; REF, reference position.

transfer axes of the robot can be represented in terms of number of unit increments of the measuring scales of these axes. The movements of the swivel and rotational axes about their centers can be expressed in units of angular increments of their decoders. The transformation of the workpiece coordinates from the camera coordinate system to that of the vision system requires trigeometric calculations. These can be quite involved with a complex multiaxis robot. For our sample robot, these calculations are straightforward.

Three aspects must be considered when a vision sensor is interfaced with the industrial robot: the physical interface, the data transfer protocol, and the coordinate transformations. To data there is no standard practice available to solve this problem. The designer of such a system therefore relies on his or her own intuition. The designer is limited by the capability of the sensor system and that of the control unit of the manipulator. A physical interface typically can be realized by an RS-232 interface.

For transformation of the workpiece coordinates from the sensor to the robot system, the following concept may be used. For each workpiece known to the sensor system, an action program is assigned by means of a program number. The grasping position of the workpiece, which usually does not correspond to its center of gravity, is calculated. This is done by bringing the gripper under hand control to the grasping position and by transferring the coordinates of this position to the sensor computer. From this coordinate axis the grasping position is calculated in reference to the sensor system. With this information the positional values are determined which are needed for the calculation of the grasping position in the work mode.

During the work mode, the control unit of the robot is waiting for the transfer of the grasp coordinates. After a workpiece is identified by the sensor, first the grasp position has to be calculated in reference to the sensor system, and then this grasp position is transferred to the robot's coordinate system. The coordinate axes obtained by this transformation are transferred to the robot control unit together with the program number of the action program which was assigned to the workpiece in the teach-in phase. The action program, modified by the new coordinate axis, can then be processed by the robot control. The geometric setup of the robot-vision system is shown in Fig. 8.58.

Coordinate Transformation for the Teach-in Phase

The teach-in phase of the coupled sensor system uses the following procedure:

1. Teaching the workpiece to the sensor system
2. Determining the grasp position with the help of the robot
3. Planning an action program for handling of the workpiece

In the following paragraphs only the calculation of a grasp position is described. The values that relate to the learning phase are marked with the subscript L. To do this we start with the setting of the robot/vision system shown in Fig. 8.58. The location of the corresponding cartesian sensor coordinate system with regard to the robot coordinate system is shown in Fig. 8.59. The sensor coordinates are described by X and Y and the robot coordinates by T and S. In the robot system the grasp position G is described by the position of the transfer axis T_G and the position of the swivel axis S_G when the

Parts Handling Assembly

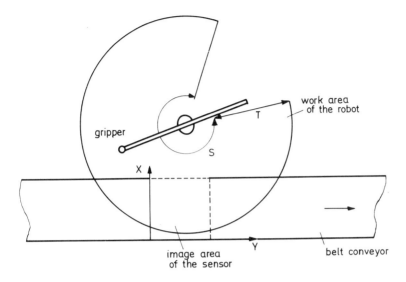

FIG. 8.58 Geometric setup of the robot/vision system.

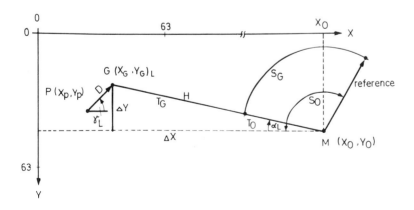

FIG. 8.59 Location of the sensor coordinate system in regard to the coordinate system of the robot.

gripper is above the grasp position. The distance H between G and M can be calculated with regard to the robot system:

$$H = T_G + T_0 \quad \text{increment units of the transfer axis } (I_T)$$

in reference to the sensor system:

$$H_S = \frac{H}{F_T}$$

The constant T_0 describes the distance between the gripper and the center of rotation of the swivel axis when the axes are at the reference point. It is measured in number of increment units of the transfer axis. The constant value F_T represents the ratio of the transfer increment units divided by the length of one pixel (I_T/PIX). The orientation of the robot arm in regard to the x axis is $\alpha = (S_0 - S_G)/F_S$ (degrees). The constant value S_0 describes the position of the swivel axis when it is parallel to the sensor x axis. The constant value F_S shows the number of increments of the swivel axis per degree of angle (I_S/deg). The coordinates of the point G in regard to the sensor-coordinate system can be calculated from

$$\Delta X = H_S \cos \alpha \quad \text{(PIX)}$$
$$\Delta Y = H_S \sin \alpha \quad \text{(PIX)}$$

Thus the following results are obtained:

$$X_G = X_0 - \Delta X \quad \text{(PIX)}$$
$$Y_G = Y_0 - \Delta Y \quad \text{(PIX)}$$

in which the coordinates X_0 and Y_0 describe the center of rotation of the swivel axis in regard to the sensor system. Thus the grasp position of the reference position of the workpiece is known. During the work phase the different positions of each workpiece have to be calculated on the basis of the reference position. For this task the distance D between the grasping position $G(X_G, Y_G)$ and the center of gravity $P(X_P, Y_P)$ and also the workpiece orientation γ are calculated during the teach-in phase. The distance D is obtained from

$$D = \sqrt{(X_P - X_G)^2 + (Y_P - Y_G)^2} \quad \text{(PIX)}$$

and the workpiece orientation is

$$\gamma = \arctan \frac{Y_P - Y_G}{X_G - X_P} \quad \text{(degrees)}$$

Parts Handling Assembly

For the calculation of the grasp position and for its transformation to the robot coordinate system, the following additional values have to be stored:

1. Distance between the grasp position and the center of gravity (PIX)
2. Orientation of the workpiece γ_L (degrees)
3. Angle α_L (degrees)
4. Number of the action program N
5. Position of the vertical lift axis H_L (I_L increments of lift axis)
6. Position of the rotation axis R_L (I_R increments of rotation axis)

Coordinate Transformation During the Operating Phase

During the operating phase the following tasks are carried out:

1. Calculation of the center of gravity $P_A(X_P, Y_P)$
2. Identification of the workpiece
3. Calculation of the rotational position δ of the workpiece
4. Calculation of the grasp position with regard to the sensor and $G_A(X_G, Y_G)$
5. Transformation of the coordinates of the grasp position to the coordinates of the robot system

The values obtained during the operating phase are marked by the subscript A. Figure 8.60 shows the interrelation of the different coordinates and the important system parameters during the operating phase. After identification of the workpiece and after calculation of its rotational angle δ with regard to the reference position, the grasp position of the workpiece with regard to the sensor position has to be calculated.

For the workpiece orientation γ_A one obtains

$$\gamma_A = \gamma_L + \delta$$

in which γ_L is the workpiece orientation and δ the rotational angle of the workpiece, both in regard to the reference position. The grasp position $G_A(X_G, Y_G)$ is then calculated from

$$\Delta X = D \cos \gamma_A$$
$$\Delta Y = D \sin \gamma_A$$
$$X_G = X_P + \Delta X$$
$$Y_G = Y_P + \Delta Y$$

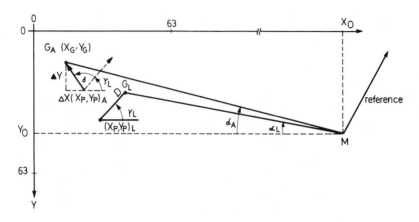

FIG. 8.60 Coordinate transformation during the operating phase.

with (X_P, Y_P) denoting the center of gravity of the object obtained during the operating phase and D the distance between the center of gravity and the grasp position.

Now the grasp position has to be transferred to the robot coordinate system. For this the following calculations are made:

$$\text{Position of the swivel axis } S_A = S_0 + F_S \arctan \frac{Y_G - Y_0}{X_0 - X_G} \quad (I_S)$$

The position of the transfer axis is calculated with the help of the formula

$$T_A = F_T \sqrt{[(X_0 - X_G)^2 + (Y_0 - Y_G)^2]} - T_0 \quad (I_T)$$

and the position of the rotational axis is obtained from

$$R_A = R_L + F_R (\delta + \alpha_A - \alpha_L) \quad (I_R)$$

Parts Handling Assembly

$$\alpha_A = \frac{S_A - S_0}{F_S} \quad \text{(degrees)}$$

The position of the vertical lift axis is

$$H_A = H_L$$

and the number of the action program is

$$N_A = N_L$$

The additional parameters of these formulas are:

R_L = position of rotational axis obtained during teach-in
F_R = number of rotational increments per degree of angle (I_R/deg)
H_L = position of the lift axis obtained from the teach-in phase

The values S_A, T_A, H_A, R_A, and N_A are then passed over to the robot control unit. Now the robot is able to move to the calculated grasp position.

Measuring the Robot Position in Regard to the Coordinate System of the Sensor

In the two preceding sections, constant values that have to be measured or calculated beforehand had been used for the calculation and transformation of the grasp position. These constant values can be defined as robot constant values and constant values of the interfaced sensor-robot system.

Robot constant values

T_0 = distance between the center of the gripper in the reference position of the robot and the center of the swivel axis; T_0 is measured in increments of the transfer axis units (I_T)
F_S = number of swivel axis increment units per degree of angle (I_S/deg)
F_R = number of rotational increment units of the gripper per degree of angle (I_R/deg)

Sensor-robot constant values

F_T = number of horizontal transfer axis increment units per pixel units (I_T/PIX)

S_0 = the angular offset of the swivel axis when it is located parallel to the sensor X axis

X_0, Y_0 = coordinates of the center of rotation of the swivel axis in reference to the sensor system

The robot constant values are measured directly at the robot. In order to obtain T_0, all axes have to be moved to the reference point. Then the distance between the center of the gripper and the center of rotation of the swivel axis T_0 is measured and multiplied by the number of increments T_R of the measurement units. The constants F_S and F_R are obtained as follows. The robot is brought to a reference position. At this position the value of the controlled variable of the swivel control loop is determined. Then the arm is rotated 180° and the new controlled variable is read. Next, the difference is taken between these two values and divided by 180. The same procedure is used to determine F_R for the rotational axis.

The sensor-robot constant values are obtained from the sensor-robot system. For this purpose both the sensor and the robot are brought to a fixed position. With the help of the robot, a disk representing an object is moved between the three points P_1, P_2, and P_3, all of which are in the field of view of the camera (Fig. 8.61). A disk is chosen because it has the following criteria:

1. The possibility of an error occurring when calculating the center of gravity is small
2. There are no geometric changes when rotated.

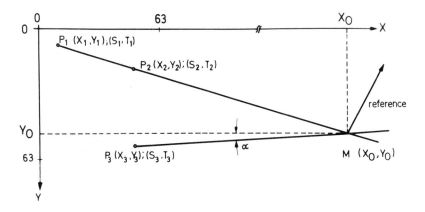

FIG. 8.61 Calibration of sensor and robot coordinate system. I_T, increments of the transfer axis; I_S, increments of the swivel axis; REF, reference position.

3. The grasp position can easily be defined as the center of gravity of the disk.

The coordinate values (S_1,T_1), (S_2,T_2), ano (S_3,T_3) are determined for each point. A prerequisite for this calibration procedure is that the position of the robot between P_1 and P_2 can be altered only by moving the transfer axis. In addition, between the points P_2 and P_3 the robot is only allowed to move about its swivel axis. Because both the center of gravity and the grasp position are identical with the disk, the grasp position with regard to the sensor system can be obtained by calculating the center of gravity at the points P_1 to P_3. Thus a fixed assignment between sensor and robot coordinates is obtained.

The constant values can be calculated as follows:

$$F_T = \frac{T_1 - T_2}{\sqrt{(X_1 - X_2)^2 + (Y_1 - Y_2)^2}} \quad (I_T/PIX)$$

In this equation the denominator is the equation of the circle:

$$(X_0 - X_2)^2 + (Y_0 - Y_2)^2 = r^2$$

and

$$(X_0 - X_3)^2 + (Y_0 - Y_3)^2 = r^2$$

with

$$r = \frac{T_2 + T_0}{F_T}$$

Thus the center of rotation of the swivel axis X_0, Y_0 with regard to the vision system can be calculated.

$$S_0 = S_3 + F_S \cdot \alpha \quad \text{with } \alpha = \arctan \frac{Y_0 - Y_3}{X_0 - X_3}$$

8.10 OPTICAL WORKPIECE RECOGNITION

Optical pattern recognition plays an ever-increasing role in the automation of manufacturing processes. An ideal recognition system for workpieces possesses the universal functions of the human visual system, enabling it to differentiate between different shapes, positions, colors, and parts. It also has the ability to store the gathered information for future reference.

Researchers concerned with the conception of universal vision systems were quickly faced with the sobering fact that identification of even simple objects presents enormous technical difficulties. To make matters worse, the exact distance, length, angular orientation, and microstructure of the object must be known in many technical processes. Because of the limited resolution capabilities of their eyes, human beings can realize this task only with optical assistance. For this reason, many vision systems were developed to obtain solutions for specific tasks. The majority of these systems are only capable of evaluating two-dimensional binary scenes. Here the picture taken is classified and processed as binary black-and-white information, the threshold of which may be determined automatically according to previously given parameters or algorithms. These methods attempt to recognize certain workpiece parameters, such as area, circumference, center of gravity, number of holes, edges, and corners. From this information the desired workpiece characteristics can be extracted. The parameters of the object under investigation must be known to the evaluation system, to which they are conveyed through teach-in and storage methods. There are also some three-dimensional vision systems being developed; however, it will take some time until they are useful for broad industrial application.

Computers are used to solve the more complex recognition tasks. In the original systems the pictures were evaluated through the use of software. The slowness of this method led to the development of hybrid vision systems in which one portion of the work is performed by software and the other by hardware. Today, the speed of completely evaluating a workpiece reaches 100 msec. This includes the task of object recognition, determination of center of gravity, and calculation of angular orientation. Since detailed three-dimensional information about the workpiece is required for quality control, the binary recognition methods do not adapt themselves to the identification of objects. Knowledge of gray levels is necessary in a majority of recognition problems in manufacturing processes. A multitide of evaluation methods have been developed to overcome this difficulty. As a rule, the picture is divided into an $n \times n$ matrix and scanned with the assistance of a submatrix in order to obtain relationships between neighboring picture points. The sum gradient, template matching, and transformation methods of converting space information into the frequency domain are used for this evaluation. For example, in the case of a picture where 128 gray values of a 128×128 matrix are investigated, 2.1×10^6 parameters must be evaluated. To date, these gray-level methods are not feasible for on-line operations because they need large computers, and will be discussed no further.

8.10.1 Tasks of a Vision System for Recognizing a Workpiece

A vision system can be used in a variety of areas, such as assisting a robot in an assembly to perform quality control tasks and defect

Parts Handling Assembly

recognition, to lead a product through a material flow system, to carry out measurements of inaccessible fast-moving parts, to adjust tools on machines, and to recognize printing errors in a printing shop. In many applications a multitude of different tasks have to be performed by the vision system. An assembly robot must, for example, first recognize the part on the conveyer belt, pick it up, assemble it under the instruction of a vision system, and finally, complete a quality control check. The variety of tasks that are expected to be met by a vision system are as follows:

1. Identification
2. Determination of position and angular orientation in relation to reference coordinates
3. Positioning, fitting, and sorting
4. Setup or adjustment of tools
5. Measurement of distances, and angles
6. Extraction and identification of specific parameters
7. Visual testing for error and completeness

The application engineer must adapt the vision system to the specific tasks to be carried out by the installation, making sure that the system is neither over- nor underqualified. In the first case, the system could be much too expensive; in the second case the required tasks would not be performed satisfactorily. One of the most difficult problems that has not yet been adequately solved is the recognition of randomly organized parts located in a container. Similarly, it is necessary to identify parts that lie overlapped on a flat surface. Generally, they will not be recognized as individual parts; instead, they will be identified as a new part. A separation of the parts is thus necessary for most vision systems, entailing a certain amount of presorting.

Figure 8.62a shows different recognition tasks that may assist robots in their work. Present vision systems are capable of performing only simple tasks (e.g., with presorted objects under special lighting conditions). Figure 8.62b presents tasks performed for quality control. Here similar problems exist. Most of the tasks can be solved to a limited extent, depending on the complexity of the part or the application [33].

When taking pictures of an object, there are many different factors to be observed. For example, a very fast speed must be used in order to obtain a clear picture of a moving object, and in extreme circumstances a flash may be required. Time plays a very important role in the ability of the system to evaluate succeeding pictures and to transfer data for remote processing. This time must be under one-half a second for assembly lines and conveyer belts; otherwise, the part flow will proceed at a faster pace than the cycle time of the vision system.

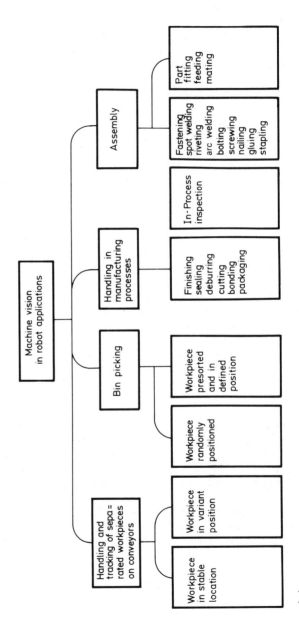

Parts Handling Assembly

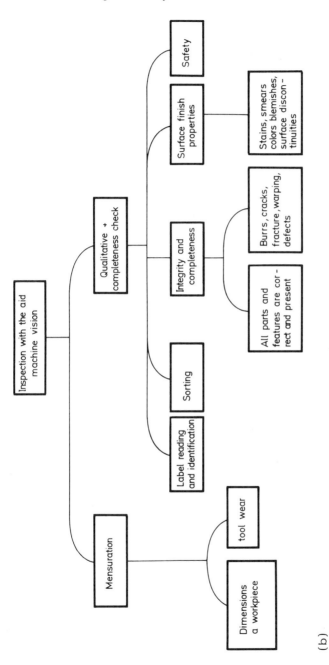

(b)

FIG. 8.62 (a) Tasks of a vision system to control the work of robots; (b) task of a vision system in quality control.

Illumination is another important factor when an identification is being performed. Workpieces are often difficult to distinguish from their background and must be placed on a contrasting or fluorescent surface to be discerned. On occasion, it may be necessary to make color corrections with filters or ultraviolet light; similarly, artificial shadows or special illumination effects can be used to bring out the shape of an object. Even the determination of an object's orientation in space can be made possible through the movement of a light plane that intersects the object.

8.10.2 Vision Systems for Simple Tasks

Manufacturing processes have long employed elementary photoelectric vision systems in the form of light beams. The oldest systems are based on incandescent lamps that emit a columnated beam of light that strikes a photoresistant detector. Any break in the light beam caused by a passing object activates either a counter or decision logic (Fig. 8.63). In principle, it is also possible to build a detector into the light source in order to measure the intensity of the reflected light (Fig. 8.64). Normal workpiece surfaces produce a diffuse reflection of weak intensity. This problem can be markedly improved by attaching to the object a retro or mirror reflector (Fig. 8.65). The disadvantage of the light-beam method is the fact that the source and the detector occupy a large area and cannot be built into a highly integrated sensor system.

Developments in semiconductor technology have made possible the manufacture of smaller light sources and detectors. The sensor system principle shown in Fig. 8.66 is constructed on the basis of TTL technology. The LED (light-emitting diode) source emits modulated infrared light that can be detected by the sensor even in a brightly illuminated room. Upon a light incident, the sensor becomes a conductor,

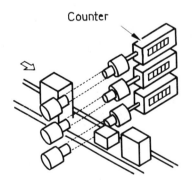

FIG. 8.63 Recognition of parts with the help of light beams.

Parts Handling Assembly 603

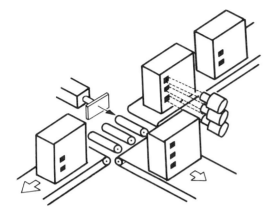

FIG. 8.64 Recognition of parts with the help of binary coding.

an increased current flows through it, and in turn it can be picked up at output A for further processing. These elements are highly sensitive and possess a greater frequency response than the conventional light beams described above. For special applications, linear and matrix arrays can be designed from these discrete building components, thereby obtaining an increase in packing density. Unfortunately, the measuring tolerances reached by the rather awkward assembly methods of arrays are too inaccurate for most measuring systems.

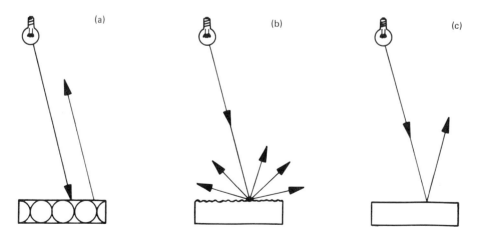

FIG. 8.65 Three different types of reflectances: (a) retro-; (b) diffuse-; (c) mirror-reflectance.

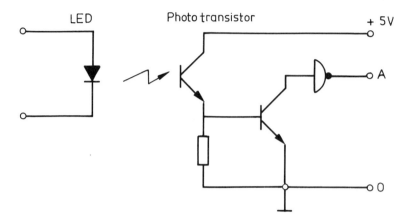

FIG. 8.66 Transmittor/sensor system based on TTL technology.

Manufacture of sensor systems with higher levels of intelligence is made possibly by the use of photoarrays. Figure 8.67 shows such an array which is employed by a vision system to read characters [34]. A matrix is used to divide the character field into segments, the number of which is dependent on the character dimension, size, and the number of characters to be identified. Approximately 9 to 30 matrix elements

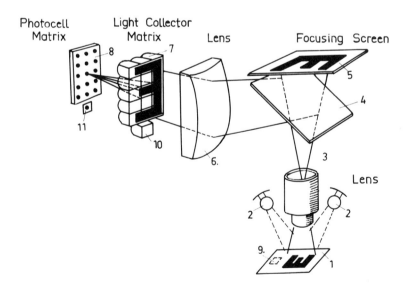

FIG. 8.67 Simple character recognition systems. (From Ref. 34. Courtesy of VDI Verlag, Düsseldorf, West Germany.)

suffice for the recognition of numeric characters and 20 to 60 elements for that of alphanumeric characters. In the system shown, the character is evenly lit by a halogen illumination system and is projected onto a detector matrix via a mirror and a lens. The light falling on the individual points of the light-conducting matrix is columnated and brought to the photocells of the recognition matrix. A ground-glass screen is used in the system to facilitate focusing on the character to be read. The matrix element voltages V_1 to V_{15} of the photocell are proportional to the intensity of light striking them (Fig. 8.68) and are amplified together with a reference voltage representing the brightness of the character's background. This reference voltage is subtracted from the individual matrix voltages by an inverter circuit, thereby compensating for varying background effects of different reading substrates. The resulting differential voltages V_1 and V_{15}, are passed on to the system for further processsing. They are also transmitted to the resistor network that stores the character features. Here the voltages V_{k1} to V_{k15} are computed as the difference of the individual differential matrix voltages and the sum voltage, their values depending on character thickness, form, and the darkness levels of the matrix elements. These differences are the inversion of the individual matrix element values as related to the values of the standard characters. The voltage values vary at the same rate as the sum voltage, eliminating diverse levels of darkness. In case a character has a different shape and darkness level than the standard character, the difference of the two is not equal to zero. Tolerance values of ±30% can be entered into the system manually to take this deviation into consideration. Similar specialized vision systems can be conceived for the identification of

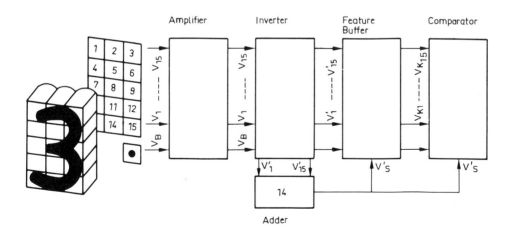

FIG. 8.68 Signal processing in a simple character recognition system. (From Ref. 34. Courtesy of VDI Verlag, Düsseldorf, West Germany.)

workpiece located on a conveyer belt; however, under certain circumstances it may be necessary that the parts retain the same orientation during the scanning phase.

Another typical example of an elementary photoelectric vision system is the use of perforated sheet metal patterns in automobile manufacturing. Here an array of holes represents, for example, the model number of a car. Such patterns are ideal for the identification of individual care to which customer-selected options must be added. The perforations are made on a part of the car body that is easily accessible. It can be scanned by an automatic vision system at specific workstations and then the desired operations may be scheduled.

Scanning devices for binary coding are another example of simple optical recognition systems. A code carrier, usually in label form, is attached to the object to be identified. Various coding methods are shown in Fig. 8.69. In the binary system it is possible, for example, to differentiate between 256 parts with eight coding bars. It is well suited for use where space is a prime consideration. Here wide bars are used for identification while the narrower ones are ignored. One coding in Fig. 8.69 represents, for example, $4 + 8 = 12$. In the case of BCD coding, four bars are used to represent the characters 0 through 9. Code labels can be designed in either a serial or a parallel fashion (Fig. 8.69). If space presents no difficulties, the 2-out-of-5 coding system with simplified error testing can be employed (Fig. 8.69). There are computer-controlled printers in existence that produce coded labels on-line for moving products. Lasers, mercury lamps, and incandescent bulbs are used as light sources, although the highest degree of accuracy is obtained with the laser. In a manufacturing installation, the depth of field, the size of the area to be scanned, and the allowable margin of error in the position of the code marker are all important factors in the selection of a vision system (Fig. 8.70). The field to be scanned can be enlarged through the use of a recognition system with a rotating scanner (Fig. 8.71). For industrial application, circular coding labels (bull's-eyes) are also practical. They are invariant to the angular position of the code carriers (Fig. 8.72).

The elementary vision systems are most applicable where objects are partially prealigned, where no specific quality characteristics are to be identified, and where the determination of a position is not necessary. Their disadvantage is the fact that a coding label must be attached to the object, thereby giving the possibility of an object mixup [35]. The object velocity may be as high as 3 m/sec when bar codes together with a laser reader are used. The code label may be misaligned by 25° in respect to a horizontal plane and 45° in respect to the vertical plane. A height difference of 65 cm is allowable; the depth of field may range from 0 to 120 cm (Fig. 8.70). The labels have the further disadvantage that they cannot be used when workpieces must be washed with certain cleaning agents or are heat treated.

Parts Handling Assembly 607

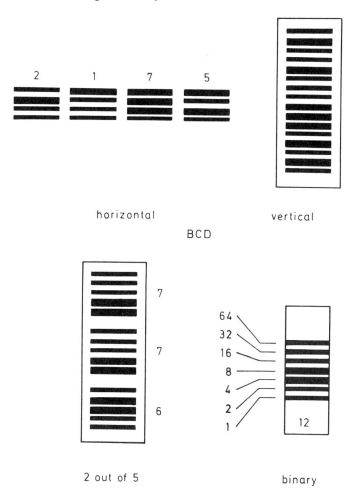

FIG. 8.69 Typical bar coding.

8.10.3 Vision Systems for More Complex Tasks

Optical systems with high-resolution capabilities are necessary for the recognition of more complex workpieces and for quality control tasks [36]. In general, processors, memory modules, and computer software are also required to evaluate the wide scope of picture data. Two-dimensional pictures can be taken by either the vector or the matrix method. In both methods the picture is divided into individual grid elements (pixels). From the varying gray levels of these pixels the binary information needed for determining picture parameters is extracted (Figs. 8.67 and 8.73).

608   Chapter 8

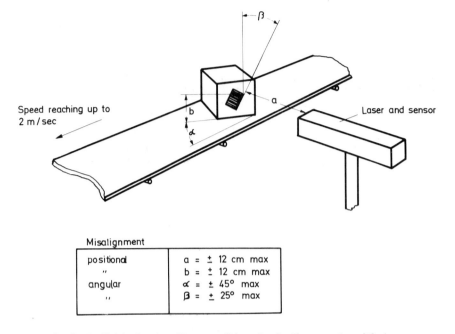

FIG. 6.70 Reliable bar coding positional misalignments with laser devices.

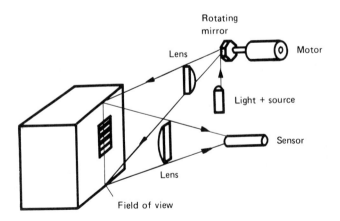

FIG. 6.71 Scanning of a bar code with a rotating light source.

Parts Handling Assembly 609

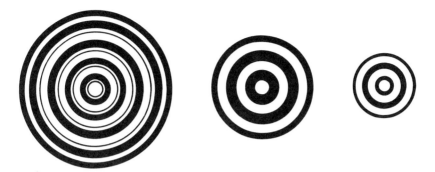

FIG. 8.72 Circular coding patterns.

The vector method requires that either the camera or the object move during the taking of the photo (Fig. 8.74). Here picture vectors of the scanned object are taken and stored at constant time intervals. Upon completion of the entire cycle, a processor evaluates the recomposed picture information and extracts the parameters of interest. This method has the advantage of a simple camera system. However, a constant object velocity is required during the picture-taking cycle, which may often be difficult to assure in the cases of conveyor belts and assembly lines. This is the only method that yields a high picture resolution with presently available cameras. For example, today there are photo vector array sensors with a maximum of 4096 pixels. This

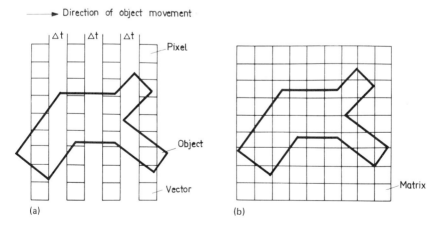

FIG. 8.73 Different methods of scanning a picture: (a) vector method (Δt can be selected); (b) matrix method.

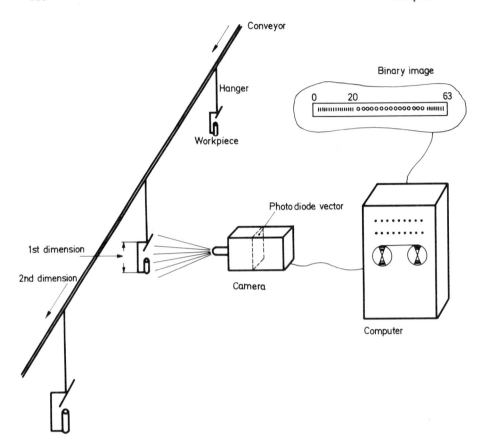

FIG. 8.74 Image recognition using a photodiode vector camera.

is in contrast to the TV camera, which is presently the matrix sensor with the highest resolution. It can evaluate 625 × 625 = 390,625 pixels.

Picture Recognition Sensors

Modern vision systems function according to different principles; thus there are the TV camera, photodiodes, and charge-coupled devices (CCDs). With increasing integration density on a semiconductor chip, however, the TV camera is becoming more and more obsolete. These principal sensor systems are discussed in the following sections.

Television camera. The TV camera is constructed on the basis of a cathode ray tube in which the picture to be evaluated is projected through a lens and onto a screen (Fig. 8.75). It is constantly scanned

Parts Handling Assembly 611

according to the raster method by an electron beam with a line density of 625. Depending on the type of camera being used, the screen has a coating of silicon, lead oxide, or antimony trisulfide. The individual particles in this coating function as tiny capacitors whose charges are dependent on the light falling on them. According to the design of the TV tube, the capacitors are either charged or discharged by the electron beam striking them. This process can be sensed on the screen with a corresponding control circuit. The individual charges are converted into a grid pattern, thereby making it possible to distinguish between 128 varying gray levels.

Self-Scanning Photoarrays

In this sensor numerous light-sensitive semiconductive elements are integrated on a chip as a photodiode vector or a matrix [37]. A depletion zone containing no mobile charges is generated on the p-n junction of the individual photodiodes by a reverse bias voltage, thereby building up a capacitance (Fig. 8.76). Upon a light incident in the junction, the photons generate positive and negative charge carriers, thus decreasing the charge. The voltage change in the junction or the junction capacity is proportional to the intensity of the light incident. Figure 8.77 shows how individual photodiodes can be combined into a single vector with the assistance of MOS switches and a shift register. It is thus possible to place the signals of the diode in sequential order to a common video input. When all switches are closed, the photodiode junctions are connected to the video output potential. The charge thus introduced will be isolated from the video output for the period of one picture-taking cycle. During this cycle the charge is changed proportionally to the intensity of the light incident. In the following scanning period, the charge change of the individual photodiodes are measured by the video control circuit and the signal is processed further. By this method, a pulse train is obtained where the levels of the individual impulses correspond to the light intensity to which the corresponding

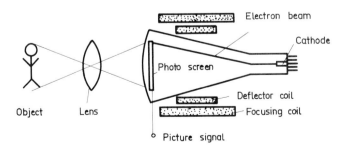

FIG. 8.75 Principle of a television camera.

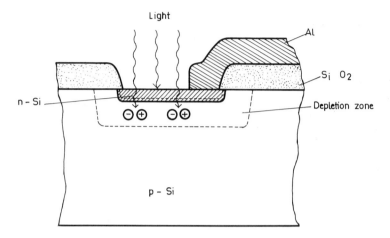

FIG. 8.76 Principle of photodiode sensors. (From Ref. 46.)

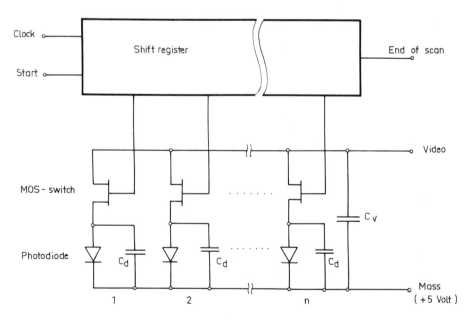

FIG. 8.77 Readout of a photodiode array with the help of a shift register. (From Ref. 37).

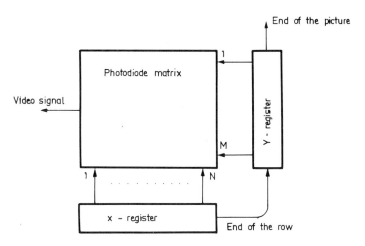

FIG. 8.78 Readout of a photo matrix.

diodes were exposed. In another method, the level of the operating voltage of each junction is polled.

By parallel connection of multiple photodiodes, a matrix is obtained. The elementary control circuit to scan the rows and columns of the matrix is shown in Fig. 8.78. During readout of the signals from each picture element, the rows are scanned first. At the end of a row the x register transfers a signal to the y register, thereby initiating readout of the next row. When the last row is completed, a picture impulse is generated that terminates scanning of the photodiodes of this matrix.

Today's linear photodiode sensors have up to 4095 picture elements in a vector array and up to 320 × 480 in a matrix array. They can be operated with a scanning frequency of 6.2 MHz. These devices are extremely reliable and due to their high degree of light sensitivity are ideal for the identification of poorly illuminated scenes. Because of manufacturing difficulties, however, a higher integration of photodiodes on a single chip has not been satisfactorily obtained to date.

Charge-coupled-device photosensors. The basic sensor element in this technology is the MOS (metal-oxide semiconductor) capacitor (Fig. 8.79). A depletion zone with reduced charge carriers is generated between the gate electrode and the substrate by the application of a voltage whose magnitude lies above the threshold level. Photons striking the surface generate minority carriers below the isolated silicon oxide zone, thereby building up a charge. This charge is integrated over a predetermined time interval and is proportional to the photon current generated by the light incident during this interval. Thereafter, the charge is read out in a two-phase clock cycle (Fig. 8.80). This is done

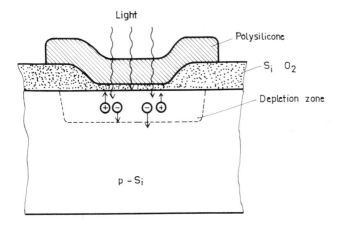

FIG. 8.79 Design of a CCD sensor. (From Ref. 37.)

according to the shift register principle, where the charge is moved from zone a to zone b and from there to zone c, and so on. Finally, it is read out as a video signal and further processed. A fraction of the charge remains in each zone, thus falsifying the original picture contents of the individual zones. This may blur edges of the picture scene. A typical linear CCD chip has 2048 elements and a matrix chip 488 × 380 elements.

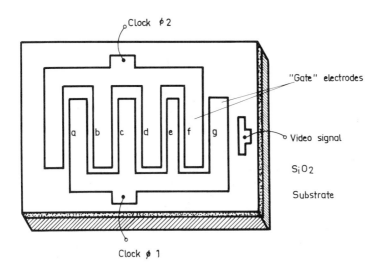

FIG. 8.80 Readout of CCD array.

Parts Handling Assembly

An improved sensor based on the principle of the charge injection device (CID) has been developed, where the stored charges of the MOS capacitors are individually read out through an electrode located in the substrate opposite the MOS capacitor. The voltage of the control electrode lies below that of the threshold voltage. This causes a contraction of the depletion zone and forces the charge into the substrate. The current necessary to move the change into the readout electrodes is proportional to the amount of light that falls on the individual cells.

Comparison of Different Technologies

The type of application for which a specific vision system is to be used determines the requirements that must be fulfilled by the system. For example, a vision system intended for quality control purposes requires a higher resolution than one for rough positioning of the parts. The following are typical parameters of picture sensors:

Resolution
Resolution time
Linearity
Geometric reproducibility
Sensitivity
Maximum scanning frequency
Dynamics
Spectral sensitivity
Cost
Aging
Reliability

As mentioned before, there are now available linear sensors with a total of 4096 photodiodes on a single chip. The highest resolution for two-dimensional tasks is obtained with the TV camera, which is equipped with 625 × 625 pixels. In the future, however, semiconductor technology will displace the TV tube. Manufacturers currently offer CCD cameras with 488 × 380 pixels integrated on a matrix array, the dimensions of which are 8.8 × 11.4 mm. The higher resolution capability of the TV camera does not offer many advantages since the possible number of picture elements to be scanned is limited by the allowable picture taking and processing time. The taking and evaluation (area, center of gravity, and angular orientation) of a picture of a moving object on a conveyer belt must be completed within 50 to 300 msec. With today's evaluation methods, these time limits can be met only when fewer than 256 × 256 pixels are to be evaluated. In addition, the TV camera is unsuited for exact measurements. A high linearity cannot be obtained in the edge zones of the picture. Special television cameras capable of reaching a linearity of 1% do exist but are very expensive. In the case of semiconductor sensors, the linearity is determined by the manufacturing accuracy. By knowing the location of the individual sensor elements on a chip, misalignments can be eliminated through software correction in the computer. These sensors are better suited for measuring tasks. In the case of moving scenes that are photographed by a TV camera, the picture can be distorted. Further disadvantages of the TV tube are the phenomenon of early aging and the noise sensitivity. One of the drawbacks of the semiconductor sensor is the possibility that individual sensor elements may fail. Typical parameters of semiconductor sensors are shown in Table 8.7.

Workpiece Identification

To be evaluated a picture must be converted into a grid composed of n × n pixels (Fig. 8.73b). In the case of the semiconductor camera, each pixel contains the information supplied by a specific sensor element. The resolution of the entire grid is determined by the number of the lines and number of points on the lines to be evaluated. In both methods the information contained in the individual picture elements is given out as an analog value, the level of which is determined by the gray level of the field of view of the individual pixel. To generate a binary picture, a threshold value must be determined below which the information to be processed is represented by a 0 and above which it is 1. This threshold is determined by either software or hardware. In practical applications, an automatic threshold adaptation is provided where, for example, the average brightness of the picture scene serves as a reference value (see the examples in Figs. 8.67 and 8.68).

The principle of an automatic workpiece recognition system using a photodiode array camera is shown in Fig. 8.81. The object to be measured is scanned by the photoarray, evaluated line by line by a processor, and then identified by a microcomputer. The angular orientation can also be determined when necessary.

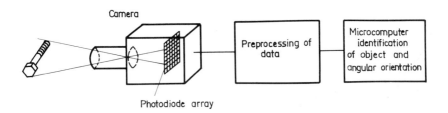

FIG. 8.81 Principle of a workpiece recognition system.

TABLE 8.7 Typical Parameters of Semiconductor Picture Sensors

Sensor element	Readout method	Maximum number of elements	Area of an element (μm^2)	Sensitivity ($\mu j/cm^2$)	Maximum scanning frequency (MHz)	Dynamic range (relative to peak)
Photodiode	Sequential read-out over video line	4096 × 1	240	0.9	10	1000:1
	Selective read-out over video line	320(H) × 244(v)	729	0.2	6	320:1
MOS charge-coupled devices	CCD shift register	1728 × 1	256	2	5	500:1
	Selective read-out over video line	244 × 188	3600	1	0.6	250:1

Source: Ref. 46.

As each picture contains a large amount of redundant information, only specific parameters of the scene need be acquired for the object's identification. The quantity of data to be evaluated is thus reduced to a minimum. This strategy has been adhered to by all methods presently available. As a rule, the picture evaluation can be performed by either hardware or by software. The latter method is slow and often yields no acceptable evaluation speed. Hardware, on the other hand, is much faster, but more expensive and less flexible. For this reason, most methods use hybrid systems where the complex and time-consuming evaluations are performed by hardware and the simpler tasks are carried out by software. The software approach is almost always used for the original development of the evaluation algorithm. Thereafter, the routines that are too time consuming for computers to solve are converted to hardware, which is often 200 times faster than software. The cost of the computer also plays a large role in the development of an on-line recognition system; thus microcomputers are used in most cases to perform the software task. In the following sections, the most important identification parameters are discussed.

Area. The information necessary to evaluate a picture matrix is contained in a fixed orthogonal grid. For further observation, we will assume this grid to be made up of n × n elements or pixels. Thus the picture can be represented by a series of gray level values H(i,j). The following relationship can be established:

$$H(i,j) = h_{ij} \tag{1}$$

where i,j = 0, 1, 2, 3, . . . , n − 1. If we limit ourselves to binary pictures, the informational contents of the individual pixels is

$$h_{ij} = \begin{cases} 1 & \text{for } (i,j) \in \text{object} \\ 0 & \notin \text{object} \end{cases} \tag{2}$$

The area can be determined as follows:

$$F = \sum_{j=0}^{n-1} \left(\sum_{i=0}^{n-1} h_{ij} \right) \tag{3}$$

Because of the finite sizes of the individual pixels, varying area values can be obtained for single objects, depending on the position of the grid and the method used to determine the threshold value (Fig. 8.82). An object's area is a unique parameter only if no other object exists that has the same area. This prerequisite is not always present in the

Parts Handling Assembly

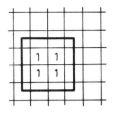
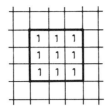

Smallest area
4 pixels

Largest area
9 pixels

FIG. 8.82 Effect of the raster size on the recognized area.

case of several workpiece projections to be evaluated; thus other parameters must often be used for the identification. One method is shown in Fig. 8.83 by which objects can be identified through parameters derived from the distribution of object elements on a scanning circle which has its center on the object's center of gravity [38]. Often, the entire object need not be evaluated; it suffices to use only selected circle segments. A knowledge of the total object area is still necessary for the following center-of-gravity calculations.

Center of gravity. The center of gravity is an important parameter of the workpiece projection. It is often used as an identification

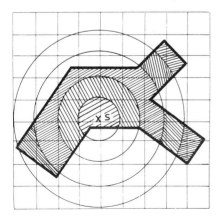

FIG. 8.83 Identification of an object with the help of circular segments about the center of gravity. S, center of gravity.

parameter itself or as a reference point for other parameters. The x_s and y_s coordinates of the center of gravity can be computed as follows:

$$x_s = \frac{1}{F} \sum_{j=0}^{n-1} \left(\sum_{i=0}^{n-1} i h_{ij} \right) \quad (4)$$

$$y_s = \frac{1}{F} \sum_{j=0}^{n-1} \left(\sum_{i=0}^{n-1} j h_{ij} \right) \quad (5)$$

The equations above require the use of a row-wise scanning mode by which the picture scene is traversed from left to right. It should be observed that the calculated center of gravity can lie within nine neighboring picture elements. This is due to the finite size of the picture elements and the location of the object (Fig. 8.84). This error must be kept in mind when considering accuracy.

Moment of inertia

1. *Axial moment of inertia of an area*: The moment of inertia about an axis can be calculated by multiplying the mass of an elementary particle of the area by the distance of this particle from the axis [39]. Thus the axial moment of inertia can be derived from the grid containing the picture information as follows:

$$J_x = \frac{1}{F} \sum_{j=0}^{n-1} \left[\sum_{i=0}^{n-1} (i - x_s)^2 h_{ij} \right] \quad (6)$$

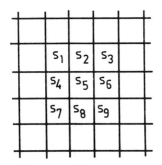

FIG. 8.84 The calculated center of gravity may lie in nine neighboring pixels.

$$J_y = \frac{1}{F} \sum_{j=0}^{n-1} \left[\sum_{i=0}^{n-1} (j - y_s)^2 h_{ij} \right] \tag{7}$$

2. *Polar moment of inertia*: The polar moment of inertia of an object is taken about an axis perpendicular to the plane of an area. The quantity is represented by J_{xy} where x and y are the coordinates of any elementary part into which the area may conceivably be divided. Thus we can calculate

$$J_{x,y} = \frac{1}{F} \sum_{j=0}^{n-1} \left[\sum_{i=0}^{n-1} (i - x_s)(j - y_s) h_{ij} \right] \tag{8}$$

Perimeter. The perimeter of an object can be obtained from the sum of individual grid elements intersected by the contour of the object [40]:

$$C = \sum_{l=1}^{N} h_l \tag{9}$$

where N is the number of elements intersecting with the perimeter. The absolute length of the contour of an object is represented by the following equation:

$$C_{abs} = \sum_{l=1}^{N} \sqrt{(x_{l+1} - x_l)^2 + (y_{l+1} - y_l)^2} \tag{10}$$

x and y being the coordinates of the contour elements.

Distance of designated object points from the center of gravity
1. *Point of intersection with scanning lines or circles*: Objects may be identified by points of intersection between specified object features and scanning lines or circles. A principle of contour determination with the help of selected scanning lines parallel to the x axis is shown in Fig. 8.85 [41]. By adjusting the vision system, the desired intersection lines can be placed through specified points of the object to be identified. They are located in the scene with the help of a CRT terminal. In Fig. 8.85, two lines, Z_1 and Z_2, intersect the picture. Each x coordinate of a designated point of intersection is normalized with respect to the x coordinate of the center of gravity. With these

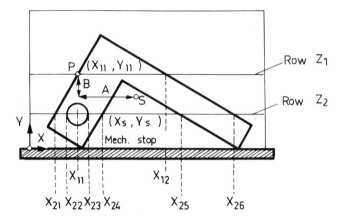

FIG. 8.85 Principle of feature extraction with the help of selected scanning lines. (From Ref. 41.)

differences, a characteristic vector for a workpiece can be obtained. Unfortunately, this method requires that during the scanning period, the workpiece lies in a defined position. The method can be improved on by employing scanning circles centered around the center of gravity of the object (Fig. 8.83). The points of intersection are invariant in regard to the principal moment of inertia. Both of these methods have the disadvantage that they are time variant depending on the number of points of intersection used. The distances between the center of gravity and specific characteristics such as edges, holes, and smallest and largest radii, may also serve as classification parameters (Fig. 8.86) [42].

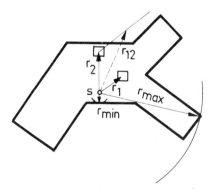

FIG. 8.86 Feature extraction by determining the distance of designated points from the center of gravity.

2. *Sequences of angles obtained with the assistance of scanning circles*: The center of this group of circles is on the center of gravity of the object (Fig. 8.87) [43,44]. The circles are positioned such that they will intersect the workpiece contour at a rather steep angle. The angles are formed by the x axis originating from the center of gravity and the lines through the points of intersection with the contour and through the center of gravity. They may be traversed either clockwise or counterclockwise. The sequence of the angles $\alpha_{k,n}$ form the characteristic vector. k stands for the number of scanning circles used and n for the running index of the angle.

In practice, multiple parameters are often combined into a characteristic vector. When in addition to the angles the coordinates of the points $P_{k,n}$ of intersection are taken, the following identification vector results:

$$M_k = (\alpha_{k,n}, P_{k,n}) \tag{11}$$

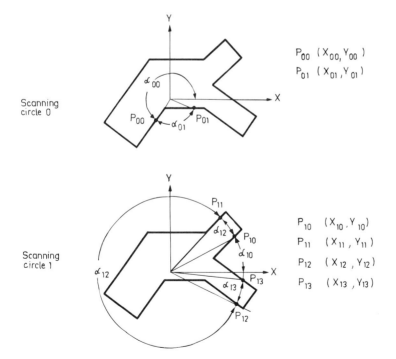

FIG. 8.87 Feature extraction with the help of polar scanning. (From Refs. 43 and 44.)

During teach-in, the workpiece is placed in the camera's field of view and intersecting circles are selected, thereby specifying a reference vector. The angle α_{k0} and the point P_{k0} need not be considered since this method is not dependent on the position of the workpiece. The angle α_{k0} is of interest, however, when the angular orientation must be determined. During the measurement procedure an identification vector is obtained from the object under investigation and is compared to the reference vector. The tolerance determined by the resolution of the picture grid and the measuring system must both be taken into consideration in the comparison. A disadvantage of this method is the fact that the scanning circles for the determination of the reference position must be adapted to the contour of the workpiece. In the case of the scanning circles approaching the object contour as a tangent at the points of intersection, large errors will result. This teach-in method may be very time consuming. Furthermore, the identification time for an object depends on the number of angles selected.

Special object characteristics. Many workpieces have numerous characteristics that may either permit identification of the object within a certain group of workpieces, or may be integrated into the overall identification process. In practice, the following characteristics are used:

Number of holes
Hole perimeter
Hole area
Position of holes
Number of corners
Position of corners

Angular orientation. In many facets of manufacturing it is necessary to determine the angular orientation of the individual workpieces before they are handled for further processing. After the position is determined, a mechanical hand can orient the workpiece properly and then transport it to another destination. The center of gravity is the basis for the establishment of a reference coordinate that can be used in determining the angular orientation. As a rule, this procedure is more difficult than identification of a workpiece. The following are the three most widely used methods for determining the angular orientation: (1) moment of inertia of the area of the workpiece, (2) correlation method, and (3) polar coding method.

1. *Angular orientation through the use of the moment of inertia of an object projection*: The calculations necessary for determining the moments of inertia J_x and J_y were described in the preceding section. Once these parameters are known, the angular orientation of the object with respect to the x axis can be calculated as follows [39]:

$$\tan 2\delta = \frac{2J_{xy}}{J_y - J_x}$$

$$\delta = \frac{1}{2} \arctan \frac{2J_{xy}}{J_y - J_x} \qquad (12)$$

The principal moments of inertia for the projection of the object can be determined in the same manner:

$$I_{1,2} = \frac{1}{2}(J_x + J_y) \pm \sqrt{\frac{(J_y - J_x)^2}{2} + J_{xy}^2} \qquad (13)$$

For objects where $J_1 = J_2$, $J_{xy} = 0$, the numerator and the denominator in the equation for the angle δ are equal to zero, making it impossible to determine the angular orientation. By observing this equation, it can be seen that the error in computing the angle or orientation increases with increasing symmetry of the object.

2. *Determination of angular orientation by the correlation method*: An overview of the various correlation methods will be given in the next section. With these methods, a correlation is made between the parameters of the stored reference value of the object and those of the picture of the object under investigation. These routines lead to long computing times when many parameters are involved. Therefore, an attempt is made to use as few parameters as possible, for example, by taking only the angles δ_i obtained by an intersection method using scanning circles (Fig. 8.87). Thus one obtains a one-dimensional function $H(\delta)$ for which the cross-correlation with respect to the reference function $H^R(\delta)$ can be expressed as follows:

$$K(\delta') = \frac{1}{n} \sum_{i=1}^{n} H(\delta_i) H^R(\delta_i + \delta') \qquad (14)$$

This function has an absolute minimum at the angular orientation $\delta' = \delta$, which is to be determined. As mentioned in the preceding section, the scanning circles must intersect the contour of the workpiece under a rather steep angle to prevent large errors.

3. *Polar coding*: With this method, the object is scanned in radially beginning at the center of gravity. Certain characteristics are recorded, such as the distance between the center of gravity and the contour as a function of an angle φ (Fig. 8.88). Thus a one-dimensional function $R(\varphi)$ is obtained which, except for a phase shift, is invariant in regard to the object's position in the field of view.

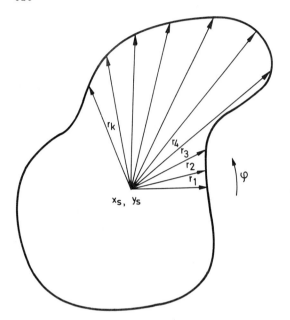

FIG. 8.88 Polar coding by radial scanning. (From Ref. 55.)

The advantage of this method lies in the fact that object information is supplied in sequential order, which makes the calculation of the angular displacement very easy. The entire object is scanned and all information obtained is used for the evaluation.

If an object contains holes or has concave contours, the function $R(\varphi)$ may assume several values, complicating further evaluation. Exact knowledge of the contour is required and extensive computations may result.

EXAMPLE 8.1 Modified Polar Coding Method Using Scanning Circles. The method presented here uses a scanning procedure by which the object information can be obtained sequentially with minimal computational requirements [45,46]. The scanning function is realized through a group of circles K_1 having a radius R, the centers of which lie on another circle K_2. The center of K_2 is the center of gravity of the object (Fig. 8.89). Circle K_2 also has the same radius R and is the geometric locus of all centers of the family of circles K_1. For an orthogonal grid, the following equation can be written:

$$K_1: (i-c)^2 + (j-d)^2 = [R + \Delta r(i,j)]^2 \qquad (15)$$

with

$$-0.5 < \Delta r(i,j) < 0.5 \tag{16}$$

The term $\Delta r(i,j)$ is the error function. The coordinates $C(o)$ and $d(o)$ are the center of the circles K_1 and are defined by the circle K_2 and the coordinates of the center of gravity x_S and y_S (Fig. 8.89). K_2 can be described by the following equation:

$$K_2: \quad (c(\varphi) - x_s)^2 + (d(\varphi) - y_s)^2 = [R + \Delta r(i,j)]^2 \tag{17}$$

The coordinates of the center of the circle K_2 are

$$c(\varphi_k) = x_s + [R + \Delta r(i,j)] \cos \varphi_k \tag{18}$$

$$d(\varphi_k) = y_s + [R + \Delta r(i,j)] \sin \varphi_k \tag{19}$$

$$\varphi_k = K \frac{2\pi}{m} + \Delta \varphi(k) \quad k = 0, 1, 2, \ldots, m-1 \tag{20}$$

where m is equal to the circumference expressed in the number of grid points.

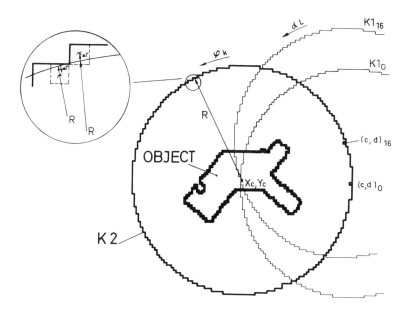

FIG. 8.89 Circular coding of an image.

Analogous to $\Delta r(i,j)$, the constraints for $\Delta\varphi(k)$ are given by

$$-\frac{\pi}{m} < \Delta\varphi(k) < \frac{\pi}{m} \qquad (21)$$

By inserting the equation $c(\varphi_k)$ and $d(\varphi_k)$ for the center coordinates of K_2 into the equation of K_1 and introducing the angle α_1, the following parameter representation is obtained:

$$i = x_s + [R + \Delta r(i,j)](\cos\varphi_k + \cos\alpha_1) \qquad (22)$$

$$j = y_s + [R + \Delta r(i,j)](\sin\varphi_k + \sin\alpha_1) \qquad (23)$$

By adhering to the above-described scanning procedure, the object elements can be obtained as a function of φ_k and α_1. In particular, φ_k determines the sequence necessary for the calculation of angular orientation.

The use of circles for picture scanning is advantageous because the circular function is invariant with regard to angular changes in the object's position. Thus it is only necessary to perform a displacement shift of coordinates c and d of the circle K_1 when leading the scanning circle K_1 along the locus K_2. In case a function other than that of a circle had been used, it would have been necessary constantly to change the angular position of the curve which the function represents while the curve is moved along the circle K_2, with an angular displacement φ_k. Here c and d are the center of rotation. The circular scanning method requires little computational power if the function of the circle is stored in the form of a table. By adding the displacement vector to these stored values, the momentary position of (c,d) would be compensated for.

Since the center of the circle K_2 is identical to the center of gravity of the object under investigation, the method described supplies the invariant position characteristic necessary to determine the angular position. When the entire bit pattern obtained from a single circle K_1 is elevated, only the characteristic values that intersect the object's image are of interest for further processing. The sum of the object elements found in one circle of K_1 are recorded. The sequence of these characteristic values then forms the function $C(k)$, representing the object's polar coding:

$$C(k) = \sum_{l=0}^{m-1} h_{kl} \qquad (24)$$

Parts Handling Assembly

The relationship between k,l and i,j is described by the equation for the center coordinates.

For the object whose angular orientation is to be determined, the function C(k) is first calculated in a reference position during the teach-in mode and stored as the reference function $C^R(k)$. The angular displacement of the object to be measured can be represented by the following equation:

$$C(k + p) = C^R(k) \tag{25}$$

In determining the phase shift p, one can either use a cross-correlation or calculate the value of the absolute difference function. The second procedure was chosen because of the low computational effort involved, no multiplication being necessary. By this method, first the differences between the corresponding values of the reference and the measured functions are calculated, and then their absolute minimum is summed over K. The resulting absolute difference function D(k') has the following equation:

$$D(k') = \sum_{k=0}^{m-1} | C(k) - C^R(k - k') | \tag{26}$$

Under ideal conditions k' = p; thus D(k') becomes zero:

$$D(k') = \sum_{k=0}^{m-1} | C(k) - C^R(k - p) | = 0 \tag{27}$$

This means that the function D(k') has an absolute minimum of zero at the position k' = p. The displacement angle the object has in regard to the reference position can be determined as follows:

$$\delta = p \frac{2\pi}{m}$$

Because of possible noise and the limited resolution of the grid structure, the shape of the measured function will not quite match that of the reference function. Thus the function D(k') will not take on the value zero for k' = p. At this point, however, if noise is not too severe, an absolute minimum is assumed, the value of which is the criterion for calculating the angular orientation.

In addition to determining angular orientation with this method, it is also possible to differentiate between upper and lower surfaces of an asymmetric object whose relationship in polar coding can be expressed as follows:

$$C^U(k) = C^L(m - k - 1) \tag{28}$$

To determine whether the upper or lower surface of the object is shown, the measured function $C(k + p)$ must be compared with $C^R(k)$ and with $C^R(m - 1 - k)$.

4. *Comparison of Various Correlation Methods*: Different mathematical methods can be employed to determine the angular orientation of an object. The method actually put into practice should be easy to implement and to use on the computer hardware. It must also be able to recognize with a high degree of accuracy the similarity between the reference picture and the measured picture. These prerequisites assure an efficient and useful recognition of the angular orientation [47].

In the case of signals with a high noise component, cross-correlation yields the best results. The method is described by the following equation:

$$K(\delta') = \frac{1}{n} \sum_{i=1}^{n} H(\delta_i) H^R(\delta_i + \delta') \tag{29}$$

This method allows the scaling factor n to be altered within a limited range. Its greatest disadvantage is the large amount of computation necessary for the multiplication.

The following simplified binary correlation can be derived from the cross-correlation. It is practical when computing is carried out by special hardware:

$$K(\delta') = \frac{1}{n} \sum_{i=1}^{n} [(\text{sign}) \Delta H(\delta_i)] [(\text{sign}) \Delta H^R(\delta_i + \delta')] \tag{30}$$

where $\Delta H(\delta_i) = R(\delta_i) - \bar{R}$, $\bar{R}$ being the mean value of the contour radius. This method is insensitive to scaling factor variations.

For extremely fast computations, the absolute difference function may be used:

$$K(\delta') = \sum_{i=1}^{n} |H(\delta_i) - H^R(\delta_i + \delta')| \tag{31}$$

It yields less exact results when noise is present. As in the case of the cross-correlation method, the following binary correlation can be derived:

$$K(\delta') = \sum_{i=1}^{n} |(\text{sign}) \Delta H(\delta_i) - (\text{sign}) \Delta H^R(\delta_i + \delta')| \quad (32)$$

Methods for Object Classification

Most recognition tasks require that the vision system be capable of differentiating between numerous workpieces or workpiece patterns. The features of the objects to be recognized must be stored in the computer so that the evaluation algorithm can classify the object under investigation. The closer the feature vectors match in the multidimensional feature space, the easier the classification becomes. Classification methods may be divided into parallel and sequential procedures.

Parallel classification procedures. In this method, the different features of the object to be identified are defined in the n-dimensional vector space by the coordinate $\underline{x} = (x_1, x_2, \ldots, x_n)$. In the following discussion, $\underline{x}^R$ represents the properties of the reference vector and $\underline{x}^M$ those of the measured vector. These individual properties must be represented invariant in relation to the position of the workpiece. Should a workpiece have numerous stable positions with varying area projections, they may be described by multiple vectors which are recognized by the identification algorithm as belonging to the same workpiece. Theoretically, the position vector $\underline{x}^M$ acquired during the measuring phase should be identical to the reference vector $\underline{x}^R$. However, because of measurement errors in the vision system, deviations in workpiece parameters, and the finite grid resolution of the picture, this is not generally the case. This error is taken into consideration by the identification algorithm and is corrected with the help of an error distribution. A multitude of methods for parallel classification of workpieces exist, but most entail large amounts of computation. Thus they are not useful in performing a real-time evaluation of the picture within a fraction of a second. Accuracy must often be sacrificed in favor of evaluation speed. Of special interest in the discussion of vision systems using mini- or microcomputers are the linear and nextneighbor classification methods, which are outlined briefly in the next few paragraphs.

For a linear classification method to be used, different object classes must be separable through linear decision boundaries [48]. When n represents the components of the feature characteristics and k the number of workpiece classes, the following equation for an n-dimensional decision space may be written:

$$F_m = a_{m0} + a_{m1}x_1 + a_{m2}x_2 + \cdots + a_{mn}x_n \tag{33}$$

$$= a_{m0} + \sum_{j=1}^{n} a_{mi}x_i \quad m = 1, 2, 3, \ldots, k \tag{34}$$

The decision criteria for the feature characteristics being examined are

$$F_m > 0 \text{ if } \underline{x}^M \text{ falls in the class } K_m \tag{35}$$

$$F_m \leq 0 \text{ if the above is not true}$$

Figure 8.90 displays the principle of this method with the help of an example using two object classes with two-dimensional workpiece characteristics. No decision can be reached for points falling on the decision boundary $F_2 = 0$. When F_2 is positive, the classifying point belongs to class k_2; if F_2 is negative, the point falls into the class K_1. Multiple object classes can be similarly represented in a two-dimensional vector space; however, if a vector has more than two dimensions, a graphic presentation is difficult. The weight factors a_{mi} must be determined for practical use of the classification procedure. Iterative methods involving the solution of matrices are well suited for this purpose. A large amount of multiplication and division is necessary to

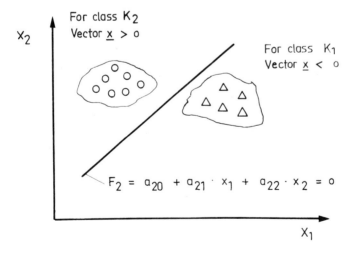

FIG. 8.90 The separation of two classes in a two-dimensional vector space.

Parts Handling Assembly 633

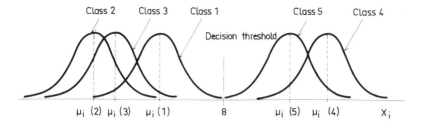

FIG. 8.91 Distributions of feature measurements. (Courtesy of Stanford Research Institute International, Menlo Park, California.)

solve problems with numerous classes and feature vectors. These calculations are extremely time consuming when microcomputers are used. Thus in time-critical recognition tasks, special hardware should be used to perform these computations.

Another classification procedure is the next-neighbor method, which uses the spatial distances between the individual neighboring classes as identification criteria. Unfortunately, this method is also too slow for real-time tasks and should be wired into hardware if used.

Sequential classification methods. With this method, the measured position characteristics $\underline{x}^M$ can be sequentially processed to reduce the computing time. Feature x_1 is examined first. Should it yield no unique identification, feature x_2 will be inspected, and so on, until a positive identification has been made. This procedure is further discussed in the following section [49].

We will again select $x_1 \cdots x_n$ to represent the values of the individual features. The mean for each class is defined by $\mu_i(k)$ and the standard deviation by $\sigma_i(k)$. Through repeated measurement of various objects belonging to the same class, a standard deviation is obtained. A typical distribution of feature parameters is shown in Fig. 8.91. If the workpieces are of a high degree of similarity, the standard deviation $\sigma_i(k)$ is equal to σ_i, which means that the distance between the means is independent of the measuring method and the workpiece itself. Different characteristics x_i of two classes K_i and K_j can be recognized by the normalized distance of the mean value.

$$r = \left| \frac{\mu_i(k_i) - \mu_j(k_j)}{\sigma_i} \right| \tag{36}$$

The error probability to differentiate between classes K_i and K_j is

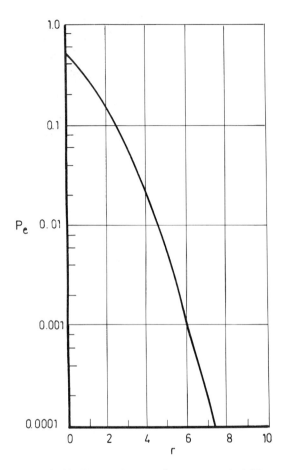

FIG. 8.92 Dependence of error probability on normalized distances between means. (Courtesy of Stanford Research Institute International, Menlo Park, California.)

$$P_e = {}_{r/2}\int^\infty \frac{1}{\sqrt{2}} e^{(1/2)x^2} dx \qquad (37)$$

This relationship is shown in Fig. 8.92. The first step to be carried out in this method is the search for a normalized distance r_{max} between two means μ_i of the class features. This value is indicated by θ in Fig. 8.91. Thus the first step of the decision process is completed. The value θ separates the feature means $\mu_i(2)$, $\mu_i(3)$, and $\mu_i(1)$ into one group, and means $\mu_i(5)$ and $\mu_i(4)$ into a second group, each of

Parts Handling Assembly

which can be further processed according to the same principle. By this method, a decision tree will be successively traversed until an exact identification of the workpiece has been made. The overlap of the distribution curves in Fig. 8.91 can lead to difficulties in determining features; therefore, it must be assured that r is as large as possible. The advantage of this method lies in the fact that when difficulties are met in reaching a decision with available featues, new features must simply be added to the evaluation algorithm. The high evaluation speed of the serial methods is associated with a larger degree of inexactness.

Figure 8.93a and Table 8.8 show the results of test runs performed with gray castings using this method. None workpieces, each having two features (for the simplification of the pictorial representation), were investigated. The greatest distance separating directly neighboring points of the characteristic x_1 is $r_{1max} = 8.1$, which is the distance between classes 8 and 9. Thus 9 is separated from the other groups. For x_2, $x_{2max} = 9.4$ is obtained as the greatest distance between the directly neighboring points 1 and 4. Since $x_{2max} > x_{1max}$, this feature is used for the first decision step. Line $x_2 = 38.6$

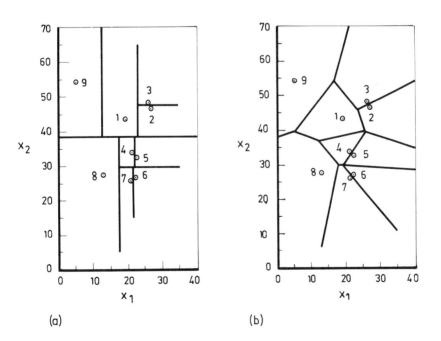

FIG. 8.93 Decision region for sequential decision-tree classifier (a) and parallel next-neighbor classifier (b). (Courtesy of Stanford Research Institute International, Menlo Park, California.)

TABLE 8.8 Feature Values of Different Castings

Class	Feature x_1	x_2
1	19.6	43.3
2	27.1	46.3
3	26.0	48.7
4	21.1	33.9
5	22.7	32.6
6	22.3	26.9
7	21.2	25.9
8	13.4	27.5
9	5.3	54.5

subdivides the problem into one group with four classes (1, 2, 3, and 9) and another with five classes (4, 5, 6, 7, 8). The successive use of this method leads to the construction of the decision tree shown in Fig. 8.94.

A comparison of the next-neighbor method and the sequential method is shown in Fig. 8.93a and b. Both have difficulty in differentiating between classes 2-3, 4-5, and 6-7. Thus additional characteristics are required to obtain a unique discrimination.

8.10.4 Examples of Vision Systems

Linear Vision System to Measure Distances

A simple vision system using a one-dimensional photodiode array sensor is shown in Fig. 8.95 [37]. With present manufacturing know-how, it is possible to produce these sensors with 4096 photodiodes on one chip, obtaining a resolution of 1%. The video signal taken from the object under investigation is brought via a video amplifier to the compensator. First, the brightness of differently illuminated picture scenes is compensated for with the help of an automatic threshold adaptor. Thereafter, the analog voltages of the individual picture elements are converted into binary information. For better adaptation of this system to the measuring task, the scanning frequency and the time interval between two scanning cycles can be varied. The binary picture will be searched for light/dark and dark/light thresholds (Fig. 8.96). The order number of the diode at which the thresholds are recorded and stored is a buffer. The actual length of the workpiece to be measured is calculated with the help of the processor. For this calculation the processor needs a comparison scale which is shown to the camera system via the teach-in mode. The length l_e of the scale is proportional to the number of diodes n_e located between the two thresholds. The processor uses the following equation to determine the scaling factor:

$$m = \frac{l_e}{n_e} \quad 1 \leq n_e \leq n_{max} \tag{38}$$

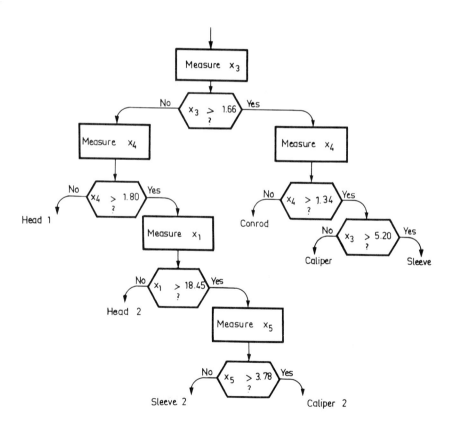

FIG. 8.94 Final decision tree for foundry parts. (Courtesy of Stanford Research Institute International, Menlo Park, California.)

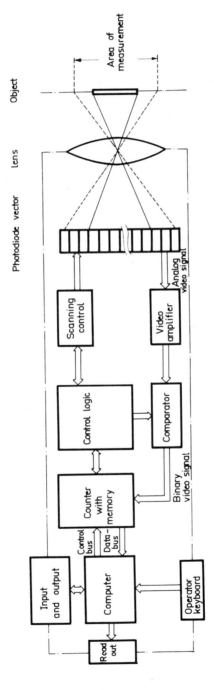

FIG. 8.95 Design of an intelligent linear camera system. (From Ref. 37.)

Parts Handling Assembly

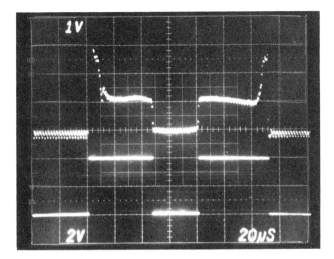

FIG. 8.96 Analog and digitized video signal. (From Ref. 37.)

In this equation n_{max} is equal to the maximum number of photodiodes of the sensor. During the measuring phase, the number of diodes n between the light/dark and dark/light thresholds are counted, making it possible to calculate the length of a part as follows:

$$l = nm = n \frac{l_e}{n_e} \qquad (39)$$

To increase the resolution of the measurement, the image of the workpiece should be located as close as possible to the camera. Figure 8.97 shows various applications for this camera.

Camera Systems to Recognize Workpieces, Their Location, and Their Orientation

EXAMPLE 8.2 The principle of using a camera system to recognize moving parts, their location, and their angular orientation is shown in Fig. 8.98 [46]. In this system the evaluation of the picture information is done both by hardware and software. While the hardware solves the time-critical tasks, the software is responsible for performing complex calculations. The sensor of the system is a 64 × 64 photodiode array integraded on one chip. The video signal to be evaluated is first sent to the preprocessor, where an automatic threshold adaptation is performed to reduce the amount of redundant information. The highest

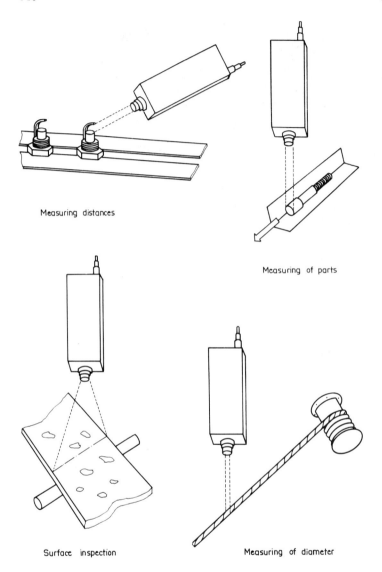

FIG. 8.97 Different uses of an intelligent linear camera. (From Ref. 37.)

Parts Handling Assembly

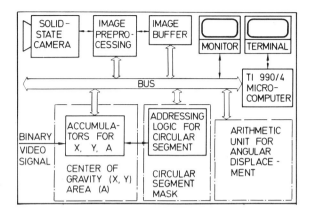

FIG. 8.98 Simplified block diagram of a real-time vision system. (From Refs. 45 and 46.)

and lowest brightness amplitudes of the picture are used for this purpose, thereby automatically adjusting the threshold to compensate for different illumination affects in the scene. An electronic circuit in this system performs an automatic frame control with the picture information. Should the picture of the workpiece intersect with this frame, the information is rejected as incomplete. If a complete picture is detected, the frame will isolate the workpiece picture on either of its ends perpendicular to the direction of the rows of the sensor elements. Thus it is possible to recognize succeeding objects as long as they are separated by the width of one row of sensor elements. The calculation of the center of gravity coordinates is done according to Eqs. (3) to (5). The summation for these equations is done with hardware accumulators during the read-in process of the individual rows of the sensed picture matrix. The division by the area is performed by the microcomputer, which is well suited for this purpose. The picture information is transferred via an interface to the computer's memory, where it is available for further processing. During the summation, the picture information is also read into two picture buffers which are later referred to for the calculation of the angular orientation.

One parameter for classification of the workpiece is its area. To increase the classification accuracy, circular segments about the center of gravity of the workpiece projection are placed over the images. The number of picture elements these segments have in common with the image serves as a second classifier (Fig. 8.83). This feature vector is independent of the angular orientation of the object. The elements of each circular segment are stored in individual buffers which are addressed by a PROM memory. For each address of a picture point it renders the corresponding number of the circular segment or its buffer address, respectively. The masks of the circular segments contained in the buffer have to be adjusted by an offset because of the random location of the object in the picture frame. This is done by adjusting the PROM addresses with an address offset about the center of gravity.

To determine angular orientation, the values of the coordinates x_s and y_s are transferred to the address control unit (Fig. 8.99). This circuit replaces Eqs. (22) and (23). During readout of the image buffer, the address control unit addresses the buffer in such a manner that at its output the picture information appears as the sequence $H(k,l)$. The circuit is designed with PROMs in which the circular function is stored. It also contains adder hardware which performs the arithmetic operations.

In the accumulator Accu 1, which is connected to the buffer, the sequence $H(k,l)$ for each k over l is summed up. Thereby the individual values of the function $C(k)$ are formed [Eq. (24)]. Since $H(k,l)$ is the binary vector of $h(k,l)$, the function of the accumulator can be realized by a simple counter.

The individual function values c_{k+p} are written into a shift register SR I. This means that after completion of the scanning procedure, the polar coding $C(k + p)$ [Eq. (25)] is available in sequential order in SR I. This reference function stored in the memory of the microcomputer is also brought sequentially into a shift register SR II. Teach-in of the reference image is done in this pattern recognition system during a special learning phase.

When the functions $C(k)$ and $C^R(k)$ are available in the desired sequence in SR I and SR II, the absolute difference function $D(k')$

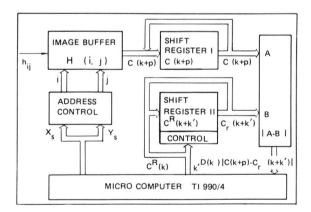

FIG. 8.99 Arithmetic unit of the angular orientation. (From Refs. 45 and 46.)

[Eq. (26)] can be calculated. The displacement variable k' specified by k causes a shift of $C^R(k)$ by k' steps. Then $D(k')$ is calculated for this cycle. During the calculations for each shift k', both shift registers are completely rotated once. From the two corresponding function values c_{k+p} and $c_{k+k'}$ the absolute difference is determined after completion of each shift and the result is summed up in the accumulator Accu II. When the entire shift cycle is completed, the individual value $D(k')$ is available at the output of accumulator Accu II. This value $D(k')$ is then clocked into the microcomputer where it is used to determine the minimum $D(k')$.

In this example the described circuit was built in bipolar TTL technology and operates with a clock frequency of 4 MHz. With the use of a scanning circle having 256 picture elements, 100 msec is necessary to determine the angular displacement. Compared to this method, a pure software system implemented on the same computer requires 30 sec for the calculation. By using this special computing circuit, a reduction in computation time by the factor of 300 was possible.

Software. The task of software in determining angular displacement can be divided into two parts: (1) performance of the teach-in phase, and (2) performance of the measuring phase (operating system).

1. *Task during the teach-in phase*: During the teach-in phase, classification and angular position characteristics of the objects are calculated and stored as reference characteristics to be used in the succeeding measuring phase. If this task were to be performed by electronic circuits, an unreasonable amount of hardware would be necessary. The teach-in is actually not time critical. Thus it is done by the microcomputer with the help of software. The software module for calculating angular displacement is able to generate the function $C(k)$ (Fig. 8.100).

The program is written in two modules to perform the following tasks: (a) calculate the coordinates of the centers of circles K_1 which are located on K_2, and adjust for the offset given by the random location of the center of gravity; and (b) calculate the functions $C^R(k)$.

(a) *Determination of the individual points of the circle K_2*: To reduce computation times, the points on the circle K_2 are not calculated with the help of the circle formula. A circle look-up table is constructed containing the x coordinates as a function of the y coordinates for the first quadrant of the circle K having a radius $R + \Delta r$. The y values correspond to the line numbers and the x values to the column numbers of the matrix. This table-look-up procedure eliminated multiplication and calculation of the square root.

When the center of gravity $S = (x_S, y_S)$ is known, the circle K_2 can be obtained from the look-up table by simple arithmetic operations (adding, subtracting). The following procedure is employed. Starting with the value $y = 0$, a loop is entered. After each pass through the loop, y is incremented by 1 and the corresponding x value is taken from the

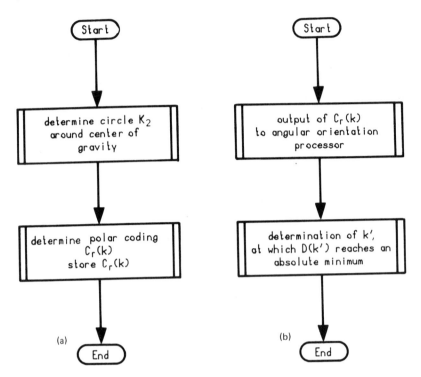

FIG. 8.100 (a) Teach-in phase; (b) measuring phase. (From Refs. 45 and 46.)

table. The equations shown below render four coordinate pairs for each of the four quadrants.

$y \in \{0, 1, \ldots, \text{final value}\}$

$x = x(y)$

Quadrant 1: $P_{Q1} = (x_s + x, y_s + y)$
Quadrant 2: $P_{Q2} = (x_s - y, y_s + x)$
Quadrant 3: $P_{Q3} = (x_s - x, y_s - y)$
Quadrant 4: $P_{Q4} = (x_s + y, y_s - x)$

These four coordinate pairs are stored in four successive quadrant tables after each pass through the loop (Fig. 8.101). When y reaches the value 33, x and y are interchanged. In a second loop, in which x is decremented by 1 after each pass, the y value corresponding to the x value is determined from the table. If x has reached the value 0, the coordinates of the circle K_2 and their relationship in respect to the center of gravity are available in the quadrant table (Fig. 8.101).

Parts Handling Assembly

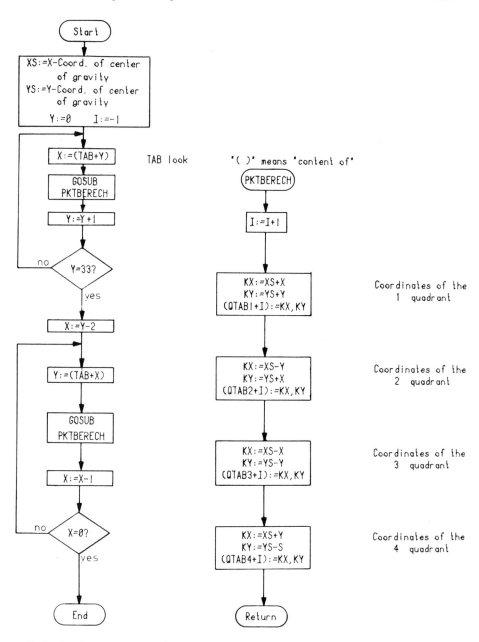

FIG. 8.101 Flowchart for the determination of the circle K_2: (a) main program; (b) table-build-up procedure. (From Refs. 45 and 46.)

(b) *Calculation of the function C(k)*: This part of the program needs more execution time, approximately 25 sec. For calculating the function, the picture buffer is searched for object points. When an object point has been found, a check is made to determine on which circle of the family of circles K_1 the point lies. For each line segment on the circle $K_{1,k}$ on which this point lies, the corresponding functional value c_k is incremented by 1 in the value table.

When all picture points have been scanned, the value table contains the complete function $C(k)$. In the last step of the teach-in phase, this value is stored in a table assigned to the individual object. This allows retrieval of the reference function $C^R(k)$ during a later measuring phase to determine the angular displacement of the object (Fig. 8.102).

2. *Task of the measuring phase*: During the measuring phase, the object is recognized with the help of its classification features. If an object is found, the angular displacement of the object can be calculated. The program for determining the angular displacement is divided into two modules: (a) output of the reference function $C^R(k)$, and (b) determination of k' at which D(k') reaches an absolute minimum.

(a) *Output of the reference function $C^R(k)$*: After recognition of the object, its angular reference function $C^R(k)$ which was established during the teach-in phase is passed on to the computing module to calculate the angular displacement (Fig. 8.99). To speed up the transfer of this function, a special parallel interface to the microcomputer was developed. This interface can be addressed similar to a memory

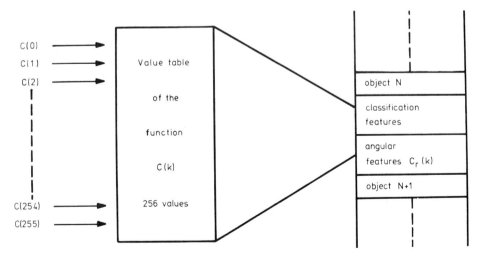

FIG. 8.102 Structure of the object record. (From Refs. 45 and 46.)

location. Thus it is possible to output the individual value of $C^R(k)$ with only one MOV instruction.

(b) *Determination of k' at which D(k') reaches an absolute minimum*: In order to find k' at which D(k') has an absolute minimum, the different values of D(k') are calculated for the points 0 to 255. These values are sent sequentially over the parallel interface to the module for determining the angular displacement. This module calculates the function value D(k'). The microcomputer reads D(k') and compares it with its preceding value. If the previous value is higher, it is replaced by the new value. The k' belonging to this new value is also stored. After all k' values have been considered, the value of k' at which D(k') has its absolute minimun is established (Fig. 8.103).

Figure 8.104 shows the function $C^R(k)$ for a small metal stamping (Fig. 8.73). The function C(k) of the rotated part is shown in Fig. 8.105 and the absolute minimum can be seen in Fig. 8.106.

EXAMPLE 8.3 This method allows the projection of an object to be recorded with a camera and stored in a 512 × 265 bit buffer [38]. As in Example 8.2, the calculation of the area and of the coordinates of the center of gravity is performed during read-in of the picture information into memory. The object traveling down a conveyor system is spotted by the camera and automatically fixed into a picture frame. This frame moves along with the conveyer belt until the coordinates of the center of gravity lie within the center of the frame (Fig. 8.107). A picture is taken of the object in this position. The frame can now be focused on the next object.

The following parameters are used for identifying the object (Fig. 8.108):

1. Area
2. Circular object segments defined by adjacent circles, whose centers coincide with the center of gravity
3. Number of points of intersection of scanning circles with the object's contour
4. Largest scanning circle that intersects the object

The user can freely select from equally spaced 128 scanning circles. They are placed over the object with the help of a CRT and should intersect the object at characteristic features. The maximum scanning circle R is automatically determined by the system, whereas the value of the angle α_{min} is supplied by the user.

The process of object identification and the measurement of the angular orientation is shown in Figs. 8.109 and 8.110. First the following independent object parameters are determined (Fig. 8.108):

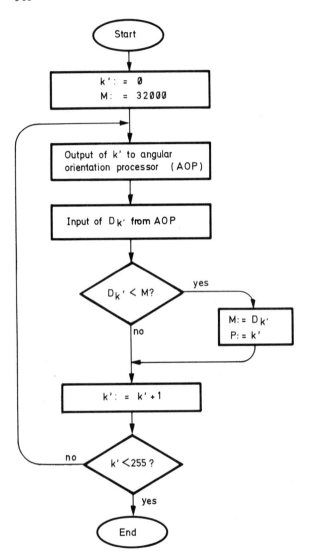

FIG. 8.103 Flowchart for determination of k', at which D(k') reaches an absolute minimum. (From Refs. 45 and 46.)

Parts Handling Assembly

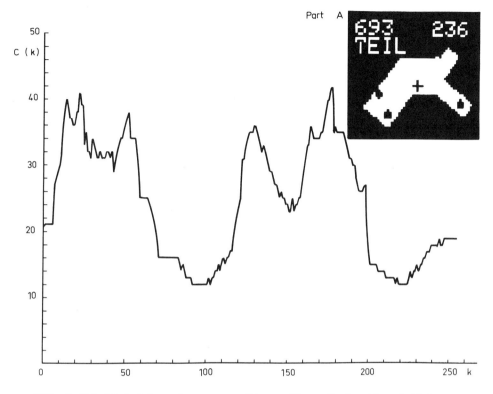

FIG. 8.104 Part A in reference position. (From Refs. 45 and 46.)

1. Area A of the projection of the object
2. Maximum object radius R_{max} about the center of gravity at a given α_{min}
3. Circular area segments intersecting the object as a function of the radius R about the center of gravity
4. Number of intersecting points of scanning circles with the object contour

In a second step, the following object-dependent calculations are carried out (Figs. 8.109 and 8.110):

1. Calculation of the autocorrelation maximum AC-MAX using the circular segments intersecting the object
2. Calculation of the cross-correlation maximum CC-MAX with the corresponding reference data

Figure 8.109 shows the reference values for eight different objects which are established during a teach-in phase. The parts to be

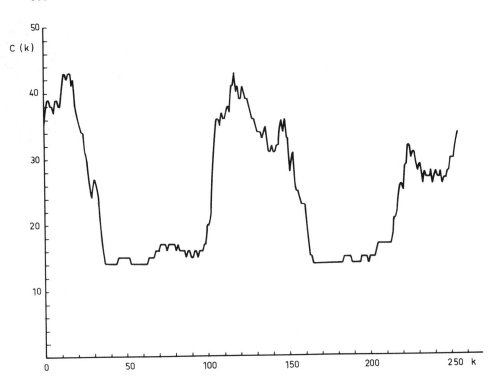

FIG. 8.105 Part A rotated 131°. (From Refs. 45 and 46.)

recognized are shown to the camera system. The number and position of the scanning circles are selected such that they assure an efficient feature discrimination. For the object to be identified during the measuring phase, the reference features are investigated in sequential order. The two fields containing $\Delta_{min}(A)$ and $\Delta_{min}(E)$ indicate that a reference object is searched whose areas A and edges E correspond to that of the measured object. Should the measuring data not be within given tolerances with data obtained during the teach-in phase, the recognition process of the object is terminated. If all parameters are uniquely identified, the object is known.

The angular orientation of the object is determined with the help of the cross-correlation method, whereby the measured data is compared with that obtained from an object in the reference position. From each of the scanning circles selected, an angle is obtained which relates the object to its reference position. Finally, a mean value is calculated from these different angles, indicating the closest approximation of the rotational angle of the object (Fig. 8.109 and 8.110). When compared with the method described in the first example, we find that this method has several disadvantages. The processing time

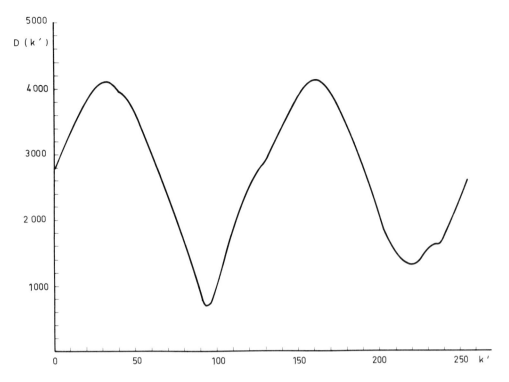

FIG. 8.106 Absolute difference function of the curves shown in Figs. 8.104 and 8.105. (From Refs. 45 and 46.)

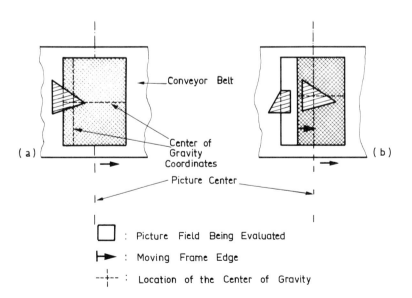

FIG. 8.107 Scanning of a picture field by an electronic frame. (Courtesy of Brown Boverie Cie, Mannheim, West Germany.)

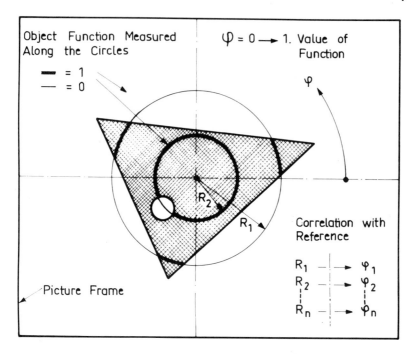

FIG. 8.108 Object-independent identification measurements. (Courtesy of Brown Boverie Cie, Mannheim, West Germany.)

of the picture increases with an increasing number of scanning circles and angles to be evaluated. Thus it is possible that the processing time of the picture could become slower than the cycle time of the conveyor system.

8.10.5 Multiple Objects in the Field of View

In many instances, more than one part may simultaneously be located in the camera's field of view. To save computing time, it is helpful to position the parts far enough away from one another so that the camera will only have one workpiece to identify at a time. In the cases of the two previous examples, it suffices if the parts follow each other at a minimum distance of one scanning line. The camera can then scan the flow in a piece-by-piece manner. If a scanning line intersects more than one workpiece at a time, a special algorithm must be used to separate the objects. If this method of separation is done by software, it is slow and difficulties may be encountered in adhering to the cycle time of the workpiece flow. It is thus

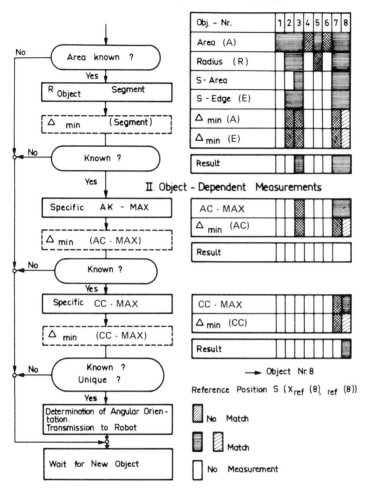

FIG. 8.109 Example illustrating the identification of an object. (Courtesy of Brown Boverie Cie, Mannheim, West Germany.)

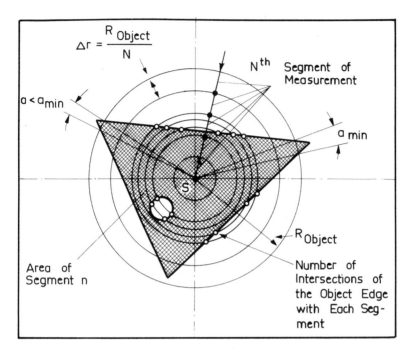

FIG. 8.110 Object-dependent measurement for identification and determination of angular orientation. (Courtesy of Brown Boveri Cie, Mannheim, West Germany.)

advantageous to make use of an algorithm that can be most easily realized in hardware. The following sections describe two methods of separating objects

Tracking of Workpiece Contours

The algorithm determines that an object must be scanned in a left-to-right, line-by-line fashion (Fig. 8.111 [48]). It is thereby only necessary to evaluate the store information from two neighboring scanning lines: information pertaining to the line presently being scanned (line K) and the line previously scanned (line K − 1). The individual lines are scanned, and points of intersection with objects O are evaluated. A vector $(x_1, l_1, r_1), (x_2, l_2, r_2) \cdots (x_n, l_n, r_n)$ for the line k_i is the result of this algorithm. If, for example, the x coordinate of the point of intersection is represented by the running index j, then the regions to the left and to the right of the intersection are described by l_j and r_j, respectively. If the object boundaries are continuous and are not touching, then for each edge (x_i, l_i, r_i) on a given line there is another edge (x_j, l_j, r_j) on the same line and $l_i = r_j$ and $r_i = l_j$ holds true.

Ascending numbers are assigned to the points of intersection on each line. The objects $O_1(K-1), O_2(K-1), \ldots O_n(K-1)$ are intersected by line $K-1$, and $O_1(K), O_2(K), \ldots, O_n(K)$ by line K. The algorithm uses two counters to assign the numbers to the points of intersection: one for the previously evaluated line, and one for the line currently under investigation. An object contour continues from one line to another if the following conditions are true (Fig. 8.111):

$$x_{j-1} < x_i < x_{j+1}$$

$$x_{i-1} < x_j < x_{i+1}$$

$$l_i = l_j$$

$$r_i = r_j$$

For these equations to hold true, the object's contour line must remain the same from one scanning line K to the next. No other contour may be intersected. If during the counting of the scanned edges, it is observed that edge i has ceased to continue as edge j, either the end of an edge has been reached or a new edge has been encountered.

When a new object appears, $x_{j+1} < x_i$ (Fig. 8.111b). The upper part of this object is identified by the line that runs through points x_j and x_{j+1} on line K. Thus two objects are now identified, $O_j(K)$ and $O_{j+1}(K)$. The point x_i on the right edge will now continue as x_{j+2}. When an edge scanned by line K terminates, the equation

$$x_{i+1} < x_j$$

becomes valid. In this case it is assumed that points x_i and x_{i+1} are combined on line $k-1$. A differentiation can be made between the following cases:

1. Termination of an object when $O_i(k-1) = O_{i+1}(k-1)$ (Fig. 8.111c).
2. Two objects have merged to become a single entity if $O_i(k-1)$ $O_{i+1}(k-1)$ $O_i(k-1)$ and $O_{i+1}(k-1)$ have the same parent (Fig. 8.111d).
3. One object encircles another (Fig. 8.111). If this is true, then:
    a. If $O_i(k-1)$ is the parent of $O_{i+1}(k-1)$, then $O_{i+1}(k-1)$ encircles $O_i(k-1)$.
    b. If $O_{i+1}(k-1)$ is the parent of $O_i(k-1)$, then $O_i(k-1)$ encircles $O_{i+1}(k-1)$.

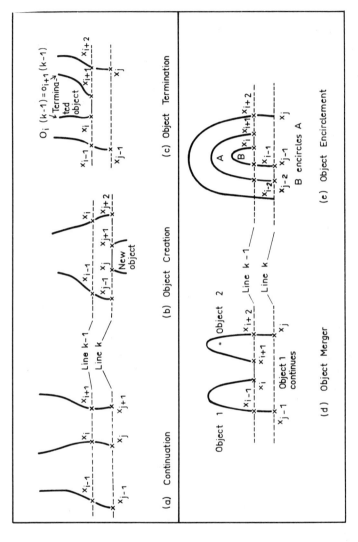

FIG. 8.111 Object tracking events: (a) continuation; (b) object creation; (c) object termination; (d) object merger; (e) object encirclement. (From Ref. 48.)

Parts Handling Assembly

In cases (a) and (b), the algorithm must keep track of the parent-child relationships of the objects (Fig. 8.112). A is the parent of B if A encircles B and B is encircled by no other object.

If a new object $O_j(K) = O_{j+1}(K)$ is identified on line K by points x_j and x_{j+1}, then $O_{j-1}(K)$ is the parent of $O_j(K)$. This algorithm can also establish a hierarchical order of the objects in the camera's field of view (Fig. 8.112). The output parameter of the algorithm is an object number that assigns the points of intersection to the contour lines. To be separated, the workpiece must be uniquely marked. Unfortunately, the workload generated by this method often slows down the computer. Thus other, simpler methods are generally used for the separation of objects.

Marking of Objects in the Field of View

There are numerous methods of performing this task. In the simplest, the object is scanned in a line-by-line fashion. As the object moves through the camera's field of view, its running length is registered and the parts of the object that are intersected are uniquely marked (Fig. 8.113). If more than one object is intersected by the line, a different mark is assigned to each. With each succeeding scanning line, the new running lengths are compared to those of the neighboring and preceding line. If neighbors are determined, both

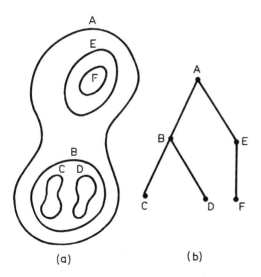

FIG. 8.112 Contour map and object hierarchy. (From Ref. 48.)

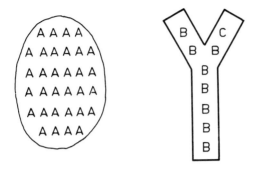

FIG. 8.113 Marking of several objects.

receive the same marking; otherwise, a new identifying mark is given. If the shape of the part is such that a line enters and leaves the object more than once, ambiguous results will be obtained. As a rule, the first mark is followed through. Upon reaching the end of the object, the entire list of markings is then examined to determine whether or not they may belong to the same object. If so, all markings are changed to the first one that was assigned to the object.

REFERENCES

1. Worldwide 30000 Are in Use, *VDI Nachrichten*, No. 5, Feb. 4, 1983.
2. J. Vollmer, *Industrial Robots*, VEB Verlag Technik, East Berlin, 1981.
3. A. S. Kondolean, Application of Technology Economic Model of Assembly Techniques to Programmable Assembly Configurations, S.M. thesis, MIT Mechanical Engineering Department, Massachusetts Institute of Technology, Cambridge, Mass., 1976.
4. J. M. Evans, et al., Computer Science and Technology, NBS/RIA Robitics Research Workshop, Delphi Forecast, National Bureau of Standards Publication No. NSBSP 500-29, Washington, D.C., 1978.
5. J. Deanvit and G. S. Hartenberg, A Kinematic Notation of Lower Pair Mechanisms Based on Matrices, *Journal of Applied Mechanics*, Vol. 22; *Trans ASME*, Vol. 11, 1955.
6. C. Blume and R. Dillmann, *Free Programmable manipulators*, Vogel Verlag, Würzburg, West Germany, 1981.
7. R. Dillmann, A Sensor Controlled Gripper with Tactile and Nontactile Sensor Environment, *Proceedings of the Conference on Robot Vision and Sensor Controls*, Stuttgart, West Germany, Nov. 1982.
8. E. Eberhardt, Development of a Functional Unit for Gripping of Objects, Master thesis, Institut für Informatik III, University of Karlsruhe, Karlsruhe, West Germany, 1981.

9. B. Faller, Design and Realization of a Sensor System Based on a Microcomputer, Master thesis, Institut für Informatik III, University of Karlsruhe, Karlsruhe, West Germany, 1981.
10. R. Dillmann and B. Faller, A Force-Torque Sensor System for Industrial Robots, *Elektronik*, Vol. 8, Apr. 1982.
11. R. Dillman, A Structured Multiprocessor System for Adaptive Sensor Controlled Assembly Robots, CAPE Conference, Amsterdam, Apr. 1983.
12. E. Freund and H. Hoyer, The Principle of Non-linear Decoupling of Systems with Application to Industrial Robots, *Regelungstechnik*, Vol. 28, Part 3, 1980.
13. R. C. Paul, Modeling Trajectory Calculation and Servoing of a Computer Controlled Arm, A.I. Memo 177, Stanford University Artificial Intelligence Lab., Stanford, Calif., Sept. 1972.
14. D. L. Pieper, The Kinematics of Manipulators Under Computer Control, Ph.D. dissertation, Stanford University, Stanford, Calif., 1968.
15. B. K. P. Horn and M. H. Raibert, Configuration Space Controls Technical Report AI-M-458, MIT Artificial Intelligence Laboratory, Cambridge, Mass., 1978.
16. R. Paul et al., Advanced Industrial Robot Control Systems, Second Report, School of Electrical Engineering, Purdue University, West Lafayette, Ind., July 1979.
17. C. Blume and W. Jakob, *Programming Languages for Industrial Robots*, Springer-Verlag, New York, 1985.
18. R. Paul, *Robot Manipulators*, MIT Press, Cambridge, Mass., 1981.
19. U. Rembold, C. Blume, R. Dillman, and K. Mörtel, Technical Requirements for Future Assembly Robots, Part 4, VDI-Zeitschrift, No. 21, Nov. 1, 1981.
20. S. Bonner and G. Kang, A Comperative Study of Robot Languages, *Computer*, Dec. 1982.
21. L. Lieberman and M. Wesley, AUTOPASS: An Automatic Programming System for Computer Controlled Mechanical Assembly, IBM Journal of *Research and Development*, Vol. 21, No. 4, 1977.
22. VDI-Proposal for Standardization, VDI 2863, Beuth Verlag, West Berlin, 1983.
23. *Computer Vision and Sensor Based Robots*, New York, 1979.
24. The Advent of Adaptable Programmable Assembly Systems, *Manufacturing Engineering*, Apr. 1979.
25. C. Blume and W. Jakob, Design of the Structured Robot Language (SRL), *Proceedings of the International Meeting on Advanced Software in Robotics*, Liège, Belgium, May 4–6, 1983.
26. C. Blume, VAL—A Robot Control System of Unimation, unpublished, University of Karlsruhe, Karlsruhe, West Germany, 1980.

27. S. Mujtaba and R. Goldmann, *AL Users' Manual*, Stanford University, Stanford, Calif., 1979.
28. K. Jensen and N. Wirth, *PASCAL—User Manual and Report*, New York, 1975.
29. R. Fikes, P. Hart, and N. Nilsson, Learning and Executing Generalized Robot Plans, *Artificial Intelligence*, Vol. 3, 1972, pp. 251–288.
30. R. Dillmann, A Graphical Emulation System for Robot Design and Program Testing, *Conference Proceedings of 13th ISIR*, Robots 7, Chicago, July 1983.
31. G. Nehr and P. Martini, The Coupling of a Workpiece Recognition System with a Robot, in *Robot Vision*, A Pugh, ed., Springer-Verlag, New York, 1983.
32. U. Rembold et al., A Very Fast Vision System for Recognizing Parts and Their Location and Orientation, *Conference Proceedings of the 11th ISIR*, Washington, D.C., 1979.
33. C. A. Rosen, Machine Vision and Robotics: Industrial Requirements. Stanford Research Institute International Technical Note 174, SRI Project 6284, Nov. 1978.
34. A. Klemt, Optical Data Acquisition, VDI-Nachrichten, No. 2, Jan. 13, 1978.
35. J. M. Hill, How to Read a Label That's Moving 17 mph, *Industrial Engineering*, Oct. 1969.
36. U. Rembold, Towards a Programmable Factory, *Europe Industrie-Revue*, Vol. 11, Oct. 1977.
37. K. Armbruster, Automatic Camera Systems for Distance Measurement, *Zeitschrift für industrielle Fertigung*, Vol. 69, No. 5, 1979.
38. R. Karg, A Flexible Opto-electronic Sensor, *Conference Proceedings of the 8th ISIR*, Vol. 1, Stuttgart, West Germany, 1978, pp. 218–229.
39. *Hütte*, Vol. I, 28th ed., Verlag von Wilhelm Ernst & Sohn, Berlin, 1955.
40. J. D. Dessimoz, Visual Identification and Location in a Multi-object Environment by Contour Tracking and Curvature Description, *Conference Proceedings of the 8th ISIR*, Stuttgart, West Germany, 1978.
41. J. Bretschi, A Microprocessor Controlled TV-Sensor for Object Recognition and Position Measurement for Industrial Robots Applications, *IITB-Mitteilungen*, 1976.
42. K. Koskinen and A. Niemi, Object Recognition and Handling in an Industrial Robot System with Vision, *Conference Proceedings of the 8th ISIR*, Stuttgart, West Germany, 1978.
43. H. Geißelmann, A TV-Sensor for Object Recognition, Positioning and Quality Control, FhG-Berichte 2, 1977.
44. O. E. Lanz, A Sensor for Position and Form Recognition, in *Automatisierung im Wandel*, Interkamongreß 1977, Springer-Verlag, Berlin.

45. G. Nehr and P. Martini, Recognition of Objects and their Orientation by Optical Sensors, *VDI-Zeitschrift*, Vol. 121, No. 10, May 1979.
46. K. Armbruster, P. Martini, G. Nehr, and U. Rembold, A Real Time Vision System for Industrial Application, *2nd IFAC/IFIP Symposium on Information Control Problems in Manufacturing Technology*, Stuttgart, West Germany, Oct. 1979.
47. P. Kammenos, Performance of Polare Coding for Visual Localization of Planar Objects, *Conference Proceedings of the 8th ISIR*, Stuttgart, West Germany, 1978.
48. R. O. Duda and R. E. Hard, *Pattern Classification and Scene Analysis*, Wiley, New York, 1973.
49. C. Rosen et al., Exploratory Research on Advanced Automation, Report 3, SRI Project 2591, Stanford Research Institute, Stanford, Calif., 1974.

9
Simulation of Manufacturing Processes

9.1 INTRODUCTION

The simulation has become an important tool when new manufacturing processes are conceived or existing ones are altered. With the simulation a mathematical model is built of the process which allows emulation of the manufacturing operation on a computer. Thus the engineer is able to observe the behavior of a process without the necessity of experimenting with the actual equipment. The engineer may want to try out different manufacturing runs, new additions, or a new layout of equipment and watch for the performance, the slowdown or speedup of cycle times, possible disturbances, and so on. There are several different simulation methods and languages available which simplify the simulation work considerably. This chapter discusses the principle of simulating manufacturing systems and gives an overview of the most important simulation tools.

9.2 SIMULATION METHODS

9.2.1 Definitions

A simulation can be described as a software tool that consists of methods, activities, and resources. In the context of this book discrete simulation is used to analyze a manufacturing process. A simulation is a model of a process that can be used to emulate it. When a practical simulation system is considered, this definition is somehow expanded. Thus a whole simulation project will contain the following subjects:

1. Planning of the simulation activity, setup of the simulation requirements, definition of the expected results, and planning of alternative simulation runs

Simulation of Manufacturing Processes

2. Problem analysis, data analysis, and data collection and processing
3. Design of the simulation model
4. Software realization of the simulation model
5. Performance of the simulation run
6. Evaluation and interpretation of the simulation results
7. Documentation of the simulation
   a. Prerequisites
   b. Used methods
   c. Simulation aids
   d. Results

A very important aspect of the simulation is to test the dynamical behavior of the model with respect to time. With this feature it is possible to investigate the stochastic behavior of the simulated object and to make visible complex system behavior. It is possible to state two basic goals for the performance of a simulation:

1. An existing process is simulated, whereby it has to be adapted or expanded to changing operating conditions.
2. A new process is investigated to avoid expensive pitfalls and to recognize the reaction to extreme operating conditions.

Under certain conditions the simulation can also be used to check performance data of the vendor specification of a new process. A requirement specification phase, a problem analysis phase, and a model phase must precede each simulation run. A simulation model is a simplified image of the real process. With it an abstract description is made of the objects and their properties. The required degree of accuracy, which is related to the simulation effort, depends on the process to be modeled and the given financial and time framework within which the study has to be performed.

9.3 ITERATIVE CHARACTER OF THE SIMULATION

A simulation study is a twofold iterative process. First, the simulation itself is done iteratively, whereby each succeeding step is a redefined or modified stage of the preceding one. Second, the evaluation of the simulation results is performed iteratively, whereby new input data are successively adapted to results of the preceding simulation run to find an optimum. Figure 9.1 shows the different phases of a simulation study; for example, when a model is being realized it is possible that a wrong simulation language was selected. In this case a step backward has to be taken to search for a new simulation aid.

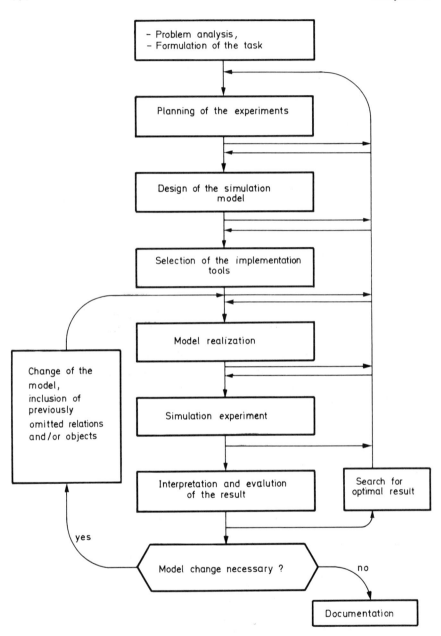

FIG. 9.1 Various phases of a simulation study.

Simulation of Manufacturing Processes

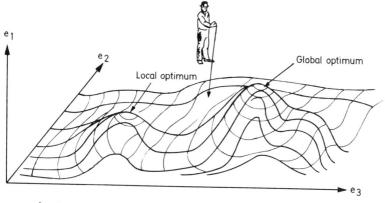

e = Input parameter

FIG. 9.2 Global and local optimum of a defined search space.

From the mathematical point of view a simulation is not an optimization procedure; despite this fact, an attempt is often made to manipulate the input data such that optimal output data are obtained. The iterative search for an optimum may lead to a problem since it will never be known if a local or global optimum has been reached (Fig. 9.2).

One can visualize the results of a search for an optimum as the activity of a navigator who is investigating the depth of an invisible ocean floor with a sonar device. The person never knows in which direction to continue the search as soon as a local optimum has been found. This problem could be resolved by systematically scanning the entire space in small increments. This may be impossible because of prohibitive cost and time factors. From a practical point of view this search can be done only on a sample basis.

9.4 ADVANTAGES AND LIMITATIONS OF SIMULATION

A simulation should be performed only when the solution of a problem is more difficult to obtain by a conventional analytical method. Simulation is not an exact method; it can be of advantage only if all prerequisites, assumption, constraints, and possible sources of errors are taken into consideration. Its most important advantage lies in the almost unlimited variations of the real process which can be investigated when the behavior of a process has to be known. This can be done independently of the process and without disturbing it. In most cases other process alternatives may be investigated with little additional effort.

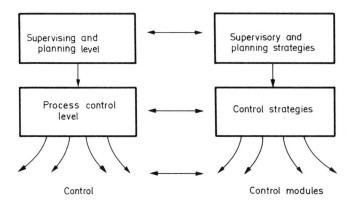

FIG. 9.3 Analogy between the reality and the levels of the simulation model.

The engineer should keep in mind that the simulation experiment is of statistical nature and that the results should be treated as such. The analogy between the reality and the simulation model holds true for all hierarchical planning and control levels (Fig. 9.3). For example, the actual process control level of the real system is designed by the control strategies of the simulation model. Simulation cannot be a sole substitute for the planning operation; it always requires a preceding design phase. This phase may be done by two different methods:

1. One starts with a very simplified model of the total system and enhances those parts in a subsequent step which are of particular interest.
2. First all critical parts of a system are investigated and then the parts are combined to a global model.

The aspect of refinement and modeling of all details may generate problems. When originally a simulation was started with a simplified model and thereafter progressively a refinement is done with the help of complex statistics, constraints, and system parameters, an economical limit can be reached at which the entire effort can no longer be justified.

Another more psychological danger may arise when the results of the simulation are interpreted. For example, it is possible to output the value of a parameter with an accuracy of five digits after the decimal point. The inexperienced user of the model may obtain a wrong sense of accuracy when interpreting the results. A good model will render output values that actually represent the accuracy of its input parameters.

Problems may also arise in case the results of the simulation are displayed as graphics. For example, a user may view the travel of a pallet on an output screen. The user may be observing the formation of queues, the operation of switches, and the interaction of system components. However, a wrongly assumed travel speed may falsify the entire simulation run. On the other hand, a well-conceived graphics may be the best means of representing complex operating conditions and the dynamic behavior of the process. This may be the most convincing argument to use a simulation.

9.5 PROGRESSION OF A SIMULATION PROJECT

In this section problems will be discussed which are characteristic of a simulation run. The simulation may be used as an aid to plan a new facility or to change an existing one. In general, the time available to perform the simulation is very short, since this activity usually precedes a facility planning phase and makes fundamental data available to it. In general, a simulation project is under time pressure because:

1. Traditional planning usually is done in a shorter time frame as is necessary for a simulation.
2. It is customary to spend more time in the later project phases rather than in the early ones.

There is a great danger that the effort required for a quick implementation of the simulation model is underestimated. Despite the fact that many standard simulation tools are available, every application has unique decision rules and a global structure which require its own solution. For example, in a material flow system, handling of pallets may be of interest. It is possible to define standard rules to transport a pallet from one machine to another. However, the decision as to which pallet has to be assigned in which order and which machining sequence has to be performed has to be stated specifically for a given project. There is no standardization possible at a global decision level.

Another problem is the starting phase of the simulation project. During planning there are many undefined problems when it is necessary to specify data and algorithms. For this reason an arbitrary state has to be set at which the simulation model is assumed to represent the process to be simulated. During simulation new parameters may have to be entered into the model because new data were obtained from a supervisory level. In this case a decision has to be made as to whether the change is needed at all and if the simulation implementation or part of it has to be repeated. Thus a simulation project always contains a risk factor which neither the designer nor the user of the program can fully anticipate. The entire effort may have been underestimated and the completion of the simulation project may be unduly delayed.

Another effect which can often be observed is that the simulation run perpetuates itself. Once the program has been completed and acceptable results are obtained, the user often tries to satisfy his or her own curiosity and simulates numerous production alternatives. The answers may require additional runs and there is no end to be seen to the simulation. For economic reasons this cycle cannot continue indefinitely and it should be terminated at a defined state.

When a simulation model is acceptance tested by a customer, problems may arise due to the fact that insufficient or assumed data were used during its development phase. This may not be the fault of either the developer or the user. There may have been parameters missing which could be detected only after the first simulation results had been obtained. Such undesirable problems often lead to an unsatisfactory user-supplier relationship.

9.6 INITIALIZATION STATE

This state defines the start of the simulation. The simulation program is either used the first time or a new simulation run is started. One distinguishes between two possibilities to initialize the system. In the first case the system is "quasi-empty"; in other words, the internal states and data have not been generated by a previous run. The simulation usually is allowed to cycle through an initialization phase until a steady state has been reached. Normally during this time there are no statistical data generated in order not to falsify the "artificial" start state. The end of the initialization phase is determined experimentally.

For practical purposes a second method is more useful. Here the system is brought into a well-defined variable start state. From it the simulation may commence. For example, the user may store a defined simulation state on a storage device and may retrieve it again to start a simulation run with this state. Of course, this leads to a higher implementation effort.

9.7 VALIDATION PROBLEMS

During the validation phase the model is compared with the relevant image of the real world. In case a manufacturing process has to be changed, it is possible to validate the prevailing state with the simulation model and to compare the results with measured data. However, changes in the system structure cannot be detected. If a new manufacturing process is to be planned, another method has to be applied.

Simulation of Manufacturing Process 669

First an attempt is made to build and test small modules of the simulation model. This is simple in case good software modularization techniques are used. The next step is to build and test the supervisory control structure. For this reason it is necessary to issue internal simulation data and states (e.g., the size of an order, the types of pallets to be selected, or the machine tools to be used). Such global strategies are advantageously tested with the help of graphics or a display screen. This allows viewing the entire process as a function of time. Microscopic problems may stay undetected, however. For example, when a material flow system is investigated, the user can follow the travel of a pallet on a conveyor or can observe the operation of a switch and the formation of a queue. Priority rules to pass a switch or the adherence to operating sequences can be checked through an output device with the help of internal data.

9.8 CONCEPTS OF SIMULATION SYSTEMS AND PROGRAMMING LANGUAGES

There are numerous aids available to build a simulation model. An essential tool to simulate material flow and manufacturing systems is the computer. Pencil-and-paper methods are suitable only for very small models. For this reason they are of no practical use.

A simulation model can be implemented with the following tools:

- Higher technical or scientific programming languages
- Instruction or block-oriented simulation language
- Problem-oriented simulation system (building module)

Before these concepts are discussed in more detail, some basic tools will be presented which are integral parts of a simulation system.

9.9 STANDARD TOOLS OF A SIMULATION SYSTEM

The following aids are needed to perform and evaluate a simulation run:

- Random number generator
- Probability distribution functions
- Statistical tests
- Statistical functions
- Time control mechanisms
- Protocol functions

The programmer has access to these functions either through special language elements or he or she uses them indirectly with the aid of system modules. A typical function is the random number generator.

9.9.1 Presentation of Stochastic Parameters

The input parameters for a simulation model are usually of a stochastic nature and must be automatically generated. These parameters can be obtained from the real manufacturing system. However, this is usually not possible, for the following reasons:

1. A real system must be in existence before the values of the parameters can be measured.
2. A long time span may be needed to obtain the data.
3. A great number of data must be acquired and stored.
4. There is no flexibility for extended experiments with the model.

Measurements are usually done only to test the model and to compare the results with the real process and to get a hypothesis for the distribution function. The values of the system variables are generated with random numbers and distribution functions. The method is based on the hypothesis that the generated values have the same probability distribution as the real system parameters.

For these methods a sequence of random numbers with a nonuniform distribution must be generated. In this sequence the value of each succeeding number is independent of that of the preceding one and there are no rules which can relate the numbers to each other. Besides the physical methods, which are of no importance in this context, there are no other methods available which can generate real random numbers. Algorithms do, however, exist which provide a reproducible sequence of uniformly distributed random numbers. These numbers are called pseudonumbers and do not completely satisfy the above-mentioned requirements. They must, however, comply with the satistical test to be suitable for the simulation experiment.

A method most widely used is based on the linear congruence algorithm, often called the residue method. Given three constants a, c, and m, the algorithm generates the (n + 1)th number from the nth number according to the following rule:

$$x_{n+1} = (ax_n + c) \bmod m$$

The method starts with a seed value x_0 and the parameters a, c, and m. It generates a reproducible sequence of numbers. The set of numbers that can be generated is finite. Ultimately, the same number of a sequence repeats an earlier one and the entire sequence starts over again. For this reason it is necessary to select the constants a, c, and m very carefully so that long sequences are obtained. The value for m is usually a power of the numbering system being used (e.g., 10^n).

For most simulation experiments nonuniformly distributed random numbers are needed (e.g., with an exponential distribution). For

Simulation of Manufacturing Processes

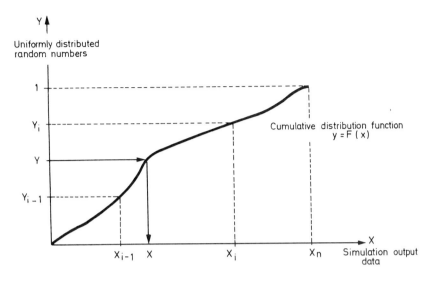

FIG. 9.4 Generation of nonuniform random numbers.

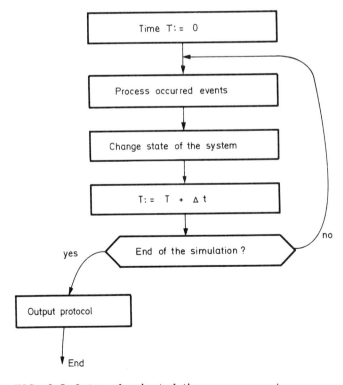

FIG. 9.5 Interval-oriented time management.

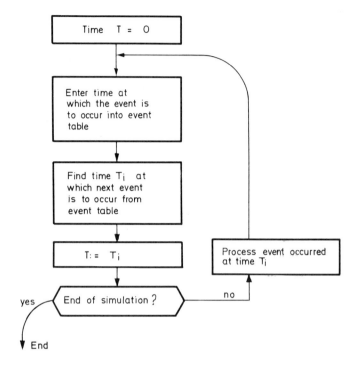

FIG. 9.6 Event-oriented time management.

this purpose the inverse of the cumulative distribution function is evaluated with a sequence of uniformly distributed random numbers (Fig. 9.4). One starts with the generation of the random number y and calculates the simulation value $x = F^{-1}(y)$.

Frequently, the distribution function is obtained as a hypothesis from experimental statistical data. To verify the hypothesis, statistical tests are used. One of the best known methods is the chi-square test. It determines the deviation of the empirical distribution from the theoretical one. The result is a measure of probability to accept or reject the hypothesis.

9.9.2 Time Control Mechanisms

Another important aspect of a simulation is the time control mechanism. The passage of time is recorded by a number that represents the clock time. It records the simulated time, the occurrence of events, and state changes. There are two basic methods for controlling the clock time. They are (1) the time-slice-oriented method, and (2) the event-oriented method. In the first case the clock is incremented by small time intervals

Δt. At each interval a check is made as to whether an event is due to occur at that time and if a system change has taken place (Fig. 9.5).

With the event-oriented method time-keeping in the model is easy; however, the selection of the time interval Δt may lead to problems. If Δt is small, long computer runs will result. On the other hand, a large Δt may indicate a system behavior which does not represent the dynamics of the real process. Since all actual event times are placed at the end of Δt, dead times will result during which the system cannot react.

This problem is avoided with the event-oriented method. Here the simulation time proceeds under variable event intervals. The length of the interval is determined by the time period at which two events succeed each other (Fig. 9.6). For the event management of a complex system with frequently occurring events, the disadvantage of this method may be long computing times and large memory requirements.

The event-driven method should be selected for a simulation where only a limited number of events occur and where the time variables differ considerably. In case there are frequent events or constant time intervals, the interval-driven method is of advantage.

9.10 HIGHER PROGRAMMING LANGUAGES

In principle every high-level programming language may be used to formulate a simulation model. For this reason they are being used very often. With them it is possible to simulate any task by any simulation method. This flexibility, however, results in a great implementation effort, in poor repeatability, and in problems in evaluating the simulation method. In addition, the programmer often has to add standard functions as they were discussed in the preceding sections.

From the software engineering point of view, modern programming languages such as PASCAL and MODULA are better suited for a simulation than the languages FORTRAN, PL/1, or BASIC. They are of particular advantage when the model needs frequent maintenance. The language SIMPAS (simulation by PASCAL) uses for the management of objects and the states a list-driven system written in PASCAL. This facilitates the definition of the initialization states and a halt of the simulation at any point. Its existing state may be written into a buffer and can be retrieved when the simulation is to be restarted from a defined point.

An event-driven time control was implemented in PASCAL by providing an event list and a corresponding list management method. A special problem poses handling of concurrent events. In the simplest case it suffices to process these events in the same sequence as they have been entered into the event list. However, in case two events pertain to the same object, for example to a switch, another solution has to be found and implemented. For this special case the standard modules, such as random number generator, statistical distribution, and protocoling, are implemented as PASCAL procedures.

9.11 STATEMENT OR BLOCK-ORIENTED SIMULATION LANGUAGES

The development of the simulation languages GPSS, SIMSCRIPT, and SIMULA started in the 1960s. They were special application-oriented languages or programming systems. The concept of these languages is based on a syntax which represents a world view of a specific class of applications and describes a defined system structure. Examples of the system features are unit, activity, transaction, class, or process. The user first has to learn the terminology in order to translate his or her application to the world view of the language. Within the framework of these languages the systems consist of (1) temporary facilities and (2) permanent ones. Temporary facilities are objects that enter a system, reside there for a certain time, and then leave it again. For example, with a manufacturing plant they are orders, workpieces, or disturbances. Permanent facilities are objects that reside in the system as long as it exists. In respect to a manufacturing plant, such facilities are machine tools, conveyors, or control elements. Permanent and temporary facilities have properties that identify and describe objects and their states. Examples of these properties are code number, type, performance, hourly rate, machine load, and sequences.

Simulation languages may be classified according to different concepts or world views, for example:

Event oriented
Process oriented
Translation oriented
Activity oriented

In the following sections three representative simulation languages of the many available ones are discussed:

GPSS: General Purpose Simulation System
SIMSCRIPT: Simulation Scripture
SIMULA: Simulation Language

Each of these languages is based on a different concept.

9.11.1 GPSS as an Example of a Transaction-Oriented Modeling System

With GPSS the temporary facilities are called transactions. They are static elements which simulate the dynamic behavior of the system by following the flow of transactions from one equipment unit to another. The equipment is represented by different units of fixed properties. For example, there is the permanent unit "facility" which will be occupied by a transaction and released after use. Another unit is the storage to which several transactions may have access in parallel. A special unit is the queue or user chain, which is automatically built up for each station or which may be implemented by the user. The general features of the equipment are:

1. *Station*: occupancy status, load, number of transactions, and their mean residency
2. *Storage*: actual contents, number of vacant places, load, mean and maximum contents, and mean residency
3. *Queue*: length, mean and maximum number of queues, and number and residency of transactions

The properties of a transaction can be defined by up to 100 numerical parameters, including priority rules, residency time, and logical attributes.

With the GPSS model the temporary and permanent facilities only appear indirectly. They form a data model in the background by which the user describes a system. The model itself is constructed from block commands, each of which performs specific functions which are unique to the language. The blocks form transactions and advance them in a flow pattern through the simulation at a specified sequence. The transactions may be changed or deleted. Figure 9.7 shows the modeling elements of GPSS. Each block type might have a name, a symbol, or a number attached to it. A specified time is consumed when a transaction is processed. With the help of the blocks the user is able to devise a flow diagram which models a process. A block can handle one facility or multiple items simultaneously; for example:

1. *Originate*: creates a transaction
2. *Queue*: creates storage space and gathers queueing statistics
3. *Split*: creates multiple transactions from one transaction
4. *Hold*: stores a transaction for a defined time period
5. *Terminate*: deletes a transaction
6. *Delay*: delays a transaction

A model is coded as a program, whereby the blocks are described with instructions.

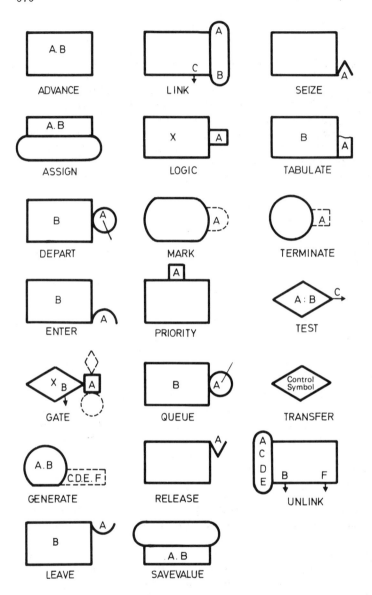

FIG. 9.7 GPSS block symbols. (After G. Gordon, *System Simulation*, Englewood Cliffs, New Jersey, 1978.)

Simulation of Manufacturing Processes 677

GPSS is an often used block-oriented language which allows the construction of small models in a short time. It is not very useful for modeling large systems in which a multitude of variables have to be investigated. GPSS is well suited to handle inventory, material flow, and queueing-type problems. The simulation runs are usually very long.

9.11.2 SIMSCRIPT as an Example of an Event-Oriented Modeling System

With SIMSCRIPT the user defines the model units needed for a simulation. This is contrary to GPSS, where all function modules are predefined. Examples of such modeling entities are a magazine, an engine block a workpiece, or a machine tool. Attributes are assigned to the entities similar to the type declaration used by high-level programming languages. During the simulation run duplicates of a prototype can be generated and may be used as a variable, as is done with programming languages.

The individual modeling entities are declared in a declaration part which is called a PREAMBLE of the program. This is similar to PASCAL or ADA. The sequence of the definition is arbitrary, whereby a distinction has to be made between (1) a permanent unit, (2) a temporary unit, and (3) a system property. By convention the unit has the following syntax:

   – EVERY <unit> HAS A <attribute> {, A attribute}

   [AND A <attribute>]

The programmer can freely select the identifier for entities and attributes, for example:

   EVERY robot HAS A joint_number AND A tool_identification.

Since the modeling entities are internally represented by the data type, the attributes are declared as:

   DEFINE joint_number, tool_identification AS INTEGER VARIABLES.

The form of the declaration is the same for temporary and permanent units. The generation of permanent units can be done by two methods. In the first case a CREATE instruction generates via a loop a number of units:

   FOR i = 1 TO n CREATE A machine.

Internally, a group counter is set up with the designator

   N. machine

With the help of the group counter, the generation of the units can be done implicitly:

   LET N.machine = n

   CREATE EACH machine

The results are realized internally by a dynamic array for each N. unit. The selection of a unit is done with an index, for example:

   LET machine(5) = ∅

Contrary to this the temporary units can be accessed only by pointers. The declaration is called TEMPORARY ENTITY, for example:

   TEMPORARY ENTITY

   EVERY object HAS A name AND A weight.

For the generation of a temporary unit, a pointer name is normally included. Otherwise, the name of the unit is used as pointer, for example:

   CREATE AN object CALLED pump

   CREATE AN object

Figure 9.8 shows the generation of temporary units, which contrary to the permanent unit, may be deleted with the instruction DESTROY, for example:

   DESTROY object CALLED pump.

Thereby the storage space for the array pump is released.
   SIMSCRIPT offers a powerful list concept to handle queues or other entities. A list consists of chained elements and a list header. The

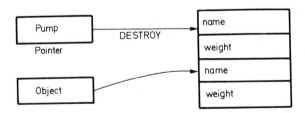

FIG. 9.8 Generation and deletion of temporary units.

Simulation of Manufacturing Processes

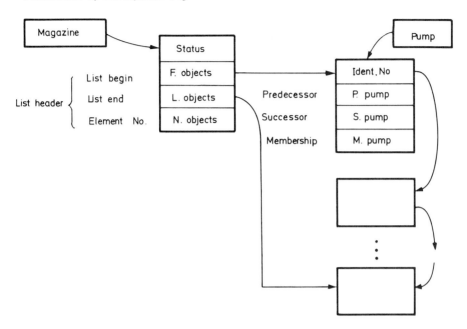

FIG. 9.9 Setup of chained lists in SIMSCRIPT.

latter is assigned to a unit and contains pointers to the list beginning and the end and to the element number (Fig. 9.9).

Elements of a list contain explicitly defined attributes, pointers to their predecessor and successor, and a membership specifier. The list structure shown in Fig. 9.9 was created by the following instructions:

PERMANENT ENTITY

EVERY magazine HAS A status AND OWNS objects

TEMPORARY ENTITY

EVERY pump HAS A identification_number AND MAY BELONG TO objects

. . .

CREATE A pump

FILE THE pump TO objects (magazine)

SIMSCRIPT includes the concept of event-oriented time control. Events are defined by the user. For each event the state change and the resulting events are recorded. Thus it is possible to represent changes and their interactions and the propagation of a change through the entire system. Internally, an event list is kept which is called a timekeeping calendar or timetable. The user declares the events and may assign different attributes to them, for example:

EVENT NOTICES

INCLUDE motor_on, drill_end AND feeder_off

The actions generated by an event are defined with event or interrupt routines. From the outside they look like subroutines. However, the triggering event is the header:

EVENT feeder_off

REMOVE FIRST object FROM feeder

SCHEDULE motor_on AFTER 4

RETURN

END

Scheduling of the event is done internally by the system according to the instructions of the user (SCHEDULE instruction). Thus the programmer may trigger events and actions at absolute or relative times. The call to an event routine during the simulation run is done by the internal timetable (Fig. 9.10). After the user has initialized the main program with the start instruction START SIMULATION, the time control assumes the command and supervises the simulation run. It fetches the text entry from the timetable, initializes the system time, and executes the corresponding event routine. In case there are no entries left in the timetable the simulation run is ended.

SIMSCRIPT contains similar control structures to control a program run as FORTRAN. Thus the user is able to formulate loops and conditional branches. There are numerous system functions for random number generation and statistical calculations. In general terms it can be stated that SIMSCRIPT is very flexible and that it is suited for large models. In addition, it is readable and self-documenting.

The modeling effort increases greatly with the size of the problem, since there are several model components necessary for the representation of a system component.

9.11.3 SIMULA as an Example of a Process-Oriented Modeling System

The program control structure and data definition of SIMULA are basically derived from the concept of the problem-oriented language ALGOL 60. With this capability the user can utilize the powerful constructs of

Simulation of Manufacturing Processes 681

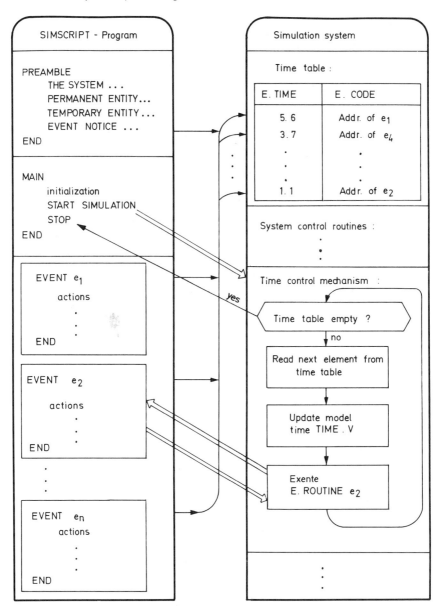

FIG. 9.10 Time control mechanism of SIMSCRIPT.

FIG. 9.11 Graphical display of a material flow system with pallets (from SIMFLEX/2 description, PSI Company).

a modern programming language. In addition, the language contains simulation specific facilities. All together this gives the language a high degree of flexibility.

Similar to SIMSCRIPT, the user designs his or her own modeling elements. Since SIMULA is based on a process-oriented modeling concept, the static attributes and the interactions between the individual objects are uniformly described by active system objects. Following a hierarchical process structure the user gradually decomposes the

Simulation of Manufacturing Processes 683

system into subsystems that form process classes. Thus it is possible to model cyclical, concurrent, or time-variant processes and to synchronize them. In SIMULA a process is defined by the prefix PROCESS, which is related to a class specification. The class concept is similar to the RECORDS of PASCAL; it allows the implementation of any desired list. A PROCESS contains automatic predefined operations for timekeeping. After the static attributes have been specified with data definitions, the dynamic behavior of a process is described in form of algorithms. Handling of time control functions is done with the operations

    HOLD (time range)
    PASSIVATE
    ACCTIVATE process (specifications)

HOLD stops the active process for a defined time and PASSIVATE transforms the process into the state "waiting" from which is can be activated only by another process. This is done by ACTIVATE, whereby different modifications may be used, such as time of activations (absolute and relative, including a priority indicator, or running before or after a concurrent process. The general state of a process is recorded by the predefined state descriptor IDLE. The instruction PASSIVATE puts the process into the state IDLE = TRUE; otherwise, IDLE = FLASE holds true.

A SIMULA model consists of individual blocks (BEGIN . . . END), which themselves may contain blocks and which may represent the processes. The outer most block is the abstract model of the dynamic system. It has the following structure:

SIMULATION

BEGIN
    declaration of global model attributes;
    declaration of classes or processes as model elements;

initialization;

HOLD (simulation time);

END SIMULATION;

During the initialization part list structures are generated and values are assigned to the attributes; then the individual processes are started.

The outermost block is also called the simulation block; it sets the time frame within which the entire simulation is done.

The timing mechanism is based on an event-oriented time schedule as in SIMSCRIPT. Once activated, every process contains a pointer to its own event entry field, which is an element of the event list and which contains the event time as well as a pointer to the corresponding process. The contents of the event list are sorted by the time control mechanism in the order of ascending event times. The process at the lower end of the list is marked as the currently active one. The user hardly has to be concerned with this timing mechanism.

The static modeling units or their attributes are described by a record of attributes which is called CLASS. Similar to the RECORD concept in Pascal, a class consists of a class header containing the name of the class in which the descriptions for the attributes are specified:

CLASS motor;

BEGIN

    INTEGER number;
    REAL assembly_time;
    INTEGER motor_type;
    REF (motor) next;

END;

With this example a motor is described by a number, the necessary assembly time, and its type. In addition, a pointer is declared which points to the next motor in the list.

The individual components of a class are declared as data types, such as standard data types INTEGER, REAL, . . . , and pointer REF. A class object and its attributes are accessed with a pointer. The object is in existence as long as it is referenced by the pointer:

Pointer:	REF (motor) object;
Generation:	part: − NEW motor;
Access:	part.number := 1517;
	part.assembly_time := 21.5;
	part.next:− NEW motor;
RELEASE:	part:− NONE;

To implement sorted lists and to manage these, SIMULA uses the predefined class HEAD. The attributes of this class are:

EMPTY:	state of a list
CARDINAL:	number of list elements

FIRST: access to the first element
LAST: access to the last element

Elements that may be entered into a list are declared by the prefix LINK. They can be entered into the list with the system procedures INTO and removed with OUT. To generate model data, SIMULA contains the corresponding distribution functions.

Protocoling and data acquisition for the evaluation of the model have to be programmed by the user, for which aids are available. SIMULA is a powerful language of high flexibility. With its help, various complex, well-structured models can be built. Learning of the languages requires a certain time and educational prerequisites.

9.12 SIMULATION OF A FLEXIBLE MANUFACTURING SYSTEM

A simulation model for a flexible manufacturing system (FMS) can be conceived as a four-level hierarchy (Table 9.1).

Level 1: basic components of the FMS
Level 2: machine control
Level 3: scheduling
Level 4: order processing and planning

The lowest level contains the basis components or building blocks of the FMS, such as conveyors, robots, machine tools, and buffers. At this level the instructions of the higher level are executed, and in rare occasions decisions are made. The elements of this level are standard modules; for example a buffer may be represented by a programmed routine to control pallets or to separate parts.

On the higher control level, decisions are made on the basis of goals and current events. The level is responsible for specific classes of actions which may be standardized. The aims of the machine control level are defined at the scheduling level; for example, load the pallet x by robots y or z. Depending on the current position of the robots, the control selects one of them and the machine control level executes the corresponding task.

The scheduling level is responsible for planning future actions and for making supervisory decisions. It also distributes orders among machine tools and conveyors, for example, on the basis of piece rate or machining time. The scheduling procedure usually is straightforward; it may contain error recovery or the determination of alternative manufacturing routes with the help of limited intelligence. A

TABLE 9.1 Different Simulation Levels of a Flexible Manufacturing System

Level	Activities	Remarks
Order processing	Generation of orders Order processing Termination of completed orders Order cancellation Order statistics	Oriented toward the field of application; can be partially standardized (the data volume and order structure may differ)
Order scheduling	Machine assignment Recovery strategies Generation and evaluation of alternative machine assignments	Programming is done; problem oriented
Control level (e.g., buffer assignment)	Decisions resulting from defined goals Structure of this level is plant specific	Problem oriented; may be standardized
Low-level control (e.g., block-oriented control)	Control modules, (e.g., conveyors, machines, magazines)	Basic control level; can be fully standardized

Simulation of Manufacturing Processes

special proglem is the change of an order or the insertion of an additional one. In this case the system has to reschedule and to generate new control instructions. For this reason the system must have at any time access to information on all orders in process and to information about the actual state of the manufacturing process. Two methods may be used to route a part through the system. They are:

1. Processing or goal information is assigned to the part and travels with it through the plant, or
2. The manufacturing equipment assigned to machining a workpiece contains the information about the part. For example, every switch contains instructions for routing a part.

In an ongoing manufacturing process it is not possible to obtain all necessary information from the different control levels. This may be possible, however, with a well-designed simulation model. At the highest level the model handles all order processing and planning activities; they include:

1. Order generation, with the aid of order data and distribution functions
2. Order buffering and retrieval
3. Order processing
4. Order completion
5. Order cancellation
6. Insertion of new orders
7. Gathering of statistics about the order (e.g., on piece rate, manufacturing times, etc.)

Order processing is problem oriented and can be partially standardized. It is possible to conceive modules for statistical calculations to generate, cancel, and terminate orders. However, data streams and the structure needed to process an order are very specific to a factory organization; they cannot easily be handled by modules. This problem may be solved with the help of lists. In general, only parts of a simulation system can be standardized to handle specific repetitive problems; otherwise, a model has to be tailored to the application.

The scheduling level causes most of the difficulties. For example, simple machine loading algorithms can be preprogrammed. However, the majority of the control strategies of plants are so different that it is almost impossible to provide the user with a set of tools which can easily handle a specific layout of conveyors and machine tools. Individual components such as a conveyor switch, its entrance and exit, its switching time and load can be programmed by standard modules.

At all levels of a simulation various statistics are accumulated. Standard data for acquiring statistical information are:

1. Order-specific processing times
2. Loading of the system components, such as machine tools, robots, and conveyors
3. Maximum and average contents of a buffer

A simulation experiment may help to solve questions such as those concerning system reserves, sensitivity to failure, wear, and process alternatives. Experiments may also be conducted to search for a more efficient machine layout, to test the effect of component failures, and to observe traffic congestion due to different buffer sizes.

10
Quality Control

10.1 INTRODUCTION

Quality control has long been recognized as an important function in a manufacturing organization. This activity is needed to correct the deviation of the quality that was actually obtained for a product from the quality that was specified. In free enterprise the standards for quality are set by the customer. Therefore, all quality standards needed to produce the parts of a product and to perform its assembly have to be specified in such a manner that the customer's expectations are met. There are a large number of quality control functions to be done at different levels of a manufacturing process. The corresponding control activities performed at each level and the data to be collected and evaluated differ considerably. In future manufacturing facilities hierarchical computer systems will be employed to supervise an integrated quality control system. The ever-increasing complexity of the products, the large model diversification, and the increasing sophistication of the manufacturing processes will lead directly to the use of these distributed systems; otherwise, quality control will be the bottleneck of an organization. To date, not all the building blocks for an integrated quality control system are available. However, there are many ongoing developments which eventually will help to build this concept. Integrated quality control functions are possible only with the availability of powerful computers. The high packing density of semiconductor circuits on one chip will make such computing equipment available. Thus it will become possible to place a large amount of computing power and low-cost intelligence into quality control equipment right down on the production floor.

There are also bus systems being developed which will simplify the interconnection of measuring equipment and computers via standard interfaces, using standard communication procedures. A considerable amount of research still has to be performed to develop modular software and data banks for quality control. Methods have to be found to integrate quality control procedures close down into the fabrication process to gain more flexibility in manufacturing. There is also evidence that many well-known statistical methods to evaluate quality control data have not yet become common working tools, because they are too cumbersome to be used by hand. Computers make these tools very valuable. The cost of quality control will steadily increase and it will become necessary to optimize the quality control operation and to find methods to allot their rated share to the different manufacturing operations.

10.2 SYSTEM CONCEPT FOR QUALITY CONTROL

In Chap. 3 we discussed the fact that an efficient quality control network can be obtained only by the integration of all quality control activities into one system (Fig. 3.40). From the consumer side, data about the customer's quality preception and the field performance of the products are entered and processed. This information is needed by engineering to conceive and design a quality product. It is the manufacturing engineer's responsibility to assure that the specified quality is obtained by the use of good manufacturing methods and practices. The quality data of the different activities are used to set quality objectives and standards. It is the goal of a firm to capture, with the help of these standards, a large share of the market for its products. However, a problem may arise when the costs of quality control become unacceptably high. In other words, a completely integrated quality control operation can be quite expensive. Careful consideration has to be given to the fact that testing still has to be economical and that the entire profit of a company will be absorbed by testing. Figure 10.1 shows how the cost of testing a product increases with increasing test effort. As the result of the increased test effort, the costs to repair a product in the field are reduced. If one adds these two curves, a minimum can be found showing the most economical test effort for a given product. If properly designed, a computer system should be able to help to establish these curves for each product. As with other organizational activities, quality control is usually structured in a hierarchical fashion. Thus the supporting computer system will also be of hierarchical design (Chap. 4). Each tier of the computer system supports a corresponding level of quality control activity.

Decentralized hierarchical computer networks have been in existence for quite awhile. However, most systems were tailored toward a specific application. A considerable amount of basic research still has to be performed to develop methods to design and install such

Quality Control

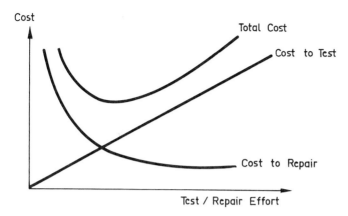

FIG. 10.1 Minimizing quality control cost.

systems systematically. Unfortunately, quality control practices and procedures performed in different factories are not the same, even in similar operations. As with other manufacturing activities, a very systematic approach has to be taken to find common building blocks from which a quality control operation can be formed. When this has been brought to a common denominator, it will be possible to design a computer-integrated quality control system. The current trend toward model proliferation makes it necessary to change production runs, often several times a day. In the automobile industry, for example, one can hardly talk about mass production any more. Cars are custom ordered and different parameters have to be checked for each car coming from the production line. Thus the quality control system has to be dynamically responsive to changes in production.

The design of the data processing system for quality control also has to take into consideration the different requirements at different hierarchical levels. General and common data from field returns, for example, should be kept on a master file in the executive computer. Here, in general, a large amount of data has to be manipulated. To eliminate bottlenecks, frequent data transfers over long distances, as required in paging operations, should be avoided. On the other hand, the local quality control operator at the end terminal at the lowest hierarchical level needs very specific information for an acceptance test to be performed. The quality control operator's data should be stored decentralized. Here the complicated question of how to structure a distributed data base arises. A considerable amount of research still has to be performed to find an optimized distributed data base in which data manipulation and communication will be kept to a minimum. Updating of files presents another problem. How often and when data should be transferred between different hierarchical levels must be determined. If updating is not done properly, the system designed can be subject to severe difficulties when a computer or part of the system fails. In this case it is possible for important data to get lost. One may also want to add recovery procedures that automatically restore the quality control system after a disturbance.

It is commonly believed that, to date, factories have not utilized the present state of statistical know-how to perform quality control functions. Formerly, many statistical methods had not been applied, and failed because without electronic computer they were too cumbersome to use. A dynamic quality control system can be taught to find its own meaningful test tolerances. It is also capable of determining automatically the number of samples to be tested and to change these requirements periodically according to the prevailing quality in the factory or of parts received from vendors. Good methods have to be found to handle mixed production runs. Here it is often very difficult to obtain large-enough sample sizes to be meaningful for statistical calculations. The problem of incomplete or wrong data is not very easy to resolve by automatic computing equipment. In conventional operations, adjustments are usually made with knowledge obtained from experience.

10.3 QUALITY CONTROL PLANNING

Quality control planning is a function similar to process planning. For each workpiece the measurement parameters, tolerances, and test sequences have to be determined. In addition, it is necessary to establish sampling plans and to conceive control methods for process capability studies. Further, the amount and type of quality data to be stored have to be established. There are two approaches to perform this planning operation: the generative and variant methods. The generative approach resembles the decision procedure used by the quality control planner. Here planning will be done with the help of functional primitives that are properly assembled and will render the quality control plan. This method is very difficult to computerize because it requires a considerable amount of experience and intuition to set up such a plan. The variant method uses a table-look-up procedure. Here quality control plans are derived from plans used for similar workpieces. Although the variant method is ideal for computerization, it is often cumbersome to implement and to maintain. Scheduling of the available measurement instruments will be an additional task of the computer. For this purpose all information about the existing measuring instruments are stored in a database. By describing the measurements to be performed with a workpiece, the batch size, and the manufacturing deadlines to be met, the system will automatically

Quality Control

assign the test equipment. In case a deadline cannot be met, an alternative schedule will be issued.

10.4 FUNCTION OF QUALITY CONTROL

Quality is the sum of all attributes and characteristics of a product or an activity which contributes to the usability of these to perform a specified function. From this definition, it is clear that quality control is a regulatory process through which actual quality performance is measured, compared with standards, and if necessary, corrective action is taken. High quality is important to help to maintain the competitiveness of a firm. With the availability of computing equipment, it is possible to systemize the quality control endeavor and to automate the work highly. The backbone of the quality control system are the company's internal, national, and international standards, which become a contractual object between the manufacturer and the customer. Adherence to these standards is the aim of a quality assurance system.

When a product has a defect, the following questions are of interest:

Type of defect
Place of occurrence
Cause of defect
Number of defects
Seriousness of defect
Cost of defect

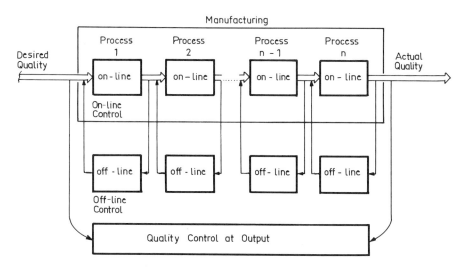

FIG. 10.2 Distribution of quality control functions in manufacturing.

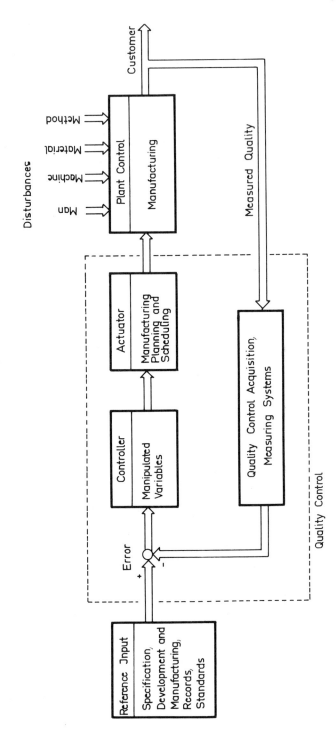

FIG. 10.3 Principle of integrated quality control.

Quality Control 695

The cost determines how much work should go into a quality test (Fig. 10.1). An effort must be made to locate all defects while the product is being manufactured and assembled. To repair a product in the field costs approximately 10 times as much as repairing it in the factory.

A quality control system can be viewed as a complex adaptive control loop. With the help of sensors, all properties and the performance of the product are identified. This is done at many places in a factory, such as (1) at receiving, (2) at the location where parts originate in the manufacturing process, (3) at subassembly stations, and (4) during the final acceptance test of the finished product. Figure 10.2 shows a typical distributed structure of a quality control system. When the cause of a defect and its place of occurrence are known, measures for its correction or possible elimination are to be initiated. Corrective actions can be taken directly at the level of the manufacturing process or may reach far into product design.

An integrated quality control system operates as follows (Fig. 10.3):

1. The manufacturing equipment is the process to be controlled and is part of the control loop. Disturbances acting on the process derive from man, material, machine, and methods.
2. The quality control acquisition system determines the actual value of the variable controlled.
3. The controlled variable is compared with set points obtained from predefined quality specification or standards.
4. To correct any deviation from a quality standard, a new manipulated variable is calculated, and if necessary, a corrective action is taken to adjust the manufacturing process.
5. The function of the controller is done by manufacturing planning, scheduling, or an automatic actuation device.

An important aspect of quality control is the reaction time, which is the time between the instant of recognizing the defect and the instant of correcting it. With the help of a coordinated effort and by using microelectronics in conjunction with larger supervisory computers, the reaction time can be optimized and the quality assurance system can be automated.

10.5 QUALITY CONTROL METHODS AND TEST PROCEDURES

When computer-controlled test methods are introduced, it may not be a good practice to duplicate conventional tests. These tests usually were developed to perform limited repetitive functions. Adaptive decision making as well as large data storage capabilities were not available in the measuring equipment. With the aid of the computer it is possible to conceive new and improved tests which are impossible to perform with conventional equipment. Thus whenever a computer is introduced, a possible redesign of the test method and test apparatus should be investigated.

Tests used by engineering, receiving inspection, and for on-line product testing are often very similar. For this reason it is of advantage to start with the buildup of a computerized test system in the engineering department. This may even coincide with the development of a new product, and thus a well-conceived test system may be engineered. The aid of engineering may be of particular importance when manufacturing has no prior computer knowledge. The laboratory usually has a more quiet working environment and allows more thorough development of the new test. Possible equipment breakdown and faulty tests may not be a severe problem, because the system developed still can resort to a manual test system. Only when the new test and the computer technology are well understood should the equipment be taken out to the factory. New and improved test methods must be developed together with the user of the equipment; otherwise, it will not be accepted and properly applied. It is also of importance to assure that the new test will be recognized by institutionalized or governmental test agencies.

With a systematically and well-developed computerized test system, many benefits may be achieved (Fig. 10.4). This figure depicts results obtained from an industry survey in which approximately 170 computerized test system users where asked about the benefits they had experienced with this equipment; 94 users responded. The most important benefits are improved test productivity, more objective quality decisions, better and improved measurements, and improved throughput. In the following sections we discuss several subjects that should be taken into consideration when a computerized test system is being developed.

When several of the same test systems are implemented, there should be only one or a very limited number of test objects assigned to one computer. Small test systems are easy to implement and to program. If necessary, they will be tied together through a larger computer at a higher level. The addition or deletion of test systems is easy to perform. Most tests need special instruments and fixtures which often have to be discarded when the test or test object are changed. With the computer it is possible to conceive programmable test systems. When a similar product is introduced or an old one is altered, writing new test programs is all that need be done. The initial installation cost for flexible test systems may be quite high. However, the equipment may pay for itself within a few years after several model changes have been made. The new tests can be implemented quickly and may need only little rebuilding of the test apparatus, if any.

An effort should be made to divide the product spectrum into similar groups or variants and to design for each of these groups a

Quality Control

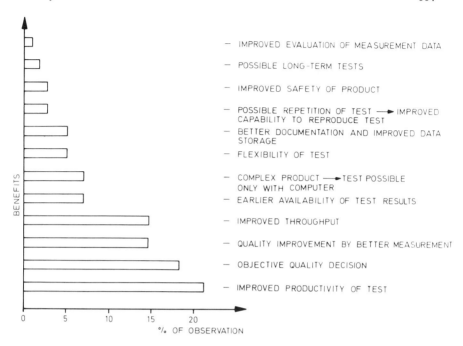

FIG. 10.4 Frequency distribution of benefits of a computer-controlled test system (average from 94 systems). (From Ref. 1.)

universal test system. Typical groups are motors, switches, thermostats, and actuators. These universal test systems should be self-sufficient so that they can be used by different departments. They also have an advantage when they are connected to a computer network. When the main computer fails, the individual test systems can still be operated.

The computer can be programmed so that the measuring system is checked at different time intervals. Thus a failure or an inaccurate measurement can be detected early. Similarly, it may be of advantage to calibrate all transducers before a test is initiated. The computer can also do automatic correction of test results (e.g., to compensate for temperature drifts) (Fig. 10.5).[2].

The configuration of the measuring system and the type of test to be performed depend on the product and on the rate at which it is tested. For this reason, the test concept and the test system must be integrated into the manufacturing process. The steps necessary to perform the test are as follows:

Recognition of the product
Positioning of the sensor

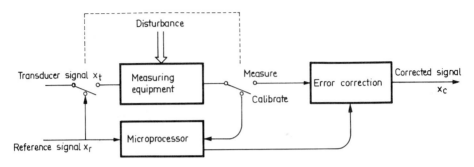

FIG. 10.5 Automatic calibration of measured signal.

    Acquisition of the test parameter or variables
    Evaluation of the test result
    Calculation of a quality index
    Output of the test result

The degree of mechanization of these individual components depends on the desired degree of automation for the entire system. For a simple product with a low production rate it may only be necessary to perform automatic data acquisition. A high production rate may require that all functions be completely automated. In accordance with these two extremes there will be different requirements for every test system. Automation of the recognition function can be done with the help of light barriers, binary coding, or machine vision. These aids are also of importance to robots. They are discussed in Chap. 8.

    For many measurements accurate contact positioning of the sensor on the test object is necessary (e.g., for temperature or vibration measurements). This may be an awkward task for moving objects such as commonly encountered on conveyors. Very complex fixtures may be necessary for positioning. To circumvent this problem, noncontact sensors are often used. This subject is discussed in a later section of this chapter.

    Acquisition of test parameters, evaluation of test results, calculation of the quality index, and output of the test results should be done by computer, as it can perform these functions quickly and very efficiently. Well-suited output peripherals are the graphic display screen and plotters.

    Automatic test systems are more efficient the more they are automated. Most errors during testing are introduced by human beings. For this reason a high degree of automation may be desirable for a test system.

10.6 ARCHITECTURE OF COMPUTERIZED MEASURING SYSTEMS

The principle of a computer-integrated quality control system was shown in Fig. 10.3. The system can be visualized as a closed control loop whereby the value of the parameter measured is returned as shown in Fig. 10.3. The system can be visualized as a closed compared with the reference input signal and the new manipulated variable is calculated. This signal actuates the control element and corrects the process variable. With quality control systems feedback may be done automatically or by the interaction of an operator. To convey the process parameter to the computer and to perform a corrective action with the manipulated variable, several signal conversions have to be performed (Fig. 3.42). Usually the measured parameter has to be converted to a voltage signal by the sensor interface. In the case of a pressure transducer, for example, the pressure may elongate a metallic bellow. This bellow moves via a mechanical linkage the center wiper of a potentiometer across a variable resistor of a voltage divider. The resistance change between the wiper and one end terminal of the voltage divider is sensed as a changing voltage signal. This parameter is related to the pressure change. An external power supply provides a standard voltage across both terminals of the potentiometer. Since the signal voltage may be of very small magnitude, it may have to be amplified by an amplifier. A voltage signal is easily affected by factory noise; thus it may be necessary to convert it to a current signal with a signal transformer. This current signal is then sent over a transmission line to an A/D converter of the computer interface. Here it is again changed to a voltage signal and then converted to digital information. The output signal of the computer may have to go through a similar chain of conversions in order to drive an actuator of a control element of the process. The exact function of these interface elements and that of the process interface of the computer are not the subject of discussion of this book. The reader interested in interfacing practices and instrumentation should consult the literature on these topics [3].

10.7 EQUIPMENT CONFIGURATION FOR QUALITY CONTROL

Depending on the desired degree of automation, quality control activities can be supported by different computer configurations (Fig. 10.6). In a simple application, probably one involving a low-volume production run, quality control (QC) observations are recorded by an inspector on a QC form. The contents of this form are then transferred to a punched card and read into the computer. Data are

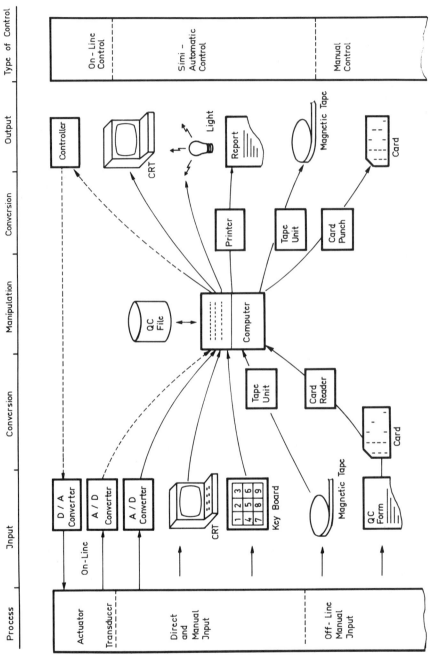

FIG. 10.6 Various equipment configurations for quality control. (From Ref. 4.)

evaluated and stored on a central QC file. The operator is provided with a report generated with the help of a printer. If data storage is required, the data can be placed on punched cards. The operator will react on the report and will manually control the process.

The other extreme is the installation of a completely automatic quality control system which senses process parameter changes with a transducer. The information is conveyed to the computer via an analog-to-digital (A/D) converter. The computer evaluates the data and stores important information. It also sends to the controller the new process parameters that are required to obtain the desired quality. Feedback information is returned to the process via a digital-to-analog (D/A) converter. An actuator performs the desired process corrections. In general, this extreme solution is very expensive (Figs. 3.6 and 3.7). For this reason, in practical installations, a less complex quality control system with some manual interaction is often selected. However, as mentioned earlier, manual interaction may be a source of problems. In addition, in quality control there are many activities that cannot be completely automated because adequate sensing equipment is not available (i.e., the detection of scratches or impurities on a coated surface).

10.7.1 Measuring Methods

Quality control information from a process may be acquired by different methods. It can be entered either by hand or by a sensor that is directly observing the process. Direct measurements are more desirable since they eliminate errors caused by a human operator. In principle, the following three measuring methods may be applied to enter a parameter into the measuring system:

Contact measurement
Noncontact measurement
Manual input

In the first two cases the process parameter (e.g, dimensional change, temperature, pH value, etc.) is converted to a voltages current, or resistance reading and sent to the analog peripheral of the computer. The input of the A/D conversion equipment requires a voltage signal with a maximum value of approximately 10 V. Thus current and resistance readings have to be converted to a voltage signal. Special sensors are discussed in more detail in Chap. 7.

10.7.2 Contact Measurements

To obtain readings of high accuracy, the contact measuring method is preferred. Typically, temperature, vibration, dimension, pressures, and so on, are measured using this method. Length measurements play a predominant role in part manufacturing. For this purpose various computer-controlled measuring machines are available. The operating principle of a coordinate-measuring machine is discussed in the following paragraphs.

Coordinate-Measuring Machine

Figure 10.7 shows a two-column universal measuring machine of the Carl Zeiss Company. The measuring table, all guideways, and the crossbar are made of good-quality granite for high dimensional

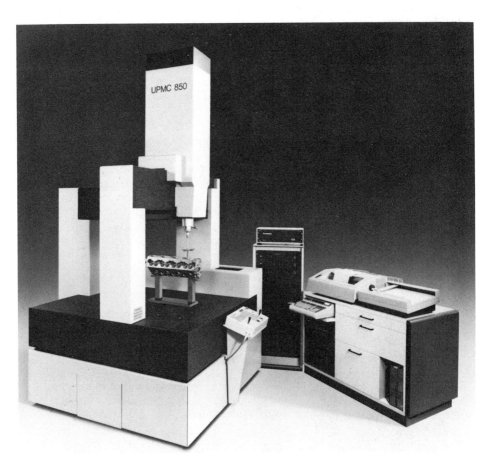

FIG. 10.7 Two-column universal measuring machine. (Courtesy of Carl Zeiss, Oberkochen, West Germany.)

Quality Control

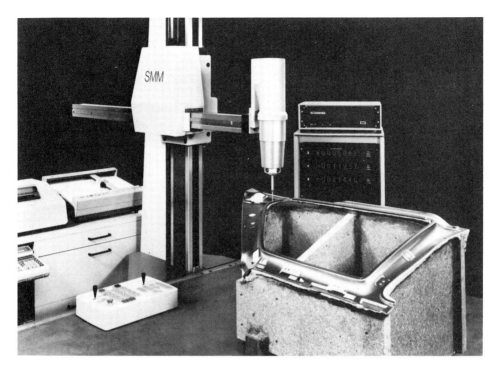

FIG. 10.8 Cantilever-type machine. (Courtesy of Carl Zeiss, Oberkochen, West Germany.)

stability. A universal three-dimensional probe measuring head is attached to the crossbar by means of low-friction air bearings. It can be moved, under program control, along the surfaces of a workpiece. During this movement the dimensions of the workpiece are recorded with the help of a probe and an induction or optical measuring system. A computer records these measurements and calculates the dimensions in reference to a given coordinate system. Depending on the application and the size of the workpiece, there are various types of measuring machines available. Figure 10.8 shows a cantilever-type machine that can cover a large work space. In principle, the measuring head and recording mechanism can be the same as that of the machine discussed above. A wide spectrum of workpieces can be handled with these types of machines. In Fig. 10.7 the measurement of an automotive engine block is shown, and Fig. 10.8 demonstrates the dimensional check of a complex sheet metal part. Figure 10.9 shows an application whereby the dimensional accuracy of a gear is recorded.

FIG. 10.9 Measuring a gear. (Courtesy of Carl Zeiss, Oberkochen, West Germany.)

Quality Control

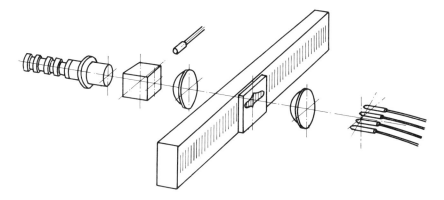

FIG. 10.10 Principle of an optical measuring scale. (Courtesy of Carl Zeiss, Oberkochen, West Germany.)

The gauging mechanism which measures the travel of the measuring head in either of the three main axes may operate on inductive, optical or mechanical measuring principles. Figure 10.10 shows the operation of an optical measuring system. The movement of the measuring head is recorded by light pulses which are generated when a light beam is interrupted by the graduation marks on a glass scale attached to the head. The accuracy of present glass scales is 0.1 µm.

With this measuring principle a reference cartesian machine coordinate system is used, with respect to which a probe head moves. The spatial reference point of the probe head may be the center of the probe tip (Fig. 10.11). The illustration shows a machine coordinate system X_M, Y_M, Z_M and the workpiece coordinate system X_W, Y_W, Z_W. The test object can be placed at any location and orientation on the table; thus the two coordinate systems do not have to coincide. The exact position of the workpiece is determined by scanning several points of its surface. For example, the surface normal is determined by measurements 1, 2, and 3. The direction of the X axis is defined by points 4, 5, and 6 and that of the Y axis by point 7. With this measurement the location of the object is mathematically defined and known with respect to the reference coordinate system. During a measuring procedure the computer will compensate all measurements for the offset between these two coordinate systems.

There are different types of probes used in the measuring head (Fig. 10.12). Theoretically, the diameter of the probe should be infinitely small in order to be able to reproduce the actual dimensions of the workpiece. Such probes, however, cannot be manufactured.

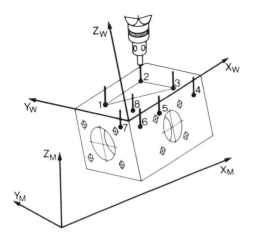

FIG. 10.11 Three-dimensional measuring principle. (Courtesy of Carl Zeiss, Oberkochen, West Germany.)

Probe types	Structural shapes	
mechanical probes	A	
probes with resistance circuit	B Closing an electrical circuit through probe and work piece during contact in arbitrary direction	
probes with built-in switches	C Opening the built-in electric switch during deflection of the probe pin (which is freely movable in several axes) during contact in the free directions	
1 D probes (mostly with inductive measuring system)	D1	D2
2 D probes (mostly with inductive measuring system)	E1	E2
3 D probes (mostly with inductive measuring system)	F1	F2

FIG. 10.12 Various structural shapes of mechanical probe systems. (Courtesy of Carl Zeiss, Oberkochen, West Germany.)

Quality Control

Actual mechanical probes are designed as shown in Figure 10.12. The computer of the measuring machine is programmed to compensate for any offset that is due to the finite dimensions of the probe tip. The problem of offset compensation was discussed in Chap. 7 under NC part programming.

The measuring probe is connected to a one-, two-, or three-dimensional measuring system (Fig. 10.12). For universal measurements the probe head may contain several probes extending in different directions (Fig. 10.13). The calibration of the probe is done with the help of a high-precision spherical standard (Fig. 10.14). For this purpose, the diameters of all probe tips and their position in regard to the machine coordinate system are determined and made known to the computer. During actual measurements the workpiece is scanned by the probe tip, thus displacing the measuring head (Fig. 10.15).

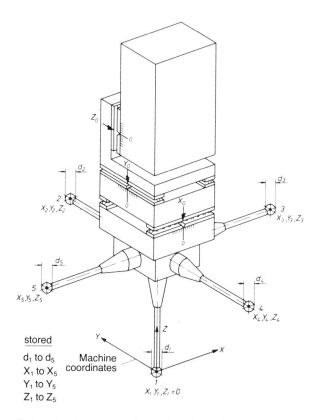

FIG. 10.13 Three-dimensional probe head. (Courtesy of Carl Zeiss, Oberkochen, West Germany.)

FIG. 10.14 Calibration of the probe with the help of a calibration sphere. (Courtesy of Carl Zeiss, Oberkochen, West Germany.)

The machine control system tries to move the measuring system back to its zero point. At this location the coordinates of the measuring machines are entered into the computer. Since the dimensions of the probe tip and the tip position are known to the computer from the calibration procedure, the computer can calculate the actual position of the workpiece contour in reference to the origin of a predefined coordinate system. A simplified control loop of the coordinate measuring machine is shown in Fig. 10.16. In this case the probe is brought into contact with the workpiece under manual control with the help of

Quality Control 709

X, Y, and Z position control sticks. After contacting the object, the hand control becomes inactive and measurements can be taken.

With a coordinate measuring machine it is also very easy to measure contours of cams and irregular surfaces. The object to be measured is rotated about an axis (i.e., with the help of an indexing table). Measurements can be taken at different angular increments and compared against a mathematical curve describing the contour of the object or against a curve obtained from a master cylinder. These types of test are very time consuming when done with conventional measuring devices. Under computer control the test time may be reduced by the order of several magnitudes. Hand operation of the machines is very time consuming when complex workpieces are measured. Figure 10.17 shows the example of a curved gear surface. For such tasks CNC-controlled measuring machines are available. The probe is guided by a program that describes the contour of the workpiece along its surface. The actual measured coordinates are compared

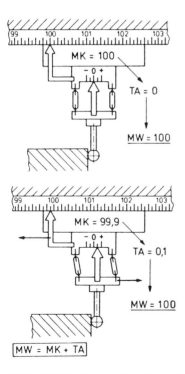

FIG. 10.15 Measuring the displacement of a probe tip. (Courtesy of Carl Zeiss, Oberkochen, West Germany.)

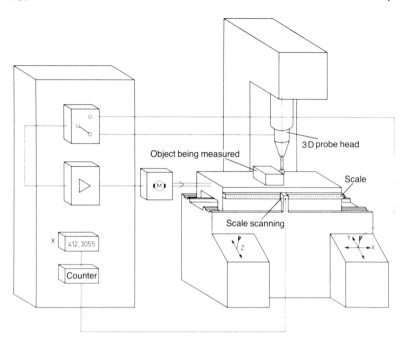

FIG. 10.16 Principle of position control by manual machine operation. (Courtesy of Carl Zeiss, Oberkochen, West Germany.)

with those of the described contour and the deviations are determined. The results may be plotted for visual inspection (Fig. 10.17).

Dimensional Measurement with the Machine Tool

The positioning scales of the machine tool in connection with a two- or three-dimensional probe head may also be used for dimensional measurements. This is done with special measuring peripherals and fixtures. The following functions are possible:

1. Supervision of operating limits (tool interference, collision, etc.)
2. Wear compensation
3. Determination of workpiece position with the help of reference surfaces
4. Compensation of dimensional changes with the help of reference points (may be necessary due to heat distortion or other external influences)
5. Automatic dimensional checks, which may be necessary due to misalignment of parts or tools or due to warpage

Quality Control

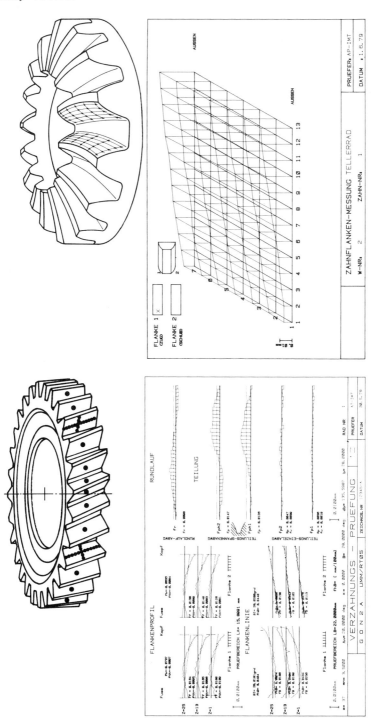

FIG. 10.17 Measuring the contour of gear teeth. (Courtesy of Carl Zeiss, Oberkochen, West Germany.)

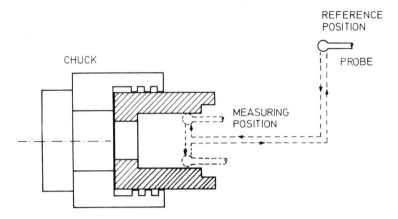

FIG. 10.18 Automatic dimensional check of workpiece in lathe. (From Ref. 6.)

A method for automatic measurement of a workpiece chucked in a lathe is shown in Fig. 10.18. The probe is guided under computer control from a reference position to the measuring points. The measurements are taken and recorded by the computer, after which the probe is automatically retracted.

Figure 10.19 shows a method of measuring the wear of a boring tool. First the tool is brought to a standstill, and then the measuring device is moved from its hold position to contact the cutting edge of

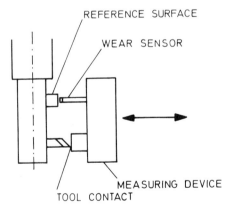

FIG. 10.19 Machine-independent wear measurement of a boring tool. (From Ref. 6.)

Quality Control 713

the tool. Then the wear sensor takes a reading on a reference surface of the tool holder. The dimensional changes between the cutting edge and reference surface of two successive readings indicate the tool wear. The measuring procedure and the evaluation of the measurement are done under computer supervision.

Measuring Robot

High-throughput production equipment often requires special or even custom-designed measuring systems. Figure 10.20 shows how a precision robot can be used for dimensional checks of complex workpieces. The measuring probes are attached to the robot arm and can be moved under program control along the contour of the workpiece. Parameters are automatically read and transferred to a computer system. As with measuring machines, the computer relates the dimensions of the object to a reference coordinate system and adjusts for the dimensions of the measuring probes and the finite diameter of the probe tip. Measuring by both robots can be done in parallel. The modularity of this system in conjunction with the mobility of the robot makes the equipment highly adaptable to a large workpiece spectrum. Measuring sequences and programs can easily be changed on-line under computer control.

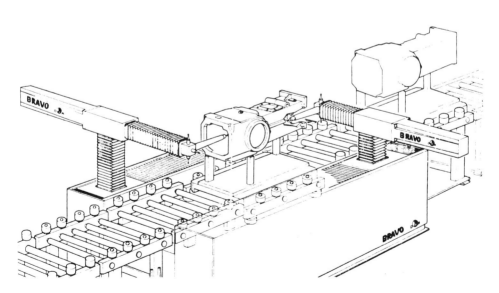

FIG. 10.20 Two measuring robots in action. (Courtesy of Digital Electronic Automation GmbH, Frankfurt, West Germany.)

10.7.3 Noncontact Measurements

Most measurements require positioning of a probe on the measuring object. This procedure is often very cumbersome and time consuming. To increase the productivity of the measuring process, many remote-measuring noncontact methods were developed. In general, instruments of this type which are presently available have low accuracy and are very complicated. In the future there will be a need to develop low-cost remote measuring devices with high accuracy. Advanced laser technology may be used for dimensional measurements and for the investigation of the surface integrity of parts. Figure 10.21 shows the principle of a laser telemetric system to measure on-line the diameter of a rod. The laser light beam hits a rotating mirror and is projected as a light sheet with the help of a collimating lens toward the object to be measured. A second lens collects the residue light that passes the object. The intensity of this light is recorded by a photodetector. With this laser principle, an accuracy of 0.0005 mm can be obtained.

Another principle of a noncontact dimensional measuring system using a linear photodiode array is shown in Fig. 10.22. The object to be measured must be clearly distinguishable from its background. For this purpose it is illuminated either by directly shining light on it or by a translucent backplane. The light emitted from the illuminated objects is collected by a lens and projected on the photodiode array. To distinguish the object from its background, a threshold is set automatically or by hand. The image from the photodiode array is digitized and stored for further processing (see Fig. 8.95). The accuracy of this measuring device depends on the number of pixels of the linear array and the distance at which the object is located. For calibration purpose an object (e.g., a metal strip) with a known width is located at the same distance at which the object under investigation is to be measured. Through a discrimination procedure the edges of the

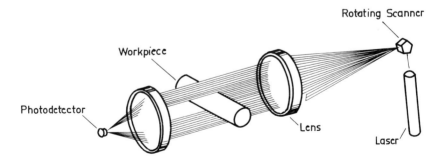

FIG. 10.21 Thickness measurement by laser. (From Ref. 7.)

Quality Control

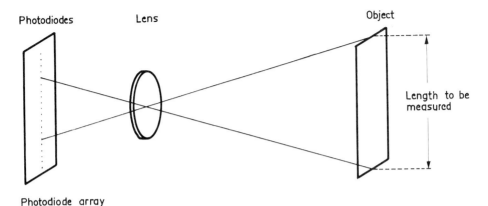

FIG. 10.22 Principle of length measurement with a photodiode array.

metal strip are located by the camera and the pixels of the photodiodes array showing these edges are marked by the computer of the camera. Now the computer counts the number of pixels located between these two marks. Their number is proportional to the width of the strip. When an unknown object is to be measured, a picture is taken and again the number of pixels between the detected edge marks are counted. The actual dimension is proportional to the number of pixels and can be calculated with the help of the information obtained from the standard by simple arithmetic.

The use of a computer is advantageous for both of the methods discussed above. The computer will retrieve the information about the stored image from memory and calculates, with the help of the known threshold, the distance between the two edges of the object. Such computerized devices are capable of performing over 500 measurements per second.

The vision systems described in Chap. 8 can also be used for quality control. However, for most visual inspection problems their low resolution is inadequate.

Acoustic signature analysis is another typical noncontact quality control tool. It is applied, for example, in situations where the vibration characteristic of a product is of interest. Vibration measurements require the sensor to be in direct contact with the object under investigation. Vibrations are also transmitted as coustic signals and can be picked up by a microphone. Figure 10.23 shows an on-line acoustic measuring system which may be under computer control. The object passes by the microphone and its sound is being monitored. At its peak value the computer initializes the reading. Infrared technology offers the possibility for remote temperature

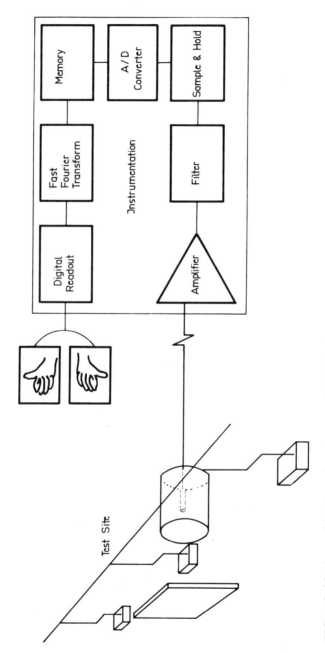

FIG. 10.23 Assembly line signature analysis test.

Quality Control

measurements. If, for example, the surface temperature of a mass-produced product must to be determined, an infrared sensor can be placed near the conveyor on which the product is moved. The sensor facing the surface under investigation will pick up the temperature. Under computer control it is possible to check for local hot spots or to record an average temperature reading.

10.7.4 Manual Input

There are many quality control parameters that cannot be measured directly or which are very difficult to acquire with instruments presently available. These parameters have to be recorded by an operator and entered into the computer by means of a keyboard or another manual device. Typical defects of this type are scratches on painted surfaces and bent sheet metal parts. In the past these defects have usually been recorded on paper and then entered into the computer. When data acquisition systems became available, passive data-entry terminals were used. With the development of low-cost intelligent terminals, manual parameter entry will be greatly facilitated. The terminal will be able to guide the operator through a quality control procedure, do validity checks on data, and immediately perform process capability calculations. The results of this calculation can be made visible on a low-cost display device. Thus it will be possible at the workstation to observe process drifts, wear of measuring devices, operator interaction with the process, performance trends between different shifts, and so on. The intelligent terminal can be connected to the factory bus system (Chap. 4) and thus will be able to communicate its quality control data to the host computer. The manual input stations for quality control are usually identical to those used in factory data acquisition systems.

Voice communication with the computer is another enhancement of a data acquisition terminal. The operator is able to call out the defects to the terminal. It, in turn, recognizes the instruction and records the defect entries in a file. To date these devices have a limited vocabulary and are very selective in regard to the master's voice by which they were trained. In case of a shift change, the voice recognition system has to be familiar with voice of its new master. Also, physical and mental stresses may change the tone of a voice and cause a recognition problem.

10.8 HIERARCHICAL COMPUTER SYSTEMS FOR QUALITY CONTROL

The necessity for a hierarchical control system was discussed in the introduction of this chapter. This hierarchical control concept is also applied to computer architectures used for quality control operation.

Principal aspects of hierarchical computer system are covered in Chap. 4. In an advanced quality control system there are often three levels of computers employed. Typical tasks to be performed by each level can be outlined as follows.

At the executive level general administrative quality control functions are performed to monitor the quality acceptance of a product by the customer and to compare it with the quality planned by the firm. Functions at these levels are listed in Table 10.1.

The computer at the next-lower level supervises the functional integrity of the quality control equipment, performs statistical calculations, and aids programming of the computers of the lowest level. These functions are listed in Table 10.2.

At the lowest level the quality control function is directly concerned with the product. For this reason it is almost impossible to conceive a universal computer system at this level, since the requirements vary considerably. Fundamentally, there are four different functions to be performed to control a test system:

System control functions
Input/output of operator and control data
Operation and control of the test program
Processing of measurement data

All four functions can be done efficiently by microprocessors (Table 10.3). They add to a high flexibility of the test system. The test program can quickly be adapted to accommodate the test object and the instruments. With local processing of the test data the results are immediately available and a higher throughput is obtained. Typical tasks of a custom-designed quality control system at the lowest hierarchical level are shown in Fig. 10.24 [8]. In this case the quality of a workpiece produced by a machine tool is supervised.

TABLE 10.1 Typical QC Functions of the Execution Computer

Monitoring of quality for each product and component

Calculation of cost of quality for each product and component

Administration of a master file for short- and long-term tests from product development

Monitoring of reliability data of competitive products

Performance of process capability studies for manufacturing

Generation of quality reports

TABLE 10.2 Typical Functions of the Host Computer

Automatic loading of programs on satellite computers
Coordination of measuring devices
Functional testing of measuring equipment
Recognition of operating errors of test equipment
Reporting system errors
Automatic change of test procedures
Storage of test data
Automatic creation of test tolerances
Periodic supervision of test tolerances
Automatic trend analysis
Optimization of test parameters
Calculation of quality indices
Calculation of nonmeasurable parameters
Comparison of test results obtained from parallel tests
Recognition of errors of different test personnel
Recognition of test data drift
Accumulation of statistical data
Automatic determination of number of tests to be performed
Location of manufacturing difficulties
Determination of most frequent and expensive defects
Storage of data for governmental regulations
Data communication with satellite (micro)computers
Data communication with other host computers
Data communication with higher hierarchical computers
Program development for satellite (micro)computers

TABLE 10.3 Function of the Microcomputer in Test Applications

System control functions

    Activation and deactivation of test run
    Central interrupt processor and bus controller
    Automatic operating mode switching
    Automatic calibration of instruments
    Automatic diagnosis of instruments

Input/output of operator and control data

    Driver for data peripherals and alarm indicators
    Manual operation program
    Operation of test program
    Editor for programs
    Monitor for additional participants

Operation and control of the test program

    Instrument driver
    Instrument initialization
    Setup of instrument parameters
    Control of the measuring process
    Handling of measurement data

Processing of measurement data

    Comparison with test limits
    Plausibility test
    Filtering of test data
    Averaging of data
    Trend analysis
    Drift correction
    Linearization
    Normalization
    Addition and multiplication with constants
    Calculation of nonmeasurable parameters
    Addition and subtraction of curves
    Differentiation of curves
    Integration of areas

Quality Control

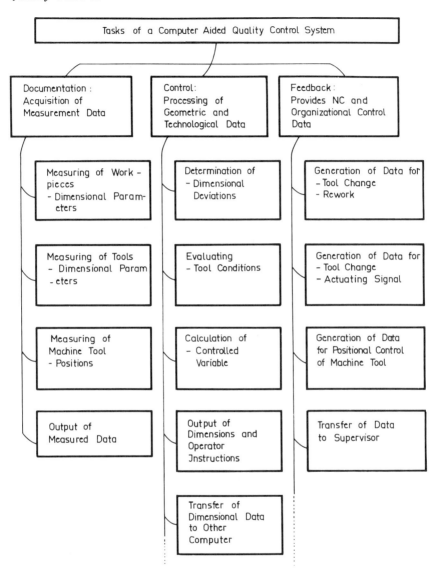

FIG. 10.24 Tasks of a computer-aided quality control system for a machine tool. (Courtesy of Kernforschungszentrum Karlsruhe GmbH, Karlsruhe, West Germany.)

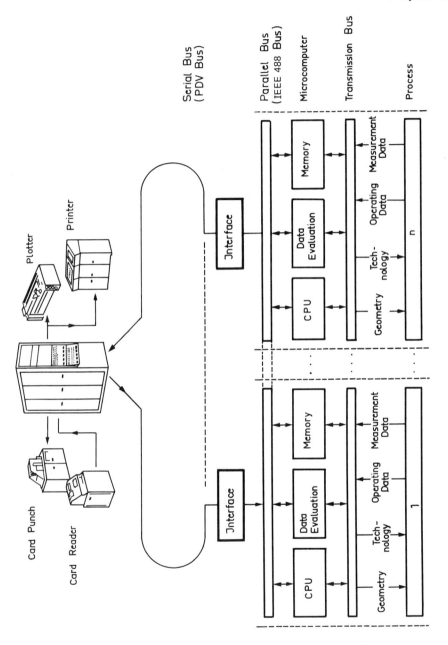

FIG. 10.25 Hierarchical quality control measuring system using a ring bus for data transmission.

Quality Control 723

If too many functions are placed on one microcomputer, there may be problems that it gets quickly overloaded (Table 10.3). This usually results in a slowdown of the equipment and in very difficult programming problems. There are various methods used to handle this situation:

1. To process measuring data by the instrument. Many new instruments contain their own microprocessor and can handle limited mathematical tasks.
2. To distribute functions among several parallel processors, as is commonly done with controls.
3. To relocate some of the system or data processing functions to a higher hierarchical level.

With hierarchical test systems it is possible to connect several self-contained groups of instruments via a communication bus to a host computer (Chap. 4); Fig. 10.25). At the lowest level a certain amount of preprocessing of data is done. The results are used by and stored on the host computer. The modules of the lowest level contain their test program for the product to be tested from the host. It may also be possible to initiate and control the test runs of modules or individual instruments remotely through the host.

10.9 IMPLEMENTATION PROBLEMS WITH QUALITY CONTROL COMPUTER SYSTEMS

There are many different methods of implementing a computerized quality control system. In general, these systems are so complex that they are installed over a long time frame. In some cases either the top-down or bottom-up approach may be the most practical method; in other cases, the system may grow together from both directions (Chap. 2). Also, the question of whether or not to have individual computers, that is, a distributed or a centralized system, may be raised. General problems associated with these computer architectures were discussed in Chap. 4. Figure 10.26 shows the most commonly used computer architectures for quality control. The major advantages and disadvantages of these systems are shown in Table 10.4. Whenever a quality control computer system is designed, the topics entered in this table should be discussed. The selection criterion depends on many factors, such as management philosophy, sampling speed, reliability, and flexibility.

Independent parallel computer-controlled tests should be considered only when there is no connection between the tests and when centralized storage and evaluation of test data are not required. The centralized system usually leads to difficulties when there are different

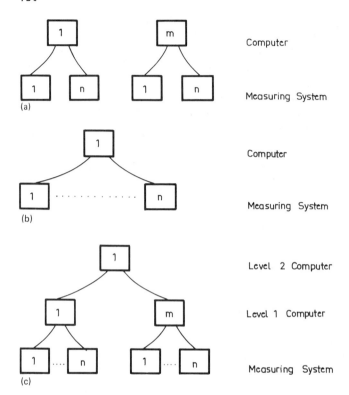

FIG. 10.26 Various computer-controlled measuring systems: (a) independent parallel control; (b) central control; (c) distributed control with supervisory functions.

types of tests and when the test locations are scattered throughout the plant. In this case the software overhead becomes excessive and the reliability of the system is poor. Usually, the distributed architecture is preferred; however, there are still situations for which the centralized computer renders the most economical solution.

In most manufacturing and assembly operations it is necessary to observe the quality of a product from the beginning of an operation to its end. Since the quality at an early processing station will affect the quality at a later station, frequent comparison of data has to be done. These quality control operations require distributed control with supervisory functions. Coordination of the test and evaluation of test data are done by the supervisory computer. This control is also necessary for mixed-model operations or situations in which products are frequently changed. On request, the centralized computer will distribute the new QC programs to the test computers.

TABLE 10.4 Advantages and Disadvantages of Different Computer Architectures for Quality Control

	Test system					
	Independent parallel computer		Centralized computer		Distributed control with supervisory functions	
	Test applications in plant are:					
	Same	Different	Same	Different	Same	Different
Cost						
Equipment	Low	Medium	Low	Medium	High	High
Software	Low	Low	Medium	High	Medium	High
Implementation	Easy	Easy	Medium	Difficult	Medium	High
Wiring	Short	Short	Long	Long	Short	Short
Software						
Overhead	Low	Low	Medium	High	Medium	Medium
Programming	Easy	Easy	Medium	Difficult	Medium	Medium
System						
Complexity	Simple	Simple	Medium	Complex	Medium	Medium
Modification	Easy	Easy	Medium	Difficult	Easy	Easy
Expandability	Easy	Easy	Difficult	Difficult	Easy	Easy
Implementation	Easy	Easy	Medium	Difficult	Medium	Medium

TABLE 10.4 (Continued)

	Test system					
	Independent parallel computer		Centralized computer		Distributed control with supervisory functions	
	Same	Different	Same	Different	Same	Different
System (continued)						
Reliability	Good	Fair	Poor	Poor	Good	Fair
Redundancy	Good	None	None	None	Good	Fair
Supervision	None	None	None	None	Good	Good
Speed	Fast	Fast	Slow	Slow	Fast	Fast
Flexibility	Good	Good	Poor	Poor	Good	Good
Peripherals						
Data	Expensive	Expensive	Inexpensive	Inexpensive	Inexpensive	Inexpensive
Process	Expensive	Expensive	Expensive	Expensive	Expensive	Expensive

Quality Control

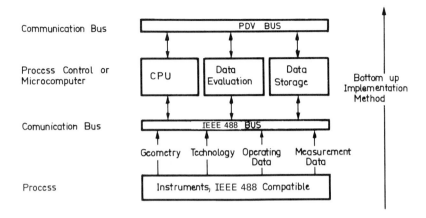

FIG. 10.27 Computer-controlled measuring system.

Computerized quality control systems are usually expensive. For this reason it may be practical to start with a conventional measuring system and introduce the computer during a later implementation phase. In this case the measuring equipment should be computer compatible. For example, they should be provided with an IEEE 488 interface. This may increase the costs of the instruments; however, the computer can be added very easily to the system at a later date (Fig. 10.27).

10.9.1 Quality Control Stations for Moving Product Lines

In mass production it is often necessary to test a product while it is moving on a conveyor or an assembly line. If several tests are to be performed on the same unit at different test stations, all data pertaining to this unit must be positively stored in the same quality file. This may lead to very difficult synchronization problems between test stations (Sec. 3.6).

Synchronization may be complicated by start-and-stop operations of the assembly line or by swinging hangers of a chain conveyor. Figure 10.28 shows as an example the various tests that have to be conducted to measure the performance and integrity of a washing machine. The supervisory computer has to track each unit from one test station to another and to store its performance parameters, taken at different intervals, in individual quality files. The time intervals are determined by the pace of the conveyor carrying the product.

When a product to be tested is moving on an assembly line, difficulties may be encountered in transmitting the test data to the computer. With a long test track and many tests to be done, it becomes

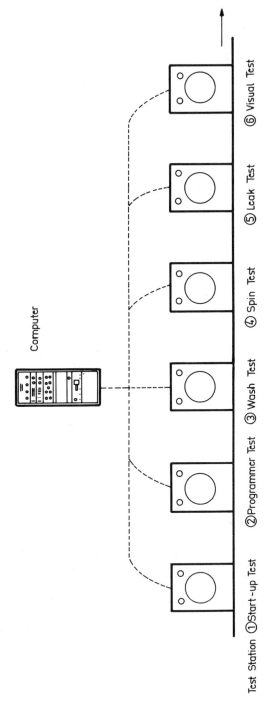

FIG. 10.28 Synchronization of test data with the test unit to measure the performances of a washing machine on a moving conveyor line.

Quality Control

impractical to connect instruments and data acquisition terminals to a stationary central computer because of the signal communication lines which have to be pulled along with the product. Depending on the type of test to be performed, there are several different solutions to this particular problem.

1. *Instant tests*: Here the performance of the test unit is time invariant. In this case it is necessary only to synchronize the storage of performance parameters taken at the different test stations.
2. *Transient test*: The test unit may have a typical pulldown characteristic, as shown in Fig. 10.29. This figure shows that tests are taken at different times and different locations. It is the task of the supervisory computer to reconstruct the pulldown curve.
3. *Test at steady-state condition*: In this case the unit has to be monitored constantly until the performance variables have reached a steady-state condition. Then a reading is taken.

Figure 10.30 [9] shows different test configurations to solve moving-line tests. The main problems are encountered when the test signals are sent to the computer. The easiest solution is to connect the sensors directly to the computer. This would require pulling the test lines along with the moving unit. In most cases this is impossible (Fig. 10.30a). The next solution would be the use of sliding contacts to transmit the signals from the moving line. Because these signals are

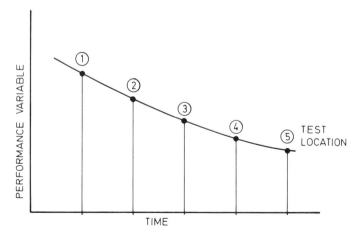

FIG. 10.29 Typical pulldown curve monitoring the transient behavior of a test unit on a moving conveyor.

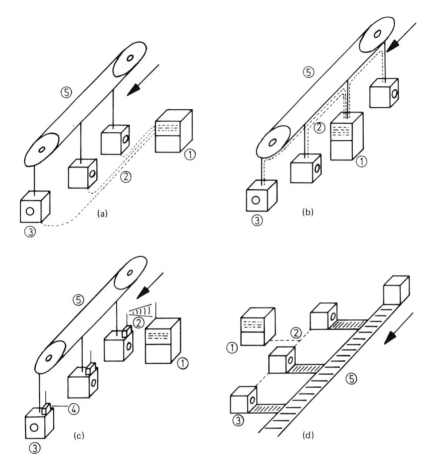

FIG. 10.30 Different test configuration for moving production lines: (a) stationary computer, test unit pulls signal cables; (b) mobile computer; (c) mobile test computer, stationary master; (d) stationary test stalls. 1, computer; 2, signal cables; 3, test unit; 4, mobile test computer; 5, conveyor.

usually very weak, they will be falsified by poor contacts or by bouncing contacts.

Figure 10.30b shows a solution where the computer travels with the test units on the conveyor system. In this case only the power to the computer has to be transmitted via sliding contacts. If the computer is provided with adequate power buffering, short interrupts of the power originating from the sliding contacts may be eliminated.

Figure 10.30c shows a test setup where each test unit is accompanied by its own battery-powered data acquisition computer. At the end of the test track the central computer receives the test data by radio or infrared communication. The test data will be evaluated by this more powerful computer. Upon completion of the test the data acquisition computer is placed on a new test unit.

Figure 10.30d shows a test system whereby the test units are pulled off the main product stream and monitored in a stationary position. This solution will render the best test data; however, it is expensive and interrupts the flow-line process. Figure 10.31 shows an engine test system that operates on this principle.

10.9.2 Self-Adapting Test Systems

The computer can be programmed to learn the upper and lower test limits on its own. The performance specifications for each model to be tested may be obtained by different methods:

Method 1: For each test model a fixed set of good units is tested (i.e., 50 units). These data are averaged and used as a performance index. The test limits are updated after another 50 units have been tested. This procedure is constantly repeated, always obtaining current test limits from the total aggregate data of good units. Figure 10.32 shows how control limits may change with this procedure. It can be noted that the allowable tolerances tend to widen as the aggregate sample size increases. This situation eventually stabilizes. A stable condition is reached only if the test parameters follow a gaussian distribution; otherwise, the tolerances will either widen or narrow. The method has another problem: It does not take into account the learning effect whereby the quality of the product increases the longer it is being built.

Method 2: With this method the tolerances are obtained from a set of good units tested previously (i.e., from 50 units). Now these tolerances are applied to the succeeding units until another 50 good units have been tested. The parameters of these new good units are used to calculate the new tolerances. The old tolerances obtained from the previous set of good units are discarded. As can be seen in Fig. 10.33, the tolerances make sudden jumps, whereby a unit may be accepted with tolerances previously used and rejected with new tolerances, or vice versa.

Method 3: Figure 10.34 shows what might be expected with a procedure that updates the tolerances every time an acceptable unit is tested, using the latest aggregate data of, say, 50 acceptable units. In this procedure the oldest set of data is discarded. This seems to be the best and most effective procedure. It

FIG. 10.31 Computerized engine test system with stationary test stalls. (Courtesy of Daimler Benz AG, Stuttgart, West Germany.)

Quality Control

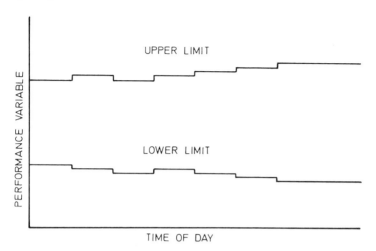

FIG. 10.32 Control limits using total aggregate data for updating.

eliminates sudden shifts or changes or limits. It also provides a trend in product quality and compensates for the learning effect.

If the performance test is carried out in a factory environment and the variables measured are affected by changing ambient conditions, temperature or other correction curves have to be incorporated in the test tolerances.

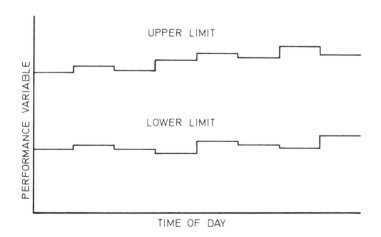

FIG. 10.33 Control limits using data of predetermined number of units for updating.

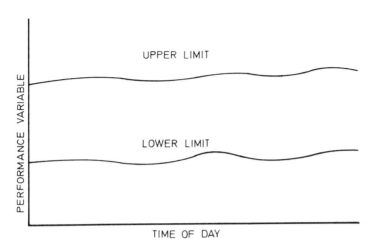

FIG. 10.34 Continuously updated control limits using latest data of predetermined number of units.

10.10 COMPUTER LANGUAGE FOR TEST APPLICATIONS

The performance of a computer-controlled quality test is a rather complex process and involves the steps described in Sec. 10.5: recognition of the product, positioning of the sensor, acquisition of test parameters, evaluation of test results, calculation of a quality index, and output of the test data. If all these operations are done completely automatically, the programming language must have the capability to control the test equipment, supervise the input/output of data, and perform statistical calculations. The first two groups of these functions require instructions that can easily manipulate individual bits, whereas statistical calculations necessitate a language that can readily do mathematical calculations. Unfortunately, there is no language that can handle both of these requirements efficiently. Therefore, the selection of the programming language is determined by factors such as the type of test, the size of the available computer, execution time of the program, required portability for the software, ease of program changes, and programming staff. Figure 10.35 gives an overview of the different programming languages used for computer-controlled test systems. Their most important attributes will be discussed briefly.

Machine-oriented languages are mainly used on small computers, such as microcomputers, which do sequential control and simple integer arithmetic. This group of languages is machine dependent and requires good programming skills. The macro and block languages are more user oriented and may reduce considerably programming

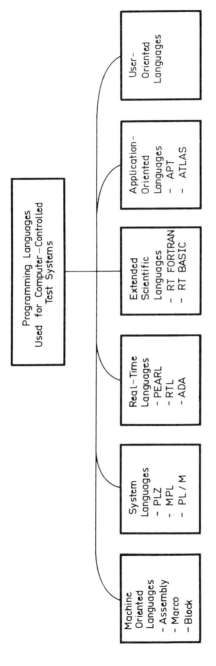

FIG. 10.35 Family of programming languages used for computer-controlled test systems.

effort and time. The machine-oriented languages offer a major advantage when memory cost has to be minimized and fast execution times are important. Their great disadvantage is that they are cumbersome to use for complex arithmetic operations, as they are needed for the execution of statistical or control algorithms.

With the help of system programming languages, more abstract formulation of the problem to be solved is possible. Thus they are more suitable for the implementation of complex algorithms. They allow conception of clearly formulated well-structured programming modules which are also self-explanatory. Their major disadvantage is that they do not support programming of I/O operations and real-time requirements.

The higher real-time languages have been specifically designed for computer process communication. They support real-time computer process interaction, task communication, simple input/output programming, and easy implementation of complex mathematical operations. Their main application area is process control, and for that reason they are of more generalized design and may lead to unwieldy programs when used for test applications. However, they are easy to learn and easy to use and will render satisfactory solutions for most test installations. Their main disadvantages are slower execution times compared with the machine and system languages and they also have large memory requirements.

The extended scientific languages are very popular since FORTRAN and BASIC are well known to engineers. For real-time applications they have been supplemented with subroutine calls to support computer process input/output communication and task scheduling and to perform manipulation of bits and bit patterns. These languages are best suited for less sophisticated test systems when the real-time requirements are not very stringent. They also occupy a large memory space and have slow execution times.

There are several languages that use APT-type constructs for measuring applications. They may be of advantage when the program that is used to manufacture a workpiece on a NC or CNC machine tool can also control the coordinate measuring machine for checking the dimension of the part. The program sends the probe along the surface of the workpiece contour and directs the measuring machine to take dimensional measurements at consecutive contact points. These points may be predetermined or selected automatically according to the results obtained from previous measurements. The APT language for this application has to be an extended version of the basic APT which includes measuring functions. Some typical functions are as follows:

1. Coordinate transformation.
2. Coordinate selection, displacement correction, shift of workpiece position, conversion of spatial axes.

Quality Control

3. Selection of measuring angle.
4. Selection of stepping increments of indexing table.
5. Selection of number of contact points along a circle, ellipsis, plane, cone, and so on.
6. Calculation of lines and radii of surface contours from measuring points.
7. Mean-value calculation.
8. Reconstruction of basic geometric elements from probed points and calculation of deviation form drawing (Fig. 10.17).
9. Calibration of probe tips with a calibration sphere and precise definition of their position, including the tip geometry. Thus a workpiece can be measured by different probes and from different positions without reclamping.
10. Numerical output, such as measuring data, dimensional deviations, and extreme-value parameters.
11. Graphical output of measurement results and deviation from the drawing.
12. Determination of statistically significant number of measurements.

Manufacturers of coordinate measuring machines supply such programming aids with their product. There are usually several levels of programming comfort from which the user may select.

10.10.1 The Language ATLAS

The language ATLAS (Abbreviated Test Language for All Sytems) was originally designed for testing avionic systems and was later extended to be capable of handling almost any test problem. It is a 17-year-old language concept and is being used by almost all well-known test system suppliers. The basic ATLAS is recognized as an IEEE standard. There are also several subsets being used which contain only the language capabilities needed for specific applications. In addition, several manufacturers have created their own version of ATLAS. A partial list of ATLAS capabilities is shown in Table 10.5 [10]. In addition to the ATLAS compiler, a programming system usually contains an editor and fault-isolation routines.

10.10.2 Special Languages

User-oriented languages in general consist of program modules that can be freely combined depending on the application. In principle, they are based on the macro concept, in which the parameters are of symbolic nature. The parameters are filled in by the user, and the macros are combined in the complete program using a linker or with the aid of a macroprocessor. It is estimated that about two-thirds of all application programs can be constructed with this method. The following are typical modules for test applications:

TABLE 10.5 Typical ATLAS Capabilities

BEGIN	Program initiation statement; needed for every test program; title and number may be defined
TERMINATE	Ends the test; no additional statement follows
Preamble section	First part of a test program; it precedes the procedure part and contains all EQUATE and subroutine definitions
Subroutine definition statement	Defines the start of a subroutine; a maximum of 25 subroutines are allowed; they cannot call themselves
END	Defines the end of a subroutine
EQUATE	Equates the computer channel number to the terminal number of the test station
Procedure section	Contains a series of statements that describe the individual test steps to be performed
PERFORM	Used to call a subroutine or another ATLAS program
CALCULATE	Used to perform the following arithmetic operations: + Add − Subtract * Multiply / Divide ** Exponentiate
COMPARE	Permits comparison of results with specified limits
VERIFY	Used to grade test results: GT Greater than LT Lower than EQ Equal to FAIL HIGH/LOW UNIT FAIL HIGH/LOW UNIT PASS
GO TO BRANCHING	Used for unconditional and conditional branching

TABLE 10.5 (Continued)

REPEAT	States number of times to be repeated
MONITOR	Reads and displays reading until operator interferes
WAIT FOR	Time delay until operator resumes data
PRINT	Print test data
DISPLAY	Output on CRT terminal
RECORD	Identifies and saves measurement data
READ	Reads analog data and retains them in MEASUREMENT register
FINISH	End of test; places equipment in quiescent state
CONNECT/DISCONNECT	Connects or disconnects test instruments
SETUP	Program tests instruments
APPLY	Compares functional test pattern with reference pattern
MASK	Can be used to ignore indeterminate outputs

Control of test cyclus
Counters
Supply of test limits
Measurement data acquisition
Parameter limit control
Plausibility control of measured parameter
Calculation of mean values
Trend analysis
Digital filtering
Logic operations
Conversion from physical to technical values
Output of control information
Writing of test protocol
Graphic presentation of data

These languages may simplify the use of a test system considerably. There is no need for learning a complex programming language. A test setup can be implemented and changed in a very short time. Their

FIG. 10.36 Output terminal displaying pictorial data. (Courtesy of KOMEG, Riegelsburg, West Germany.)

big disadvantage is the rather inflexible structure of the program, which makes the language only useful for the specific application for which it was designed. With this method programs also require considerable memory space, and in addition the execution times are slow.

In general, any test system should be user friendly. This may be obtained by several aids, such as user guidance, easy-to-use input/output pheripherals, graphic displays, and menus to select pictorial data and graph presentations on an output terminal (Fig. 10.36).

REFERENCES

1. J. Stehle, University of Karlsruhe, Test Systems, Master Thesis, University of Karlsruhe, West Germany.
2. G. Baner, The Application of Microprocessors for Adapting Measuring Systems PDV-Bericht, GfK-PDV 101, Kernforschungszentrum Karlsruhe GmbH, Karlsruhe, West Germany, Jan. 1977.
3. U. Rembold, K. Armbruster, and W. Ülzmann, *Interface Technology for Computer-Controlled Manufacturing Processes*, Marcel Dekker, New York, 1983.
4. W. Steinchen, Quality Control by Computer, *VDI Nachrichten*, No. 46/12, Nov. 1969.

5. K. Herzog, Zeiss Multi-coordinate Metrology, Hardware-Software-Application, reprint from *Zeiss Information 91*, Carl Zeiss, Oberkochen, West Germany, 1980.
6. A. Fürst and W. Vollaard, NC Machine Controls Itself, *VDI Nachrichten*, No. 7, Feb. 1982.
7. C. Wick, Automatic Electro-optical Inspection, *Manufacturing Engineering*, Nov. 1979.
8. I. Bey, Computer Aided Quality Assurance in Industrial Production, PDV-Bericht KfK-PDV 170, Kernforschungszentrum Karlsruhe GmbH, Karlsruhe, West Germany, 1979.
9. P. C. Chen. U. Rembold, and J. Weinstein, On-line Computerized Product Testing, IEEE Transactions on Industrial Applications, Jan./Feb. 1973.
10. ATS 961, *Automated Test System Programmer's Manual*, Texas Instruments, Inc., Dallas, 1978.

11
The Programmable Factory

11.1 INTRODUCTION

Hard automation has made possible the mass production of very complex products at a reasonable cost. These manufacturing methods, however, could not be used efficiently in low- and medium-volume production runs. It needed the development of NC technology to improve the productivity of batch production. Despite its inherent advantages, numerical control has penetrated metal cutting by only 5% over a period of 25 years. This is due primarily to the fact that for many manufacturing organizations, it is still very difficult to devise good justification procedures to determine the economics of this technology. In particular, the intangible benefits which are often associated with the introduction of new tools are not fully understood.
In most cases production personnel become aware of these benefits only after several NC machine tools have been operated for a longer period. This trial-and-error procedure discourages many organizations from assuming the technological risk connected with the introduction of sophisticated production equipment. In addition, the justification phase for the investigation of complex manufacturing technology is very time consuming and costly because it usually involves the investigation of horizontal integration across the factory. New machine tools will also make it necessary to change the present production organization so that alternative ways of doing business will have to be conceived.

The DNC technology was devised by manufacturers who used a great number of NC machine tools. In this technology both hardware and software drive several machine tools simultaneously. The operation is coordinated by one or several computers. This approach to centrally supervised but rather loosely connected production machines led to the development of flexible manufacturing systems (FMSs). The total number of DNC systems presently installed is also very low since they usually constitute an enormous investment. Their number is less than 200 worldwide. The justification difficulties mentioned above also impose a problem with the introduction of DNC and FMS technology.

11.2 GENERAL ASPECTS OF FLEXIBLE MANUFACTURING SYSTEMS

A flexible manufacturing system consists of a group of machine tools and/or production equipment interconnected by an automated material handling system. A computer, or usually a hierarchical computer system, plans, executes, and controls the production process. Workpieces are loaded on a carrier and are sent progressively from one machining operation to another. The production sequence can be selected according to the manufacturing plan of the workpiece spectrum. The part is often mounted on the carrier or a pallet serving as the machining table. It does not need reclamping when being transported from one operation to another. Special tools and fixtures may also be automatically scheduled and changed under computer control. In addioion, quality control may be done on an automatic measuring machine. Direct labor is kept to a minimum and may only be needed for loading and unloading workpieces on or off the manufacturing system.

Flexible manufacturing is economical for medium-size production runs of 200 to 20,000 parts per year (Fig. 11.1). These production runs account for 50 to 75% of the value of all parts manufactured. In Fig. 11.1 simple and complex parts are differentiated. The workpiece spectrum consists of spherical parts made from steel, cast iron, aluminum, or magnesium having a weight of 1 to 1000 kg. Typical operations include turning, drilling, boring, milling, tapping, and grooving. Figure 11.2 shows various complex workpieces made on FMS.

To date, these medium-size production runs are done either on NC machine tools or by mass production methods. Neither of these processes renders economical results. Eventually, industry will learn how to justify and use FMS. When this happens their growth potential should be enormous. The objectives in employing FMS are as follows:

1. Use NC technology for medium-size production runs.
2. Provide a manufacturing facility that can machine certain types of part families and can easily be reprogrammed for other part families.
3. Provide a self-contained manufacturing facility that can automatically schedule, machine, and inspect production runs.

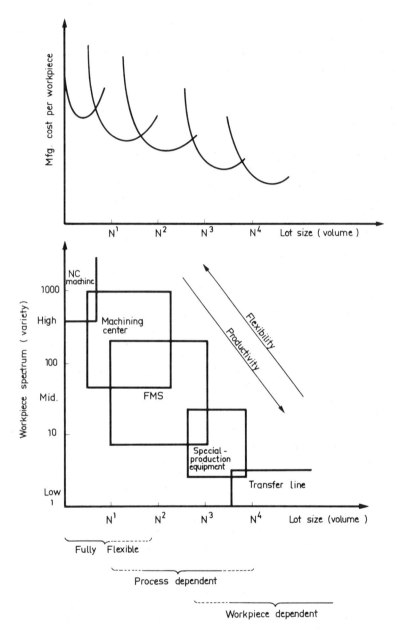

FIG. 11.1 Flexible manufacturing concept. Lot size N varies with complexity of part. Typically N = 1 for complex and N = 10 for simple parts. (From Ref. 1.)

The Programmable Family

FIG. 11.2 Typical complex workpieces produced on flexible NFS.
(Courtesy Maho Werkzeugmaschinenbau, Pfronten, West Germany.)

4. Provide a supervisory function to control and supervise production operations and the manufacturing equipment.
5. Construct independent machining modules that can easily be assembled to a production unit, permitting unobstructed reconfiguration when necessary.

There are many benefits that can be expected from FMS. Figure 2.1 shows that for an average part, approximately 95% of its residence time in a factory is spent waiting, 2% of its time is used loading and unloading it from machine tools, and 3% is used to actually perform machining operations. If one assumes that a conventional machine tool can be operated 24 hours a day 365 days a year, its productive time is only 6% (Fig. 11.3). Equipment geared for highly automated production has similar problems (e.g., its calculated optimal utilization time typically is 80%, but in practice it is only 59%). This problem is analyzed further in Table 11.1. If these losses are restructured as in Table 11.2, it can be seen that 64% of the losses are related to managerial problems. A similar study was made for NC machining centers (Table 11.3). In this case active machining occupies only 23% of the total available time.

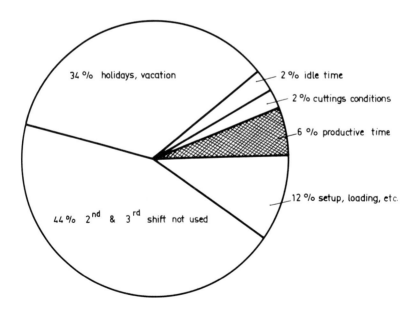

FIG. 11.3 Distribution of lost capacity in machine tools. (From Ref. 2.)

TABLE 11.1 Contributing Factors to Productivity Loss

Equipment failure	42%
Machine in wait mode	34%
Workforce control	16%
Others	8%

Source: Ref. 3.

TABLE 11.2 Management-Related Productivity Losses

Skilled trade factors	14%
Machine in wait mode	34%
Workforce control	16%
	64%

Source: Ref. 2.

TABLE 11.3 Equipment Utilization in Machining Centers

Metal cutting	23%
Positioning and tool changing	27%
Gauging and loading	18%
Setup	5%
Waiting and idling	14%
Repairing and others	13%

Source: Ref. 3.

Flexible manufacturing systems are designed to minimize the idle time inherent to conventional manufacturing systems. Possible benefits include the following:

1. *Operate equipment around the clock*: During the two day shifts, the equipment will be supplied with raw parts and properly maintained by personnel. Scheduling and supervision is done by the computer under operator observance. During the night shift (ghost shift) a computer supervises the operation independently and if necessary, turns the system off when problems arise.
2. *Minimize direct labor*: Machining, tool changing, fixturing, measuring, and material moving and handling are controlled automatically by the computer. Labor may be used during day shifts for observance functions, loading, unloading, and maintenance.
3. *Minimize lead time*: This is performed by the computer, which knows the production schedule and the machine status.
4. *Reduction of in-process inventory*: Since a FMS operates on the flow line production principle, in-process inventory buffers are reduced to a minimum. They may only be maintained to provide parts during a possible equipment failure.
5. *Reduce tools and fixture requirements*: Since FMSs produce a larger part spectrum, universal tools and fixtures are used. This results in shorter retooling and setup times.
6. *Obtain a high flexibility*: The part spectrum for which a FMS is conceived has a major influence on the flexibility and utilization of the equipment. Future product variants, engineering changes, and manufacturing methods should be anticipated when such a system is conceived.

11.3 CLASSIFICATION OF FMSs

Since the conception of flexible manufacturing systems, basically three different types have been designed. Common to all of them is the machine tool, a material handling system, and supervising control equipment. A typical representative of each type will be shown. The various concepts of machine tools and conveyance and control systems employed will also be discussed.

11.3.1 FMS with Complementing Machine Tools

With this principle several NC machine tools are interconnected by a material handling system. The machine tools complement each other. The workpiece enters the system via a loading station. From here it

The Programmable Family

is sent under computer control from one workstation to another to be processed by consecutive machining operations. The number of operations done by each machine tool depends on the workpiece design and machining process. Usually, with this FMS the path of the workpiece through the system is fixed. However, different machine tools may be used for different parts. Figure 11.4 shows the principal layout of such a system. The workpiece is clamped on a pallet in the setup area and from there it is channeled via a roller conveyor system to the different machining operations. The machine tools consist of a NC mill, a duplex turning station, an NC lathe, and two head indexer boring machines. Any machine tool may have provisions to select tools automatically and to dispatch these to the spindle on demand.

This type of FMS in general is very economical; however, during equipment failure the entire system may be idle. Our example shows two machine tools that could be used as backup. These are the second turning and the second head indexer boring machine.

11.3.2 FMS with Substituting Machine Tools

Possible standstill of the entire system due to equipment failure may be eliminated by the use of substituting machine tools. Figure 11.5 shows a FMS constructed from eight machining centers, a storage system, and a shuttle-operated material flow line. In this case workpieces can be brought to any machining center that has the proper tool to produce the part. The computer has in its memory the status of each machine tool and assigns machining capacity upon release of the workpiece for processing. Each machine tool is equipped with an automatic tool changer and selects the correct tools on demand. A given machining center may perform part or all of the machining operations. This FMS has the great advantage that upon equipment failure only part of the system is idle.

11.3.3 Hybrid FMS System

In practical installations a strict adherence to either the complementing or substituting principle is seldom seen. Often, both principles are intermixed. The actual system shown in Fig. 11.4 is of hybrid design.

11.3.4 Special Features of FMSs

In most of the FMS designed to date, the material handling system is the center of the installation. The machine tools are located such that they can be serviced readily with parts. An alternative system, where a centralized tool conveyor plays the predominant role, is shown in Fig. 11.6. This system employs three special-purpose machining

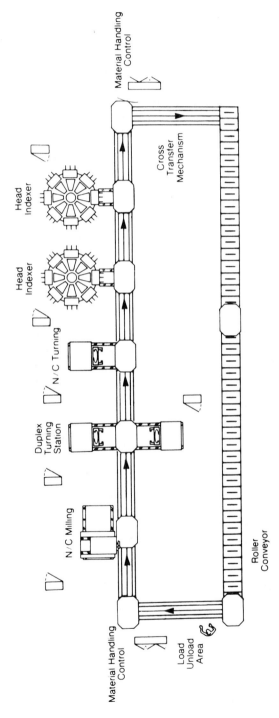

FIG. 11.4 FMS with compensating machine tools. (Courtesy of Kearney & Trecker Corp., Milwaukee, Wisconsin.)

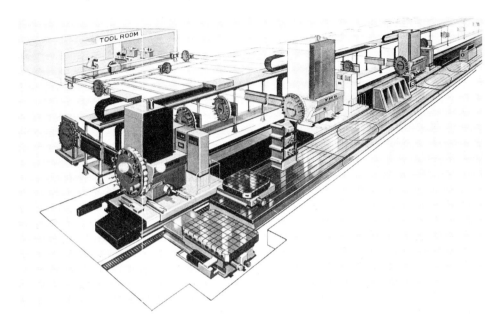

FIG. 11.5 FMS with substituting machine tools. (Courtesy of Yamazaki Machinary Works, Ltd., Nagoya, Japan.)

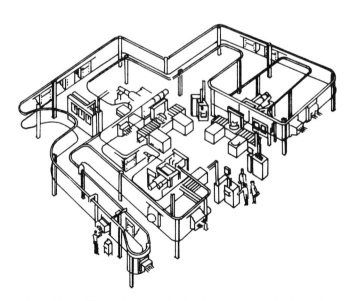

FIG. 11.6 Flexible manufacturing system designed around a centralized tool conveyance system. (From Ref. 5. Courtesy of Society of Manufacturing Engineers, Dearborn, Michigan.)

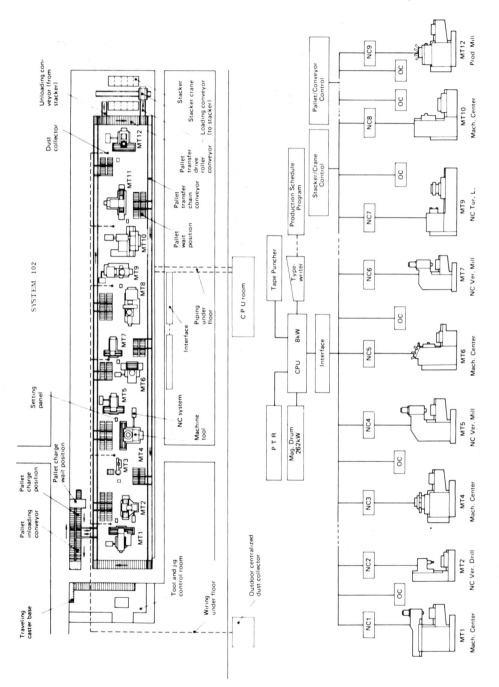

The Programmable Factory 753

centers, on either of which a workpiece may be clamped. The tools are located in tool heads and are centrally stored. Upon assignment of a workpiece to a machining center the control system selects the proper tool and sends it to the workplace. Here the tool is connected to the machining center to perform its machining operation.

11.4 COMPONENTS OF A FMS

As mentioned earlier, the FMS consists of the machine tool, material handling system, and the control computer. In addition, a considerable amount of software is needed to operate the equipment. The complexity of the system is determined by the user. It depends on the part spectrum, the number and types of machining operations needed to fabricate the part, and the amount of horizontal and vertical integration of the FMS into its manufacturing environment.

Theoretically, it is possible to conceive the FMS as an independent small factory, whereby full integration of computer-aided design, process planning, production scheduling, and production control would be feasible. However, this will be done in only a small number of installations. The average FMS will be part of a larger manufacturing facility which is the direct user of the parts manufactured. The integration of the FMS into the manufacturing environment will play an important role when such a system is conceived and installed. Particular attention has to be paid to common data processing procedures and protocols and the compatibility of the computers and the communication equipment. General components that will make up a programmable factory were discussed in Chap. 3. In the following section we discuss those components that are of particular interest to FMSs.

11.4.1 Machine Tools

A FMS may be constructed from conventional NC or CNC machine tools such as lathes, milling, and boring machines (Chap. 3.5; Fig. 11.7). In this case the flexibility must be designed into the software. Thus the software and the supporting computer system will be quite complex. This requires that the software and the computer be highly modular to be useful for different types of installations. There is, however, a trend toward the use of more powerful machine tools. Newer installations often employ machining centers and multiple-spindle boring machines (Figs. 3.18 and 3.19).

FIG. 11.7 FMS with conventional NC machine tools. (Courtesy Hitachi Seiki Ltd., Tokyo, Japan.)

Chapter 11

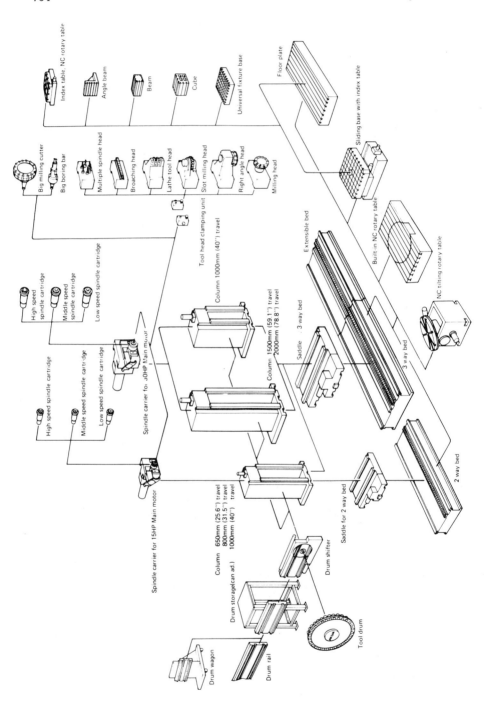

The Programmable Factory 755

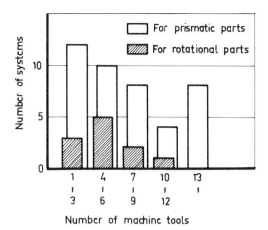

FIG. 11.9 Number of machine tools in the system. (From Ref. 8.)

Another concept that will have a considerable impact on FMS technology is the convertible modular machine tool (Fig. 11.8). Here the machining system is assembled from standard modules and can be tailored to suit different machining requirements. In this case the flexibility resides mainly in the hardware. There will be less emphasis on computing equipment and software. Of the two alternatives, the first offers more flexibility. However, it is more difficult to conceive, implement, and maintain.

In a recent study in Japan, 53 flexible manufacturing systems were investigated; 42 of these were designed for machining of prismatic parts and 11 for rotational parts [8]. The number of machine tools in these systems is shown in Fig. 11.9 and the types of machine tools in Fig. 11.10. Of particular interest is the great number of machining centers used. The distribution of additional important parameters, such as maximum part size, number of different parts, processing time, and average lot size is shown in Table 11.4. Workpiece transfer and loading were automated in most cases. Automation of the tool routing was automated in only four installations. No system had both the flow of the workpieces and that of the tools fully automated. Fourteen systems for prismatic parts and one system for rotational parts could be operated without attendance during the night shift.

To obtain versatility the machine tools of a FMS will have to have automatic tool head changers and tool buffers (Fig. 11.11). Tool

---

FIG. 11.8 FMS with convertible modular machine tools. (Courtesy of Yamazaki Machinery Works Ltd., Nagoya, Japan.)

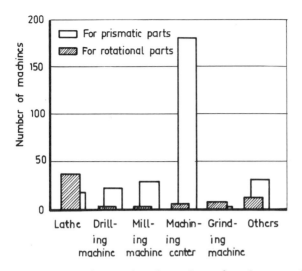

FIG. 11.10 Accumulated number of each type of machine tool. (From Ref. 8.)

TABLE 11.4 Distribution of Parameters of Installed FMS for Prismatic Parts

Parameter	Minimum	Maximum	Mean
Maximum part size (mm)	200	1000	600
Number of different parts	2	100	30
Processing time for a part (min)	2	90	25
Average lot size	1	501	40

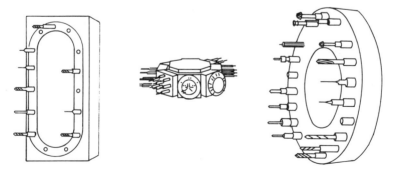

FIG. 11.11 Different tool holder concepts.

The Programmable Factory 757

selection and changes are made under program control. A great number of tool-changing concepts are available. Usually, each machine tool manufacturer has its own design.

A FMS may be constructed from several machining modules (Fig. 11.12). Each module is self-sufficient; the machanics, electronics, and the electrical interface are integrated into one unit. Control of the machining operation and supervision of the machine function should be done autonomously. The interface to other FMS components is of particular importance in order to be able to connect modules with each other and to a supervisory controller. It is essential that the same communication protocols, connectors, pin assignments, and logic signal levels be used. Standardization is also needed for the mechanical interfaces.

11.4.2 Materials Handling Systems

The material handling equipment ties the individual machine tools together to form a flow pattern production system. There are several

FIG. 11.12 Modular turning system. (Courtesy of Pittler Maschinenfabrik AG, Langen be: Frankfurt/M., West Germany.)

material handling concepts in use. Typical representatives are the roller, shuttle (Fig. 11.5), and towline (Fig. 11.13) systems (see also Fig. 3.30). These systems are designed to move palletized parts to and from the machine tool and to secure them in a fixed position for machining. Typical flow patterns for the system layout are shown in Fig. 11.14.

Usually, the material handling system contains a work setup area (Fig. 11.4). Here the part is clamped on a pallet and the pallet becomes the actual worktable. For efficient operations an attempt is made to perform clamping operations only once during machining since reclamping adds to manufacturing cost. The pallet will be delivered to the conveyor system and the identity of the workpiece will be made known to the computer. The computer determines the routing of the part and sends it from one machining station to the other. For machining the pallet will be clamped securely to the worktable of the

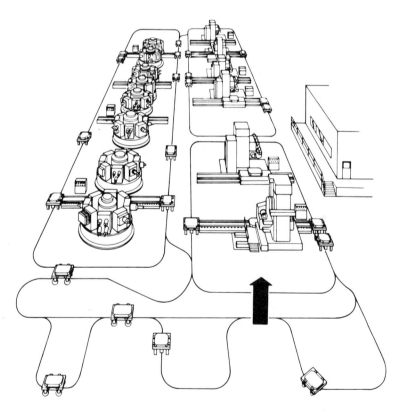

FIG. 11.13 Towline material handling concept. (Courtesy of Kearney & Trecker Corp., Milwaukee, Wisconsin.)

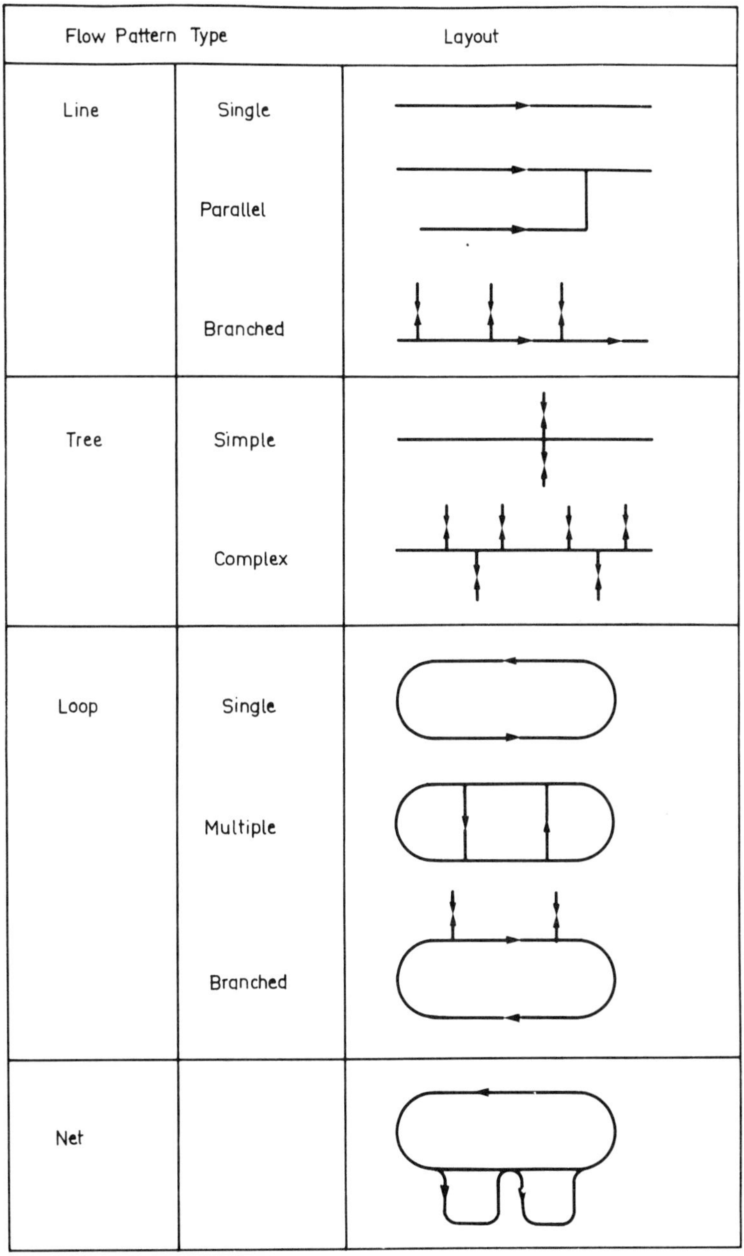

The Programmable Factory

Flow Pattern Type		Layout
Line	Single	
	Parallel	
	Branched	
Tree	Simple	
	Complex	
Loop	Single	
	Multiple	
	Branched	
Net		

FIG. 11.14 Different types of flow patterns of a FMS. (From Ref. 8.)

processing equipment. Upon completion of the operation the pallet is released and sent to the next destination. During its travel through the plant the computer must keep track of the part to know its whereabouts. There are several principles that can be used in FMS to track parts; they were discussed in Chap. 3.

11.4.3 Control of FMS

Modern control systems employ a hierarchical control concept (Chap. 4). In principle, the computer can perform both production and machine control. Production control is usually done at a higher hierarchical level. Both of these functions are closely interwoven and extensive data transfer between the different levels of computer is necessary. Figure 11.15 shows the flow of data and control information in a FMS. Such a machining installation is usually part of a comprehensive manufacturing system (Fig. 2.22). For this reason it may need only limited control functions, such as human-machine communication, NC machine control, material flow control, machine monitoring, and quality control.

A manufacturing task starts with the definition of a long-range goal. This information is entered into the computer at the highest level and decomposed to smaller tasks which are executed at that level. The output of each task is either an instruction for an operator or one for a manufacturing process. The instructions are handed down to the next level and further decomposed to subtasks. The process is repeated until all participants of the control system have received their assignments. From each lower level feedback is sent back to the higher level, assisting the decision-making process. This will complete the individual assignments of each computer. Communication between levels takes place with the help of a mailbox service and a common memory. Only one user at a time can write into a mailbox; however, if necessary, several users can read it. Mailboxes are needed for commands, control and status information, and data. Updating is done when a production process to which the mailbox is assigned changes its status. Thus the contents of all mailboxes represent the status of the entire system.

A FMS will be designed as a virtual machining process. Upon presentation of a part description, the tasks to make the parts are decomposed and the part route, machine tools, tools, fixtures, and quality test will be assigned to the part. Users have the illusion that they have at their service a fixed set of resources.

When a part spectrum is presented to the computer, first an analysis will be made as to whether the job can be done by the FMS. The decision logic must contain knowledge of all the constraints of the system and its capabilities. An economic analysis on the production run and economical batch sizes should also be conducted. The next step

The Programmable Factory

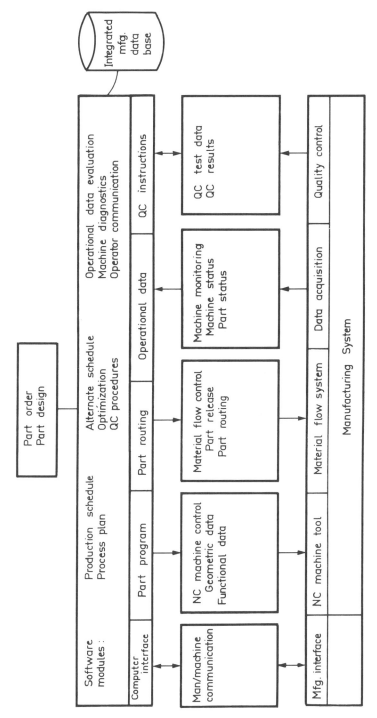

FIG. 11.15 Flow of data and control information in a FMS.

is the selection of the part families. For this the parts are usually divided into groups that have similar processing requirements (Chap. 6). However, other criteria, such as geometric shapes, common tools or fixtures, production cost, and material composition may also be used. Information about the part, its design, and the required completion dates is entered from a higher hierarchical level. The computer then produces the process plan and production schedule. It may also go through an optimization procedure.

In case of a bottleneck, rescheduling of another part may be necessary. If there is still excessive queueing of parts, it may be necessary to use decision rules to minimize waiting time. Each part has its own route file. Upon release of the part the part program, part routing plan, and quality control program are activated. Thus processing of the part can commence.

During processing, operational and quality control data are gathered and sent to a higher level for evaluation. In case the production process proceeds normally, the operational data are filed for historical evaluation. However, if a problem is detected, the computer will try to locate it and may even attempt a system recovery procedure. All these functions make it necessary that the computer have human knowledge about the manufacturing process and machine control. The system architect must be able to structure this knowledge so that it can be used efficiently by the computer.

Role of the Computer in a Flexible Manufacturing System

The most important functions of computers at different hierarchical levels were discussed in Chap. 4. A flexible manufacturing system will have hierarchical control employing two or three control levels. The administrative level will probably be located in a central plant computer which supervises the FMS and other entities of the plant. Figure 11.15 depicts the flow of data and control information in a FMS. A supporting computer system will be laid out according to the control requirements of the entire installation. The functions of the various control levels are described next.

Administrative Control

The tasks at this level are production scheduling, production control, and maintenance of the management data base supporting the FMS. Table 11.5 and Fig. 11.15 show a more detailed breakdown of the required functions. This level interfaces with supervisory control of the FMS.

Supervisory Control

At this level the production plan will be executed and the operation of the various production facilities, such as that of machine tools,

The Programmable Factory

TABLE 11.5 Administrative Tasks to Support a FMS

Scheduling	Control
Process planning	System startup/shutdown
Equipment scheduling	Order routing
Part family assignment	Production supervision
Part programming	Operator interaction and guidance
Resource configuration	System recovery
Rescheduling	Reporting
Order dispatching	Productivity
Production optimization	Parts manufactured
Production simulation	Quality
Maintenance scheduling	Scrap
	System problems

material storage, conveyor system, tool setup, and quality control, will be coordinated and supervised. The tasks to be performed are scheduling and controlling functions (Table 11.6 and Fig. 11.15). The computer will obtain a daily production schedule and commit the resources needed to perform the production runs. Upon release of an order, the raw part is retrieved from storage and clamped on a pallet in the setup area. From here it is routed under computer control

TABLE 11.6 Supervisory Tasks to Support a FMS

Scheduling	Control
Daily production runs	Equipment synchronization
Machine tools	Operating data evaluation
Tools	Part flow control
Material handling	Operator support
Quality Control	Operator guidance
NC programs	
NC program distribution	

through the virtual manufacturing system which was configured for production of the part. The flow of the part from one machine tool to another is synchronized, and conflicts with the production of other parts are avoided. There must be a model of the virtual manufacturing process in the computer memory which simulates the flow of the part through the facility. Upon completion of the part a test program and procedure will be dispatched to the measuring equipment to check if the quality standard is met.

An important function at this level is operator support. The operator can query the system to have the production schedule or the status of an order displayed and if desired may receive information on machine utilization and productivity. Quality control functions can also be supervised interactively. Upon request, a display terminal may show inspection procedures, tolerances, deviations, necessary compensation for tool wear, and historical quality data. Interactive functions will also be supporting the tool setup procedure used to prepare a new production run. In addition, a simulated machining operation can be observed for a workpiece to assure that it can be produced with the tools provided and that the tool path is free of any collision. There will also be notification about a necessary change of worn tools. In case a change is necessary, the operator can query the system about the life expectancy of other tools. If there is a change scheduled for a tool in the near future, the system will inform the operator as to whether it is economical to replace it now.

Machine monitoring will be essential with a highly automated production equipment such as a FMS. The running time should be close to its maximum rated value. Therefore, all functions, such as movements of actuators, relays, valves, and cylinders, will be monitored closely. Any problem or slowdown of equipment will be recorded. Interactive failure diagnostics procedures will help the operator to locate the problem. The operator will obtain information on the probable cause of a failure and instructions for repair. For this purpose the computer will have a failure catalog and decision rules to derive a failure diagnosis of abnormal machine behavior. When a problem occurs, a failure code will be issued together with information on the cause of the problem, repair instructions, and the accumulated failure history of the machine. If a less important failure occurs, the computer will decide wether to continue the operation. In any event, it will issue a report on the problem to initiate a repair that can be made during a work break.

Machine Monitoring and Control

At this lowest control level a multitude of simple control functions will be performed. The FMS may contain a part, tool, and chip conveyor system, robots, several machine tools, a workpiece setup

The Programmable Factory

area, and an automatic measuring machine. One or several computers will be installed to control and monitor each of these operations. The basic functions of a computer at this level are explained in Table 11.7 and Fig. 11.15. Basically, the tasks are control and monitoring. There are provisions for manual and automatic operations and initialization of the individual system components.

The controller for each machine must be capable of starting and completing an entire machine cycle autonomously and of monitoring the operation. Synchronization between machines will be done according to the handshaking protocols commonly used with computer communication. NC and measuring machines will have aids for program debugging and correction as well as trace routines and provision for adjusting to a possible offset of the reference coordinates.

Example of Hierarchical Computer Architecture for a FMS

Figure 11.16 shows the layout of a FMS installed by the Renault Company. The system produces gearboxes for trucks. There are four machining centers: one boring and facing machine and two head changers interconnected by wire-guided battery-powered carts. The substitution of machine tools can be randomly scheduled and the tools accessed by carts through a complex transportation network.

Figure 11.17 depicts the hierarchical computer system used to control the equipment. Supervisory control is carried out by two minicomputers operating in duplex mode. At the next lower level there are 13 microcomputers, 25 programmable controllers, and 5 numerical controllers. The following are special features of the control system:

1. Decision making is done centrally by a supervisory computer.
2. Control information is processed locally.
3. Each machine can be operated independently without the supervisory computer.
4. Correction can be done locally without interfering with the entire system.

The supervisory computer coordinates the operation of the individual subprocesses. Some of its important missions are the following:

1. To initialize and control a 24-hour production plan
2. To optimize machine utilization
3. To coordinate the parallel operation of the machines and the transportation system
4. To control the transportation system
5. To activate standby equipment, if necessary
6. To control the tooling system and to monitor tool wear
7. To issue production records and statistics

TABLE 11.7 Control and Monitoring Functions of a FMS

	Machine tool	Conveyor	Robot	Workpiece setup	Measuring machine
Path control					
Workpiece	×				
Measuring probe					×
Control of:					
Actuator	×	×	×	×	×
Motor	×	×	×	×	×
Relay	×	×	×	×	
Pump	×		×		
Valve	×		×		×
Monitoring of:					
Actuator	×	×	×	×	×
Motor	×	×	×	×	×
Relay	×	×	×	×	×
Pump	×		×		
Valve	×		×		
Limit switch	×	×	×	×	×
Light barrier	×	×	×	×	×
Bar code		×	×		

The Programmable Factory

	×	×		×	×	×		
			×		×	×		
		×	×		×	×		
		×	×		×	×		
			×				×	×
	×	×	×	×				
								×

Tool breakage
Production rate
Scrap
Downtime

Measurement of:
  Performance
  Dimension
  Diagnosis
  Tool wear
  Temperature
  Chatter

Workpiece
  Identify
  Tracking

Correction of:
  Tool chatter
  Tool wear
  Spindle error
  Heat distortion
  Setup error

768   Chapter 11

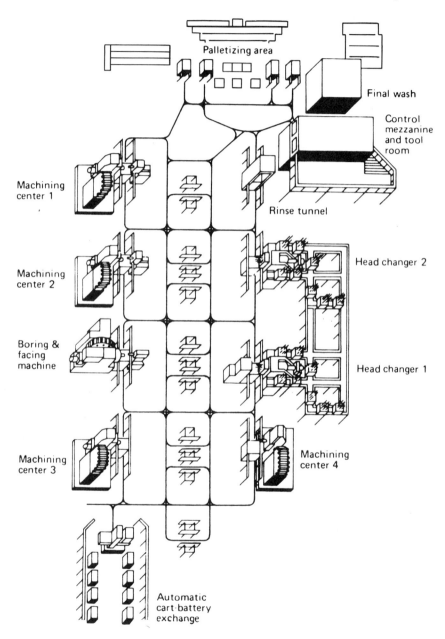

FIG. 11.16 Layout of the Bouthéon FMS. (From Ref. 10.)

The Programmable Factory

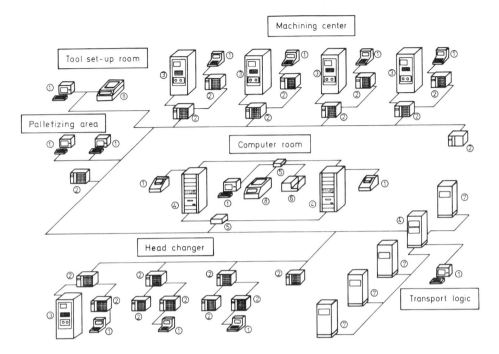

FIG. 11.17 Hardware configuration of Bouthéon FMS: 1, terminal; 2, programmable controller; 3, control computer (machine tool); 4, master computer; 5, interface; 6, card reader; 7, microcomputer; 8, printer.

Each machining center, as well as the facing and boring machines, have its own independent numerical controller and two programmable controllers. One programmable controller identifies the workpiece upon arrival of a pallet and initiates the NC program to be used. The second programmable controller monitors the tool wear. There are three programmable controllers for each head changer.

The material handling system has its own independent microcomputer controllers. The transportation network is divided into four control sections, each supervised by a microcomputer. The local control logic processes the transportation instruction issued by the supervisory computer by means of a transportation control logic. An important function performed by the transportation logic is the creation of a system map and its transfer to the supervisory computer. In addition, it is responsible for cart control.

770   Chapter 11

Each cart contains its own control logic, which communicates with the local control logic via dialogue devices embedded in the floor. The cart control obtains its destination address from the supervisory computer and conducts the car along its chosen path. In addition, an optimal path is selected.

11.5 A FACTORY OF THE FUTURE

A model for a fully automated factory was developed by the Japanese and made public in 1972. In this plant machine tool components were to be manufactured with little human interaction. Within the following 5 years this factory of the future had matured to the concept discussed in this section. In principle, all essential functions of this factory have been discussed in this book. For this reason only a brief description of the factory will be given (Fig. 11.18) [11].

The core of the plant consists of a warehouse section, a machining area, and an assembly facility. All these functions are interconnected by a material handling system and controlled by a computer. Emphasis was given to high flexibility, reliability, safety, and low consumption of resources and energy.

The warehouse section is serviced by computer-controlled stacker cranes performing the storage and retrieval operations. Raw workpieces

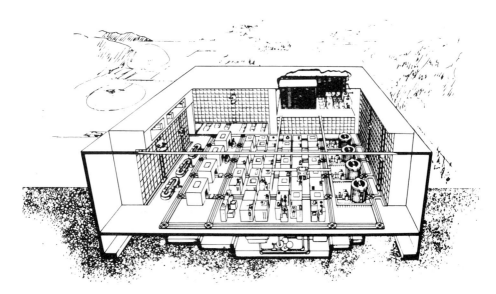

FIG. 11.18 Unmanned factory. (Courtesy of North-Holland Publishing Company, Amsterdam, Holland.)

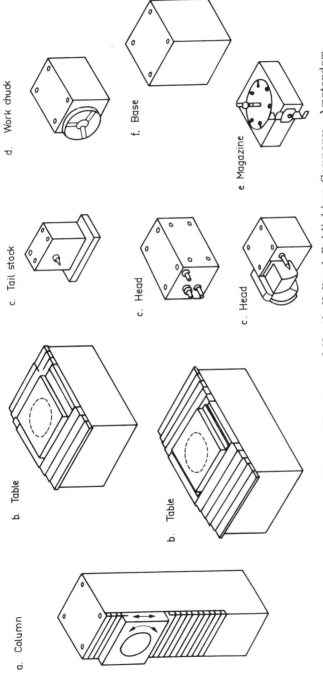

FIG. 11.19 Machine tool modules. (Courtesy of North-Holland Publishing Company, Amsterdam, Holland.)

TABLE 11.8 Capsules and Their Machining Functions

	Capsule type				
	A: Horizontal boring	B: Vertical lathe	C$_1$: Horizontal lathe	C$_2$: Cylindrical grinder	D: Chuck lathe	E
			Module			
	CTHM	CTHM	CTHM	CTHW	CTHB	CT
Milling	×	×			×	×
Boring	×					
Drilling	×	×	×		×	×
Tapping	×	×	×		×	
Reaming	×	×				
Surface grinding	×					×
Screw threading	×	×	×		×	×
Turning			×		×	
Center drilling			×			
Grooving			×			
Cylindrical grinding					×	
Screw grinding				×		
Internal grinding						×

[a]Symbols C, T, H, M, W, B identify the machine tool components used by a module.

The Programmable Factory

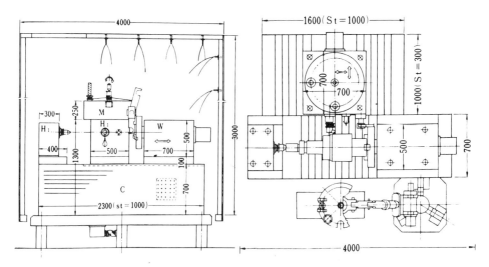

FIG. 11.20 A C-type machining capsule. (Courtesy of North-Holland Publishing Company, Amsterdam, Holland.)

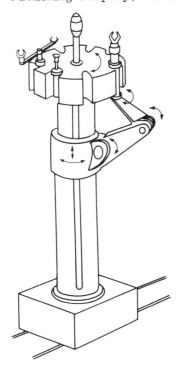

FIG. 11.21 Movable factory floor robot. (Courtesy of North-Holland Publishing Company, Amsterdam, Holland.)

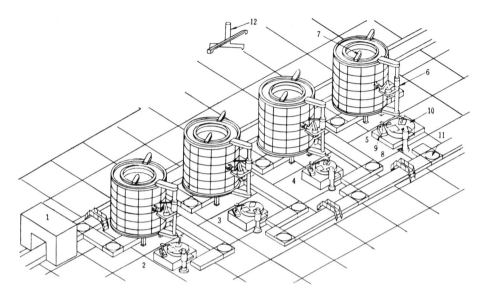

FIG. 11.22 Assembly area: 1, washing station; 2, assembly stage 0; 3, assembly stage 1; 4, assembly stage 2; 5, assembly stage 3; 6, stacker crane; 7, receiving-dispatching system; 8, assembly robot; 9, working machine; 10, working table; 11, turntable; 12, capsule transportation crane. (Courtesy of North-Holland Publishing Company, Amsterdam, Holland.)

are brought to the setup area and channeled to different machining capsules. The machine tools in these capsules are of modular design and assembled to perform specific functions (Fig. 11.19). The six capsules that can be constructed from these modules are described in Table 11.8. A C-type capsule is outlined in Fig. 11.20. Loading and unloading is done by a standardized manipulator which runs on rails laid throughout the factory (Fig. 11.21). Finished parts are transported either to main storage or to the rotational storage system of the assembly facility (Fig. 11.22). From here a robot can pick the parts for an assembly order and present these to the assembly capsule (Fig. 11.23). A workstation for assembly consists of a rotary index table with tilting fixtures, robot arms, and vision systems. The assembly is done by one or several robots and supervised by cameras.

A conventional factory manufacturing the same type of product would employ about 700 people. With computer-controlled automation as described, only 10 operators would be needed. Software, especially expertsystems for knowledge processing, would play an important role in this factory and would amount to over 50% of the installation cost.

The Programmable Factory 775

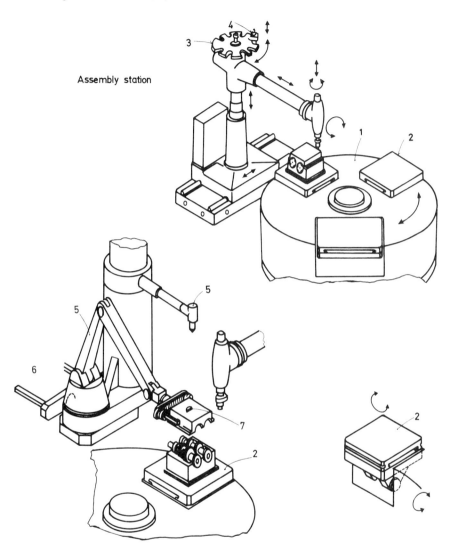

FIG. 11.23 Example of assembly capsule: 1, indexing table; 2, tilting platform; 3, magazine; 4, automatic wrench; 5, assembly robot; 6, pallet fork; 7, identification mark. (Courtesy of North-Holland Publishing Company, Amsterdam, Holland.)

REFERENCES

1. KT's World of Manufacturing Systems, Kearney & Trecker Corporation, Milwaukee, Wis., 1980.
2. Machine Tool Utilization, Internal Report, University of Karlsruhe, Karlsruhe, West Germany, 1983.
3. C. F. Carter, Towards Flexible Automation, *Manufacturing Engineering*, Aug. 1982.
4. C. Dupont-Gatement, A Survey of Flexible Manufacturing Systems, *Journal of Manufacturing Systems*, Vol. 1, No. 1, 1982.
5. R. N. Stauffer, Flexible Manufacturing Systems, Bendix Builds a Big One, *Manufacturing Engineering*, Aug. 1981.
6. W. Maßburg, Automation Brings Progress to Japan, *VDI-Nachrichten*, No. 7/13. Feb. 1981.
7. Japan Knows What the Customer Wants, *VDI Nachrichten*, No. 50/12, Dec. 1980.
8. T. Ohmi, Y. Ito, and Y. Yoshida, Flexible Manufacturing Systems in Japan—Present Status, *Proceedings of the First International Conference on Flexible Manufacturing Systems*, Brighton, England, Oct. 20–22, 1982.
9. Machining module, Sales literature, PITTLER Maschinenfabrik AG, Langen near Frankfurt/M., West Germany.
10. C. Dupont-Gatemand, A Real Time Control System for Unaligned Flexible Manufacturing System, *Fourth IFAC/IFIP Symposium on Information Control Problems in Manufacturing Technology*, Gaithersburg, Md., Oct. 26–28, 1982, published in the proceedings of this meeting.
11. H. Yoshikawa, Unmanned Machine Shop Project in Japan, Procedings of the 3rd International IFIP/IFAC Conference on Programming Languages for Machine Tools, PROI, North-Holland Publishing Company, Amsterdam, Holland, 1976.

Index

A/D converter, 701
ACC type controller, 446
ACO type controller, 446
Acoustic programming, 555
ADA programming language, 112
Adaptive control, 443
Administrative
 calls, 189
 control, 24, 102, 762
AL robot programming language, 566
ALGOL 60, 680
Algorithmic decision process, 450
Analysis
 of a part spectrum, 230
 method, 665
Angular orientation of a workpiece, 598, 624, 642, 650
APT language, 472, 736
Artificial intelligence, 334, 501
Assembly
 of parts, 6, 46, 83, 501, 505, 524, 546, 549, 553, 557, 584
 code, 112
 facility, 507, 770
 line operation, 379

[Assembly]
 machine modules, 83
ATLAS programming language, 737
Attendance reporting, 39
Automation, degree of, 698
Automation criteria, 48
AUTOPASS programming system, 559

Back scheduling, 358
Balancing strategies, 379
Balancing techniques, 371
Bar code, 607
BASIC, 112, 736
Basic software system, 144
Backup system, 749
Behind tape reader mode, 495, 497
Bill of lading, 13
Bill of materials, 3, 14, 16, 34, 55, 61, 293, 318, 319, 330, 358
 block type, 320
 intended, 319
 single level, 319
Binary code, 606
Binary vision, 598, 616
Bit-serial data transfer, 116

777

Block diagram for controls, 431
Block-oriented language, 677
Block-type language, 112
Bottom-up implementation of QC, 723
Boundary presentation of objects, 213, 265
Business computer, 2, 241

CAD applications, 236
  electrical engineering, 236
  mechanical engineering, 236
  physics, 236
CAD
  software/hardware, 240, 243
  system, 235, 236
    COMPAC, 245
    GRIN, 250
    PADL, 262
    three-dimensional, 238
    TIPS, 271
    two-dimensional, 238
CAD/CAM
  integration, 753
  systems, 391
  technology, 113
Canonical pattern, 271
Capacity planning, 365
Capital equipment planning, 32
Cartesian
  coordinate system, 512, 515, 551, 580, 585
  interpolator, 414
  trajectory, 414
Center of gravity, 598, 619
Central
  computer, 241
  data base, 10
  robot controller, 548
  scheduler, 122

Centralized system for QC, 723
Character generator, 153
Charge-coupled devices, 610, 613
Chi-square test, 672
Circular DDA interpolation, 421
Class concept of SIMULA, 683
Classification of workpieces
  design-oriented, 299
  process-oriented, 299, 302
  benefits, 293
  by features, 302
  by visual inspection, 301
  code matrix, 323
  manual, 292
  methods, 301
  number, 298
  parameters, 292
  system, 292, 302, 323
Classification of FMSs, 748
CLDATA concept, 469, 534
Clipping of pictures, 179, 180
Clossed-loop
  control, 428
  search mode, 450
Communication network, 98
CNC
  architecture, 485
  control, 485
  machine tool, 753, 736
  part program, 489
Code
  chain type, 299
  hierarchical, 300
  number, 299
  number system, 298
  reader, 71
  system, 304
Coding of parts, 287, 323, 330
Communication
  horizontal, 96
  vertical, 96
  in the graphics system, 177
  link, 131
  network, 114

Index

[Communication]
  protocol, 28
COMPAC/CAD system, 203, 205
Compiler, 20, 580, 582,
  583, 584
Components of a robot, 510
Composite component concept,
  312
Computer
  architecture, 96, 119
  control, 53
  hierarchy, 96
  numerical control, 18
  -aided design, 2, 16, 149
  -aided manufacturing, 44
  -aided part programming,
  467
Computer equipment, 102
Contact
  measurement, 701
  sensors, 698
Continuous-path control, 501
Continuous process control,
  27
Contour pattern, 271
Control
  computer, 2, 18, 91, 718
  information flow, 760
  layer of plant, 24, 28, 109
Conveyor system, 758
  fixed type, 76
  monorail, 76
  power and free, 76
Coordinate
  measuring machine, 702
  measuring time, 736
  system, 512
  transformation, 417, 514,
    515, 534, 578, 580,
    590, 593
  transformation level, 546
CORE/CAD system, 243
Corporate
  level control, 21, 102
  management, 21

Correlation method, 630
Cost
  accounting, 13
  center, 13
Cross-correlation, 629
Cubical
  parts classification, 327
  workpiece, 325
Customer
  order, 10, 357
  order servicing, 32
Cut force sensor, 458
Cutting speed, 346

D/A converter, 701
Data
  acquisition, 136, 366, 371, 383,
    698, 717
  base,
    centralized, 14, 16
    collection of independent, 14
    distributed, 14, 18
    independent, 15
    interfaced, 14, 17
    solitary, 14
  collection system, 39
  files
    direct access, 187
    linked list, 187
    sequential organization, 187
  formats, 28
  flow, 760
  flow machine, 147
  format, 10, 21
  link layer, 118
  peripherals, 159
  processing
    computer, 91
    protocol, 753
    procedure, 753
  structure for graphics,
    185
DDC algorithm, 437

Decision
  rules, 334
  strategy, 448
  table, 34, 336
Definition phase of a project, 49
Degree of automation, 13, 27
Degrees of freedom, 512
Delivery date, 364
Denavit-Hartenberg matrix, 515, 516, 538, 551
Design, 6, 33, 141
  and development, 55
  family of parts, 288
  phase, 33, 252
    detailed drafting, 227
    functional analysis, 227
    layout, 227
    product definition, 227
    types of, 227
  variants, 352
Dialogue calls, 189
Dictionary for an expert system, 144
Digital differential analyzer DDA, 418
Digitized circuit design data, 113
Digitizer, 159, 161
Dimensional image, 588
Dimensioning tree, 205
Direct
  measurement of parameters, 701
  numerical control, 18
  wear measurement, 461
Discrete
  process, 27
  simulation, 662
Display
  file for CAD, 179, 180
  screen for CAD, 180
  tube for CAD, 151
Distributed computer system, 2, 19, 96, 723

DMA, 93
DNC
  machine tool, 63
  system, 392, 492
Drafting machine, 171
Drawing, 34, 319, 330
  facet method, 205
  fill-in-the blanks method, 209
  generative principle, 199
  organizational contents, 196
  table printout method, 209
  technological contents, 196
  three-dimensional primitives, 203
  variant method, 209
  from simple form elements, 205
Due dates of orders, 358
Dynamic programming, 357

Economic order quantity, 360
Editor for robot language, 594
Effector (robot hand), 500, 513, 514, 517, 519, 526, 538, 546, 549, 566
  control, 546
EIA code for NC, 465
Electrical motor, 532, 537
Electronic mail, 138
Element
  table for CAD, 221
  time for line balancing, 376
Engineering, 33, 44, 293
EPOS software engineering system, 114
Equipment specification, 293
Error diagnostic routine, 20
Ethernet, 116
  bus, 117
Event-oriented simulation method, 672
EXAPT/NC programming system, 475
  basic, 475

Index

[EXAPT]
  1, 475
  1.1, 475
  2, 475
Expediting of parts, 39
Expert system, 762
External sensor for robots, 523

Fabrication release card, 381
Facet method for CAD, 256
Facility monitoring, 138
Facility planning, 32, 293
Factory flow analysis, 315
Failure diagnostic procedure, 764
FAPT TURN programming system, 480
Feature extraction for vision, 622, 623
Feed rate, 346
Feedback control system, 429
Feed and speed selection, 324
Fifth generation computer, 140
Figure list for CAD, 188
Financial modeling, 59
Financial planning, 61
Finite element method, 55
Firmware, 113
Fixed type of automation, 53
Fixture file, 346
Fixture selection, 344
Flexible automation, 53
Flexible manufacturing system (FMS), 1, 48, 505, 685, 742, 748
  components, 753
  computer control, 760, 762

[Flexible manufacturing system (FMS)]
  hybrid system, 749
  with complementing machine tools, 748
  with substituting machine tools, 749
  features, 749
Flow line production, 83, 288, 291
Flying master principle, 101
Force-torque sensor, 524
Forecasting, 17, 31
Form elements in CAD, 232, 245, 248, 350
FORTRAN, 112, 680, 683, 736
Frame concept for robots, 526, 538, 544, 546, 549, 550, 551, 553, 570, 585
Frequency distribution of parts, 230
Future robots, 509

Gauging method, 705
Generation of
  drawings, 199
  robots (historical), 504
Generative
  method of planning QC, 692
  programming of NC, 484
Geometric primitives in CAD, 199
GKS (graphic kernel system), 243
GPSS simulation language, 674, 675
Grammar for problem solving, 144
Graphic
  calls, 189
  data processing, 150
  description of workpieces, 216

[Graphic]
  display, 740
  language, 140
  peripherals, 159
  programming, 398
  software, 173
  tablet, 161
  terminal, 163
Grasp by robot
  coordinates, 590
  position, 590
Gray value in vision, 598
GRIN CAD system, 250
Gripper for robot, 520, 522, 529, 532, 549, 550
Ground vehicles, material flow, 78
Group technology, 7, 287, 288, 323, 333, 762
Guided vehicles in material flow, 78

Handshake procedure, 21
Hardwired
  interpolators, 417
  logic circuits, 393
  logic controllers, 394
  programming, 398
Heuristic methods H for line balancing, 379
Hierarchical
  computer control, 2, 19, 21, 26, 99, 119, 743, 760, 765
  computer network, 10, 690, 717
  data acquisition, 126
Hierarchy of
  control, 100
  management, 100
  organization, 100
High
  volume production, 63, 228

[High]
  level inquiry language, 145
  level programming language, 673
  rise warehouse, 81
Homogeneous coordinates, 516
Horizontal integration of workpieces, 325
Host computer, 717, 719
Hydraulic drive, 533
Hypothesis testing, 670

Identification
  number, 298
  task, 446
IEEE 488 bus, 116
Image data base, problem solving, 144
Implicit programming for robots, 140, 544, 573
In-process inventory, 366
Incremental transducer, 458
Indirect wear measurement, 463
Inference, 141
  machine, 145
Information
  flow, 12
  processing network, 141
Infrared sensor, 524
Initialization phase of a simulation, 668
Inoperative time element in scheduling, 367
Integrated
  manufacturing system, 563
  quality control, 689, 695
Integration of CAD/CAM, 283
Intelligent
  interface, 140
  interface machines, 145
  sensor, 509
  terminal, 130, 717
Interactive
  CAD, 183

Index 783

[Interactive]
  process communication, 764
  symbolic programming, 478
Interfaces for software systems, 10, 21, 28
Interfacing
  of FMS components, 757
  of vision system with a robot, 588
Internal sensor of robot, 523
Interpolation method, 414, 538, 547, 566, 569, 574, 580
Interpreter for robots, 529, 574, 578, 582
Inventory, 7, 46
  carrying cost, 362, 375
  management, 42, 61
  master file, 358
Investment planning, 347
Invoice, 13
IRDATA concept for robots, 534, 561
ISO code, 465
Iterative process in simulation, 663

Job assignment, 39
Joystick, 159

Kernel language for CAD, 145
Keyboard for CAD, 159
Kinematics for robots, 510, 512, 534, 574
Knowledge base for problem solving, 141
Knowledge-based management, 140

Labor
  cost, 347

[Labor]
  ticket, 383
Ladder diagram for circuits, 402
Language for QC programming, 734
Laplace transform, 433
Large volume production, 291
Laser telemetric system, 714
Lathe, 55, 63
Lead time for production, 347, 358, 366
Library routines for CAD, 179
Light pen, 163
Light-emitting diode, 602
Line
  balancing, 376
  balancing sheet, 34
  list for CAD, 188
  production, 376
Linear
  arrays, 603
  camera, 640
  congruence algorithm, 670
  DDA interpolation, 418
  interpolation, 538, 541
  programming, 357
  vision system, 636
Load balancing, 370
Local area network, 136
Logic
  network, 393
  operations control, 93
  programming of computer, 145
Long range planning for manufacturing, 3, 31
Look-up tables, 324, 336
Low volume production, 63, 228, 291, 334, 742

Machine
  assignment, 39
  axes, 63
  control, 138, 760, 764
  control unit MCU, 408

[Machine]
  coordinate system, 707
  language, 736
  load profile, 39
  loading plans, 34
  monitoring, 760, 764
  tool, 63, 119, 288, 318, 386, 748, 753
  balancing, 293
  master file, 339
  matrix file, 324
  selection, 324, 339
Machining
  centers, 755
  elements, 336
  facility, 770
  instruction, 46
  module, 63, 333
  primitives, 334
  sequence, 46, 325
    file, 339
Magnetoelastic transducer, 460
Mailbox communication principle, 760
Maintenance of machine, 43
Make-or-buy decision, 6
Management information system, 2, 8, 71, 138
Manual
  input of QC data, 701, 717
  part programming, 465
  programming, 554, 578
Manufacturing
  control, 6, 59, 61, 141, 330, 381
  control of the future, 25
  cost master file, 350
  cost, 349, 350
  data base, 13
  document, 13, 34
  family concept, 288
  function, 29
  lead time, 325
  monitoring, 39, 330

[Manufacturing]
  planning, 6, 55, 59, 141
  process planning, 34
  resource planning, 362
  resources, 356
  route, 383
  scheduling, 46, 357
  system, 55
    components, 55
    time, 324, 346
    file, 349
  control, 13, 39
Marketing, 34
  research, 29
Mass production method, 743
Master production
  plan, 357
  schedule, 356, 358, 366
Master-slave programming, 555
Material
  card, 383
  control, 138
  flow, 39, 71
    control, 6, 760
  handling, 39, 501
    system, 748, 749, 757, 769, 770
  master file, 336
  movement, 3, 76
  receiving, 40
  requirement, 61
    file, 336
    planning, 357
  storage, 6, 71, 78
Measurement
  automatic, 712
  dimensional, 710
  displacement, 458
  noncontact, 714
Measuring
  head, 703
  methods, 701
  phase of a vision system, 644
  robot, 713
  system, 699
Medium
  range planning, 39

Index

[Medium]
  volume production, 228,
    291, 742, 743
Meta interference system,
  144
Methods for object classification, 631
MICLASS classification system,
  307
Milling machine, 55, 63
Mirror call in CAD, 189
Modeling of solids, 209
Modified polar coding for
  vision, 626
Moment of inertia calculation
  for vision, 620
Monorail conveyor, 76
Move
  statement for robots, 569,
    580
  ticket, 381
MPST hierarchical control
  system, 119
Multiple
  objects in the field of view,
    652
  spindle technology, 66
Multitasking, 136, 567

Natural language, 140, 145
  programming, 572
NC (numerical control), 1,
  6, 389
  engine lathe, 55, 63
  machine control, 408, 760
  machine tool, 53, 63, 501,
    736, 742, 748, 753
  machining, 391
  machining centers, 746
  part programming, 34, 39,
    240, 464, 465
  programming language, 391,
    472, 475, 480, 736
  punch press, 71

[NC (numerical control)]
  punched tape, 55, 465
Noncontact sensors, 698
Nonlinear load balancing, 371
Nonrotational parts, 292

Object classification for vision, 631
Object presentation
  boundary presentation, 213
  primitive instancing, 211
  quasi-disjoint, 213
  sweeping, 213
  wire frame, 211
Offset compensation, 707
On-line quality control system, 89
Open-loop
  control, 428
  search mode, 450
Operating
  budget, 44
  system, 19, 20, 113, 136, 173,
    578, 582
Operation
  plan, 34
  sequence, 364, 365, 366
Operational factory data, 129
Operations research, 287
Operative level of control hierarchy, 104
Operator dialog, 132
Optimization of manufacturing
  processes, 350, 371, 379
Order
  completion time, 366
  dispatching, 381
  identification, 383
  monitoring, 381, 383
  planning, 39
  processing, 14, 687
  release, 13, 381
  release planning, 362, 365
Organizational planning, 29
Overtime determination,
  375

PADL(CAD programming system), 262
Palletizing of manufacturing parts, 81
Pancake motor, 533
Parallel
 arithmetic processor, 528
 classification procedures, 631
Part
 families, 34, 241, 288
 identification, 71
 list, 381
 program (NC programming), 34, 63
 routing, 315
 spectrum, 350
Part demand, 362
PASCAL, 112, 566, 677, 683, 684
Path control for robots, 536, 580
Payroll, 17
 calculation, 383
PDV bus, 116, 117
PEARL programming language, 112
Perforated paper tape, 465
Phase object in CAD, 213
Photo array sensor, 604, 609
Photodiode, 610
 array camera, 616
 matrix, 588
Photoelectric
 sensor, 602
 transducer, 460
Physical layer communication, 118
Picture parameters, 607
Piezoelectric crystal, 460
Planning
 algorithms, 358
 control, 44, 104
 manufacturing, 6
 module, 570
 process, 365
 processor, 336
 sheet, 322

Plant
 floor, 109
 maintenance, 44
 management, 24
Pneumatic drive, 533
Point
 list for CAD, 188
 to-point control, 538, 580
Polar
 coding for vision, 625
 moment of inertia for vision, 621
Polynomials for interpolation, 417
Positioning wheel for CAD, 159
Postprocessor, 473, 490
Power and free conveyor, 76
Pragmatical analysis in problem solving, 143
Preoriented parts, 86
Primitive instancing presentation, 211
Primitives in CAD, 183, 245, 251, 262, 265
Priority
 interrupt, 20
 rules, 369
Problem
 domain, 144
 solving, 140, 573
  language, 145
  machine, 145
Process, 682
 boundary, 339, 341
 capability file, 339
 equipment file, 34
 grid for part families, 339
 master file, 336
 model, 350
 monitoring, 330
 optimization, 93
 peripherals, 112
 plan, 34, 330, 762
  output, 324
 planning, 3, 6, 13, 138, 241, 287, 296, 318, 322

Index 787

[Process]
 [planning]
  generative, 330, 333, 350
  hybrid, 330
  variant, 330, 350, 352
  selection, 324, 336
  sheet, 34
Processing
 times manufacturing, 324
Product, 509
 design, 13
 group, 230
 life cycle, 8, 29
 planning, 356
 simulation, 55
Production
 bottleneck, 375
 control, 296, 760
 high-volume, 63, 228
 low-volume, 63, 228, 298, 334, 742
 medium-volume, 228, 291, 742, 743
 monitoring, 39
 scheduling, 102, 138, 297, 762
Profit maximizing, 357
Programmable
 controller, 93, 398, 401, 769
 factory, 742
 test system, 696
Programming
 language, 112, 144, 472, 531, 549, 553, 557, 558, 559, 561, 566, 573, 585, 674, 680, 683, 734, 736
 for graphic data processing, 188
 methods, 554
 system, 114, 203, 236, 238, 243, 245, 250, 262, 271, 464, 475, 549, 559, 674, 677, 682
Project phases, 49

PROLOG, 145
Protocol, 10, 21
 for communication, 10, 21, 114
 logical, 114
 physical, 114
 transmission, 114
Proximity sensors, 463
Pseudo-display-file compiler, 179
Pseudo numbers, 670
PTP control, 538, 541
Punch press, 63
Punched paper tape, 410
Purchasing, 13, 40, 298

Quality
 assurance, 102
 control, 46, 86, 138, 296, 760
  computer architectures, 723
  equipment, 699
  functions, 689, 693
  implementation problems, 723
  instruction, 34
  methods, 695
  on moving product lines, 727
  planning, 692
  system concept, 690
Quasi- disjoint decomposition, 213
Queueing, 366
 theory, 357
Quotation, 13

Random
 access, 110
 numbers, 670, 672
Raster display, 152
Raw material selection, 324, 336
Real time
 clock, 93,
 control system, 107

[Real time]
  language, 736
  vision system, 641, 647
Refresh memory, 157
Regular algorithmic dicision for control, 449
Relational algebra machine, 147
Relationship matrix, 223
Relative frames, 587
Remote center compliances, 86
Residue method simulation, 670
Resource planning, 357
Return-on-investment, 50
Ring
  bus, 116
  structured list, 187
Robot
  axis, 512
  calibration, 585
  control, 538, 578
  controller, 537, 546
  coordinate, 514, 515, 580, 586
Rotation command, 179, 189, 195
Rotational parts, 292, 304, 325, 327, 339
Route
  file, 762
  sheet, 762

Sales forecasts, 29
SARS specification system, 114
Satellite computer, 107, 110
Scaling command, 179, 180
Scheduling, 6
  algorithms, 366
  alternative equipment 371
Scientific computer, 91

Segmentation for picture display, 184
Self
  -adapting test system, 731
  -scanning photoarrays, 611
Semantical analysis, 143
Semantics of a phase object, 216
Semaphore, 570
Sensor, 457, 504, 505, 510, 566
  522, 523, 529, 534, 537,
  549, 554, 561, 566
  coordinates, 544, 551
  data, 532, 534
  signals, 558
Sequential
  classification method, 633
  control, 93, 407
  decision-tree, 635
  parameters, 509
Service, 298
  programs, 174
Servo control level, 546
Servomotor, 533
Setup time, 346
Shape primitives, 334
short-range planning, 3, 39
Signature analysis, 715
SIMPAS, 673
SIMSCRIPT, 674, 677, 682
SIMULA, 674, 680, 682
Simulation, 574, 582, 662
  aids, 59
  experiment, 666, 670, 688
  languages, 674, 677, 680, 682
  model, 663, 669, 677
  project, 667
  run, 667, 669, 677
Software, 107, 173
  cost, 48, 51
  development aids, 113
  engineering, 336
  for CAD, 175
  overhead, 19
  phase, 50
Solid geometry for CAD, 265
Solitary computer system, 20

Index

SRL, compiler, 573
  language for robots, 559, 566, 567, 570, 571
Stacker crane, 770
Stand-alone system, 582
Standardization, 757
Standby control, 108
Starting phase, in simulation, 667
Stepping motor, 533
Strain gauge, 460
Strategic
  goal setting, 29
  management, 102
Supervisory
  computer, 748, 769
  control, 762
  level, 102
Syntax
  of a CAD language, 235
  of a phase object, 216
System
  control, 91
  programming language, 736

Table-look-up for variants, 692
Tablet, graphical system, 165
Taylor equation, 349
Teach-in, 549, 557, 580, 584, 585, 586
  method, robots, 549, 555, 557, 580, 584, 585, 586
  phase of a vision system, 643
Technological planning, 29
Teleoperator, 500, 501
Test
  language, 737
  performance, 697
  procedures, 695
Text generator, 324

Textual
  language, 140
  programming, 398, 501, 555, 557, 558, 580
Three-dimensional
  models, 251
  vision, 598
Threshold value, vision, 616
Time
  control in simulation, 680
  control mechanism, 672, 684
  measurements, 369
  scheduling, 29
  slice-oriented method, 672
TIPS CAD system, 271
Tool
  card, 383
  change time, 346
  design, 298
  file, 346
  life, 346
  selection, 344
Top-down implementation, 723
Touch sensor, 553
Touch-sensitive terminals, 165
Trace routine, 179
Trackball, graphical system, 159
Tracking of workpieces
  direct, 71
  indirect, 75
  of workpiece contours, 654
Trajectory
  calculation, 513, 515, 534, 536, 541, 555, 566, 578
  control level, 547
  interpolation level, 546
  planning, 537, 574
Transaction
  interpreter, 132
  language, 132
Transfer
  behavior of control, 432
  buffer of synchronization, 124
  function, 432

Transformation
　matrices, 517
　routines, 179
Translation of a displayed
　　figure, 180
Tree-structured list in CAD,
　　187
Truth tables in circuit design, 393
TV camera, 610

Ultrasonic sensor, 524
UNIX operating system, 108
User guidance by computer,
　　740

VAL language for robots,
　　566
Validation phase in simulation,
　　668
Variant
　design, 334
　method, 692
Vector
　display, 155
　screen, 156
Vendor performance, 46
Vertical integration of
　　workpieces, 325
Virtual machining process,
　　760
Vision, 504
　for complex tasks, 607

[Vision]
　for simple task, 602
　parameters, 615
　sensor, 589
　systems, 71, 86, 544, 584,
　　585, 636
Voice communication, 717
Volume primitives, 251

Warehouse, 46, 81, 770
Wear measurement, 461
Wire frame model in CAD, 211
Word processing, 138
Work
　center identification, 383
　elements, 376
　queues, 367
Workpiece
　analysis, 230
　description, 199, 232
　drawing, 199, 256, 318
　features for vision, 618, 624
　identification, 616
　perimeter, 621
　recognition, 597
　system, 639
　spectrum, 230
World
　coordinates of robot, 514, 515,
　　534, 544
　model, 544, 570, 573, 584

Z-transform, 443